FOLSOM

FOLSOM

NEW ARCHAEOLOGICAL INVESTIGATIONS OF A CLASSIC PALEOINDIAN BISON KILL

DAVID J. MELTZER

with contributions by

MEENA BALAKRISHNAN

DONALD A. DORWARD

VANCE T. HOLLIDAY

BONNIE F. JACOBS

LINDA SCOTT-CUMMINGS

TODD A. SUROVELL

JAMES L. THELER

LAWRENCE C. TODD

ALISA J. WINKLER

UNIVERSITY OF CALIFORNIA PRESS
Berkeley   Los Angeles   London

University of California Press, one of the most distinguished university presses in the United States, enriches lives around the world by advancing scholarship in the humanities, social sciences, and natural sciences. Its activities are supported by the UC Press Foundation and by philanthropic contributions from individuals and institutions. For more information, visit www.ucpress.edu.

University of California Press
Berkeley and Los Angeles, California

University of California Press, Ltd.
London, England

Library of Congress Cataloging-in-Publication Data

Meltzer, David J.
    Folsom : new archaeological investigations of a classic Paleoindian bison kill / David J. Meltzer; with contributions by Meena Balakrishnan . . . [et al.].
        p. cm.
    Includes bibliographical references and index.
    ISBN 0-520-24644-6 (case : alk. paper)
    1. Folsom Site (N.M)   2. Folsom culture—New Mexico—Colfax County.   3. Folsom points—New Mexico—Colfax County.
4. Excavations (Archaeology)—New Mexico—Colfax County.
5. Animal remains (Archaeology)—New Mexico—Colfax County.
6. Colfax County (N.M.)—Antiquities. I. Balakrishnan, Meena. II. Title.
    E99.F65M45 2006
    978.9'2201—dc22

                                                    2006002482

Manufactured in the United States of America
10 09 08 07 06
10 9 8 7 6 5 4 3 2 1

The paper used in this publication meets the minimum requirements of ANSI/NISO Z39.48–1992 (R 1997) (Permanence of Paper). ♾

Jacket illustrations: Looking east across the South Bank, where the first Folsom point (inset) was recovered in 1927. Background photo by David J. Meltzer; inset photo courtesy Denver Museum of Nature and Science.

In memory of George McJunkin and Carl Schwachheim,
who thought Folsom worth telling others about,
and to Joseph & Ruth Cramer,
whose extraordinary generosity made that telling possible

CONTENTS

CONTRIBUTORS LIST

Meena Balakrishnan
Department of Geological Sciences
Southern Methodist University
Dallas, Texas

Donald A. Dorward
Department of Anthropology
Southern Methodist University
Dallas, Texas

Vance T. Holliday
Department of Anthropology
University of Arizona
Tucson, Arizona

Bonnie F. Jacobs
Environmental Science Program
Southern Methodist University
Dallas, Texas

David J. Meltzer
Department of Anthropology
Southern Methodist University
Dallas, Texas

Linda Scott-Cummings
Paleoresearch Institute
Golden, Colorado

Todd A. Surovell
Department of Anthropology
University of Wyoming
Laramie, Wyoming

James L. Theler
Archaeological Studies Program
University of Wisconsin – La Crosse
La Crosse, Wisconsin

Lawrence C. Todd
Department of Anthropology
Colorado State University
Ft. Collins, Colorado

Alisa J. Winkler
Department of Geological Sciences
Southern Methodist University
Dallas, Texas

PREFACE

I began in archaeology as a teenager working on the Thunderbird Paleoindian site in the Shenandoah Valley of Virginia. It was a spectacular site. Located near a major stone (jasper) outcrop, it was a camp for Late Glacial hunter-gatherers who cycled through the area to refurbish their toolkits and, in the process, left literally tons of debris from the manufacture of their fluted projectile points and tools. The artifacts at Thunderbird were scattered about well-defined living floors, had been little disturbed, even to the point of preserving the outlines of knapping activity, and were buried by gentle overbank silting, thus neatly sealing those remains from intrusions of materials from higher levels. The record there of stone tool technology was nothing short of superb.

Yet, it was thought the site was inadequate, because we hadn't found any mammoth bones. This was supposed to be an eastern Paleoindian site. As it was well known that Paleoindians were big-game hunters, their sites ought to contain the remains of their prey. So we looked for mammoths. I well remember watching as a bulldozer tore apart the floodplain of the valley upriver of Thunderbird, to expose a deeply buried Late Glacial backwater channel, searching for the mammoth skeletons that surely went with all our fluted points.

And I vividly recall the palpable excitement when, at Thunderbird itself, we found what appeared to be a mammoth vertebra sitting atop the Paleoindian surface. We worked late in the night to get that precious fossil out of the ground (Hurricane Agnes was bearing down on us and would soon flood the site). Once safely in the field laboratory, we gathered around as the piece was carefully cleaned, a painstaking process that finally revealed . . . a piece of rotting quartzite, doing an astonishingly good imitation of a mammoth vertebra. Everyone felt just awful.

The episode made a big impression on me, though I scarcely understood it at the time. What I later came to realize was just how much our expectations of the archaeological record had colored our views of that record. As Binford and Sabloff (1982:139) rightly observe, archaeologists are often "unaware of how their traditionally held paradigms influence their views of the past." I subsequently tried to understand where those big-game expectations had come from, a search that led me into the history of archaeology, and straight to Folsom.

For it was at Folsom that a decades-long, exceedingly bitter controversy over human antiquity in America was finally resolved. For a variety of reasons, largely having to do with the difficulty of determining the age of archaeological remains in those pre-radiocarbon days, resolution required a kill site like Folsom, where projectile points were embedded between the ribs of a now extinct species of bison. Once found, Folsom provided a model and method for subsequent discoveries, and in the decade following that discovery a battery of comparably aged and older (Clovis) sites with associated bison and mammoth remains were found on the western Great Plains and in the Southwest. There is a simple reason all these sites had large mammal bones: like Folsom, they were initially fossil discoveries. The skeletons of megafauna are much more visible in the ground (even from a distance) than the odd Folsom or Clovis point, so the search was on in the 1920s and 1930s for those large bones, which were then followed up to see if they would lead to artifacts. "Boneless" Folsom and Paleoindian sites were rarely found in those years, largely because no one was looking for them (Meltzer 1989a).

That repeated association of Paleoindian points with the bones of extinct megafauna and the apparent scarcity of nonkill sites, led naturally to the inference that Paleoindians were big-game hunters: which is why we were looking for and expecting to find evidence of a kill at Thunderbird. Had the history of resolving the antiquity dispute decades earlier been different, had prehistory not been made at Folsom in

Part of the Folsom crew on site, July 1998, from left to right: Brent Buenger, Todd A. Surovell, Nicole M. Waguespack, David J. Meltzer (in surveyor flagging tape lei), Jason M. LaBelle, Russell D. Greaves, John D. Seebach, and Pei-Lin Yu. (Photo by D. J. Meltzer.)

the way that it was, North American Paleoindian studies would have been very different. But it wasn't, so they weren't.

In teasing apart the historical development of Paleoindian studies (Meltzer 1983, 1991b, 1994), I spent considerable time at Folsom in the 1920s—metaphorically speaking, of course. In doing so I was struck by how much we knew about what happened there in the 1920s, yet how little we knew about what happened there in Late Glacial times. Like many Paleoindian archaeologists, I'd visited the Folsom site, and perhaps as others had been, I was curious about whether more remained of the original deposits, and if additional fieldwork might fill in some of the very large gaps in our knowledge of the site.

That curiosity remained largely academic, however, for I was busy with fieldwork and research on other sites, Paleoindian and otherwise. But in the mid-1990s, through a series of fortunate events, the opportunity to conduct fieldwork at Folsom presented itself. I grabbed it, excavating there from 1997 to 1999 and doing collections research and laboratory analyses then and since. This book is the result of that effort and is, in a real sense, the last piece of the puzzle that bewildered me so many years ago.

One racks up many debts in doing a field, analytical, and archival project as large as this one, and the satisfaction of finally finishing it is only surpassed by the pleasure that this occasion allows of publicly thanking all those who helped along the way.

Investigations at the Folsom site were supported by the Quest Archaeological Research Fund, an extraordinarily generous endowment to support studies of North American Paleoindian occupations on the Great Plains, established in 1996 by Joseph L. and M. Ruth Cramer. Under the auspices of the Quest Fund I have been able, with students and colleagues, to conduct field investigations at a number of Paleoindian sites, Folsom among them, and explore a variety of issues and problems related to Late Glacial environments and human adaptations. Summaries of this work are available on the web at www.smu.edu/anthro/faculty/dmeltzer.htm and www.smu.edu/anthro/QUEST/home.htm, both of which provide access to many of the publications that have emerged from these studies. Those published works, and this book dedicated to the Cramers, are but small recompense for their generous legacy.

In-kind support for the Folsom fieldwork came in the form of the best field quarters I've ever experienced, the beautiful Trinchera Pass Ranch, which we were able to use thanks to the gracious hospitality of Leo and Wende Quintanilla. Their ranch fully surrounds the site and they allowed us the run of the place, enabling us to survey the area for that (still) elusive Folsom encampment. Even before I planned fieldwork at Folsom, I met Leo and Wende in a Kevin Bacon–like connection, through Grant Hall via Kay Hindes, who arranged for all of us to meet at the ranch in November of 1996. Although the Folsom site is not on his property, I wanted to have Leo's permission to work there since we'd clearly be spending a great deal of time on his land. He kindly agreed, and with that green light, fieldwork preparations commenced.

Our yearly planning and field logistics were aided by Fred Owensby, longtime steward of the site, who helped often with equipment, tools, mechanical matters, and the thousand-and-one unexpected problems and challenges that arise in the course of a field project. He and his late wife Jane were always pleased to share their lifetime of knowledge of the area's history, natural and cultural. Their son and daughter-in-law, Stuart and Sue Owensby, stepped in on many occasions to help as well.

Additional financial support for the research and writing came from the Potts and Sibley Foundation, Midland, Texas, and a Research Fellowship Leave from Southern Methodist University, for which I would like to thank Robert Bechtel and Dean Jasper Neel, respectively. The National Science Foundation (DIR-8911249), the National Endowment for the Humanities, and the Department of Anthropology at the Smithsonian Institution supported my research into the archives and history of the human antiquity controversy and Folsom's role in it. Publication of this volume was supported in part by an anonymous donor.

Fieldwork at Folsom was conducted under Archaeological Excavation Easement Permits AE-74 (1997), AE-78 (1998), and AE-83 (1999) from the State of New Mexico. For help with the permit process and discussions about the site and its preservation, I would like to thank David Eck, Daniel Reilley, and Norman Nelson.

I was fortunate to have with me all three seasons a skilled, energetic, and hardworking crew. The 1997 group was comprised of Michael Bever, Jason M. LaBelle, and Joseph Miller, joined later by Luis Alvarado, Douglas Anderson, Krystal Blundell, Elizabeth Burghard, Virginia Hatfield, Jemuel Ripley, and Pei-Lin Yu. The 1998 crew consisted of Jason M. LaBelle, Joseph Miller, John D. Seebach, Todd A. Surovell, and Nicole M. Waguespack, who were joined for various shorter stretches by Kathy Bartsch, Krystal Blundell, Brent Buenger, J. David Kilby, Ethan Meltzer, Christine Ponko, Jemuel Ripley, and Pei-Lin Yu. The 1999 field crew consisted of Krystal Blundell, Brent Buenger, Robert Godsoe, J. David Kilby, Jason M. LaBelle, Jason A. Meininger, Allison Mittler Cheryl Ross, John D. Seebach, Joy R. Staats, Todd A. Surovell, Nicole M. Waguespack, and Christopher Widga, joined for briefer periods by Thomas Loebel, Ethan Meltzer, and Pei-Lin Yu. Dr. Russell D. Greaves—Rusty—deserves special thanks for serving as field director in 1998 and 1999 and bringing to the task his considerable energy, extraordinary work ethic, logistical skill, and expertise in all things related to archaeological fieldwork.

Many colleagues came to visit at Folsom and ended up working there and offering valuable help and advice, including Stephen Durand, Edward Hajic, Grant Hall, C. Vance Haynes—who shared his firsthand knowledge of the site as well as his field notes and photographs (the latter dating back to his initial visit in 1949), Bruce and Lisa Huckell, Robert L. Kelly, Daniel H. Mann, and Richard Reanier. Vance T. Holliday and Lawrence C. Todd, with whom I collaborated in chapters 5 and 7, helped throughout the fieldwork, analysis, and writing, going above and beyond the call of co-author obligations. Holliday's participation in the project was supported in part by the National Science Foundation (EAR-9807347).

Roger Phillips, Doug Weins, and Patrick Shore arranged and oversaw the seismic and resistivity surveys on the site. Access to Bellisle Lake was provided by Frank Burton. Pat Fall and Steve Falconer supplied the raft parts and other critical equipment and helped in the coring effort in the summer of 1999, as did Steve Durand and Renata Brunner-Jass. Thanks to the good offices of Robert S. Thompson, Joseph Rosenbaum and Jeff Honke of the U.S. Geological Survey came down one cold day in February 2001, with the necessary coring expertise and equipment for our successful winter coring. Jason LaBelle and Kent Newman provided much needed assistance on that occasion. Joe Rosenbaum also provided helpful comments on a draft of the Bellisle discussion.

Folsom site artifacts and skeletal remains excavated in the 1920s are in various museums, and I am indebted to the curators who made access to that material possible. These include, at the American Museum of Natural History, David Hurst Thomas and Lori Pendleton (Anthropology) and Mark A. Norell and John Alexander (Vertebrate Paleontology). Alexander also provided his transcription of Barnum Brown's (1928b) talk at the International Congress of Americanists. At the Denver Museum, the visit was facilitated by Ryntha Johnson and James Dixon (Anthropology) and Russell W. Graham and Logan Ivy (Paleontology); at the Louden-Henritze Museum, by Loretta Martin; at the University of Pennsylvania Museum, by Lucy Williams and William Wierzbowski; and at Capulin Volcano National Monument, by Abbie Reeves and Margaret Johnston. I would also like to thank Tony Baker, Anna Brown, Darien Brown, and Bill Burchard, who provided access to Folsom site materials in their collections, as well as Jack Hofman and Phillippe LeTourneau, who early on made available their records on the artifacts. So too did Adrienne Anderson, who first took me to the site in 1990. Ryan Byerly provided a characteristically thorough rechecking and inventory of the faunal remains recovered during our excavations.

Chapter 2 was originally written for a 1990 SAA symposium on Folsom archaeology. I am especially grateful to the late Dorothy Cook Meade, daughter of Harold Cook, for reading and commenting on that manuscript, and for a most memorable afternoon of conversation with her and the late Grayson Meade at the Cook family ranch in Agate, Nebraska. Tom Burch and Emily Burch Hughes generously shared the diary and photographs of their uncle, Carl Schwachheim, that I might see clearly his contributions to the work at Folsom, of which they are justly proud.

After the last major field season at Folsom, we moved on to fieldwork at other sites, and the business of working up the Folsom material began. Wanting to be in the field, but also wanting to keep the Folsom research and writing moving along, I experimented: I stayed in each morning to work on this book, while my crews were out on the site. I then joined them for lunch and an afternoon of fieldwork. The experiment was a great success, largely because of the ability and expertise of the students who have worked with me over these last few years and the good judgment of my advanced graduate students—Brian Andrews, Jason LaBelle, and John Seebach—in charge of the fieldwork at those sites. They made my constant presence unnecessary. Indeed, I probably fool myself into thinking I was needed at all, but they at least were polite enough not to say otherwise when I showed up at midday. That the crew could never quite fathom why I would choose to be indoors writing in the cool morning hours, and then be outside in the blazing afternoon heat of, say, a West Texas sand dune, is perfectly understandable. It's not a schedule any rational person should keep.

The preparation of this monograph has benefited from the wise counsel, comments, and advice of Stanley Ahler, Tony Baker, Lewis R. Binford, Paul Goldberg, Donald K. Grayson, Jack Hofman, Amber Johnson, Dan Mann, Tony Marks, Paul Matheus, Garth Sampson, Thomas Stafford, and Crayton Yapp. Their help is much appreciated. I am especially grateful to Michael B. Collins, Bob Kelly, and Todd Surovell, each of whom provided, at different points in the process, careful readings of the entire manuscript.

Bob Kelly, when asked by the press whether there were any books that would be competing with this one, replied, "No, and there won't be unless the site's original excavators, Cook, Figgins, et al., come back from the grave." I'd like to think that won't happen. But I'd also like to think that if it does, they won't object to what I have done with their work.

David J. Meltzer
Dallas, Texas

ONE

Introduction

The Folsom Paleoindian Site

DAVID J. MELTZER

The Folsom site, in Colfax County, New Mexico (29CX1, LA 8121), is one of the most widely known archaeological localities in North America. It is routinely mentioned in archaeological texts, regularly appears on maps of notable American sites, and, of course, served historically as the type locality for the Folsom Paleoindian period—a slice of time and a distinctive archaeological culture dating from about 10,900 to 10,200 radiocarbon years before the present (hereafter, ^{14}C yr B.P.).[1] Folsom is on the National Register of Historic Places (Murtaugh 1976:481), as well as being a National Historic Landmark, and a New Mexico State Monument.[2]

All this because excavations there from 1926 to 1928 uncovered finely made fluted projectile points—now called Folsom points—lodged between the ribs of a species of bison that went extinct at the end of the Pleistocene. That these animals were hunted at Folsom demonstrated for the first time and after decades of controversy that American prehistory began at least in late Pleistocene times, bringing to an end—at least for a time—one of the most bitter disputes in American archaeology (chapter 2; also Meltzer 1983, 1991b, 1994).

But while Folsom is one of the best known sites in American archaeology, it is also one of the least known sites in American archaeology—scientifically speaking.

The major purpose of the 1920s excavation was to recover bison skeletons for museum display, and once the site's archaeological significance became known, to document the association of the artifacts with the bison skeletons, and determine the site was indeed Pleistocene in age. That was done well, much to the relief of an archaeological community anxious to have the ugly controversy over human antiquity in the Americas put behind them, and keen to have the deep past that Folsom provided (Kidder 1936; Kroeber 1940; see chapter 2).

As the decades passed our knowledge of the Paleoindian period to which Folsom bestowed its name grew considerably, with the discovery on the Great Plains of dozens of other sites of the same age and cultural tradition (fig. 1.1). Yet, our knowledge of the type site lagged behind, largely because of the narrow goals of the original excavations, the field and analytical methods in place at the time, and the meager number of publications that emerged from that earlier work, many of which were merely abstracts or discussion comments (e.g., Brown 1928a, 1929, 1936; Bryan 1929, 1937; Cook 1927a, 1928b; Figgins 1927; Hay and Cook 1928, 1930). Ironically, the articles by Cook (1927a) and Figgins (1927) that are routinely cited as the breakthrough publications on Folsom were largely polemical pieces, written over the winter of 1926 to 1927, well before any projectile points had been found in situ, and with a gambler's eye on several other sites they felt provided better evidence of a much older human presence in the Americas than Folsom could muster (chapter 2).

As a result of these historical circumstances, the most basic questions about the Folsom site—its age; its geological history; the environment at the time of the occupation; how the Folsom hunters may have used the landscape to reduce the hunting risks; how thoroughly and in what fashion they butchered the animals; the nature of and variability in their artifacts; how long they may have lingered at the site; what else may have occurred at this locality, besides a bison kill; where these groups came from—and where they headed afterward; the structure, scale, and subsequent taphonomic history of the bonebed—all remained unanswered. In effect, the Folsom type site revealed little about Folsom period adaptations.

That situation was only partly rectified in the 1970s in work done by Adrienne Anderson, then a doctoral candidate at the University of Colorado, and C. Vance Haynes of the University of Arizona, under the rubric of the Folsom Ecology Project. But a detailed investigation of the site, and particularly the Paleoindian bonebed, had not been

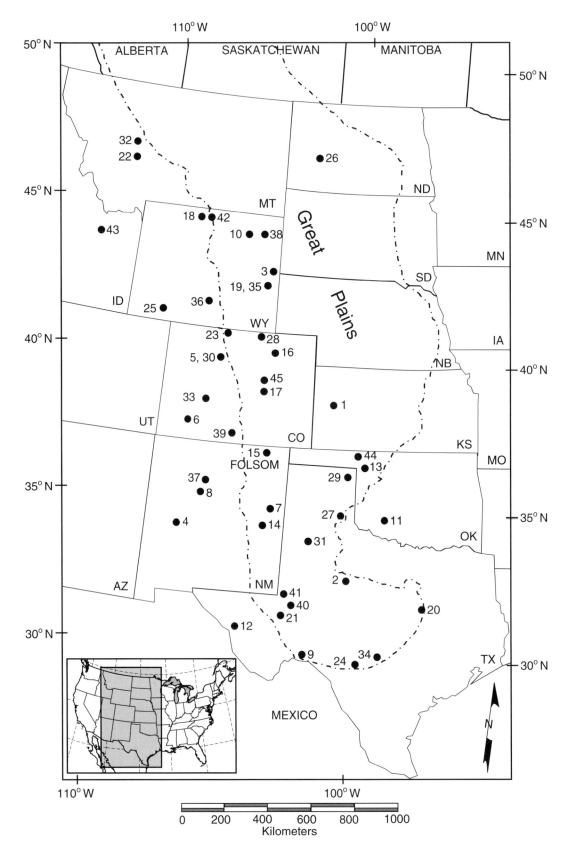

FIGURE 1.1 Folsom sites on the Great Plains and in the Rocky Mountains. Key to site numbers: (1) 12 Mile Creek, KS; (2) Adair-Steadman, TX; (3) Agate Basin, WY; (4) Ake, NM; (5) Barger Gulch Loc B, CO; (6) Black Mountain, CO; (7) Blackwater Locality No 1, NM; (8) Boca Negra Wash, NM; (9) Bonfire Rockshelter, TX; (10) Carter/Kerr-McGee, WY; (11) Cedar Creek, OK; (12) Chispa Creek, TX; (13) Cooper, OK; (14) Elida, NM; (15) Folsom, NM; (16) Fowler-Parrish, CO; (17) Hahn, CO; (18) Hanson, WY; (19) Hell Gap, WY; (20) Horn Shelter, TX; (21) Hot Tubb, TX; (22) Indian Creek, MT; (23) Johnson, CO; (24) Kincaid, TX; (25) Krmpottch, WY; (26) Lake Ho sites, ND: Big Black, Bobtail Wolf, Young-Man-Chief; (27) Lake Theo, TX; (28) Lindenmeier, CO; (29) Lipscomb, TX; (30) Lower Twin Mountain, CO; (31) Lubbock Lake, TX; (32) McHaffle, MT; (33) Mountaineer, CO; (34) Pavo Real, TX; (35) Powars II, WY; (36) Rattlesnake Pass, WY; (37) Rio Rancho, NM; (38) Rocky gciift WY; (39) San Luis Valley sites, CO: Cattle Guard, Linger, Reddin, Zapata; (40) Scharbauer, TX; (41) Shifting Sands, TX; (42) Two Moon Cave, WY; (43) Wasden, ID; (44) Waugh, OK; (45) Westfall, CO.

FIGURE 1.2 Looking west up Wild Horse Arroyo toward Johnson Mesa, 1997. (Photo by D. J. Meltzer.)

undertaken since crews from the American Museum of Natural History left the site in October of 1928.

In an effort to enhance our understanding of the site, an interdisciplinary field project under the auspices of the Quest Archaeological Research Program at SMU was initiated at Folsom in 1997. Fieldwork continued over portions of the 1998 and 1999 seasons. The work, which subsequently included extensive analyses of archaeological and faunal remains recovered during the 1920s, focused on those basic questions just noted, and more (as iterated below). A few very brief preliminary notices and a longer interim article were published on the SMU/QUEST work (Meltzer 2000; Meltzer, Holliday, and Todd 1998, 1999; Meltzer, Todd, and Holliday 2002).

This book presents the full data and analytical results of that investigation. But it is more than that. Because the data from the original investigations were never systematically or completely analyzed, let alone fully published, this is also a report on the 1920s excavations based on our analysis of curated faunal and archaeological collections. The reanalysis of that material was an integral component of our recent fieldwork and investigations, for it was apparent that our understanding of what remained of the site would require—and also be considerably enhanced by—embedding our data and results in the much larger sample of artifact and faunal remains recovered in the 1920s. Similarly, making sense of what came out of the site in the 1920s requires putting these data in the context of the broader understanding of the site's geology and stratigraphy, taphonomic history, paleoenvironmental setting, and archaeology that our recent investigations provide.

All of that is explored in detail in the chapters that follow. It is useful at this point to provide a summary account of the 1920s work, and of the several brief field stints that followed, as a prologue to those chapters and to situate the questions and goals of the reinvestigation that began in 1997.

A Synopsis of Earlier Work

The Folsom site is located in the far northeastern corner of Colfax County, New Mexico, in the Raton Section of the Great Plains physiographic province. The site itself straddles Wild Horse Arroyo, a northwest-southeast trending tributary of the Dry Cimarron River. Both Wild Horse Arroyo and the Dry Cimarron have their headwaters on nearby Johnson Mesa, a prominent regional landform just west of the site (fig. 1.2). This is an area that is relatively dry, receiving an average of just ~41 cm (~16 in.) of precipitation per year, most of which comes during summer thunderstorms, one of which played a critical role in the discovery of the Folsom site.

On the evening of August 27, 1908, a late summer rain began on the eastern side of Johnson Mesa. The storm grew violent and rapidly expanded, and as an estimated 38 cm (~15 in.) of rain fell into the night, the normally placid Dry Cimarron River, which heads on Johnson Mesa just above the site, rose quickly out of its banks. A frantic alarm was telephoned from the Crowfoot Ranch, just below the Mesa, to warn residents of the small village of Folsom, less than a dozen miles downstream, of the advancing tide of water. By the time the floodwaters of the Dry Cimarron reached the town, they were "half a mile wide and . . . at least five feet deep." Rolling walls of water swept through town, destroying property, carrying away livestock, and killing 17 men, women, and children (Guyer 1988:32; McNaghten 1988:33). It was a pivotal event in village history.

FIGURE 1.3 Jesse Figgins (left) and Fred Howarth having lunch near the Folsom site, March 1926. (Photo by H. J. Cook, courtesy of Denver Museum of Nature and Science.)

So too in the history of American archaeology, for the August 1908 storm triggered or accelerated the head-cutting of Wild Horse Arroyo, incising the channel more deeply than it had been before.[3] Sometime after—but how long after, no one knows—George McJunkin, foreman of the Crowfoot Ranch, who tended cattle and broke horses in this area (hence the arroyo's name), spotted large bones eroding out of the arroyo wall ~2 m to 3 m below the surface.[4] As archaeological lore has it, he surmised that bones at that depth were probably old and, on closer inspection, recognized them as slightly mineralized, from a bison, and apparently from a form larger than modern bison. Whether he found artifacts with them has been the subject of much speculation, even some speculative history (e.g., Folsom 1992). But there are no facts bearing on the question. All we really know, and this is fully to McJunkin's credit, is that he must have recognized the bones as being of interest. Otherwise, they simply would have been ignored.

McJunkin spoke of the bones on one of his trips into Raton, New Mexico, when he stopped at the home of blacksmith Carl Schwachheim, who had a kindred curiosity in natural history (T. Burch and E. Burch Hughes, personal communication, 1997; Steen 1955:5). Schwachheim was intrigued by McJunkin's report, and repeated it to Fred Howarth, a Raton banker and amateur naturalist, who often joined Schwachheim on natural history and fossil hunting outings. Yet, it was not until December 10, 1922, sadly, after McJunkin died, that Schwachheim and Howarth, accompanied by several others, first visited the site (appendix B; fig. 2.8).

Intrigued by the fossils, they made several unsuccessful attempts to interest the State of New Mexico in excavating the site; they then went looking for another institution to take an interest (T. Burch, personal communication, 1997). Howarth knew rancher and paleontologist Harold Cook, then Honorary Curator of Paleontology at the Colorado Museum of Natural History in Denver,[5] so arranged to visit the Museum with Schwachheim in January of 1926. There they saw Cook and met Jesse Figgins, the Museum Director (appendix B; Steen 1955:5–6), and told them about the bison remains, some of which they subsequently shipped to Denver. Cook identified them as coming from a previously unknown and apparently extinct species of bison. Their interest piqued, Cook and Figgins joined Howarth and Schwachheim (fig. 1.3) at Folsom on March 7, 1926, deciding on the spot to excavate there with the aim of "supplying a mountable [bison] skeleton" for display at the Museum (Figgins to Taylor, June 21, 1926, JDF/DMNS; appendix B).[6] Importantly, when excavations began in May of 1926, this was considered a "bison quarry," and not an archaeological site (Cook to Barbour, February 15, 1926, EHB/NSM).

The Folsom excavations started on the South Bank of Wild Horse Arroyo and were conducted largely by Schwachheim, with help from several individuals, including Frank Figgins, Jesse's son. A field camp was established on the North Bank, just across from the excavations (appendix C). By mid-June bison bones were being uncovered, and in mid-July the first artifact, the distal end of a Folsom fluted projectile point (DMNS 1391/3), was uncovered, though not in situ (appendix B; July 14, 1926). The discovery was reported to Figgins in Denver, who urged the crew to be more careful and try to find a point in place—then notify him immediately so he could examine the find (Figgins to Howarth, July 22, 1926, DIR/DMNS). Unfortunately, none were.

FIGURE 1.4 The first Folsom point recovered in situ, South Bank, August 31,1927. (Photo courtesy of Denver Museum of Nature and Science.)

That fall, Figgins and Cook wrote their oft-cited papers on the site (Cook 1927a; Figgins 1927); but these came on the heels of the 1926 season, a year before any artifacts were found and examined while still in place within the bonebed, and before the human antiquity controversy came to an end (chapter 2).

During the 1927 field season, the excavation area was expanded but the techniques were the same. Only this time, as a result of an exchange Figgins had with the formidable

Aleš Hrdlička in Washington that spring (chapter 2; also Meltzer 1983, 1994), the crew was explicitly instructed to watch carefully for artifacts, and leave unexcavated any that were spotted in place. Finally, on August 29, 1927, a Folsom point was found in situ, squarely between two ribs of the extinct bison, a fossil snapshot of a hunter's killing thrust (fig. 1.4).

The point was carefully guarded while various individuals and institutions were alerted to come see the evidence in

FIGURE 1.5 Looking west across the South Bank excavations at Folsom, July 1928. Lud Shoemaker (black hat) is working the mule team; Carl Schwachheim stands with his shovel on the right. Note the backdirt berm behind them. (Photo by N. Judd, courtesy of National Anthropological Archives, Smithsonian Institution.)

the ground. Responding to the call were Barnum Brown, vertebrate paleontologist at the American Museum of Natural History; Frank Roberts, archaeologist at the Bureau of American Ethnology, Smithsonian Institution; and A. V. Kidder of the Carnegie Institution of Washington and the leading American archaeologist of his day. All agreed that the point and the bones of the extinct bison were contemporaneous (Meltzer 1983:35–37, 1994), testimony in those preradiocarbon days that humans were in America by at least the latest Pleistocene. That discovery profoundly changed the face of American archaeology (chapter 2; also Kidder 1936; Kroeber 1940; Meltzer 1983).

In order to expand the sample of artifacts and bison remains, and resolve more precisely the age of the site—it was known that the Folsom bison was extinct, but just when that animal went extinct was not (Meltzer 1991b)—the American Museum of Natural History joined the excavations in 1928. Barnum Brown took overall charge, but much of the actual fieldwork (fig. 1.5) was under the immediate supervision of Peter Kaisen, the crew again including the peripatetic Carl Schwachheim.

The 1928 fieldwork substantially expanded the 1926–1927 excavations on the South Bank and, also, stretched across the arroyo the open a small area on the North Bank. Brown's crew was joined by geologist Kirk Bryan of Harvard University, there at the behest of the Smithsonian Institution, who corroborated Brown's assessment of the site's antiquity: Folsom

was late Pleistocene in age (Brown 1928a, 1928b, 1929; Bryan 1929:129).

The faunal remains from the 1926–1927 excavations are housed at the Denver Museum, which until recently displayed the mounted bison they had recovered seven decades earlier. The American Museum also had a mounted skeleton, and between these two institutions several thousand individual bison bones are curated (chapter 7). The stone artifacts from the original investigations are curated at several institutions, including the Denver Museum, which preserves the sediment block displaying the first Folsom point found between the bison ribs; the American Museum of Natural History; the University Museum of the University of Pennsylvania; and several private and smaller museum collections.

The details of the Folsom discovery and original excavations are discussed in a variety of sources. The primary published literature on the 1926–1928 work at the site includes papers by Brown (1928a, 1928b, 1929), Bryan (1929, 1937, 1941), Cook (1927a, 1927b, 1928; also Hay and Cook 1928, 1930), and Figgins (1927, 1928). Also important are the rich archival materials from the 1926–1928 excavations, including letters to and from the field, sketchy field notes and maps, unpublished reports and manuscripts (e.g., Cook 1947, 1952; Hay 1927), and the diary of Carl Schwachheim, the relevant portions of which are presented here as appendix B. These are housed with the archives of the principals, a listing of which is given in the References Cited.

There are also secondary sources on the history and results of the earlier excavations at the site, many of which pay particular attention to George McJunkin's role in the discovery—though not always agreeing on just what that role was (e.g., Agogino 1971; Folsom 1992; Hewett 1971; Hillerman 1971, 1973; Preston 1997; for more general overviews, see Meltzer 1983, 1991b, 1994; Roberts 1939, 1940, 1951; Wormington 1957).

However, none of these papers or reports provides detailed analyses—let alone any of the data—on the artifact or faunal remains from the site. Indeed, the first inventory of the bison remains excavated in the 1920s would not be conducted until nearly 70 years had passed, and then just of the material at the American Museum of Natural History (Todd and Hofman 1991). Only a part of that study was published (Todd, Rapson, and Hofman 1996).

Save for sporadic visits after 1928 (chapter 4), there was no further fieldwork at Folsom for many decades, though Cook had Howarth visit Folsom in July, 1933, to collect charcoal in hopes of getting a tree ring date (Cook to Libby, December 7, 1949, HJC/AHC). Cook submitted the charcoal for radiocarbon dating in 1949, soon after Willard Libby of the University of Chicago invented the technique that would garner his Nobel Prize (Cook to Libby, December 7, 1949; Libby to Cook, December 15, 1949, HJC/AHC). As it happens, however, the charcoal Howarth collected and Cook submitted was not actually from the site (chapter 5), nor did it prove to be Paleoindian in age, much to the consternation of archaeologists in the 1950s eager to know how old the Folsom type site was (e.g., Roberts 1951).

In the early 1970s, Adrienne Anderson returned to the Folsom area to undertake an intensive archaeological site survey and develop a paleoenvironmental record for the region. Limited testing was carried out at Folsom, aimed at determining "(1) the remaining extent of the Folsom-bearing deposit, (2) the feasibility of additional excavation, and (3) the presence of diatoms, snails, pollen, and other information enabling paleoenvironmental reconstruction" (Anderson 1975:19). As a part of that effort samples were also collected for radiocarbon dating by C. V. Haynes, partly with an eye on correlating the site deposits with what was then thought to be the relatively recent eruption of nearby Capulin Volcano (Anderson 1975:39; Anderson and Haynes 1979; Haynes et al. 1992:83–84). At about the same time, local avocationals and a group from Trinidad State Junior College salvaged a relatively large *Bison antiquus* cranium along with several other elements eroding out of the North Bank of Wild Horse Arroyo.

Why Go Back to Folsom?

These brief stints of fieldwork notwithstanding, there was no significant or sustained excavations at Folsom after 1928. There are likely many reasons for this, not least the impression that there was nothing left to excavate after the American Museum had finished its work. By late August of 1928, Kaisen was convinced they had exhausted the site. As he put it, it "look[s] like we got around the Indians Buffalo hunt" (Kaisen to Brown, August 29, 1928, VP/AMNH). The suspicion that virtually nothing remained was frequently repeated (e.g., Brown 1928b; Cook to Jenks, January 11, 1929, HJC/AHC; Howarth to Figgins, October 12, 1928, DIR / DMNS; Brown to Figgins, October 12, 1928, DIR /DMNS). This was apparently a common ploy of Barnum Brown's, intended to keep other paleontologists from jumping his claims (L. Jacobs, personal communication).

Still, with encouragement from both Brown and Figgins, archaeologist A. E. Jenks made a halfhearted stab at mounting a field project at Folsom in 1929, but nothing came of it (Jenks to Cook, April 29, 1929, HJC/AHC; Cook to Howarth, March 12, 1929, HJC/AHC; Figgins to Brown, January 26, 1929; Brown to Figgins, February 1, 1929, DIR/DMNS). Jenks knew, because Harold Cook told him, that such a project would be an expensive gamble given the amount of overburden that had to be removed for what might possibly be very meager archaeological returns (e.g., Cook to Jenks, March 31, 1929, HJC/AHC).

Certainly the same cost/benefit concern gave me pause when, in the mid-1990s, I considered reinvestigating the site. However, there were tantalizing hints in the archives that intact deposits might still be found on the North Bank and along the western margin of the AMNH excavations on the South Bank (e.g., Cook to Jenks, March 31, 1929, HJC/ AGFO; Brown to Figgins, February 1, 1929, DIR/DMNS). Moreover, there were also clues that the western margin of the site might have been where the bison processing had taken place. The prospect of finding intact deposits, particularly ones that might inform on bison butchering, site structure, and activity areas, was appealing.

After all, the 1920s archaeologists and paleontologists were so keen to affirm the association of the artifacts with bison remains, and a Pleistocene bison kill was so novel, that little more was learned at Folsom than that humans were in America for a very long time. In the decades since the discovery of the site, many Folsom-aged bison kills on the Great Plains were excavated and carefully analyzed and studied in detail (reviews in Frison 1991; Hofman and Graham 1998; Jodry 1999a; Stanford 1999). As a result, considerable and valuable information was learned of Folsom hunting strategies, including a measure of the scale of the kills, season of predation, numbers of animals involved, herd composition, bison butchering patterns, carcass utilization, site and bonebed taphonomic processes, etc. (e.g., Frison 1991; Hofman 1994, 1995; Jodry 1999b; Johnson 1987; Todd, 1987a, 1987b, 1991; Todd, Hofman, and Schultz 1990, 1992; Todd, Rapson, and Hofman 1996). Yet, knowledge of the type site did not keep apace with these developments, making it difficult to evaluate how or whether it fit into the larger adaptive patterns marking this period.

Then there was the matter of what else may have occurred here. A much broader picture of Folsom activities and adaptations would surely have emerged were there preserved traces of the areas beyond the confines of the kill, where the hunters lived and worked during their time there (e.g., Stanford 1999). Unlike bonebeds, with their archaeological singleness of purpose and activity, habitation/camp areas have the potential to reveal additional activities, such as the secondary and more intensive processing of the bison carcasses, meat and hide preparation, and refurbishment of weaponry, evidence of other prey species, and construction of structures, all of which holds the promise of revealing a greater diversity and representation of tool forms and debitage, a measure of raw material patterning, lithic tool production, technological organization, settlement scale and mobility, and other habitation activities (e.g., Amick 1995, 1996, 1999a, 1999b; Bamforth 2002; Hofman 1991, 1992, 1999a, 1999b; Hofman, Amick, and Rose 1990; Ingbar 1992, 1994; Ingbar and Hofman 1999; Jodry 1999b; LeTourneau 2000; Sellet 2004). There had been surveys—beginning in the 1920s under Clark Wissler at the American Museum—to locate Folsom-age camp or habitation areas that might have been nearby. None was ever found. But none of those surveys involved excavation beyond the bonebed. Thus, even if the bonebed proved to be exhausted, there was still a potential reward in finding an associated camp.

And if neither a camp nor intact portions of bonebed remained, it nonetheless seemed likely that a field investigation would have a payoff. Geological mapping and paleoecological sampling would surely yield important data on the site's geology, antiquity, paleotopography, stratigraphy, depositional history, and Late Glacial climatic and environmental context, little of which was known, but all of which could help provide insight and understanding of the remains recovered in the 1920s.

There was still another reason to return to Folsom, and that was its historical importance. This was the place that forever changed American archaeology, and while I reopened investigations there with a long agenda of scientific goals and questions, detailed below, not far beneath the surface of that research plan was the inherent interest and challenge of understanding the site where, arguably, American archaeology in the early twentieth century was born.

In the end, when the opportunity arose to work at the site, to have a chance to gain a better understanding of its archaeology and, in so doing, perhaps bring our knowledge of what happened there in Late Glacial times up to the level of our knowledge of what happened there in the 1920s, the opportunity proved impossible to resist. Indeed, there seemed no choice but to go back to Folsom.[7]

A Framework for Reinvestigation

Given the considerable analytical attention that has been paid to Folsom archaeology and Late Glacial environments over the last nearly 80 years, it is perhaps not surprising that there are significant differences of opinion regarding the nature of hunter-gatherer adaptations during this period, differences that cut across virtually all aspects of the Folsom

record. Those issues currently in play range from the role of bison hunting in Folsom subsistence strategies, particularly whether Folsom groups were specialized bison hunters, to what role other animal and plant resources may have played in the diet, to the degree to which Folsom groups aggregated or hunted communally, to whether they were residentially mobile foragers who followed bison herds from kill to kill, or logistically mobile collectors who sent out specialized task groups that would make kills, to the question of whether or to what degree bison or perhaps other prey/food resources structured Folsom group mobility and technological organization, to the use and significance of exotic raw material in Folsom lithic assemblages, to the meaning of technological and stylistic differences in projectile point assemblages, to the effects of Younger Dryas climates on these Late Glacial hunter-gatherers—to name just some of the larger issues on the table (e.g., Ahler and Geib 2000; Amick 1994, 1995, 1996, 1999a; Bamforth 1985, 1988, 1991, 2002; Boldurian and Hubinsky 1994; Bradley 1993; Frison 1991; Frison and Bonnichsen 1996; Frison and Bradley 1990; Frison, Haynes, and Larson 1996; Hofman 1991, 1992, 1994, 1999a, 1999b; Hofman and Graham 1998; Hofman and Todd 2001; Ingbar 1992, 1994; Ingbar and Hofman 1999; Jodry 1999a, 1999b; Kelly and Todd 1988; LaBelle 2004; LaBelle, Seebach, and Andrews 2003; LeTourneau 2000; MacDonald 1998, 1999; Sellet 1999, 2004; Stanford 1999; Todd 1987b, 1991; Tunnell and Johnson 2000).

These issues are not settled here. *Caveat lector.*

There are a couple of reasons why. First, and most obviously, this is an intensive study of a single site, and not a study of the archaeology of the Folsom period or a synthesis of Folsom adaptations and environments. As a result, the questions that can be asked and answered are by design narrower in scope. That does not mean, however, that the evidence from this site has no bearing on some of those larger questions regarding Folsom adaptations. It only means that when we presume to speak to those issues, the warranting arguments and analytical linkages need to be made explicit, so it is clear how and in what way the data from this single site are relevant.

It must also be borne in mind that the subsistence, mobility, technological, and organizational strategies of Folsom groups may have varied considerably over space and time, for the Folsom range extended from the Rocky Mountains into the western margins of the Mississippi Valley and spanned 700 radiocarbon years (Haynes et al. 1992), perhaps a millennium or more calendar years. The data from this site may fail to support a particular hypothesis, but that by itself would not necessarily falsify the hypothesis and might be far more interesting for what it might reveal of adaptive variability in Folsom times.

Second, and perhaps less obviously, large bison kills like Folsom, despite their visibility and profound influence on our interpretations of Folsom period adaptations, represent less than 5% of all known Folsom localities (LaBelle, Seebach, and Andrews 2003; also Bamforth 1988; Frison,

Haynes, and Larson 1996; Hofman and Todd 2001). Many of the questions about Folsom adaptations that have arisen over the years are a result of the increasing realization that the archaeological record of this period is not just composed of large bison kills, but is dominated by hundreds of smaller kill sites, quarry localities, lithic scatters, and isolated fluted point finds (Amick 1994; Blackmar 2001; Frison, Haynes, and Larson 1996; Hofman 1999b; Jodry 1999a; LaBelle 2005; LaBelle, Seebach, and Andrews 2003; Largent, Waters, and Carlson 1991; LeTourneau 2000). Under the circumstances, data and evidence from the kind of site that is part of the interpretive problem may not be altogether useful in building an analytical solution.

Still, it is also the case that interpretive myths (sensu Binford 1981) can emerge around certain classes of sites, and therefore clarifying the precise nature of the evidence from the Folsom bison kill—which is, after all, the type site—can perhaps help clear away some of the haze.

In this section, then, let me explore the major analytical questions that guided the investigations and that I hoped to answer—if only in part—at the Folsom site, highlighting both those that bear strictly on the archaeological record of this particular locality and then those that potentially can inform on some of those larger issues. These are only the broader questions; many narrower ones pertaining to specific data sets will be found in the relevant chapters. I begin this discussion at the site-specific level and range upward from there, though with frequent slides back down the analytical scale.

Following the archaeological questions and issues, I turn briefly to analytical questions regarding the history of the archaeological work at Folsom site and the site's signal role in the resolution of the human antiquity controversy in North America. I return to all of these in chapter 9.

Folsom Paleoindians: Open Questions and Unresolved Issues

ARE THERE INTACT ARCHAEOLOGICAL DEPOSITS REMAINING AT THE FOLSOM SITE?

On its face, this is the most mundane of questions, but it has to be asked and answered straight off, for if Kaisen was correct about having "got around the Indians Buffalo hunt," there would be significant limits on what a reinvestigation of the site might accomplish. There would be far less to learn about bonebed structure, butchering and processing areas, bonebed taphonomy, other possible aspects of Folsom subsistence strategies, or whether any evidence of an associated camp or habitation remained, were it necessary to rely entirely on the results of the 1920s excavations. Answering this question initially required determining the limits of the original excavations, and identifying and mapping the extent and distribution of Folsom age deposits—matters which were at the top of the research agenda in 1997 when we began work

at the site. Fortunately, we quickly discovered that Kaisen was wrong (chapter 4).

WHAT IS THE GEOLOGICAL HISTORY AND CONTEXT OF THE FOLSOM SITE?

The original geological work at Folsom by Brown, Bryan, and Cook was done in the bonebed on the South Bank of Wild Horse Arroyo, in and around the original excavations. There was some discrepancy in their interpretations. Cook (1927a:244) described the deposits in which the bonebed occurred as swampy and marshy, a muddy bottom in which freshwater invertebrates occur. Yet, Brown (1928b) identified those same gastropods as "pulmonate land shells" and the sediments enclosing them and the bonebed as being aeolian in origin—albeit filling an old stream course.

The more recent stratigraphic studies and radiocarbon dating (e.g., Haynes et al. 1992) were conducted primarily on the North Bank—where the precise position of the Folsom bonebed was not known, and where it was suspected that redeposition of materials had occurred. Although this study helped refine the age of the occupation, in 1997 much remained to be done (Haynes et al. 1992:87). Little was known of the geological processes affecting the Folsom site, before, during, and after the occupation; it was unclear whether or how the geological history of the North and South Banks of the site differed; the stratigraphic context of the bonebed and any other potential Lake Glacial age surfaces needed to be better understood, if only to help determine how and why the site formed where it did and how it did and what might remain of it. These are matters taken up in chapter 5.

WHAT IS THE AGE OF THE FOLSOM BISON KILL?

Libby's effort to radiocarbon date the Paleoindian occupation at the site failed (Roberts 1951). The failure cannot be laid at the doorstep of radiocarbon dating, even though the technique was still very much in its early phase of development. The age, as it happens, was likely valid (chapter 5)—just not relevant to the Paleoindian occupation. Users of the new technique were themselves not up to the task of the importance of selecting samples in a way that they bore on the archaeological event of interest.

Later efforts to date the site proved more successful, in that they produced ages within the known temporal range of the Folsom period (Anderson and Haynes 1979; Haynes et al. 1992; see Holliday 2000b). Yet, there was a dichotomy in the resulting ages: A sample run on bison bone collagen had yielded an age of 10,260 ± 110 ^{14}C yr B.P. (SMU-179), while a cluster of ages from charcoal fragments had yielded an average age of ~10,890 ^{14}C yr B.P.—a difference of 600 radiocarbon years (Haynes et al. 1992:87). Even granting a 2σ variation, these represent significantly different ages. Haynes et al. (1992) believed the older cluster was more accurate, and reasonably so, given concerns at the time about the reliability of bone dating. However, there was also a suspicion that the dated charcoal might be unrelated to the human occupation of the site. Moreover, it was obtained from the North Bank, and not directly from the bonebed on the South Bank, making its association with the archaeological event unknown. Obviously, if reliable ages on bison bone using modern techniques (e.g., Stafford et al. 1987, 1991) could be obtained, they could more directly pin down the age of the Paleoindian kill (chapter 5).

WHAT WAS THE CLIMATE AT THE TIME OF THE FOLSOM SITE OCCUPATION?

Virtually nothing was learned of this matter in the 1920s, as it was not then customary to ask such questions. Brown (1928b), however, observing the aeolian character of the sediments, inferred that they must have accumulated "during a long period of little or no rainfall."

More recently, Holliday (2000a) suggested—based on evidence from the age and distribution of aeolian sediments and variations in stable carbon isotopes—that the Southern Plains was subjected to significant, rapid fluctuations in temperature and moisture in the last millennia of the Pleistocene, and that the Folsom period was one of episodic drought. However, Haynes (1991) earlier argued that the Folsom period was a time of a net increase in effective precipitation, the result of reduced evaporation. He puts a drying period in the preceding Clovis times. Both may be correct about the climate of Folsom times, for the issue is one of scale and variability, both temporal and spatial.

We now know the Folsom period corresponds neatly with the Younger Dryas Chronozone, a geologically rapid reversal of Pleistocene deglaciation that brought a brief—thousand-year—return to cold glacial conditions (Allen and Anderson 1993; Clark, Alley, and Pollard 1999, Clark et al. 2001, 2002; Clarke et al. 2003; Mayewski et al. 1993, 1994; Peteet 1995; Severinghaus and Brook 1999; K. Taylor et al. 1997; Teller, Leverington, and Mann 2002; Yu and Wright 2001). The Younger Dryas is a dramatic example of rapid climate change, at least on a geological timescale. It is further assumed to have been detectable on a human timescale, with the result that the climatic changes of the Younger Dryas are increasingly invoked as an explanatory mechanism for culture change, from broad transitions such as the advent of agriculture (Richerson, Boyd, and Betinger 2001) to finer-grained cultural changes such as the disappearance of fluting in lanceolate projectile points of New England (Newby et al. 2004). Leaving aside the analytical challenges of linking climate and cultural change (Meltzer 2004), recent research has shown that the effects of the Younger Dryas were not uniform across space or through time, or uniformly severe (e.g., Shuman et al. 2002; Williams, Shuman, and Webb 2001; Yu and Wright 2001).

There are apparent Younger Dryas effects recorded in pollen and sediment cores from localities in the Rocky Mountains several hundred kilometers north of and at elevations nearly 1000 m higher than Folsom (e.g., Fall 1997; Markgraf and Scott 1981; Reasoner and Jodry 2000), but how this climatic episode played out at Folsom, the scale at which it played out, and how it might have affected these hunter-gatherers, or the resources on which they depended, are not known. What is known, using models of hunter-gatherer adaptations (Binford 2001) based on the present climate and environment of the Folsom area (chapter 3), is that this is an area capable of supporting logistical hunting forays, but not necessarily long-term forager residence or—because of heavy snowfall and the nature of the resource base—overwintering by hunter-gatherers. Whether that was also true of Late Glacial times is examined in chapter 6.

WHAT BIOTIC RESOURCES WERE AVAILABLE TO HUNTER-GATHERERS AT THE TIME OF THE FOLSOM SITE OCCUPATION?

The largest mammal on the Folsom landscape today is elk or wapiti (*Cervus elaphus*). Bison are today absent, save for a small boutique herd, but were recorded historically in this area of northeastern New Mexico, though this was considered the western margin of their range (Bailey 1931). Bison were obviously present on this landscape in the Younger Dryas. But was that herd here year-round, or were they present on the landscape only seasonally, say, in the summer? If bison did inhabit the region in winter, what were they feeding on, and what might that reveal of winter temperature and precipitation, given bison tolerances for cold and snow (Guthrie 1990; Telfer and Kelsall 1984)? Was their presence predictable, and were their numbers abundant?

What other fauna—and, for that matter, floral resources—might have been available for exploitation by hunter-gatherers? This is not an area that, at the moment, provides sufficient plant resources for long-term occupation (chapter 3); but were conditions different in the Late Glacial? Knowing something of the structure of the biotic community as well as the climate can shed light on what Paleoindians may have been doing in this area—and how long they may have lingered.

Ultimately, answers to the questions regarding the climate and environment at the Folsom site will have to be obtained from relevant data acquired at the site or in the surrounding region. And because different paleoecological indicators—such as pollen, macrofossils, soils, gastropods, bison, and stable isotopes—sample and record the climate and environment at different spatial and temporal scales, a reliable reconstruction requires suites of converging evidence.

Once obtained, however, those data must still be linked to the occupation at the Folsom site, a matter requiring continued attention to matters of scale, for evidence in the geological record and that projected by climate models provide patterns that are time-resolvable in the very best of circumstances to decades, and more commonly to centuries and millennia. Hunter-gatherers and the resources they exploited adapt to daily, seasonal, or annual changes in the weather, changes occurring on a much more rapid, human timescale.

The data on Folsom climate and environment, and the warranting arguments linking those data to the Paleoindian occupation, are detailed in chapter 6.

DID FOLSOM GROUPS OCCUPY PROTECTED FOOTHILLS AND INTERMONTANE BASINS DURING THE COLD SEASON?

The specifics of the climate and environment at Folsom tie in to a larger debate over Folsom settlement systems. Based on an extensive analysis of surface assemblages, Amick (1996:413; also Hofman 1999b) argued that Folsom groups in the Southern Plains and the Basin and Range Provinces seasonally varied their subsistence and settlement strategies, corresponding to annual changes in resource availability. Behind his argument is the assumption that the key subsistence resource of Folsom groups—bison—would themselves have favored protected foothills and intermontane basins in the cold season, like the area around Folsom, and then shifted to the grasslands and marshes of the open Plains during the warm season. Accordingly, Amick argues that Folsom groups, mapping on to their prey, would have shown a similar annual and seasonal pattern of movement.

Located in a small valley just below Johnson Mesa, the Folsom site is particularly well situated to provide a test of Amick's (1996) hypothesis about the use of such topographic settings during the cooler months of the year. Initial seasonality estimates put the Folsom occupation in the late fall or early winter (Frison 1991:159; Todd, Rapson, and Hofman 1996), potentially putting people and bison in a position to overwinter in this area. Testing this hypothesis would require asking, and if possible answering, the questions of whether bison could and did overwinter here; whether other animal resources, such as elk, deer, and bighorn sheep, or plants suitable for human consumption were available; and whether this was a single kill or a series of winter-long, multiple, closely spaced kills of bison and other animals as at Agate Basin, for example (Frison 1982a; Hill 2001:249); as well as whether any camp or habitation area associated with the kill exhibits structures, storage structures, meat caches, or other evidence of long term cold-weather habitation (Binford 1993; Frison 1982b). These matters will be taken up in chapters 6 and 7.

DID FOLSOM HUNTER-GATHERERS EXPLOIT FAUNAL RESOURCES OTHER THAN BISON?

Embedded within the overwintering hypothesis is the assumption Folsom groups were not strictly bison specialists but had a broader diet breadth, which included the use of seasonally available resources, possibly including medium and small mammals as well as plants, the latter increasing in importance as the average search and transport time for the larger bodied resources increases (Cannon 2003:12). Animals such as pronghorn antelope, for example, can maintain higher fat levels throughout the year and, thus, provide a critical balance to a bison-dominated diet (Hill 1994:125). Pronghorn occur, in fact, in the Folsom levels at

Agate Basin (Frison 1982a; Hill 1994, 2001), among other sites (Wilmsen and Roberts 1978). Admittedly, those occurrences are relatively rare (Bamforth 1988).

The possibility that Folsom subsistence strategies might have been broader than supposed is a provocative hypothesis, and one that implies adaptive strategies more in keeping with what is generally known of human forager behavior (Kelly 1995). This hypothesis remains largely untested, however, partly because it has had to rely on indirect evidence of subsistence and adaptation inferred from tool stone distribution, and mostly because few sites have provided the kind of detailed evidence needed to test it (LaBelle 2004). The attention to bison kills has overshadowed other potential aspects of Folsom adaptations and biases interpretations of their diet and hunting tactics (e.g., MacDonald 1998). Arguably, the role of large-bodied prey in Folsom diets (and those of Paleoindians generally) may be overrepresented relative to their actual importance (see also Cannon and Meltzer 2004; O'Connell, Hawkes, and Blurton-Jones 1992:338–339).

This is especially intriguing in this instance, not just because Folsom was the defining Paleoindian bison kill site, but because the 1920s excavations also yielded the remains of "five other species of smaller mammals" (Brown 1928b; Hay and Cook 1930). While most of those were isolated jaws of burrowing animals, and thus not likely associated with human activities, fragments of deer bone were also recovered. Excavation techniques being what they were in the 1920s, it is not clear whether the deer were Paleoindian prey. Determining whether this or other species, if any, were exploited, will require close examination of any additional nonbison species recovered from the site or from any associated camp areas, issues addressed in chapters 5 and 7.

WHAT MIGHT BE INFERRED OF THE TACTICS AND STRATEGIES OF BISON HUNTING AT THE FOLSOM SITE?

There is no doubt that bison were a food source in Folsom times, but leaving aside the question of just how important this resource was, consider a more practical issue. Hunting large mammals—and Late Glacial *Bison antiquus* were very large mammals, perhaps 20% larger than modern Bison bison (Hofman and Todd 2001:206)—is risky (Hawkes, O'Connell, and Blurton-Jones 2001). Risk is used here in two senses: the economic risk of uncertain returns and the more colloquial reference to the physical danger of hunting large mammals (e.g., Bamforth and Bleed 1997; Binford 2001; Christenson 1982; Hawkes and Bleige-Bird 2002; Lee 1979; Meltzer 1993; Nelson 1969; Silberbauer 1981; Smith 1991).

Folsom hunters were extremely adept at preying on bison, using, as Frison (1991:155) observes, "great ingenuity" in their hunting strategies and tactics to reduce the elements of risk. More than 30 bison were killed at the Folsom site, and early on there was speculation about how the hunters may have used aspects of the topography to accomplish that (Brown 1928b; Wissler 1928) but little hard evidence. Had they used impermanent features of the landscape, such as brush or snow banks, to help maneuver the bison herd, it would not, of course, be visible archaeologically. But if they used the natural topography to advantage—and Frison (1991:156; also Frison, Haynes, and Larson 1996:214) argues that headcuts on dry arroyos were an oft-used landscape feature—and if such could be discerned at Folsom, then that might shed light on how this group of hunters mediated the danger of the hunt.

The archaeological difficulty faced here, of course, is that arroyos are actively eroding features of the landscape and may not be visible 10,000 years later, or, if visible, may not preserve the remains of the kill that occurred within them (Frison 1991). As Albanese put it, "Any preserved kill sites are geological oddities" (Albanese 1978:61). This raises two specific questions: first, What is the setting within which the kill occurred, and how might it have been used by the Folsom hunters; and, second, What geological mechanisms led to the preservation of the site? Chapters 5 and 7 explore these matters.

All this might also shed light on the related question of whether this was an ambush/intercept or an encounter hunt (Binford 1978a, 1978b; Hofman 1999b). Among ethnographic hunters the former tend to occur on landscapes where the appearance of prey is relatively predictable, perhaps owing to localized resources or perennial water holes; the latter are less tied to features of the place (Binford 1978a, 1978b; Hawkes, O'Connell, and Blurton-Jones 2001). Some have suggested that ambush kills were a common strategy among Folsom groups (Stanford 1999:300); others, that encounter hunting was more likely (LaBelle, Seebach, and Andrews 2003). Knowing the geology and paleohydrology of the Folsom locality, particularly if this was a period of aridity, as well as what food resources might have been available to bison populations in the area (chapter 6), might shed light on this matter, or at least how it played out in this particular locality.

A related issue is whether the kill at Folsom was the product of a communal effort and, more particularly, whether groups aggregated for the purposes of communal hunting (e.g., Bamforth 1985, 1988, 1991; Hofman 1994; Jodry 1999a:330). The two concepts—communal and aggregation—are often used interchangeably, yet I think it is important to keep them separate. Communal hunting merely implies a cooperative effort among a group of hunters. Aggregation bespeaks the coming together of otherwise seasonally dispersed groups for a variety of purposes, including the exchange of information, resources, and mates and, of course, pooling labor toward a large and labor-demanding common task—like procuring an ample store of bison meat (Bamforth 1988:24–25; Hofman 1994; Jodry 1999a:262). For that matter, the archaeological literature rarely provides guidance as to what distinguishes "communal" from noncommunal hunting, usually only contrasting communal with individual or small-group hunting. In this regard, as Kelly (1995:218; also Binford 2001) suggests, the appropriate

question is not whether foragers hunt individually or communally, but the optimal size of the foraging party. The answer, of course, depends on the nature and structure of the resources (Binford 2001) and on the return rates a forager could expect if working alone or in a larger group (Kelly 1995; Smith 1991). The size of the foraging group at Folsom—optimal or otherwise—will likely elude us.

Still, it might be possible to assess whether an aggregation occurred here. Aggregation among hunter-gatherers occurs at different intervals, with smaller events taking place on a seasonal or annual basis, and larger ones on a multiyear basis (Binford 2001). Given the distances groups may have had to travel to aggregate, it would have been difficult to carry large stores of food, so aggregation sites were likely selected for the richness of its food resources rather than, say, the presence of abundant lithic raw material. Sufficient food would offset the costs of aggregation and allow communal foraging (Kelly 1995:219–221). Several have argued that seasonal aggregation for bison hunting was carried out by Folsom groups (e.g., Bamforth 1988, 1991; Fawcett 1987; Greiser 1985; Jodry 1999a; Wilmsen and Roberts 1978), but such claims have been received with some skepticism (Hofman 1994). Obviously, before claims of any general patterns are made, each site needs to be evaluated for the specific evidence it provides of aggregation, as opposed to the apparent presence of a "communal" labor pool no larger than what might be expected of a logistical task group (Binford 2001).

IS THERE EVIDENCE FOR MORE THAN ONE BISON KILL AT FOLSOM?

Most Folsom kills are single events (Bamforth 2002:57; Frison 1991; Stanford 1999), although there are apparent exceptions (Bement 1999b; Frison 1982a). There are arguments as to why the dominant pattern occurs, which are rooted in several assumptions about Folsom mobility and hunting strategies (LaBelle, Seebach, and Andrews 2003). These assumptions are, first, that Folsom groups more commonly practiced logistical as opposed to residential mobility (LaBelle, Seebach, and Andrews 2003)—that is, small task groups moved to/from resources, rather than the entire band or local group (Binford 1980; Kelly 1995); second, that Folsom hunting was based more on an encounter strategy rather than an ambush-intercept strategy (LaBelle 2005; LaBelle, Seebach, and Andrews 2003); and, finally, that logistical mobility and encounter hunting are a response to a resource base assumed to be patchily distributed on the landscape, making the use of logistical forays by hunting parties a more efficient foraging strategy.

A contrasting situation—Folsom groups moving in a residential fashion from bison kill to bison kill over vast areas (Kelly and Todd 1988)—would produce multiactivity sites in which camps are associated with kill and processing areas, as at Cattle Guard (Jodry 1999b) and Shifting Sands (Hofman, Amick, and Rose 1990). These sites are less common (LaBelle, Seebach, and Andrews 2003). More common are large residential sites in which multiple prey taxa are represented, and there is evidence for a range of activities taking place over a relatively long-term occupation—as in a winter camp. These include the sites of Agate Basin (Frison 1982a), Blackwater Locality No. 1 (Hester 1972), and Lubbock Lake (Bamforth 1985; Johnson 1987). Most of these occur in areas where stable resources are found in conjunction with favorable landscape features— such as water sources, quarries, and lookouts (Hawkes, O'Connell, and Blurton-Jones 2001).

Rarest of all are Folsom sites in which only a single activity—bison killing—takes place repeatedly at the same locality. Such sites were common in Late Prehistoric times (e.g., Head-Smashed-In, Alberta; Vore, Wyoming), where groups undertook large-scale communal hunts at strategic spots on the landscape where large herds of bison could be anticipated and that had topographic features well suited to disadvantaging animals—like the sinkhole at Vore and the jump at Head-Smashed-In—and where other resources to sustain a groups' residence in an area were present. The only possible Folsom example of this is the Cooper site (Bement 1999b), but whether those same topographic, ecological, and archaeological conditions obtained there has been questioned (LaBelle 2000).

Thus, the question of whether there was one or more than one kill at Folsom has broader entailments and may inform on the nature of mobility—whether logistical or residential—the length of time spent at the site, or whether the area had resources or topographic features that especially suited it for bison hunting or other activities.

How many kills took place here was not known from the earlier work at the site (Frison 1991:159). Compounding the ability to answer the question is the limited window of most archaeological excavations, including the one at Folsom, and the correspondingly large area around a kill site where associated archaeological camp or habitation debris might be found (O'Connell 1987; O'Connell, Hawkes, and Blurton-Jones 1992). Before any conclusions are drawn about Folsom fitting a particular pattern, the stratigraphic context of the bonebed must be examined, and testing must be conducted in the areas outside the bonebed, to see what other remains might occur (chapters 4 and 5).

WHAT WAS THE NATURE OF THE BUTCHERING AND PROCESSING OF THE BISON AT THE FOLSOM SITE?

The preceding issues raise the question of whether the Folsom bison remains display a "gourmet" butchering strategy or more intensive processing. The 1920s excavators merely reported the presence of bison skeletons and an area of the site in which bones were "more or less mixed." The impression given from both published and unpublished sources was that whole, articulated bison skeletons dominated the faunal assemblage (e.g., Brown 1928a, 1928b; Bryan 1937:141), suggesting that very little processing had taken place here.

While such a pattern of "light butchering" is present at some Folsom sites (e.g., Bement 1997:158–161), in other Folsom localities there is considerably more processing and often removal of selective elements, such as high-utility parts (Frison 1991; Todd 1991). In many of the latter instances, it appears that bones were removed from the site as "complete limb units rather than as segmented subsets," suggesting that bulk processing took place beyond the kill area, perhaps in nearby camps or more distant habitation sites (Todd 1991:224, 229). Variations on such patterns are well documented ethnoarchaeologically (e.g., Binford 1981; Cannon 2003; Hawkes, O'Connell, and Blurton-Jones 2001; O'Connell, Hawkes, and Blurton-Jones, 1990, 1992).

Then there is the more specific issue of whether bison fat was specifically targeted by hunters. Fat plays a critical role in the human diet (Driver 1990; Frison 1982b; Speth 1983; Speth and Spielmann 1983; Todd 1991; Wandsnider 1997), particularly during the winter months when bison, whose meat is otherwise not noted for its high fat content (Brink 1997; Wandsnider 1997:20), are at their leanest. Although we would expect Folsom groups to maximize all the fat resources available from their kills, that does not appear to be the case, at least in most of the relatively large and representative Folsom-age bison kills (e.g., Bement 1999b; Frison 1982b; Todd 1991; Todd, Hofman, and Schultz 1992). Todd (1991:230) and others (e.g., Jodry 1999b) attribute this apparent "indifference" of Folsom hunters to fat resources— which contrasts dramatically with bison processing in later time periods—to less strongly seasonal Late Glacial climates, which would have reduced the length of time for which bison were resource-stressed. But there are also hints that this general pattern may mask subtle seasonal variation in carcass utilization (Hill 1994:126, 139; but see Hill 2001:104–105), and perhaps a kill made in the early fall when animals were at their prime (Frison 1982b:201), the season the Folsom kill may have occurred, could shed light on the use of the fat of these animals.

How intensively and in what manner these groups were processing animals and preparing them for transport has a bearing on their degree of settlement mobility (Cannon 2003:12; Metcalf and Barlow 1992). Among modern hunter-gatherers, residential camps are sometimes moved to a kill, rather than the products of the kill being moved to a camp. Kelly and Todd (1988:236) believe that Folsom hunters would have moved residential groups "from kill to kill." Which strategy is adopted and when depends on many circumstances, among them the number and size of the carcasses, the number of available carriers, the distance to an offsite camp, and the climate/season when the kill took place (Bartram 1993:121; Cannon 2003:4; Emerson 1993:139–140, 150; Lee 1979:220). Transport distance and cost are especially relevant, since the full weight of the prey (bison) is greater than the heaviest load a pedestrian forager can transport any significant distance, which is probably in the range of 20 kg to 45 kg (Cannon 2003:6; O'Connell, Hawkes, and Blurton-Jones 1988, 1990). Still, large animal kills permit transport decisions based more on body part utility than do small animal kills; one can afford to be selective with regard to what is transported when the animals are large and abundant (Cannon 2003:4; Emerson 1993:139). If the kill is not made by a residential group, then carcasses have to be processed for transport. The more time a hunter spends "processing a carcass in the field, the more parts of low food value should be removed from the load that is taken home, so that the utility of that load, measured in calories per unit weight, is increased. If more time is spent field-processing carcasses as transport distance increases, then a smaller proportion of low utility parts should be taken home when more distant patches are used" (Cannon 2003:4). The climate/season comes into play in this equation, insofar as temperature and precipitation influence how easily groups could have butchered the animals, removed the meat from the bones, and dried it (thereby lightening the load) for transport (Bartram 1993:121, 131–132; Cannon 2003; Frison 1991). Those discarded low-utility parts should, of course, be left at the kill.

Knowing the patterns of carcass processing at Folsom would provide further insight into the nature and context of the activities there; a gauge, perhaps, of the time spent at the site and whether Paleoindians overwintered here; the degree to which elements of different kinds were transported off-site; and how much may have been transported rather than left behind. LaBelle, Seebach, and Andrews (2003) observe that there is an element of equifinality to the equation, in that "gourmet" butchering could reflect residential foraging groups moving rapidly from kill to kill or the presence of smaller task groups who, for lack of sufficient labor to process and transport the meat, had to leave much of it behind. To the degree that these matters can be resolved, it will require detailed analysis of the extant collections, careful inventory of elements left behind in the kill area, and examination of any clues about the spatial patterning within any portions of the bonebed that might still exist, all of which are taken up in chapter 7.

WHAT IS THE TAPHONOMIC HISTORY OF THE FOLSOM BISON BONEBED?

Of course, before any far-reaching conclusions are drawn about the character and degree of carcass processing and transport, careful attention must be paid to the postdepositional taphonomic processes effecting the faunal remains (e.g., Binford 1981; Frison and Todd 1986; Frison, Haynes, and Larson 1996; Kreutzer 1988, 1996; Todd 1987a). Otherwise, one runs the risk of attributing patterns in skeletal element occurrence and frequency or surface modifications to cultural activity, which might actually result from carnivore action, weathering and exposure, fluvial transport, or some other nonhuman agency.

Taphonomy was not a matter of concern or attention during the original Folsom site excavations; the concept was scarcely known then. But faunal material collected during those investigations can still be examined for evidence of taphonomic processes, insofar as the bone is reasonably well preserved and curated. Such a study, however, also requires simultaneous attention to the geological context in which the remains were recovered, to understand the mechanisms that might be operative in a particular setting and that might account for the observed patterns. Much of the attention in chapter 7 focuses on Folsom bonebed taphonomy.

WHERE OR HOW DID THE FOLSOM HUNTERS PROCURE THE STONE FOR THEIR TOOLKITS?

Early on in the investigations at Folsom, Brown and others recognized some of the fluted points as being manufactured of Alibates agatized dolomite from the Texas Panhandle several hundred kilometers southeast of Folsom. That realization was not accorded particular notice, save for Figgins' suggestion that someone ought to go to Alibates to look for where the Folsom group had camped (chapter 8).

Nowadays, of course, the source(s) of the stone used in Paleoindian assemblages is examined more closely, for the information it provides of the scale of their mobility (e.g., Amick 1995, 1996; Hofman 1990, 1991, 1999b). Routinely discarded in Folsom sites are projectile points made of high-quality stone often acquired from sources hundreds of kilometers distant. Even if one makes the unrealistic assumption that they moved in unerringly straight lines from source to site, Folsom groups were still, arguably, among the most widely traveled pedestrian hunter-gatherers in prehistory (Amick 1996:441). Hofman (1991; Hofman, Todd, and Collins 1991), for example, identified several points in the Folsom site assemblage as having been made of Edwards chert, the nearest outcrop of which is 575 km away (but see chapter 8).

Moreover, in recent years distinctive geographic patterns in the use of particular lithic raw material have been observed in Folsom sites and assemblages (e.g., Amick 1994; Hofman 1991; Hofman, Todd, and Collins 1991; Jodry 1999b; MacDonald 1999; Stanford 1999; Wyckoff 1999). This includes directional trends to the movement of Edwards chert, Alibates agatized dolomite, and Tecovas jasper, which dominate Folsom assemblages on the Southern Plains (Hofman 1999b:406; Wyckoff 1999). These patterns are interpreted by Stanford as possibly defining "boundaries of traditional areas of exploitation by independent Folsom bands" (Stanford 1999:303; also Amick 1996; Hofman 1994; MacDonald 1999; Hester and Grady 1977). Others have suggested that they may also reflect the wide-ranging search for mates by individuals in regions of highly dispersed and low-density populations or small-scale exchange for the purpose of maintaining alliances (MacDonald 1998:227). Still others have attributed the acquisition and discard patterns to the unpredictability of movement on the Folsom landscape (Bamforth and Bleed 1997:133).

Each of these hypotheses may be correct in whole or in part, and though the data from this particular site cannot fully test these notions, it can nonetheless provide additional evidence of the geographic space across which this particular group traveled and the points on the map at which stone was acquired (chapter 8). But that evidence must be tempered by an awareness of the disparity that often occurs between the quality of stone used in the manufacture of projectile points and that used in the production of less formal tools, a disparity that is often congruent with the use of exotic versus local lithic raw material (Bamforth 2002). Looking only at projectile points—and a bonebed assemblage often yields scarcely more than that—may reveal little of how other artifacts were incorporated into the toolkit.

It is also important to keep in mind the prospect that the stone was acquired via exchange, and not directly at the rock outcrop by the group who used it (e.g., Hayden 1982). If that occurred, it would badly skew our interpretation of the scale of mobility (Kelly 1992). For a variety of reasons, I am skeptical that exchange played any significant role in lithic raw material procurement at this early stage in North American prehistory (Meltzer 1989b; also Bamforth 2002; Hofman 1992; G. Jones et al. 2003). Nonetheless, the possibility bears watching, especially since by this time in prehistory there were other populations and groups—not all of whom were Folsom—on the North American landscape.

So too does the diversity of the lithic raw material, for it may provide vital data on the question of whether an aggregation occurred here, on the supposition that dispersed groups converging from different points on the landscape would come equipped with different kinds of stone (Bamforth 1991; Hofman 1994; Jodry 1999a). Sorting these various processes raises a challenging problem of equifinality, for direct procurement, indirect procurement/exchange, and aggregation can all leave similar archaeological products (Meltzer 1989b; also G. Jones et al. 2003; Kelly 1992). But the problem may not be wholly intractable (chapter 8).

HOW DID THE FOLSOM HUNTER-GATHERERS ORGANIZE THEIR TECHNOLOGY?

Relying on stone acquired from distant sources requires that hunter-gatherers solve the twin problems of resource incongruence—the disparity between where stone is acquired and where it is put to use and time stress—the gap between when stone is acquired and when it is put to use (Amick 1994:22; G. Jones et al. 2003). Much fruitful discussion has taken place on these matters over the last decade (e.g., Amick 1994, 1996, 1999a; Bamforth 2002; Bamforth and Bleed 1997; Bradley 1993; Hofman 1991, 1992; Ingbar 1992, 1994; Ingbar and Hofman 1999; Jodry 1999b; G. Jones et al. 2003; Kelly 1992; Kelly and Todd 1988; LeTourneau 2000; Sellet 1999, 2004). But not all of it has a bearing on the Folsom case, since, so far at least, the assemblage is comprised almost entirely of projectile points. Projectile points from kill sites are, again, a narrow technological window within which to view larger organizational patterns

(Bamforth 2002:62). Nonetheless, and until such time as a wider variety of tools is recovered, this assemblage can be used to probe aspects of Folsom technological organization, starting with the strategy of toolkit provisioning, and the related matter of the timing and role of fluting.

Put in its barest terms, mobile foragers can opt to provision a toolkit continuously, replacing tools as they break or wear out, or they can replace it wholesale in "gearing-up" bursts in which a surplus of tools is produced for future needs (Binford 1979; Hofman 1992; Ingbar 1992, 1994; Kuhn 1989; Sellet 1999, 2004). The former strategy requires a somewhat regular supply of stone, its success obviously depending on the abundance and distribution of sources in the region through which a group is moving (Bamforth and Bleed 1997; Sellet 2004), and/or the ability to judiciously recycle tools to extend their use-lives (Hofman 1992). Gearing up, in contrast, demands that a large amount of stone be available all at once, as at a quarry or outcrop.

Gearing up and gradual replacement are situational strategies that were assuredly used by the same group at different times of the year (Sellet 2004). Nonetheless, they may be correlated with certain foraging and mobility conditions, and possibly intersect the question of discerning ambush versus encounter hunting. Hofman (1992:198), for example, argues that Late Prehistoric bison hunters may have geared up once or twice a year, because it was usually known in advance when and where a bison hunt would occur (e.g., Reher and Frison 1980). In contrast, he believes that Folsom groups, because they "may have lived from a relatively steady sequence of kills throughout the year," had to be constantly maintaining their toolkits (Hofman 1992:198). As the locations of those kills were unpredictable and the demands of hunting might not have allowed for special visits to a quarry, they had to maintain their toolkits using the stone they had available—which is why, as that supply dwindled across the long stretches of the Plains landscape where stone sources were rare, tactics for toolkit maintenance were brought into play (Hofman 1992).

As Sellet suggests (2004), the matter is probably more complicated than that, not least because Folsom groups may organize themselves differently at various points in time and space—examples of both gearing up and gradual replacement are evident among Folsom sites—but also because of the energy/time demands of each strategy. Gearing up, he suggests, is ideally a task taken up in winter when a group is relatively less mobile for long periods, food resources are close at hand, and the supply of stone is abundant. Since stone supplies are not necessarily located in a spot that might be suitable for overwintering, putting in that supply may require a special procurement trip to an outcrop (also Bamforth and Bleed 1997:127).

In contrast, gradual replacement would more likely occur when a group is on the move during the warm months and using their tools, and thus it takes place in a very different organizational context: in response to actual and immediate needs, not merely anticipated ones. In this instance, the demands of subsistence dictate the need for tool replacement and, more critically, determine access to raw material replacement (Binford 1979; Sellet 2004:10). If the former (subsistence needs) override the latter (raw material access), as they often do, groups would have to make do with the supply of stone already available to them.

Correlates aside, the strategy used at the Folsom site can perhaps be gauged through the raw material patterning in the projectile points in this assemblage, which, as highly curated items, have a longer "past" than most tools and, in some measure, record the history of stone acquisition and tool provisioning (Ingbar 1994; G. Jones et al. 2003; LaBelle, Seebach, and Andrews 2003; MacDonald 1999; Meltzer 1989b).

This brings up the matter of fluting. Driving a long flute from the face of points as thin as these required considerable skill and could easily result in breakage or failure if the force was misapplied. Fluting failure among Folsom knappers is variously estimated as 25% to 50% (Amick 1995, 1999a; Bradley 1993; Bamforth and Bleed 1997; Flenniken 1978; Frison 1991; Ingbar and Hofman 1999; Sellet 2004; Winfrey 1990). Given those costs, one would expect that fluting would occur close to a stone source or, at least, when there was ample stone in camp during a time of gearing up, if only because failure would then have lower costs. Yet, despite this, fluting did not always occur at the stone source. Instead, production was aimed at "intermediary forms"—bifaces, primarily—that put the stone into easily transportable packages, which could then subsequently be modified into a number of different tools (Ingbar and Hofman 1999; Kelly 1988), including, of course, fluted points. In many instances points were fluted at sites well away from a stone supply (Amick 1994; Ingbar and Hofman 1999; Sellet 2004).

The apparent willingness of Folsom knappers to gamble on successfully fluting a projectile point in the face of potential failure and loss of valuable stone has led to multiple hypotheses about possible, nonutilitarian motives for fluting, as well as about other organizational strategies that might have mitigated this risk (e.g., Bamforth and Bleed 1997; Bradley 1993; Frison 1991; Ingbar 1992; Ingbar and Hofman 1999; Sellet 2004). At times, however, Folsom groups apparently elected not to gamble, for certain assemblages contain unfluted points. Some of these unfluted forms are technologically and morphologically identical to Folsom points; others diverge sufficiently from the Folsom type that they have been separately designated Midland points and have been the subject of longstanding debate (Agogino 1969; Amick 1995; Judge 1970; Wendorf and Krieger 1959). Hofman (1992) has argued that the frequency of these pseudofluted or unfluted forms should be roughly proportional to the hunter-gatherer group's distance in time and space from the stone source and the number of kill/retooling events that occurred in the interim (chapter 8).

Knowing the incidence of fluting in the Folsom assemblage, and whether fluting or even point manufacture occurred at this site, warrants attention, given that this assemblage appears to be dominated by exotic rather than local stone sources (chapter 8). One caveat: Save for a few comments in the context of the historical section in chapter 8, I steer well clear of how Folsom points are fluted. That issue has been amply discussed elsewhere (e.g., Ahler and Geib 2000; Crabtree 1966; Flenniken 1978; Frison and Bradley 1980; Roberts 1936; Sellet 2004), and as there are no preforms or manufacturing debris at Folsom, there are few data from there that can contribute to the discussion.

WHAT TACTICS DID FOLSOM HUNTER-GATHERERS USE TO MAINTAIN THEIR TOOLKITS?

How a toolkit was provisioned and the decision to flute or not to flute are but two of the organizational tactics for maintaining the viability of a toolkit across time, space, and anticipated activities (Ingbar 1992). There are still other aspects of the organizational technology that may have helped minimize the logistical disparity between where stone was procured and where it was put to use. Many of those are invisible in an assemblage comprised almost entirely of finished and discarded/lost projectile points, as Folsom's is. Even so, clues can be found in the degree to which tools are resharpened and reworked prior to disposal, whether broken projectile points ended their use-lives as other tools, and how well raw material is conserved (e.g., Ahler and Geib 2000; Hofman 1991, 1992; Ingbar and Hofman 1999).

Maintenance tactics might also be manifest in the way these points were hafted. Soon after the first Folsom points were out of the ground, speculation began about the purpose of fluting and how these points would have been attached to spears. Fluting was first likened to the groove on a bayonet (Cook 1928b), and fluting and hafting were thought to be unrelated. But once it became apparent the flute would have been buried within the haft and could not have served a blood-letting function, attention turned to how the two might relate (chapter 8).

Once turned, consensus was quickly reached that fluting was not necessary for hafting—unfluted points can be hafted just as readily—nor was that its sole purpose (e.g., Bradley 1991, 1993; see also Ahler and Geib 2000; Amick 1999a; Collins 1999; Ingbar and Hofman 1999; Osborn 1999; Roberts 1935; Wilmsen and Roberts 1978). But fluting certainly enhanced the anchoring of the point to a spearshaft. The actual mechanics of hafting seemed straightforward: The base of a point was placed between splints or a slotted piece of wood or bone, with perhaps a bit of mastic applied, then tightly wrapped (Crabtree 1966). There seemed to be ample data to support this model (e.g., Judge 1973), but it had one puzzling aspect. Folsom flute scars routinely extend beyond the base nearly to the tip, and thus each flute would have continued well beyond the hafted area. Why risk the cost of fluting failure for a superfluous result?

The answer, Ahler and Geib (2000) argue, is that the conventional model of hafting is wrong. In their view, the hafting of Folsom points was more akin to a modern-day utility knife, wherein only a small portion of the blade is exposed at any one time, and as that blade dulls or breaks, the unexposed portion is slid forward for use. So too, fluted points were hafted nearly their entire length, and as they broke or dulled, the haft was unbound and the point was slid forward, then rehafted. Fluting allowed this forwardly adjusting friction haft and, in their view, was "designed foremost to conserve raw material and vastly extend the use-life of a given projectile" (Ahler and Geib 2000:806).

The paradox that the costly business of fluting was intended to save stone can be examined with the well-traveled assemblage from Folsom, for it contains points jettisoned in use and, thus, provides snapshots of how hafting appeared in points at different stages in their use-lives (chapter 8).

WHAT IS REVEALED IN THE BREAKAGE AND DISCARD PATTERNS OF THE ARTIFACTS?

Not all artifacts could be maintained indefinitely in the toolkit, of course: sooner or later they broke, were worn to exhaustion, or were simply lost. How they broke can reveal something of how they were used, though I would hasten to add that, broadly speaking, there is little mystery here. The bulk of the Folsom assemblage is comprised of projectile points propelled with considerable force into very live, very fast-moving, very large animals. But how they were propelled, whether by thrusting spears or thrown spears, is less transparent. So too are the questions of how often they broke, if patterns of breakage correspond to raw material types, and whether it was the case, as Roberts (1935:17) suspected early on, that the process of fluting was a fundamental design flaw that virtually guaranteed these points failed in use (also Ahler and Geib 2000; Crabtree 1966).

And what of the points found here? Were all these specimens lost—an understandable enough prospect, had they been embedded in a "mass of meat and gore" as Wheat (1979:95) put it—or were some of them merely discarded; and, if so, why? Are there detectable thresholds in the size of the broken or heavily worn specimens that can provide a gauge of the overall stone supply and, thus, insight into the decision to salvage and recycle or discard points from a kill? Is it possible to tell the difference between points abandoned and points lost?

More broadly, what does the number of points and tools found on a site represent, and how much is that number a consequence of, say, the size of the excavation area, where the excavations occurred, or perhaps geological processes of transport or erosion (Frison, Haynes, and Larson 1996:213; Hofman 1999a:122; Jodry 1999b)? If not, or even if so, is it possible to infer from the number and density of points and butchering tools in a bonebed something of the relative dispersal of animal prey, the mechanics of the weaponry that was used, or the nature of the hafting system (Hofman 1999a; Wheat 1979)?

Although I hesitate to use the term "stone tool taphonomy" since stone tools were never alive (though they did

have use-lives), and they do not undergo diagenesis (though they can be affected by a variety of site formation processes), I think it is important to attempt to understand their taphonomic history and the various processes that account for its character and appearance in the archaeological record of this assemblage (chapter 8; also Frison, Haynes, and Larson 1996:213).

IN WHAT MANNER DO THE POINTS FROM FOLSOM VARY, AND IS THAT VARIATION A RESULT OF RAW MATERIAL, TECHNOLOGY, FUNCTION, OR STYLE?

There has emerged in the last decade considerable discussion about the morphological variation one might expect to see in an assemblage of fluted points, and about the cause(s) of that variation (Amick 1999a; Bolduan and Cotter 1999; Ingbar 1992; Ingbar and Hofman 1999; Hofman 1999a, 1999b; see also Hawkes and Bleige Bird 2002). That discussion is not altogether congruent with Judge's (1973) realization, since confirmed by others, that there is also considerable standardization in fluted point morphology, perhaps related to the costs and demands of hafting devices. This and other models of point production (e.g., Ahler and Geib 2000) have very specific implications for point morphometrics, hafting technologies, and patterns of use-life, and these can be tested with this assemblage.

Exploring the manner in which these points vary also sheds light on the morphological consequences of use, breakage, and reworking relative to raw material types, and allows a glimpse of stylistic variability in what is otherwise a dominantly functional form. The detection of stylistic variability is of particular interest, insofar as it can help address the question, raised earlier, of whether an aggregation of previously dispersed groups occurred at Folsom or whether perhaps some of the points used at the site were obtained via exchange (Bamforth 1991; Hofman 1994; Meltzer 1989b).

ARE THERE ANY ASSOCIATED CAMP/HABITATION AREAS AT FOLSOM?

Ethnoarchaeological (e.g., Fisher 1992; O'Connell, Hawkes, and Blurton-Jones 1992) and archaeological (Frison 1996; Hofman 1996:56, 62; Hofman, Amick, and Rose 1990; Jodry 1999b; Jodry and Stanford 1992; LaBelle, Seebach, and Andrews 2003; Stanford 1999) studies have shown that kill sites are frequently accompanied by camps, though the latter are far less visible archaeologically, in part because they may be situated 10 m to 70 m from kill areas. The Mill Iron camp, for example, is 25 m away from bonebed (Kreutzer 1996:101; also Fisher 1992:73; Hofman, Amick, and Rose 1990; Jodry 1992; O'Connell, Hawkes, and Blurton-Jones 1992). The nature and function of those camps vary. Some are long-term habitations, others butchering/processing locales, and still others are some combination of these and/or other activities. Such areas have the potential to yield a broader complement of stone tools, provide evidence of other activities, and help put a kill site into a broader adaptive context (Amick 1996:413; LaBelle, Seebach, and Andrews 2003).

There is good reason to suppose that a camp ought to have once been present at Folsom: More than 30 Pleistocene bison were killed and butchered here (chapter 7). Ethnoarchaeological evidence suggests at least 2 hr of processing time per animal, depending, among other things, on the size of the animals, the distribution of the carcasses, the extent of the butchering (in this case, reasonably thorough), the size of the labor force that is working on the task, and even the temperature—when it is cold, the group can afford to take more time, since putrefaction is delayed, but then it becomes harder to work the carcasses (L. R. Binford, personal communication, 1998). Regardless of the precise details, the killing/butchering must have taken several days, and the hunters were likely nearby most of that time.

Yet, the original Folsom excavations did not encounter any associated habitation areas, though those excavations also did not extend any distance away from the bonebed; indeed, the full extent of the bonebed was not known in the 1920s. Although earlier investigators sought evidence of an associated camp, none was found. But given the potential importance of such a discovery, further efforts were made to locate a camp at Folsom (chapter 4).

Probing the History of Archaeology

Any volume on the Folsom site attempting to be reasonably comprehensive, as this one is, cannot neglect the historical aspects of the work at this site and its role in the history of American archaeology. Much of that is well-trod ground, and need not be repeated here (see Meltzer 1983, 1991b, 1994). But an analytical overview of that history is in order, and is provided in chapter 2, along with an exploration of several additional questions that can provide a deeper understanding of how the work at Folsom unfolded. This effort is critical to understanding Folsom's reception in the scientific community, how and why the longstanding controversy over human antiquity in the America's was resolved here and not at another site, how matters of status and rank played out within the scientific community, what impact Folsom had on the evolution of American archaeology as a discipline, and the post-Folsom trajectory of Paleoindian studies in North America. These questions too can be answered—though in archival rather than archaeological records.

WHY FOLSOM, BUT NOT ANY OF THE OTHER SITES PREVIOUSLY CHAMPIONED AS PLEISTOCENE IN AGE?

Folsom was the last in a very long line of sites, stretching back to the mid-nineteenth century, offered as evidence of a Pleistocene human presence in the Americas. Unlike all those other sites, however, Folsom was accepted. Moreover, in the queue of rejected sites were localities, like Lone Wolf Creek (Texas), that later proved to be Paleoindian in age. Why were they rejected, and Folsom accepted? One answer offered to that question is that there was a "paradigm bias"—or perhaps just plain ignorance—on the part of archaeologists of the time (e.g., Alsoszatai-Petheo 1986; Rogers and Martin 1984, 1986, 1987; Schultz 1983). But as I have argued elsewhere, the historical matter is far more complicated than this simplistic rendering suggests (Meltzer 1991b, 1994). An examination of the published record, and particularly the archival material available from this period, reveals clearly why Folsom was accepted, while the others were not (chapter 2).

WHY WAS CREDIT FOR RESOLVING THE HUMAN ANTIQUITY CONTROVERSY GIVEN TO OTHERS, NOT COOK AND FIGGINS?

Folsom caused a sea change in the way archaeologists viewed the prehistory of the Americas. Within a decade of the 1927 site visit, there were seven major symposia devoted to human antiquity in the Americas. Yet neither Cook nor Figgins, both of whom had done so much to bring about this sea change, were invited to participate in these meetings. On only one occasion were they even invited to attend as audience members. Speaking on those occasions about the Folsom evidence were, instead, half a dozen archaeologists, geologists, and vertebrate paleontologists, most of whom had only visited the site briefly, and a few who had never even been there at all.

That Cook and Figgins received little, if any, public acclaim for their work at Folsom reveals a great deal about how science works, how controversy is resolved, who is deemed competent to evaluate the meaning of discoveries, and who gets to judge when resolution is achieved and controversy is over.

WHAT MADE FOLSOM SO IMPORTANT TO AMERICAN ARCHAEOLOGY?

One of the curious aspects of the high drama that played out at Folsom in September of 1927 is that A. V. Kidder was there. The representatives of the American Museum of Natural History (Barnum Brown) and the Smithsonian Institution (Frank Roberts) had both come in response to invitations from Figgins. Kidder arrived four days later at Roberts' behest (Meltzer 1983). It is perhaps not entirely surprising that Roberts asked Kidder to join him; they had just been together at the first Pecos Conference, and Roberts knew Kidder from his Harvard days. But it is surprising that Kidder, who otherwise had little interest and no previous participation in the debate over human antiquity, would be so anxious to visit Folsom or feel compelled to announce his opinion of its antiquity and importance at a public forum at the Southwest Museum just a few weeks later (Kidder 1927). And unlike Roberts, who was so inspired by what he'd witnessed at Folsom that he almost immediately abandoned his work on the Southwestern Late Prehistoric sites and went on to become a major figure in twentieth-century Paleoindian studies, Kidder was content to continue working in the Southwest and, later, the Mayan region.

Even so, Folsom was extremely important to Kidder (1936), who for a variety of reasons was relieved to have the "chronological elbowroom" this discovery afforded (145). Kidder was hardly alone in seeing the profound implications of Folsom for American prehistory, as well as for the practice of American archaeology (also Bryan 1941; Kroeber 1940 Nelson 1928b, 1933; Roberts 1940). Why that was so is apparent in the state of American archaeology before Folsom, and in the sharp turn the discipline took afterward (chapter 2).

To be sure, the Folsom discovery also set the foundation for North American Paleoindian studies, in ways that are both obvious and subtle. Not only did Folsom teach archaeologists how to find more sites like it (Meltzer 1989a), but also it helped create an inferential basis of Paleoindian adaptations that would soon harden into fact (chapter 2), and it exposed conceptual and methodological problems in the manner in which concepts like "type" and "culture" were applied to archaeological remains (chapter 8).

The SMU/QUEST Folsom Project:
A Brief Summary

Our reinvestigation of Folsom was conducted over portions of three summer field seasons from 1997 to 1999. It involved a range of activities (chapter 4), including surveys of the modern ecological, climatic, and geological context (chapter 3), extensive geological and geophysical studies (chapter 5), sampling for radiocarbon dating and paleoenvironmental and paleoclimatic indicators—including sediment coring of a nearby lake (chapters 5 and 6), archaeological surface survey and intensive excavations in a remnant of the bison bonebed (chapters 4, 7, and 8), and even historical archaeology in the areas of the 1926–1927 and 1928 field camps (appendix C). During the "off-season" and afterward, there was laboratory analysis of the recovered remains (chapters 5–8), archival and historical work on the correspondence and field notes from the original investigations (chapters 2 and 4), and examination and analyses of the extant museum and private artifact and faunal collections from the site (chapters 7 and 8). The archaeological fieldwork was completed in 1999, but annual visits have been made to the site since, to monitor erosion and map items that surface in the interim. In addition, an extensive study of the Late Glacial and Holocene fluvial geomorphology of the Upper Dry Cimarron was initiated in 2002 under the direction of Daniel Mann (2003, 2004). As this work is ongoing it will be reported elsewhere; however, some early results are incorporated here (chapters 5 and 6).

Plan for the Volume

Because this volume is not just our work, but also aims to publish the results of the 1920s investigations, it presents an organizational challenge: What is the best way to bring out the results of two very different projects, done for very different reasons, using very different methods and techniques, operating under very different understandings of the archaeological and Paleoindian records, and separated by nearly three-quarters of a century?

Dividing this volume into two separate parts, or even writing two separate volumes, one on the 1920s work and the other on our investigations, is an option that would—in theory—have the virtue of maintaining a very clear division of what was done, when and why, and what came out of it. However, such an approach would be impossible to maintain in practice, since the earlier results cannot be easily understood without knowing what we know now, and vice versa; it would be difficult to execute; and it would hamstring the larger effort to unite the results of those investigations and provide a fuller understanding and synthesis of the Folsom site. But then the flip side, ignoring the historical aspect of the investigations in a strict effort to integrate the various investigations, would risk misunderstanding the reliability, representativeness, context, and completeness of the data recovered and the conclusions reached during the various projects on the site, and would inevitably fail to fully credit the earlier investigators and investigations. Finally—and this is a small but not insignificant point—ignoring the historical context would drain much of the color and understanding from a discussion of what is, after all, one of the historically pivotal sites in American archaeology.

As a compromise strategy, then, I sought a middle ground: Where appropriate, each chapter discusses the approach, data, and results of the earlier investigations, reports on the results of our work, and then synthesizes the whole. Like all compromises, this one has some awkward organizational moments, and may invite invidious, if unintended, comparisons between the different projects. But on the whole this seems the best way, if not the only way, to do archaeological and historical justice to the task at hand. The format is put into effect in chapter 4 and characterizes most of the chapters that follow.

Before that, however, chapter 2 takes up the historical questions raised above, as well as others surrounding the larger intellectual context of the excavations at Folsom. Chapter 2 also helps explain the archaeological context within which the work was done, which, in turn, clarifies aspects of why the fieldwork evolved as it did between 1926 and 1928.

From the Folsom site's historical context, attention turns in chapter 3 to a summary of its natural context: the geology, hydrology, topography, and present-day climate and environment of Folsom and the surrounding region. As such summaries can tend to be almost-generic in scope and content, I have tried to limit the discussion to those aspects directly relevant to the site and the Paleoindian occupation or those central to an understanding of its archaeological context. I then use that summary and what it might suggest to draw a series of hypotheses about the climate and environment at the time of the Paleoindian occupation, hypotheses that are tested in later chapters.

The primary data for those tests were derived over several seasons of fieldwork in the 1920s and in the 1990s. To understand the nature, limitations, and potential biases of those data, chapter 4 describes the procedures and techniques of the 1920s excavations, and those of the brief stints of fieldwork that followed, and then summarizes in detail our research design, methods, and results. As an experienced consumer of archaeological monographs, I can only sympathize with the reader who would happily skip such oftendreary recitations of what was done when, and where, for the good stuff coming later. I usually head for the artifacts and radiocarbon dates myself. If you're so inclined, they're in chapters 8 and 5, respectively. But before you do, a few words.

Knowing the details of what had been done at the Folsom site prior to our investigations is critical in several ways, not least in framing the research questions we took to the investigations, in knowing what kinds of data would be needed to answer those questions, and in having fewer illusions of how many of these data might be left at the site—indeed, of how much of the site might be left. The review in chapter 4 details the prior work with the larger goal of using it to set the stage for the subsequent discussion in that same chapter of the research strategies and tactics we brought to bear in attempting to answer the archaeological questions discussed above. Knowing the details of how the data were recovered—both in our work and in the earlier investigations—will help in understanding and assessing the evidence presented in the subsequent chapters.

For the reader's sake, however, the very driest parts of this necessary discussion, such as the particulars of our excavation grid and level systems, excavation methodologies, and recording formats, are presented separately in appendix A. As a supplement to the historical record of excavations, appendix B presents the relevant portions of Carl Schwachheim's diary, which he kept throughout his seasons at Folsom and which provides a first-hand and almost-daily account of the excavations. In appendix C, Donald Dorward and I take an archaeological look at the Colorado (1926–1927) and American Museum (1928) field camps, reporting on the results of our metal detector surveys and mapping in 1997 of the surface material from those camps, located on the North Bank of the Folsom site.

In chapter 5, Vance Holliday and I examine the site geology, focusing on the paleotopography, stratigraphy, and geochronology. We follow here in the footsteps of Barnum Brown and Kirk Bryan, who anticipated several of our conclusions about the geology and age of the site, this despite working decades before the advent of radiocarbon dating. The goals of our analysis are several, including to understand the shape of the now deeply buried landscape, to gain a sense of how a herd of more than 30 Pleistocene bison were maneuvered to the kill, to understand the geomorphic processes operating in this setting that served to both preserve the site and modify it in critical ways, to gain control

over the age of this site, and to assess whether or where camp or habitation areas might exist beyond the bonebed area. An ancillary study to this chapter as well as chapter 7 was a Fourier-transform infrared spectroscopic (FTIR) analysis of sediment mineralogy, presented here as appendix D, by Todd Surovell.

Chapter 6 provides a fine-grained paleoclimatic and paleoenvironmental history of the Folsom site and area, an endeavor in which I am joined by several collaborators. Little was done on this topic during the original investigations. Our efforts involved a variety of kinds of data, including pollen, macrofossils, land snails, and fauna; an analysis of stable isotopes—carbon (δ^{13}C) primarily, but also oxygen (δ^{18}O) and nitrogen (δ^{15}N); and efforts to understand Folsom within the larger context of Late Glacial climates, notably what it reveals of the Younger Dryas in this portion of North America. The Folsom site, not surprisingly, proves to have looked very different in Late Glacial times than it does today—but not in ways we had anticipated at the outset of our research.

In chapter 7, Lawrence Todd and I provide a detailed analysis of the bison remains. Our focus there is to understand the structure and characteristics of the herd of bison killed at the site, the patterns of Paleoindian butchering and processing of these animals, and the postoccupation taphonomic and diagenetic processes that modified the character of the bonebed. Because this is the site at which this genus and species of bison was first defined as *Bison taylorii* (see Hay and Cook 1928, 1930), and then frequently and rapidly redefined, to the consternation of later taxonomists (e.g., MacDonald 1981; Skinner and Kaisen 1947), we provide as well a guide through this taxonomic thicket. Both ours and the original investigations yielded non-bison remains. Despite earlier suspicions, none of these prove to be related to the Paleoindian occupation on site; they too are described in this chapter, albeit briefly.

The artifact assemblage from Folsom, the subject of chapter 8, contained a handful of tools but over two dozen projectile points. This was, of course, the site after which points of this type were named, and its salient attributes first identified. Chapter 8 explores the early efforts by Figgins, Brown, and others to describe these specimens, including patterns in their raw material, and also attempts to reconstruct for subsequent analytical purposes just how many points were found at the site and where. Like the debate over bison taxonomy that followed the Folsom discovery, so too there was a debate over how best to define the Folsom projectile point type, a task made no easier by the rapidly increasing tally of lanceolate points, fluted and otherwise, from across the continent. Understanding that debate is important to this book, though not to this chapter, so exploration of the decade-long effort to resolve the "Folsom-Yuma problem" is relegated to appendix E. The second and larger part of chapter 8 explores the morphometric variability, patterns in lithic raw material use, technology, and life history of the projectile points and other tools from this artifact assemblage. This is done with an eye toward addressing the questions of Folsom mobility, technological organization, and patterns of toolkit use, breakage, and maintenance, as well as what these data may reveal about where this group of hunters came from and where they may have been headed afterward.

These many and varied threads are then woven together in chapter 9, which summarizes and synthesizes what we know of what occurred at the Folsom site in Late Glacial times, using as a framework the archaeological and historical research questions iterated above. As forewarned, not all of those questions are or can be answered, but they certainly provided a challenge to think as broadly and as deeply as we could about the data and evidence from this site.

We will likely never know as much about what happened there 10,000 years ago as we do about what happened there in the 1920s, but I would venture the immodest claim that Folsom is no longer just one of the best-known sites in American archaeology; it is also now reasonably well understood.

Notes

1. By convention and to avoid confusion, ages are provided here in radiocarbon years. When matters get to the nitty-gritty of assessing the ages of specific samples (chapters 5 and 6), radiocarbon ages are also calibrated into secular calendar years.

2. The site's National Register citation reads "Paleo-Indian (ca. 9000 B.C.) Kill site. Excavations between 1926 and 1929 by paleontologists J. D. Figgins and Barnum Brown of Colorado Museum of Natural History uncovered flint spear point embedded between the ribs of an extinct bison species. Important milestone in history of American archaeology; confirmed theories favoring man's early advent into the Americas" (Murtagh 1976:481). A few factual errors aside, that is not a bad 50-word summary of the site's history and importance.

3. There are historical reports and supporting geological evidence, discussed more fully in chapters 4 and 5, that

Wild Horse Arroyo in the vicinity of the site had hardly eroded prior to the early twentieth century (Owen 1988:27).

4. McJunkin's story is a compelling one, though poorly known. Born a slave in pre–Civil War Texas, he was befriended at an early age by the plantation owner, Jack McJunkin, whose name he took and who may have taught him to read and supplied him with books. In his teens, George moved west, took a series of ranching jobs on the Southern High Plains, and by the late nineteenth century was working and living in northeastern New Mexico. Intensely curious about the world around him, George McJunkin became, apparently without benefit of formal education, an astute naturalist. It was American archaeology's good fortune that this intellectually curious cowboy was checking the Crowfoot's fence lines and cattle after the Folsom flood. Any cowboy would have done the same, but few would have noticed or appreciated the significance of the bones jutting out of the bottom of the newly incised arroyo. The ex-slave who rode out of Texas and into history has attracted several biographical efforts (e.g., Folsom 1992; Hillerman 1973; Preston 1997) which, given how many long stretches of McJunkin's life will forever remain unknown, understandably vary somewhat in the telling.

5. The Colorado Museum of Natural History later became the Denver Museum of Natural History. It is now the Denver Museum of Nature and Science. To avoid the historical anachronism, the original name is used throughout in the text where reference is made to the work in the 1920s. Where reference is made to archival collections housed at this institution, the current name is used.

6. Here and throughout the work (especially in chapter 2), citations to unpublished archival materials use acronyms that refer to collections and the institutions where they are housed. Acronyms are defined in the Note on Archival Sources provided with the "References Cited."

7. That Folsom happens to be located in a breathtakingly beautiful, well-watered, and green mountain valley teeming with game was an incidental benefit to one who had spent much of the prior decade working in parched, treeless, rattlesnake-infested, blazing-hot sand dunes on the Southern High Plains. I believed I had earned a site with shade.

TWO

Folsom and the Human Antiquity Controversy in America

DAVID J. MELTZER

Folsom played a pivotal role in the development of American archaeology. Most everyone knows this. What may be less well known is why this particular site, alone among dozens of localities championed since the mid-nineteenth century, including several bison kills, finally established that humans were in the Americas by late Pleistocene times (see Meltzer 1991b). What may not be known at all is why, in the decade after the breakthrough at Folsom, the site's investigators—Jesse Figgins and Harold Cook—were completely excluded from professional discussions of the site and North American Paleoindians.

As it happens, those issues are linked in ways that reveal much about the history and context of research into human antiquity in America, and about the nature of scientific controversy and its resolution. This chapter explores those issues, but two brief comments on what this chapter is not: (1) it is not intended to be a strict narrative of the history of fieldwork at Folsom (the necessary parts of that are given in chapter 4) but, rather, aims more broadly at this and other archaeological and paleontological localities being investigated in the 1920s, to show how events and actions elsewhere set the stage and influenced the work—and the perceptions of the work—here at Folsom; (2) this chapter is also not intended to be an overview of the human antiquity controversy, although it necessarily requires a brief summary of that long and bitter dispute in order to establish the intellectual backdrop against which the research at Folsom was inevitably set and the gauge with which the evidence from this site would be measured (see also Meltzer 1983, 1991b, 1994).

This chapter explores just what made the Folsom site so important, why it mattered, and what it meant for the discipline and those involved, by seeking to answer—invoking the spirit of Groucho Marx—a deceptively simple question.

Who's Buried in Grant's Tomb?

The Folsom discovery in 1927 triumphantly resolved a dispute over human antiquity in the Americas that reached back to the mid-nineteenth century. Today, we credit Figgins and Cook with the "breakthrough" at Folsom (e.g., Daniel 1975:275; Fagan 1987:50–51; Willey and Sabloff 1980:121; Wilmsen 1965:181; Wormington 1957:23–25). We do so for seemingly good reasons.

After all, it was Cook, then an Honorary Curator of Paleontology at the Colorado Museum of Natural History (CMNH), whose report on the Lone Wolf Creek (Texas) site evidently spurred Fred Howarth, a Raton, New Mexico, banker, to see the potential of the Folsom site and bring some of the deeply buried bison bones from the site to the Colorado Museum. It was field parties under Cook and Figgins—the latter then Director of the Museum—that in 1926 conducted the initial excavations at the site and discovered, though not in situ, the first Folsom projectile points.

It was Figgins who traveled east in early 1927 to show the Folsom artifacts to various skeptics, Smithsonian anthropologists Aleš Hrdlička and William Henry Holmes among them, in an unsuccessful effort to convince them of the age and association of the find (Wormington 1957: 23). And it was Figgins who again sent crews to Folsom the following summer, where his faith was rewarded on August 30, 1927, when the crew found a projectile point, this time embedded between the ribs of an extinct species of bison. It was Figgins who sent telegrams nationwide, inviting the scientific community to come view the Folsom artifact in position and confirm its age and context (Meltzer 1983, 1993).

In looking back at this episode we routinely lump the Folsom discovery and discoverers with the resolution of the human antiquity controversy. Cook and Figgins' dual 1927 publications in *Natural History* (Cook 1927a; Figgins 1927a) are routinely cited as marking this turning point in American archaeology—when our discipline finally found itself in possession of deep time (e.g., Willey and Sabloff 1980:121).

But consider this: Four months after the electrifying news from Folsom, the American Anthropological Association devoted one of the four symposia at its December 1927 annual meeting to "The Antiquity of Man in America" (Hallowell 1928). There Nels Nelson of the American Museum of Natural History (AMNH) spoke about the long-standing controversy over human antiquity in the Americas and the implications of the Folsom discovery for that dispute, while Frank H. H. Roberts of the Bureau of American Ethnology (BAE) and Barnum Brown of the American Museum addressed the site's archaeology and paleontology, based on what they had seen there when they visited Folsom in response to Figgins' telegrams that September of 1927 (Hallowell 1928:543).

Nelson and Brown were, of course, established figures in 1927: Brown was a well-known vertebrate paleontologist, while Nelson was an archaeologist with an involvement in the human antiquity issue that reached back over a decade (e.g., Nelson 1918, 1928a). Both of them had been following events at Folsom from the outset, but Nelson had not visited Folsom that September to see the point in situ (Nelson to Figgins, September 13, 1927, JDF/DMNS). Brown had, but then he had had little prior experience with Pleistocene faunas, and none with an archaeological fauna. In December of 1927, Roberts was a newly minted Harvard Ph.D. who had been at BAE only a year, and whose prior work was on Late Prehistoric sites in the Southwest (Judd 1967). He was not the Paleoindian archaeologist he would ultimately become—his reputation-building excavations at Lindenmeier were still years away. When he visited Folsom that September on behalf of the Smithsonian, his degree was scarcely two months old, and by December his sole experience with Paleoindian materials generally or Folsom in particular was the three days he spent visiting the site.

The AAA meetings that December in Andover were a triumph. The artifacts from Folsom held center stage, and "all the anthropologists and archaeologists present accepted the authenticity of the find, saying they established a definite landmark in the history of prehistoric man in America" (Brown to Figgins, January 10, 1928, VP/AMNH; Jenks to Figgins, January 4, 1928, DIR/DMNS). Yet, it was Brown, Nelson, and Roberts who spoke at that session, not Cook or Figgins. Neither of them was even asked to participate—only to loan photographs and specimens (Brown to Figgins, December 8, 1927, DIR/DMNS).

In fact, from 1927 to 1937 there were seven major symposia devoted to human antiquity in the Americas. In these, the fast-emerging fundamentals of Paleoindian chronology, artifacts, and faunal associations were being hammered out. Each of these discussions took place on national and even international stages. The symposia were held at the 1927 American Anthropological Association meetings (Hallowell 1928); the 1928 meeting of the New York Academy of Medicine; the 1928 Geological Society of America meeting, held jointly with the American Association for the Advancement of Science (AAAS); the 1931 AAAS meetings (Danforth 1931); the 1933 Fifth Pacific Science Congress (Jenness 1933); the 1935 meeting of the American Society of Naturalists (Howard 1936); and, finally, the 1937 International Symposium on Early Man sponsored by the Academy of Natural Sciences of Philadelphia (MacCurdy 1937).

Taking their turns at the center of these stages speaking about Folsom archaeology and geology, including the type site, were Kirk Bryan, Edgar B. Howard, John C. Merriam, E. H. Sellards, and Chester Stock, along with Brown, Nelson, and Roberts, among others. In almost every case, Hrdlička was invited to present his views on the subject and the site, and on several occasions he did so (e.g., Hrdlička 1928, 1937; see also Dixon to Hrdlička, November 27, 1927; Boas to Hrdlička, April 21, 1931; Howard to Hrdlička, November 30, 1936, all in AH/NAA).

Yet, in all the planning that went into the selection of participants for these meetings, I have found only two occasions when Cook or Figgins was even suggested as a possible speaker. In both cases the suggestions were ignored (see Gregory to Boas, April 27, 1931, FB/APS; Howard to Merriam, November 16, 1935, JCM/LC).

Cook did receive an invitation to attend the 1937 International Symposium on Early Man in Philadelphia, where, had he gone, he could have listened to Bryan and Roberts discuss the Folsom site. But the press of business prevented his attending (Howard to Cook, December 31, 1936; Cook to Howard, March 15, 1937, HJC/AGFO), and perhaps that was for the best. Otherwise, he would have heard Bryan describe the Folsom finds as "discovered" by Figgins and Cook, but "confirmed by the masterly excavation of the site by Barnum Brown," and heard Gladwin give sole credit for Folsom to Brown and not even bother to mention Cook or Figgins (Bryan 1937:139–140; Gladwin 1937:133).

Cook and Figgins' complete absence from the postdiscovery discussions of Folsom, let alone of the human antiquity issue, and the fact that their contributions were completely ignored, if not devalued, by their peers—despite their continued hand in Paleoindian research (e.g., Figgins 1933a, 1934, 1935)—are rather surprising, at least given how we look back on the Folsom episode today. Yet, I think their absence and the contemporary measure of their contribution show that the Folsom *discovery* and the subsequent *resolution* of the human antiquity controversy were very separate events involving very different participants. It also shows, when probed deeper, the sharp boundaries of scientific status and rank, and a clear lesson about the nature of the scientific enterprise. And, finally, it provides a cautionary tale of what might have been, had Figgins not visited Hrdlička at the Smithsonian Institution in the spring of 1927 and showed him the first points recovered from the site; for Folsom was not the only locality championed as evidence of a Pleistocene human presence in the Americas, but it was the only one that was accepted.

Elsewhere, I detail the history and issues involved in the nineteenth- through early twentieth-century dispute over human antiquity in the Americas that provide the larger context for the Folsom discovery (e.g., Meltzer 1983, 1991b, 1993, 1994, 2006), and that ground need not be covered again, save in summary fashion.

Background to Controversy

The possibility that the arrival of people in the Americas might be geologically ancient was only seriously considered after 1859/1860, when the Old Testament barrier was finally broken in Europe (Grayson 1983). At Brixham Cave, England, and in Abbeville in northwestern France, human artifacts were found in direct association with extinct mammals: mammoths, cave bear, hyenas, and the like (Evans 1860; Prestwich 1860; also Grayson 1983, 1990; Gruber 1965). Although the absolute age of those animal remains was unknown—this was a century before the advent of radiocarbon dating—their presence in a deposit was widely accepted as marking an earlier geological period (the Pleistocene), one which predated the modern world (Lyell 1830–1833). That human artifacts were in those same deposits meant humans too had a past beyond history, a fact with deep and profound intellectual consequences (Grayson 1983; Stocking 1987).

The shock waves of that realization quickly reached America owing largely to Joseph Henry, first Secretary of the Smithsonian Institution. He issued a set of detailed instructions (Gibbs 1862) to military officers, missionaries, and other travelers in the "Indian country" on how to search for and record archaeological evidence that might reveal "analogous stages of the mental development of the primitive inhabitants of this country and those of Europe" (Henry 1862:35; see also Hinsley 1981; Meltzer 1983).

With that, archaeological questions of the origin, antiquity, and adaptations of the first Americans emerged in a recognizably modern form. They began with the hope that there would be evidence in America of stone tools alike in form, evolutionary "grade," and antiquity as those of Paleolithic Europe. Although naïve in retrospect, in the 1860s there was good reason to expect as much, for it was generally believed that there was an "exact synchronism [of geological strata] between Europe and America" (Whittlesey 1869).

By the 1870s stone artifacts, seemingly akin to ancient European Paleolithic tools, were reported by Charles Abbott from apparent Pleistocene-age gravels at Trenton, New Jersey (e.g., Abbott 1877). He had little doubt of their antiquity. After all, "had the Delaware River been a European stream, the implements found in its valley would have been accepted at once as evidence of the so-called Paleolithic man" (Abbott 1881:126–127).

Abbott's discoveries were soon replicated by others. In the spring of 1883 G. F. Wright predicted that "when observers become familiar with the rude form of these Paleolithic implements they will doubtless find them in abundance."

He was correct. Over the next decade, many more "American Paleolithic" artifacts were reported from surface and buried contexts at other sites, and their presence was taken as proof of prehistoric Americans living here thousands, if not tens of thousands of years ago, when northern latitudes were shrouded in glacial ice. It did not matter that the precise age of these tools proved difficult to pin down by geological evidence (Lewis 1881; Shaler 1876; Wright 1881, 1888, 1889b). These artifacts so readily mimicked European Paleolithic tools of undeniable antiquity that they were assumed to be just as old (Haynes 1881:135–137; Putnam 1888:423–424). Abbott (1881) concluded triumphantly that "the sequence of events, the advance of culture, have been practically synchronous in the two continents; and the parallelism in the archaeology of America and Europe becomes something more than "mere fancy" (1881:517).

By 1889 the evidence for an American Paleolithic was almost universally accepted. If there was skepticism about it, it was well hidden. Indeed, British Paleolithic expert Boyd Dawkins (1883) himself proclaimed that the American "implements are of the same type, and occur under exactly the same conditions, as the river- drift implements of Europe" (1883:347). The last years of the 1880s saw a parade of symposia, feature articles, and books, all testifying to the veracity of the American Paleolithic (e.g., Abbott 1889, Dawkins 1883; McGee 1888; Putnam 1888, 1889; Wallace 1887; Wright 1889). For many, the only lingering question was how early in the Pleistocene humans may have arrived.

Yet, scarcely a year later the American Paleolithic was under harsh fire, sparked by William Henry Holmes' (1890) studies of stone tools at the prehistoric Piney Branch quartzite quarry in Washington, D.C. He learned there that an artifact might appear "rude" merely because it was unfinished, not because it was ancient. To explain why that was, Holmes drew on the then-popular notion in biology that ontogeny recapitulates phylogeny, such that in the evolution of a species, ancestral adult forms become descendant juvenile stages. As Holmes (1894) translated this into archaeological terms, the "growth of the individual [stone tool] epitomizes the successive stages through which the species [history of stone tool making] passed" (137). Thus, if early on in the process of manufacture a stone tool was discarded or rejected, it would appear like the "rude" and ancient stone tools of Europe (fig. 2.1).

Thus, the analogical argument that the similar form of American artifacts with European Paleoliths implied a similar age and evolutionary grade, as was routinely argued by American Paleolithic proponents, was flawed. Artifact form, Holmes (1890) stressed, had no chronological significance whatsoever. Age must be determined independently, by the geological context of the artifacts (also Holmes 1892).

Paleolithic proponents like Abbott and Wright replied that the similarity between paleoliths and the Piney Branch quarry debris was purely accidental and thus irrelevant to the antiquity issue (Abbott 1892). Critics retorted that American and even European paleoliths sometimes looked

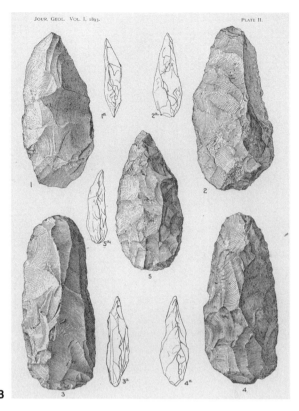

A   B

FIGURE 2.1 A. G. F. Wright's (1890) composite image of the Newcomerstown paleolith alongside a paleolithic biface (reduced to one-half size) from Amiens, France; B. Holmes' (1893b) depiction of the Newcomerstown specimen alongside "four ordinary rejects." Holmes left it to the reader to decide which of the five specimens was from Newcomerstown, and which were quarry rejects. (From Wright 1890, Holmes 1893b.)

alike only because some European paleoliths were themselves quarry rejects (Meltzer 1983). Each side recognized that the key to sorting unfinished rejects from finished paleoliths lay in whether the objects showed signs of use. But even though they examined the very same objects, they could not agree whether or not they were used.

The growing dispute over the American Paleolithic, Abbott himself would come to admit, was resolving itself as a geological matter, but geology was providing little guidance. There were questions of whether an artifact had actually come from a primary context, an ambiguity complicated by the absence of agreed-on field strategies for removing artifacts and reading their stratigraphic units and depositional contexts. Then there were questions about the age of the artifact-enclosing deposit, which in the 1890s and early decades of the twentieth century were thoroughly entangled in an increasingly contentious debate over how to recognize Pleistocene-age strata and a simmering controversy over the number and timing of the glacial periods (e.g., Chamberlin 1893b, 1903; Salisbury 1893b; Wright 1889a, 1889b, 1892).

That controversy and the American Paleolithic dispute exploded publicly in the last months of 1892, sparked by the appearance of Wright's *Man and the Glacial Period*, which advocated both an American Paleolithic and a single glacial period. Wright's book met an ugly reception at the hands of critics who, directed by Thomas Chamberlin, Chief of the

Glacial Division of the United States Geological Survey (USGS), savaged the book's archaeological and geological claims and contents (e.g., Chamberlin 1892, 1893a; McGee 1893a; Salisbury 1892a, 1892b, 1893a). The critics, in turn, were counterattacked by Wright's allies and other Paleolithic proponents (e.g., Claypole 1893a, 1893b; Winchell 1893a, 1893b; Youmans 1893a, 1893b).

Yet, the battle over *Man and the Glacial Period* was only nominally about Wright's advocacy of an American Paleolithic and the unity of the glacial period. Instead, it thinly veiled a proprietary dispute between government and nongovernment scientists (Meltzer 1991b). Long-simmering resentment about the perceived heavy-handedness of USGS and BAE scientists boiled over, which, in the wake of economic hard times brought on by the Panic of 1893, triggered a new wave of attacks on profligate federal science (Rabbitt 1980; Worster 2001). While Wright and his defenders sought redress in the press and in Congress, Holmes stayed on the attack, systematically criticizing all Paleolithic claims, Abbott's Trenton gravels included (e.g., Holmes 1893a, 1893b, 1893c; Meltzer 1991b). By the August 1893 meeting of the AAAS, the talk of the Paleolithic was fiery, and the positions on either side had hardened beyond compromise (e.g., McGee 1893b; Moorehead 1893).

Following a few years of relative quiet on the rhetorical scene, the principals separately visited Trenton in June and

FIGURE 2.2 Aleš Hrdlička examining the stratigraphy in the Gilder Mound, Nebraska, January 1907. (Photo by E. Barbour, courtesy of Nebraska State Museum.)

July of 1897, then met to confront each other at the AAAS meetings that August. Once more, their positions were irreconcilable. But this time the BAAS was meeting jointly with the AAAS, and Sir John Evans himself, the Dean of the European Paleolithic, was there. At the meetings he was shown a set of the Trenton paleoliths but dismissed them as Neolithic, not Paleolithic (McGee 1897). It was a devastating blow.

Still, by decade's end contemporaries would call the American Paleolithic battle a draw. Nonetheless, on one point the critics undeniably won: By the turn of the century new discoveries of American paleoliths had virtually ceased, and advocates of the American Paleolithic could do no more than recycle very worn evidence. The active search for a deep past continued, however, and with the discovery in December 1899 of a human femur in the Trenton gravels, the character of the evidence for deep human antiquity, and the nature of the controversy, took an abrupt turn.

The femur was turned over to Aleš Hrdlička (1902), who at that time was noncommittal as to its age and affinities and had no particular opinion about human antiquity in the Americas. That would soon change. By 1903, in the employ of Holmes at the Smithsonian, Hrdlička embarked on a thorough vetting of all skeletal remains purportedly of great antiquity and came away convinced that none was anything more than recent in age. Hrdlička's criticisms rested largely on his strongly held belief that earlier forms of the human species would slowly dissolve into a mosaic of primitive features as one moved back in time. Thus, any allegedly Pleistocene humans found in the New World ought to look primitive. So far as Hrdlička was concerned, none did. Over the next several decades, like Holmes before him, and often with Holmes at his side, Hrdlička challenged each and every claim to a Pleistocene human antiquity:

"Like Horatio at the land bridge between Asia and North America, mowing down with deadly precision all would-be geologically ancient invaders of the New World," as Hooton (1937:102) put it (also Kroeber 1940:461).

In structure, Hrdlička's argument in bone was no different from Abbott's in stone: If it is old, it should look primitive, and vice versa. But there was one significant difference—the argument worked for Hrdlička, who was fast becoming the premier physical anthropologist of his day. Few could challenge his broad and deep knowledge of human variability and evolution. To claim a human skeleton was Pleistocene in age, one had to play by Hrdlička's rules.

Over the next two decades, the discovery of human bones in apparently Pleistocene age deposits at places like Lansing (Kansas), Gilder Mound (Nebraska), and Vero and Melbourne (Florida). In each case, descending on the site to inspect the skeletal remains and their geological context was a swarm of archaeologists and geologists, Hrdlička usually leading the pack (fig. 2.2).

In each case, Hrdlička challenged the claim that the bones were primitive enough to be old. And taking a cue from Holmes, Hrdlička suggested that since humans bury their dead and because bone is so easily broken and moved in the earth, the odds were that any bone in ancient deposits had arrived there fortuitously, redeposited from younger deposits (Hrdlička 1907, 1918). To this, paleontologist Oliver Hay, who had long battled Hrdlička over the issue of human antiquity from an office just down the hall from his, tartly replied, "Perhaps we get a clue here to the reason why civilized people nail up their dead in good strong boxes" (Hay 1918:460; see also Meltzer 1983:32-33).

So it went for nearly four decades. Dozens of purportedly Pleistocene-age sites were championed—some with stone tools, others with human skeletal remains—but all were suspect, and all faced withering criticism from Hrdlička, Holmes, and others (e.g., R. Chamberlin 1917a, 1917b; Holmes 1899, 1902, 1918, 1925; Merriam 1914, 1924; for details on several of the individual sites and how they played out in the controversy, see Meltzer 1983, 1991b). In the end, all the claims were deemed unacceptable—not because of flaws in the geology of the sites, although such was criticized by others (e.g., Chamberlin 1902, 1919), but because all the remains seemed to Hrdlička (1903, 1907, 1918) to fit within the anatomical range of modern or historically known Native Americans. Hence, they had to be recent in age—regardless of the geological context, apparent fossilization, or other signs of great age.

"Facts are facts," Harold Cook had assured John Merriam (Cook to Merriam, January 1, 1929, JCM/LC), but Cook was wrong. Facts were not just facts: They were theory-laden and "controversy-laden" observations about the empirical realm (Rudwick 1985:431). The empirical evidence was never viewed in quite the same way by all who saw it.

Once again, the situation reached an angry impasse. Journalist Robert Gilder (1911), who found the Nebraska "Neanderthals," called Hrdlička a liar. Holmes (1925) snarled that the evidence from Vero was "dangerous to the cause of science." Archaeologist Nels Nelson thought it best to "lie low for the present" (Nelson to Hay, April 5, 1920, OPH/SIA). Shrewd advice, and many followed it.

Ultimately, the dispute over a deep human antiquity in America ranged widely over archaeological and nonarchaeological evidence and issues, led to fiery encounters at meetings and in print, and exposed irreconcilable differences between advocates and critics. In this wide-open field there were few rules of engagement, the controversy grew bitter, and it left lasting scars. When it was all over, Frank Roberts (1940:52) darkly admitted that "the question of early man in America [became over those decades] virtually taboo, and no anthropologist, or for that matter geologist or paleontologist, desirous of a successful career would tempt the fate of ostracism by intimating that he had discovered indications of a respectable antiquity for the indian." Or as A. V. Kidder (1936:144) put it, "[We] comforted ourselves by working in the satisfactorily clear atmosphere of the late periods."

Yet, for all its ambiguity and bitterness and after defying resolution for more than half a century, the controversy over human antiquity in North America simply vanished at Folsom in early September 1927. Paying no heed to his own advice, Roberts went there, and Alfred Kidder too. Folsom, they saw, gave American prehistory a significant time depth, "even by Old World standards," though it was no Paleolithic (Kidder 1936). Indeed, after decades of dispute, nothing ended up looking quite the way it had at the beginning.

Forerunners to Folsom: Two Creeks, One Toad

Cook and Figgins actively entered into the controversy over human antiquity not at Folsom but, earlier, at several other sites, including Snake Creek (Nebraska) and Lone Wolf Creek (Texas) sites, and, slightly later, at Frederick (Oklahoma). Snake Creek is actually a series of localities, one of which was the type site for the apparently Pliocene-aged Anthropoid ape, *Hesperopithecus haroldcookii*, and some comparably aged bone "artifacts." Lone Wolf Creek is a Paleoindian (Plainview age) bison kill. Frederick produced artifacts—including metates—in apparent association with mammoths. Both Cook and Figgins were involved at Frederick and Lone Wolf Creek; only Cook was involved at Snake Creek. These early episodes do not just mark their initial forays into the human antiquity controversy; they strongly influenced the reception accorded them at Folsom, and thus it is important to understand what came out of those sites, and what was said of them—by Cook, Figgins, and everyone else.

Snake Creek

Agate Ranch, Harold Cook's family homestead in western Nebraska, sits amidst one of the most spectacular paleontological localities in the world (part of which became, in 1965, the Agate Fossil Beds National Monument). In 1908, Cook, along with American Museum vertebrate paleontologist W. D. Matthew, discovered promising fossil exposures in the Snake Creek drainage 20 miles south of Agate. Within what they took to be Lower Pliocene units, they collected molars and premolars that showed "a startling resemblance to the teeth of Anthropoidea" (Matthew and Cook 1909:390; the Snake Creek formation is now identified as Miocene in age [see Skinner, Skinner, and Gorris 1977]). They were not altogether surprised. The associated fauna included antelope, which, in deposits of similar age in Europe, co-occurred with Anthropoid apes (Matthew and Cook 1909:390).

Cook collected at this locality intermittently over the next decade and, in February of 1922, sent a molar tooth he had found in the Upper Snake Creek beds to Henry Fairfield Osborn at the American Museum of Natural History. Cook thought the tooth "agrees far more closely with the anthropoid-human molar, than that of any other mammal known" (Cook to Osborn, February 25, 1922, cited in Osborn 1922:1; also Cook 1927b:115; see Skinner, Skinner, and Gorris 1977:277–278). Osborn (1922:1, 4) agreed, hastily identifying the specimen as "the first anthropoid ape of America" and naming it *Hesperopithecus haroldcookii*. More detailed studies by William K. Gregory and Milo Hellman at the AMNH seemed to confirm that the tooth's "nearest resemblances are with 'Pithecanthropus' and with men rather than with apes" (Osborn 1922:2; also Gregory and Hellman 1923a:14).

Osborn's identification of a new genus was not universally accepted. "Responsible critics" offered alternative suggestions of its possible affinities, ranging from bear to horse to monkey (Gregory 1927; Gregory and Hellman 1923b: 526). Obviously, more fossil material was needed.

The summer of 1922 Osborn sent Albert Thomson, his longtime field assistant, back to the "sacred ground" of Snake Creek (or so Osborn viewed it) in search of more of *Hesperopithecus*: "This animal will have a remarkable skull, and heaven grant that you may secure a bit of it" (Osborn to Thomson, June 27, 1922, HFO/AMNH, quoted in Skinner, Skinner, and Gorris 1977:281). But Thomson's efforts were stymied by an uncooperative landowner, so he went elsewhere in the Snake Creek Beds in hopes of finding *Hesperopithecus*. He had no luck (Cook 1927b:115; Skinner, Skinner, and Gorris 1977:281).

A quick surface survey in 1925 back at the type site, however, yielded another specimen (an upper molar), but again, permission to excavate was refused (Skinner, Skinner, and Gorris 1977:282). However, in 1925 and 1926 Thomson, occasionally joined by Osborn, Gregory, and Barnum Brown, collected other material, which largely convinced Gregory that *Hesperopithecus* was actually *Prosthennops*, an extinct peccary (Osborn to Cook, January 27, 1926, HJC/AGFO; Gregory 1927:580). The case was sealed the summer of 1927, when Thomson's excavations at the type site recovered abundant *Prosthennops* fossils—*Hesperopithecus* had been a gigantic taxonomic blunder (Gregory 1927:518).

Nonetheless, in 1925, just as Gregory was beginning to have doubts about *Hesperopithecus*, radically different evidence of early humans was apparently recovered from the Snake Creek beds (Cook 1927b:115–116; Skinner, Skinner, and Gorris 1977:282). Cut, shaped, and drilled bone had been found by Thomson, which Cook and Osborn believed to be genuine Pliocene artifacts. These "artifacts," Cook admitted, were found amidst a large number of broken bones, and there was no "question *at all* that a great deal [of] natural wear, breaking and erosion has also taken a hand" (Cook to Osborn, January 8, 1926, HJC/AGFO; emphasis in original). But Cook insisted that not every specimen was the result of natural causes:

> Of course Thomson has a whole lot of doubtful scrap. Some of it may be artifacts, some probably is not. But excluding all the doubtful things, he still has a nucleus of material in which the evidence, viewed without bias, is so strong that it requires a straining stretch of a negatively inclined imagination to account for the conditions found in any way but the one which says "Made by Hand." (Cook to Osborn, January 8, 1926, HJC/AGFO; also Cook to Ingalls, Febuary 27, 1928, HJC/AHC)

In structure, this argument is little different from the one used half a century later to defend the authenticity of "artifacts" from the Calico and Old Crow sites (Irving 1985; see also Irving, Joplin, and Beebe 1986:59–60; Simpson, Patterson, and Singer 1986:99).

The other paleontologists at the American Museum were "very skeptical" about the Snake Creek "artifacts." Osborn advised Cook to proceed "with extreme caution, because it will not do to take up a position and then later be obliged

to back down and apologize" (Osborn to Cook, January 27, 1926; also Osborn to Cook, May 16, 1927, HJC/AGFO).

Osborn followed his own advice. He would never publish on the Snake Creek "artifacts," but he did speak to them in Philadelphia at the December 1926 AAAS meeting and, again, the next spring at the New York Academy of Sciences. Cook was bolder, marshaling bits of *Hesperopithecus* and the Snake Creek tools into a *Scientific American* article on human antiquity in America, which took "rather sharp issue" with an article published in that same journal just five months earlier by Hrdlička (Cook to Barbour, October 3, 1926, EHB/NSM; Cook 1926; Hrdlička 1926). Cook's bravado was in part fueled by what he believed was corroborative evidence from the site of Lone Wolf Creek.

Lone Wolf Creek

In 1924 Nelson Vaughan, a fossil and artifact collector, wrote Figgins of his discovery of a "huge skeleton" of bison eroding out of the banks of Lone Wolf Creek, just outside of Colorado City, Texas (Vaughan to Figgins, April 1924, JDF/DMNS; cf. Figgins 1927a:229). Figgins was interested in obtaining the fossil for the Colorado Museum and hired Vaughn and rancher H. D. Boyes to collect the specimen. Since neither had training in paleontology, Figgins sent a letter instructing them on how to excavate and crate the specimen (Figgins to Vaughan, April 28, 1924, JDF/DMNS). The goal, as it often was for Figgins, was to "secure fossils of a character that can be mounted, that is, nearly complete skeletons," for such was Figgins' gauge of whether a fossil was valuable (Figgins to Vaughan, May 9, 1925, JDF/DMNS; Cook to Osborn, July 7, 1927, HFO/AMNH).

Figgins would soon regret his decision to supervise the excavations long distance. Vaughan and Boyes dug out the bison in the early summer of 1924 (fig. 2.3) but made a mess of it, failing to apply enough shellac to stabilize the bone. Worse, in the process of crating the fossil for shipping, Boyes took the bone-bearing slabs and

> trimmed [them] down to make them thin enough to go into the crates, instead of making deeper crates. This resulted in the processes being destroyed on a whole series of vertebra. (Figgins to Vaughan, March 31, 1925, JDF/DMNS).

So, instead of having multiple complete skeletons of what appeared to be an extinct, possibly Pleistocene-age species, the Museum "had but one and that had to be restored in many places" (Figgins to Vaughan, March 31, 1925, JDF/DMNS).

Less fortunate still, Figgins learned only afterward that three projectile points had been found with the bison skeleton. The artifacts had not been left in place or photographed, and only two of them actually reached Denver. The third was lost, or possibly stolen (Figgins to Hay, March 11, 1925, JDF/DMNS; Figgins to Vaughan, March 16, 1925, JDF/DMNS; Figgins 1927a:231). Boyes was suspected.

FIGURE 2.3 H. D. Boyes and Nelson Vaughn at Lone Wolf Creek, Texas, 1924. Note the bison remains in place. (Photo courtesy of Heart of West Texas Museum.)

Figgins grumbled that Boyes was just an "ignorant rancher" who felt he was there merely for the purpose of "getting fossils" and had failed to appreciate the site's archaeological significance (Cook to Hay, December 23, 1926, OPH/SIA; Figgins to Hay, December 21, 1926, OPH/SIA; Figgins to Vaughan, January 11, 1927, JDF/DMNS). But in truth, "getting fossils" was precisely why Figgins had Boyes and Vaughan excavating there. For that matter, Boyes may well have appreciated the significance of the artifact, as he had his own "small personal collection of arrowheads."

Figgins suspected that Boyes pilfered the third Lone Wolf Creek artifact (cf. Figgins to Hay, December 21, 1926, OPH/SIA, and Vaughan to Figgins, January 27, 1927, JDF/DMNS).

Regardless, the pressing issue facing Figgins was whether the artifacts had been associated with the fossil bison, and he found himself in the awkward position of building the case for Lone Wolf Creek's antiquity, long after the evidence was out of the ground. Neither Figgins nor Cook believed that Boyes or Vaughan had either motive or knowledge to fake an association of points and fossil bison but well knew that their trustworthiness would be a central issue (Cook to Hay, December 23, 1926; Figgins to Cook, December 28, 1926, HJC/AGFO; Figgins to Hay, December 21, 1926, Hay to Figgins, November 1, 1926, OPH/SIA). Unhappily, neither excavator had proven to be an especially useful witness about the details of the discovery, though Vaughan insisted that two of the points were "immediately associated with the bison remains as tho [sic] they had been shot there with great force" (Vaughan to Figgins, March 20, 1925, JDF/DMNS; also Figgins to Vaughan, March 16, 1925, December 21, 1926, January 11, 1927, JDF/DMNS).

With that and little else to go on, the Lone Wolf Creek claim to great antiquity centered on Cook's assessment of the geology, which he worked out in a visit to the site in the spring of 1925, long after the excavations were over. Yet, on that slender footing, Cook (1925:459) boldly announced in *Science* that Lone Wolf Creek provided "good, dependable definite evidence of human artifacts in the Pleistocene in America." In his opinion, it had profound implications for the "unfortunate controversy" over human antiquity in America (Cook to Cattell, June 25, 1925, HJC/AGFO), since

> There is no possibility of accidental intrusion of these artifacts with the bison, or of their being of later age. *They are certainly and positively contemporaneous with that fossil bison and the associated fauna of mammoths, camels and extinct horses*—of a type found elsewhere in beds of known Pleistocene age. (Cook 1925:460; emphasis in the original)

Cook may have been mostly correct—at least about the artifacts associated with the bison, which he saw as being similar to those Samuel Williston had recovered with a fossil bison at the site of 12 Mile Creek, Kansas, some three decades earlier; Barnum Brown was of a similar opinion (Cook to Cattell, June 25, 1925; Figgins to Cook, July 27, 1925, HJC/AGFO). But while his announcement rang with assurance, the circumstances of the discovery were disappointing even to proponents of a deep human antiquity (e.g., Hay to Cook, December 17, 1926, HJC/AGFO; cf. Goddard 1926). Under the circumstances, there was little reason for confidence in Lone Wolf Creek, no matter how vigorously Cook tried to promote it. And he tried very hard indeed. However, that only made the situation worse.

Three days after Cook's paper was published in *Science*, Holmes wrote to geologist John C. Merriam, President of the Carnegie Institution of Washington:

> You have doubtless seen an article by Harold J. Cook in the last issue of SCIENCE, in which the finding of arrow heads in the Pleistocene is constantly announced. The frequency of such risky announcements requires, if possible, that prompt attention be given in each case. . . . It occurs to me that you may find it possible to have Mr. [Chester] Stock or some other well-qualified person visit the Texas site and report upon the evidence. (Holmes to Merriam, November 23, 1925, WHH/SIA)

Hrdlička was equally troubled by Cook's paper: "the cocksureness of the author makes one rather suspicious" (Hrdlička to Merriam, via Gilbert, November 28, 1925, JCM/CIW). Merriam reported that unfortunately a site visit was not possible, but he made discrete inquiries of Cook's old mentor, W. D. Matthew (Merriam to Holmes, November 27, 1925, WHH/SIA; also Merriam to Matthew, November 23, 1925, JCM/LC). In just a few days, Matthew replied:

> I read Harold Cook's article in SCIENCE rather carefully and critically. Harold has, as you know, a somewhat optimistic temperament, and I find it necessary to discount his geological conclusions more or less. (Matthew to Merriam, November 27, 1925, JCM/LC)

And so Merriam, Holmes, and Hrdlička did as well.

Cook persisted. In a *Scientific American* broadside attack on Hrdlička published the next year, Cook (1926) invoked Lone Wolf Creek, *Hesperopithecus*, and the Snake Creek artifacts in support of his claim for a Pleistocene or even earlier human presence in the New World.

In Hrdlička's eyes, Cook's paper was just "another head of the hydra," and he moved swiftly to sever it.[1] As Holmes had the year before, Hrdlička wrote Merriam asking whether it was possible to have the Carnegie Institution send "some reliable man who would command the confidence of us all to look into the matter" (Hrdlička to Merriam, November 10, 1926, AH/NAA). For good measure, Hrdlička wrote again a week later, criticizing the Lone Wolf Creek claim:

> It appears [the discovery] was not even made by himself [Cook]; there has been shown no photographic record of the find; and there is not the slightest corroboration of the claims by any one in whom we all could have full confidence. (Hrdlička to Merriam, November 15, 1926, AH/NAA)

Cook, arguing vigorously on the basis of very thin evidence, was decidedly not one in whom Hrdlička had full confidence. And if Hrdlička had less than full confidence in Cook regarding Lone Wolf Creek, it was not improved by Cook's claims for the Frederick site, made early the next year.

Frederick

After reading Cook's *Scientific American* article, F. G. Priestley, a Frederick, Oklahoma, physician, wrote the magazine's editor, Albert Ingalls, to report that he had found fossils of extinct animals and apparently ancient stone tools in a commercial gravel pit. Ingalls had Priestly contact Cook, who, with Figgins, visited the gravel pit in late January 1927, admonishing all that "secrecy, secrecy is the thing" (Ingalls to Cook, January 12, 1927; Priestly to Cook, January 19, 1927, HJC/AGFO; Figgins 1927a:235).

Whatever the reason for secrecy, Cook soon took the wraps off Frederick. As he interpreted the geology, the site

FIGURE 2.4 A. H. Holloman (left) and Harold Cook in the gravel pit at Frederick, Oklahoma, January 1927. (Photo courtesy of Denver Museum of Nature and Science.)

was in an ancient river bed that, following extensive erosion around it, had been left a high remnant above the landscape (Cook 1927b:116). Found in the "undisturbed deposits of this old stream bed" were bones of mammoth, mastodon, sloth, horse, camel, and other Pleistocene fauna (Cook 1927b:117), along with projectile points and ground stone metates (Figgins 1927a:237–238). "Strangely enough," Cook (1927b:117) remarked, "these implements show a degree of culture closely comparable with that of the nomadic modern Plains indians."

Strange, indeed. The artifacts had been recovered during the quarrying, but since the quarry operators "did not appreciate their importance," many more were surely lost, either spread on the local roads or buried somewhere in the quarry debris piles (Cook to Hay, February 25, 1927, OPH/SIA; Figgins 1927a:235). Still, "not less than five unquestionable metates" were discovered in the presence of A. H. Holloman, the quarry owner (Figgins 1927a:238) (fig. 2.4).

Cook inferred the age of the deposit using the amount of time it must have taken to erode the locality to its present position: Assuming known erosion rates, he calculated about a million years. That was "undoubtedly excessive," even for Cook (1927b:117), so he took a fraction of that and concluded that the site was about 365,000 years old. Paleontologist Oliver Hay (1927), on the basis of the fauna, put its age in the "early Pleistocene, the first interglacial stage, in round numbers, 500,000 years ago." As to the presence of modern-looking artifacts at such great antiquity, Cook explained:

It now appears to me quite probable that these very early people *were* closely comparable in most ways to our modern Indian, structurally as well as in culture; and if so, this would account for the errors in interpretation as

to man's antiquity in America, fallen into by some talented scholars. (1927b:117; emphasis in original)

Talented scholars may have made "errors," but there were also gaping holes in Frederick, as even its proponents admitted. Neither Cook, nor Figgins, nor any other visitor to the site ever found any artifacts in situ (Cook to Gould, October 22, 1927, HJC/AGFO; Gould to Figgins, October 11, 1927; Figgins to Gould, October 14, 1927, JDF/DMNS; Gould to Hay, October 10, 1927, OPH/SIA). Nor were any subsequently found in place by Holloman, who had been urged to leave them that way so that "scientists of established reputation" might see them (Cook to Hay, December 2, 1927; Gould to Hay, October 10, 1927, January 24, 1929, OPH/SIA).

Again, as at Lone Wolf Creek, crucial details on what was found and where rested on the testimony of an inexperienced collector (Cook to Hay, March 16, 1927; Hay to Priestley, September 14, 1927, OPH/SIA; also Figgins to Cook, October 12, 1927, HJC/AGFO). Holloman seemed honest, but that was hardly enough for Charles Gould, the Oklahoma State Geologist:

I agree with you that Mr. Holloman is honest and that he believes that he found the various artifacts where he says he found them. At the same time, we all realize that he has very little conception of the values of things which are important and those which are unimportant. (Gould to Figgins, October 18, 1927, JDF/DMNS).

In fact, Holloman, showed "utter ignorance of all things pertaining to fossils and artifacts" (Figgins to Hay, October 22, 1927, JDF/DMNS). He had identified, with the help of the local chiropractor, glyptodon fossil fragments as "carved stone" and a mammoth tooth as a "perfect human skull." He

FIGURE 2.5 Harold Cook and "Pleisty" the toad at Frederick, Oklahoma. (Photo courtesy of Dorothy Cook Meade.)

even reported finding a live toad (fig. 2.5) encased in a Pleistocene-aged mud ball (Figgins to Gould, October 14, 1927; Gould to Figgins, October 18, 1927; Hay to Figgins, September 16, 1927, October 25, 1928, JDF/DMNS). Cook, of course, was "thoroughly of the opinion" that the toad was modern, but he did "want to know how they got into the situation in which they are found" (Cook to Hay March 31, 1928, OPH/SIA). In any event, such absurd claims as made by Holloman, Figgins unhappily admitted, were "a more or less serviceable weapon in the hands of the enemy" (Figgins to Hay, September 19, 1927, October 22, 1927, OPH/SIA).

Still, Holloman insisted that the formation in which the artifacts were found "had never been disturbed sinse [sic] it was first formed" (Holloman to Hay, September 1927, OPH/SIA). Cook, Figgins, and Hay well realized, of course, that an untrained observer picking up stones in a commercial gravel quarry might well misinterpret the stratigraphy. Nonetheless, they believed the conditions at Frederick limited the possibility, which brought the discussion back around to Cook's interpretation of the geology (Cook to Hay, March 16, 1927, OPH/SIA; Cook to Gould, October 22, 1927, HJC/AGFO; Figgins to Gould, October 14, 1927, JDF/DMNS).

Yet, even Oliver Hay was not completely satisfied with Cook's assessment of the geology (Hay to Cook, January 17, 1928, HJC/AGFO; Cook to Hay, January 25, 1928, OPH/SIA). He urged that an independent geologist be sent to the site to determine its age; Cook and Figgins agreed (Hay to Cook, July 15, 1927, HJC/AGFO; Cook to Hay, July 18, 1927, OPH/SIA; Hay to Figgins, July 15, 1927; Figgins to Hay, July 20, 1927, JDF/DMNS).

The task fell to State Geologist Charles Gould, working with Oliver Evans of the University of Oklahoma, who was especially well versed in the geology of the valleys of western Oklahoma (Gould to Cook and Figgins, October 17, 1929, HJC/AGFO and JDF/DMNS). Evans concluded that the Frederick strata were "not necessarily more than 10,000

years old, and might be somewhat younger" (Evans, October 17, 1929, JDF/DMNS; see also Evans 1930). That was a "rather ridiculous" conclusion, Cook snarled, but Gould cautioned him that Evans "*may be right*" (Cook to Gould, November 2, 1929; Gould to Cook, November 4, 1929, HJC/AGFO, emphasis in original; see also Cook 1931a). Cook quietly spread the poisonous rumor that Evans' paper had initially been rejected because of its inaccuracies and was only published because of Hrdlička's intervention and influence (Cook to Ingalls, April 3, 1928, HJC/AGFO).

Evans' assessment of the geology was almost immediately followed by a sharp critique of the archaeology by Leslie Spier, then at the University of Oklahoma. Spier had seen the site and was appalled: Any claims about artifact context were doubtful, and even if the Frederick artifacts were genuine (Spier hadn't seen them), the age Cook and Hay assigned them was utterly incongruous with known human history (Spier 1928a:160–161).

Hay had initially been pleased to learn that Spier was to visit Frederick, thinking Spier was "a good man" (Hay to Figgins, October 18, 1927, JDF/DMNS). But after Spier's criticism appeared in *Science*, Hay changed his mind. Both he and Cook quickly fired off rejoinders.[2] Frederick had real artifacts, they insisted, and Spier's a priori conclusions about what artifacts at this age might look like were entirely inappropriate, but just the sort of thing one could expect, Cook opined, from someone with "that sort of mind and training" (Cook to Hay, February 20, 1928, OPH/SIA; the rejoinders are Cook 1928a and Hay 1928; see also Cook to Brown, February 27, 1928, HJC/AHC; Figgins to Hay, November 8, 1928, DIR/DMNS):

> Dr. Spier holds that there is an incongruity in the association of artifacts, as identified by our anthropologists, with fossils bones and teeth of animals of Aftonian age. There is an incongruity, but this is the creation of the anthropologists. (Hay 1928:442)

Spier, at first reluctant to engage in public debate, chose instead to write Cook a letter of reply, which he also circulated among interested individuals, including Hrdlička (Spier to Cook, April 26, 1928, HJC/AGFO, also AH/NAA). Hrdlička was naturally pleased with Spier's response, and after urging him to publish it, which he ultimately did (Spier 1928b), Hrdlička bemoaned what he saw as another

> "epidemic" of claims by the paleontologists, and with some of these men the matter has become a regular obsession—I can find no better word with which to describe the state of affairs. The matter ought to be taken up energetically by our archaeologists, where archaeological evidence alone is concerned; [but] we almost seem to have no archaeologists. (Hrdlička to Spier, May 7, 1928, AH/NAA)

Credible or Incredible?

While Hrdlička was not persuaded by Cook and Figgins on Snake Creek, Lone Wolf Creek, and Frederick, he was equally unconvinced when he heard the story from others. In December 1926 at the AAAS meeting in Philadelphia, Osborn presented his opinion that the Snake Creek artifacts were genuine. Hrdlička, having himself just presented a paper on "Recent Findings in America Attributed to Geologically Ancient Man" (Kidder and Terry 1927), was in the audience, along with *Scientific American* editor Albert Ingalls, who later reported:

> All during [the] presentation of [Osborn's] Nebraska paper at Philadelphia I watched Hrdlička sitting there, face getting red, eyes getting glassy. At the end he rose and said of course he had never been able to accept any of the evidences but did not relish being put in a class which has shut its mind in the matter. . . . He says the find should have been seen by "some authority." I wonder *who* he means??! (Ingalls to Cook, May 10, 1927, HJC/AHC; emphasis in original)

Hrdlička, in fact, had a committee of people in mind.

For a number of years Hrdlička had advocated the establishment of a blue-ribbon panel to examine each new claim of great antiquity in the Americas. As he described it, it was to be an elite group, composed of "the foremost men in Geology, Anthropology, etc." and sponsored by the AAAS or the National Research Council. To such a panel

> . . . all cases of finds relating to man's antiquity on this continent [should] be referred to this committee as soon as possible after the finds are made and before anything has been removed or isturbed. (Hrdlička to Chamberlin, October 20, 1919, AH/NAA)

While that idea never got off the ground in that particular form,[3] Hrdlička always held to the principle that all such finds needed to be examined and evaluated on-site by recognized experts.

It was only this sort of approach that would appease skeptics, who believed that advocates of a deep human antiquity "always assume the highly unscientific attitude of endeavoring to prove the case without considering the evidence to the contrary" (Hodge to Hrdlička, June 1, 1928, AH/NAA). Even Cook's editor at *Scientific American* saw the need to keep him in check:

> You fellers [sic] get enthusiastic once in a while, you know, and kinda run ahead. I have to be the flywheel, sometimes. (Ingalls to Cook, July 11, 1928, HJC/AHC)

That enthusiasm was also too much for old friends: Matthew was "unable to take the Snake Creek 'artifacts' seriously"—so were Brown and Gregory (Matthew to Cook, August 11, 1927; Osborn to Cook, May 6, 1927, May 16,

1927, HJC/AGFO). After his May 1927 presentation on the Snake Creek artifacts, Osborn published nothing and said little more about them, apparently driven to silence by Nels Nelson's verdict that these were just "waterworn bones" (Nelson 1928a; Skinner, Skinner, and Gorris 1977:282).

Cook and Figgins paid the skeptics little mind, writing that Frederick—and, for that matter, Snake Creek and Lone Wolf Creek—pushed humanity back "by hundreds of thousands of years" (Cook 1927b:116). Cook even refused to accept Gregory's conclusion that *Hesperopithecus* was just a peccary. After all, the artifacts from Snake Creek indicated that a Pliocene or early Pleistocene human must have been present. As he saw it, "Hesperopithecus was probably the direct ancestor of our Aftonian men from Frederick, Oklahoma!" (Cook to Ingalls, February 27, 1928, HJC/AHC). Then and afterward he stuck to his guns. In talks he gave into the 1950s he continued to insist that

> among those so-called "artifacts" were a few so formed, cut, shaped, drilled or polished that it seems incredible that they could be formed by any natural agent of erosion with which we are familiar. . . .This whole question as to Pliocene man or artifacts in America is far from being settled, either way. (Cook 1952:17–18)

Yet, "incredible" as it may have seemed to Cook, as far as the archaeological community of the 1920s was concerned, the Snake Creek artifacts were not humanly made, Frederick was almost certainly a hoax, and the Lone Wolf Creek case was so shoddy it was simply unacceptable at face value (Antevs fieldbook, August 11, 1934, EA/UA; Hrdlička 1928:809; Roberts 1940:58). As Hrdlička put it, here was a trio of questionable finds that plunged the human presence in the Americas back hundreds of thousands of years, but none had "been examined except superficially by any anthropologist or archaeologist outside of those directly concerned," and all were believed only by Cook and Figgins (Hrdlička to Hodge, June 7, 1928, AH/NAA).

It was in this context that Cook and Figgins' work at Folsom emerged.

Folsom, 1908–1928

The precise circumstances of the initial discovery are poorly known (Anderson 1975:43–44). The various accounts (e.g., Agogino 1971, 1985; Cook 1947; Fagan 1987; Folsom 1992; Hewett 1971; Hillerman 1971; Little 1947; Owen 1951; Preston 1997; Reed 1940; Roberts 1935; Steen 1955; Thompson 1967) are mostly secondary, based on the recollections of those who were not there, and are often contradictory—as one might expect. Most credit the initial discovery to Crowfoot Ranch cowboy George McJunkin (fig. 2.6); others, however, grant him only bystander status, belittle his role, or simply ignore him altogether (e.g., Cook 1947; Owen 1951, Roberts 1935; Thompson 1967). This, too, is not unexpected.

FIGURE 2.6 George McJunkin at the Crowfoot Ranch. (Photo courtesy of Eastern New Mexico University.)

FIGURE 2.7 Carl Schwachheim in front of his antler fountain, Raton, New Mexico. (Photo courtesy of Carl Schwachheim Collection.)

FIGURE 2.8 Fred Howarth sitting on the floor of Wild Horse Arroyo, pointing to bison bone eroding out of deposits on the South Bank, December 1922. (Photo courtesy of Carl Schwachheim Collection.)

In one of the most bizarre renditions of the Folsom story, Cook claimed that Fred Howarth had seen his 1925 *Science* paper on Lone Wolf Creek, and while checking the Crowfoot cattle herd for the Raton bank he told the "Negro Cowboy" (McJunkin) of Cook's paper. McJunkin then apparently replied that he had spotted some bones deeply buried in an arroyo and took Howarth to see them (Cook 1947:1). There is one *literally* fatal flaw to Cook's version of the events: McJunkin died in 1922, three years before Cook's paper on Lone Wolf Creek appeared in *Science*. Cook later claimed to have met McJunkin in 1927 (see Cook to Markman, February 14, 1947, and Cook to Sellards, May 5, 1950, both HJC/AGFO).[4]

In any case, the reasonably secure facts are as outlined in chapter 1: namely, that sometime after the 1908 flood, McJunkin spotted bison bones eroding out near the base of Wild Horse Arroyo. The bones obviously piqued his interest, for he began telling others about the find (Brown to Figgins, May 28, 1937, VP/AMNH). Among those he told was Carl Schwachheim, the Raton blacksmith: One engaging scenario has McJunkin stopping by Schwachheim's home to admire a fountain (fig. 2.7) made with the antler racks of two bull elk that became entangled in a mortal contest (Folsom 1992; Steen 1955:5), whence the two learned of their shared interest in natural history. None of the entries in Schwachheim's diary, which begins in 1918, makes mention of McJunkin.

Schwachheim ultimately visited the site with Howarth and others on December 10, 1922, after McJunkin had died. Their visit was recorded in Schwachheim's Diary (appendix B) and memorialized in a photograph of Howarth sitting on the floor of Wild Horse Arroyo, pointing to some bison bone in the side wall (fig. 2.8).

After failing to get the State of New Mexico to excavate the site (T. Burch, personal communication, 1997), in late January of 1926 Schwachheim and Howarth visited the Colorado Museum of Natural History in Denver (appendix B; Roberts 1935:4; Steen 1955:5–6). Figgins and Cook were interested to hear of the site, and when Howarth subsequently shipped a package of the bison bones from Folsom, Cook was immediately struck that not only was this a previously unknown and extinct species of bison, but also it was possibly "very similar to the Texas [Lone Wolf Creek] material" (Cook to Barbour, March 12, 1926, EHB/NSM; Howarth to J. D. Figgins, February 4, 1926, DIR/DMNS). A site visit was arranged for all parties in the spring of 1926 (Figgins to Howarth, February 6, 1926, March 31, 1926, DIR/DMNS) (appendix B; fig. 1.3). Years later, Cook claimed that after that visit he recommended excavations:

First, for mountable *new* bison; Second: as we might get there geological data to date it, and; Third; as the charcoal and other evidence I noted there suggested it *might*

FIGURE 2.9 Carl Schwachheim in the excavations on the South Bank at Folsom. (Photo courtesy of American Museum of Natural History.)

be a "kill," we might find evidence here of primitive man, with primitive *bison*, as we had done in Texas. (Cook 1947:2, emphasis in original; Cook to Bailey, February 5, 1947, DIR/DMNS)

Figgins would likewise claim that he was interested in Folsom because of the "possibility that additional evidence of man's antiquity might be uncovered, and with that prospect in view . . . gave explicit instruction that constant attention be paid to such discoveries" (Figgins 1927a:232; also Figgins to Hay, November 6, 1926, DIR/DMNS).

Hindsight is always perfect, never more so than here. In fact, the contemporary correspondence shows no evidence that Folsom was excavated because Cook or Figgins thought it might be an archaeological site, or expected it to yield artifacts—though others anticipated the possibility.[5] Instead, the excavation was aimed at "supplying a mountable [bison] skeleton [and] . . . exchange material" (Figgins to Taylor, June 21, 1926, DIR/DMNS; see also Cook to Barbour, February 15, 1926, EHB/NSM; Figgins to Brown, May 26, 1926, VP/AMNH). This is why, when a projectile point was later found with the bones, it was "shocking news" (Figgins to Brown, July 23, 1926, VP/AMNH).

An initial sum of $300 was transferred to Howarth's bank so he could finance the startup of the fieldwork, which included paying the land leasee, J. L. Shoemaker of the Crowfoot Ranch, $50 for the "exclusive right and privilege [sic] of excavating and removing from his property skeletons of prehistoric animals." Howarth thought that was a steep price to pay, but then "these ranchers seem to get the idea that when a Museum wants anything like fossils that they have a large money value" (Howarth to Figgins, May 11, 1926, DIR/DMNS).[6] Of course, this was little more than a courtesy payment: Howarth well knew that the site was on State of New Mexico land, and he had applied for and received permission from the State Land Office to excavate (Schwachheim to Figgins, March 5, 1926; Swope to Howarth, May 5, 1926, DIR/DMNS).

Fieldwork began at Folsom in early May, 1926, the heavy pick and shovel work being done by Schwachheim (fig. 2.9). The work was slowed early on by snow and rain and, when Harold Cook showed up in mid-June, by conflicting instructions (Figgins to Cook, June 4, 1926, HJC/AGFO; Figgins to Howarth, June 21, July 24, 1926; Howarth to Figgins, July 5, 1926, DIR/DMNS; Schwachheim to Figgins, May 24, 1926, June 11, 1926, JDF/DMNS). In early July Figgins' son Frank arrived, perhaps to help ensure that excavations proceeded according to plan—although his unaccounted spending of Museum funds raised eyebrows (Howarth to Figgins, July 5, 1926, July 22, 1926, September 2, 1926, DIR/DMNS).[7]

In mid July, the first Folsom point came out of the bone bed (fig. 2.10). Unfortunately, it was not found in situ (appendix B; July 14, 1926). Learning of the find only afterward and by letter (again!), a disappointed Figgins, the memories of Lone Wolf Creek undoubtedly fresh in his mind, told Howarth to

try to impress the boys [Schwachheim and Frank Figgins] with the importance of watching for human remains and then in no circumstances, remove them, but let me know at once. I would want to study and collect such material in situ. This find is of double importance in that it substantiates the other finds. (Figgins to Howarth, July 22, 1926, DIR/DMNS)

FIGURE 2.10 Left, the first fluted projectile point found at the Folsom site, July 1926; right, second projectile point found 1926. Neither was found in situ. (Photo by D. J. Meltzer.)

While he instructed them to "scan every particle of dirt they remove" (Figgins to Brown, July 23, 1926, VP/AMNH), the crew found no points in situ the remainder of the field season (Schwachheim to Figgins, August 9, 1926, DIR/DMNS). However, one additional point (fig. 2.10) was found that refit to a sliver of stone adjacent to a rib (Figgins 1927a: 232–234, fig. 3 right, fig. 4; Figgins to Cook, November 16, 1926, HJC/AGFO; Figgins to Brown, November 16, 1926, VP/AMNH; Figgins to Hay, November 8, 1926, OPH/SIA).

That autumn, at the urging of Barnum Brown, Cook and Figgins wrote their *Natural History* papers, but obviously before any artifacts had been found and examined in place, let alone before Folsom's antiquity had been confirmed (Cook 1927a; Figgins 1927a; see Brown to Figgins, November 30, 1926, VP/AMNH). In fact, the *Natural History* papers were polemical pieces, repeating their claims about Lone Wolf Creek and Frederick and written in "a deliberate attempt," as Figgins boasted to Hay, "to arouse Dr. H. [Hrdlička] and stir up all the venom there is in him" (Figgins to Hay, February 23, 1927, OPH/SIA; Figgins to Brown, February 16, 1927, April 9, 1927, VP/AMNH). Unfortunately, Figgins comments also stirred up the editors at *Natural History*, who rewrote his introduction in order to remove the "personal [and] controversial matter" (Brown to Figgins, April 4, 1927, VP/AMNH).[8]

Figgins didn't particularly care whether or not he aroused Hrdlička's ire (Figgins to Brown, April 9, 1927, VP/AMNH). As he explained it to Cook:

Everyone seems to think Hrdlička will attack it and if you haven't realized that I will fight back in a two-handed manner, then watch the dust the instant Hrdlička appears in print. . . . You see, I am a free lance and without responsibility in the matter of "scientific courtesy," so if a party tears a chunk of hide off my back, in a suave and courteous manner, there is nothing to prevent my removing three upper and two lower incisors, black one eye and gouge the other, after I have laid his hide across a barbed wire fence, and that is precisely what will happen. (Figgins to Cook, December 28, 1926, JDF/DMNS)

"I am daring the whole miserable caboodle of them," he proclaimed.

Brave words, but they rang hollow. Hay plotted "a showdown" over Folsom between Hrdlička and Figgins at the Smithsonian, but Figgins back pedalled and suggested Harold Cook should go to Washington to "be the goat" (Figgins to Howarth, November 27, 1926, JDF/DMNS; also Figgins to Cook, November 26, 1926, December 21, 1926, HJC/AGFO; Figgins to Hay, November 8, 1926, November 26, 1926, December 10,1926; Hay to Figgins, November 17, 1926, December 6, 1926, DIR/DMNS and OPH/SIA).[9]

In the end, Cook was unable to serve as sacrificial animal, so Figgins traveled east in the spring of 1927 with the Folsom artifacts, although "not with the idea that I am submitting them for [Hrdlička's] opinions. I view this find a

FIGURE 2.11 Aleš Hrdlička at the Smithsonian Institution. (Photo courtesy of National Anthropological Archives, Smithsonian Institution.)

question for the paleontologists and geologists to settle and the opinion of the anthropologist is of the scantest [sic] importance" (Figgins to Hay, April 11, 29, 1927, OPH/SIA; Hay to Figgins, April 25, 1927, DIR/DMNS). By the time he arrived at the Smithsonian, he'd worked himself into a lather at the prospect of meeting the fearsome Hrdlička in his lair[10] (fig. 2.11).

Yet, to his surprise and relief, Hrdlička was courteous and extremely pleased to see the Folsom points, and only expressed his regret that none had been found in place (Figgins to Hay, July 1, 1927, September 29, 1927, OPH/SIA). They then went down the hall to see Holmes, who, initially unaware of where the points were from, dismissed them as "undoubtedly European."[11] After being told they had come from New Mexico, Holmes averred that

> he would never believe man and those prehistoric mammals were contemporaneous—that man hunted those mammals—until arrowheads were found imbedded in the bones of the latter in such a manner as to prohibit question that they were shot into them while the animal was living. (Figgins to Brown, June 8, 1927, VP/AMNH; also Figgins to Hay, July 1, 1927, OPH/SIA)

That, of course, had not happened—yet.

In the event that it did, Hrdlička offered Figgins some advice: When additional artifacts appeared in situ, the crew should immediately stop excavating, and telegrams should be sent around the country inviting "outside scientists" to come and examine the artifacts in place (Figgins to Brown,

June 8, 1927, VP/AMNH; Figgins to Hay, July 1, 1927, OPH/SIA; Meltzer 1991b:32). Figgins thought that "perfectly reasonable" advice, and came away with newfound respect for Hrdlička (Figgins to Brown, June 8, 1927, VP/AMNH; Figgins to Hay, September 29, 1927, OPH/SIA; Figgins to Nelson, August 24, 1927, DIR/DMNS; Figgins to Cook, July 22, 1927, HJC/AGFO). What Figgins hadn't appreciated was Hrdlička's motive for offering that advice. Hrdlička didn't trust Figgins or Cook and wasn't going to be convinced by anything they said about the Folsom site, its age, or any possible association of artifacts with extinct animals. He wanted that panel of blue-ribbon scientists called in when the time came.

In late May of 1927, excavations resumed at Folsom, and on August 29, 1927, as Schwachheim reported to Figgins, "I found an arrow or spear point at noon. . . . Thought perhaps some of your doubters would like to see the evidence in the matrix and in place" (Schwachheim to Figgins, August 29, 1927, DIR/DMNS). Figgins promptly wired his congratulations and instructed Schwachheim to carefully cover and secure the point (figure 1.4), and to keep his eye on it "every minute and not let any one remove it or dig around it regardless of who it is" (Figgins to Schwachheim, August 31, 1927, DIR/DMNS). Hrdlička's advice was followed: Telegrams were fired off to individuals and institutions around the country, Hrdlička included (though he was then in Alaska), announcing, "Another arrowhead found in position with bison remains at Folsom, New Mexico. Can you personally examine find. Answer telegram" (e.g., Figgins to Brown, August 30, 1927, BB/AMNH).

FIGURE 2.12 Carl Schwachheim (left) and Barnum Brown posing with the first in situ Folsom point, September 4, 1927. (Photo courtesy of American Museum of Natural History.)

Schwachheim duly awaited the arrival of "Scientists, Anthropologists, Archaeologists, Zoologists, or other bugs" (Schwachheim to Figgins, September 4, 1927, DIR/DMNS). Among those converging on Folsom on September 4th were Figgins and Howarth, as well as Barnum Brown, already out West doing fieldwork, and Frank Roberts, sent by the Smithsonian in Hrdlička's stead (Wetmore to Figgins, September 2, 1927, DIR/DMNS). Brown made a series of notes on the stratigraphy and geology, and had his picture taken next to the in situ point (fig. 2.12). Roberts had been attending the first Pecos Conference and was quite taken by what he saw at Folsom. He left at the end of that day (without having his picture taken), but returned again on September 6th, and once more on September 8th, the last time having invited along A. V. Kidder of the Carnegie Institution of Washington (appendix B; Kidder to Figgins, October 13, 1927, JDF/DMNS; Roberts 1935:5).[12]

All who visited the site agreed that the artifacts and fossil bison were contemporaneous (Brown 1928a; Kidder to Figgins, October 13, 1927, DIR/DMNS; Figgins to Granger, September 30, 1927, VP/AMNH; Roberts to Fewkes, September 13, 1927, BAE/NAA; Wetmore to Figgins, September 8, 1927, DIR/DMNS; Meltzer 1983:35–37, 1994; Roberts 1935:5). As Kidder put it:

> As an archaeologist, I am of course not competent to pass either upon the paleontological or the geological evidences of antiquity, but I have paid great attention for many years to questions of deposition and association. On these points I am able to judge, and I was entirely convinced of the contemporaneous association of the artifact which you so wisely had left "in situ" and the bones of the bison. (Kidder to Figgins, October 13, 1927, DIR/DMNS)

Within the month, Kidder announced publicly what he'd always hoped for privately (Meltzer 1993:129–130): The first Americans had arrived some 15,000 to 20,000 years ago (Kidder 1927). The announcement, elaborated on by Brown, Nelson, and Roberts at the December American Anthropological Association meeting, electrified the scientific community.

Hrdlička had been invited to attend that meeting but declined because of a previous commitment (Dixon to Hrdlička, November 27, 1927; Hrdlička to Dixon, December 2, 1927, AH/NAA). The following spring, he spoke at the New York Academy of Medicine, where he reiterated his position on the purported ancient human skeletal evidence—about which, of course, the evidence from Folsom was irrelevant. He then suggested that the problem of human antiquity in the Americas might be unresolvable—at least within his generation (Hrdlička 1928:807).

Nelson, who followed Hrdlička, politely disagreed. The problem, he suggested, might easily be solved after an hour or so of discussion (Nelson 1928:822). Brown, who followed Nelson, would not need even an hour. "In my hand," he announced with the projectile points from Folsom held high for all to see, "I hold the answer to the antiquity of man in America. *It is simply a question of interpretation*" (Brown 1928a:824; emphasis in original). Hrdlička's reaction to Brown that evening went unrecorded, but several days later, he wrote Brown asking for a copy of one of Brown's slides showing a general view of the site and a bit more information:

> I should also be glad if you would give me your views as to how the artifacts may have come into association

with the bisons [sic]. (Hrdlička to Brown, March 15, 1928, AH/NAA)

Brown replied with his interpretation of the site's geological context and then testified that

> . . . I personally removed the five inches of stratified clay that immediately covered all of the fifth point, excepting one of the barbs. . . . There is absolutely no possibility of any introduction of the points subsequent to the natural covering over of the bison skeleton. (Brown to Hrdlička, March 16, 1928, AH/NAA)

That wasn't quite good enough for Hrdlička:

> Thanks very much for your good letter. There is only one additional point on which I should like your opinion and that is, the manner in which the arrow points or darts got into these places where they were found. (Hrdlička to Brown, March 19, 1928, AH/NAA)

Brown tried again, first asking what Hrdlička thought of the photograph he had sent showing the point "lying in situ between the two ribs" that he had personally uncovered, then repeating:

> It is my opinion that, at least, three of these points were embedded in some part of the flesh of the animals when the carcasses were entombed. . . . Please write me further if I have not made the situation clear. (Brown to Hrdlička, March 21, 1928, AH/NAA)

Unable to get Brown to admit to any possibility of intrusion at the site, Hrdlička backed off:

> Thank you for your supplementary letter which is very satisfactory. (Hrdlička to Brown, March 22, 1928, AH/NAA)

Hrdlička portrayed the exchange in a very different way when he described it to F. W. Hodge a few months later:

> A curious incident happened to me with Barnum Brown. He showed, you may remember, among others a picture that gave a general view at Folsom, and as this picture showed some suggestive conditions [a "little ravine" in the middle of the site], I wanted a copy and so wrote him. In answer he sent me four other pictures quite useless, without mentioning the one I wanted; and when I wrote again specifying clearly what I was after, there was no answer. Such things naturally make one suspicious. There are many points about the Folsom find which need explanation; but they are not the points touched upon in the talks or reports on the find. Hay, Gidley, Barnum Brown, Harold Cook, Figgins, and Renaud—what a constellation of blossoms. (Hrdlička to Hodge, June 7, 1928, AH/NAA)

Hrdlička would never endorse the Folsom evidence, but he nonetheless tacitly accepted it. The evidence for that is clear enough: After nearly three decades of publicly attacking any and all claims for human antiquity in the Americas (e.g., Hrdlička 1902, 1903, 1907, 1918, 1926), Hrdlička said virtually nothing about Folsom. Holmes, when questioned about his view of the Folsom discovery, sidestepped the issue, saying he had "now dropped the matter entirely, and am perfectly willing to accept the conclusions of the skilled men who are carrying on researches with the full knowledge of the problems and the dangers" (Holmes to Sellards, March 6, 1930, WHH/SIA).

When Hrdlička (1942:54) finally broke his silence regarding the Folsom site some 15 years later, he offered the opinion that the Folsom bison may not have been extinct very long—on an *absolute* timescale. He was hardly alone in thinking so (Figgins 1934:4–5); most paleontologists, geologists, and even archaeologists of the time agreed (e.g., Antevs 1935:302-303; Bryan 1941; Colbert 1937; Howard 1936a:401, 1936b:317; Kidder 1936:144; Roberts 1936:345, 1940:104-105; Romer 1933:70). Speaking of the Folsom discovery, Kirk Bryan lamented that "one of the difficulties is that we know so little about the Pleistocene faunas" (Bryan to Wetmore, August 16, 1928, USNM/SIA). Figgins' belief that the age of Folsom could be "determined only through studies of the fossils themselves" was mistaken; he was closer to the truth with his remark that "the whole question rests upon the geological age of the formation in which the artifacts were found" (Figgins to Hay, December 21, 1926, OPH/SIA). That Folsom artifacts were not found in association with the remains of a modern species of bison made it clear nonetheles that the site was old on a *relative* timescale (Figgins 1934:5).

For his part, Barnum Brown was so taken by Folsom he arranged for the American Museum to excavate there in 1928 in a joint project with the Colorado Museum (chapter 4). That July, when points were again found in situ with bison remains, telegrams were once more broadcast to institutions across the country, and again, in response the find was seen by "several of the best men in the country" (Cook to Loomis, November 12, 1928, HJC/AGFO)[13] (fig. 2.13).

Frank Roberts came back in 1928 as well (fig. 2.14), joined by his Smithsonian colleague Neil Judd, and by Kirk Bryan, a USGS/Harvard University geologist whose credentials even Cook admitted were impressive (Cook to Hay, July 18, 1927, OPH/SIA; but see Cook to Hay, August 21, 1928, OPH/SIA). After mapping the geology in the region, Bryan put the "age of the material containing *B. taylori* and the implements [as] late Pleistocene or perhaps early Recent" (Bryan 1929:129; Bryan to Wetmore, August 3, 1928, USNM/SIA). Brown reached a similar opinion based on the bones, as did others (Brown 1928a, 1928b, 1929; Romer 1933; see also Romer to Cook, May 26, 1931, HJC/AGFO).

Cook too thought Folsom was "later Pleistocene" in age, but then he never thought Folsom "had any such antiquity

FIGURE 2.13 Visitors to the Folsom site, September, 1928. Identifiable individuals are Fay Cooper Cole and his wife on the far left. Albrecht Penck is in the foreground, left side, next to Harold Cook and his wife. Carl Schwachheim is immediately behind Cook. Fred Howarth is on the far right, back row. (Photo courtesy of Carl Schwachheim Collection.)

as the Frederick, Oklahoma and Colorado, Texas, evidence" (Cook to Hay, December 23, 1926, January 25, 1928, OPH/SIA; Cook to Ingalls, January 6, 1929, HJC/AHC; Cook to Wissler, March 25, 1929, HJC/AGFO; Cook 1928b:39).

Why Folsom?

Brown and Bryan's results, presented in late 1928, sealed the case for the acceptance of Folsom. But why Folsom, and not Snake Creek, Lone Wolf Creek, Frederick, or some of the other sites purportedly of Pleistocene age? There are several reasons (Meltzer 1991b). For one, Folsom was just what it was advertised to be. It was a late Pleistocene-age site, those were undeniable artifacts, and their context was unimpeachable. Even at a distance, Snake Creek and Frederick were suspect, even to those who had no particular stake in the outcome (e.g., Antevs fieldbook, August 11, 1934, EA/UA; Roberts 1940:58). Ultimately, Figgins himself (1935:4) publicly disavowed Frederick as a hoax.

Equally important, Folsom was a kill site where points were found lodged in the skeletons of extinct bison. Under those circumstances, there was little chance that this could be a fortuitous association, as Holmes himself had said to Figgins that spring of 1927. While at the site, Kidder was reputedly heard to utter, "The ever-present prairie dog didn't have anything to do with that!" (Cook to Matthew, September 9, 1927, HJC/AGFO).

And because this was a kill that involved multiple bison skeletons (chapter 7), the discovery of points in and among the bison bones was repeated in 1928 and witnessed again by a troop of visiting scientific dignitaries. The fact of geological contemporaneity was replicated and thus further reinforced (Meltzer 1991b:33).

But what of Lone Wolf Creek or, for that matter, the sites of 12 Mile Creek, Kansas (Williston 1902), and Meserve, Nebraska (Barbour and Schultz 1932)? All were found prior to 1927, all are today recognized as late Pleistocene to early Holocene in age, and all contain Paleoindian artifacts associated with *Bison antiquus*—the same species found at Folsom. Were legitimate late Pleistocene archaeological sites discovered before 1927 being ignored (e.g., Alsoszatai-Petheo 1986:18–19; Bryan 1986; Rogers and Martin 1984, 1986, 1987)? Much as Cook had, latter-day scholars wonder why these sites were not "decisive . . . in ending the controversy" and find it "amazing . . . what little effect these somewhat repetitious discoveries had on the archaeologists of that time" (Rogers and Martin 1986:44, 1987:82). The historical circumstances, however, are more complicated than this simple rendering suggests.

In 1902 Samuel Williston (1902) reported a projectile point found next to the scapula of a fossil bison at the 12 Mile Creek site in western Kansas (Rogers and Martin 1984). His report was largely ignored, but not because he was some provincial amateur: Williston was a Yale Ph.D. (under O. C. Marsh), who in 1902 joined Chamberlin and Salisbury on the Geology faculty at the University of Chicago. Williston himself never claimed great antiquity for the site, since he was unsure of its age:

FIGURE 2.14 Frank Roberts (left) and Kirk Bryan (right) at Folsom, July 1928. (Photo by N. Judd, courtesy of National Anthropological Archives, Smithsonian Institution.)

The [bison] bones were at first identified by me as pertaining to the extinct species *Bison antiquus* Leidy. . . . Later, however, Mr. Lucas of the National Museum, after an exhaustive study of the known species of fossil bisons [*sic*] of America, recognized the species as new, and gave to it the name *Bison occidentalis*, known otherwise only from a fragmentary skull collected in Alaska. (Williston 1902:313)

The reason that the 12 Mile Creek site never entered the debate is simply that in 1902, bison still lived on the Plains, and there was no evidence then to indicate that these particular bison remains or the associated projectile point were late Pleistocene in age. Williston did observe that "not far distant" from the site in the same "upland marl" were bones of *Elephas primigenius (sic)*, which he noted was an indicator of the *Equus* beds (Pleistocene), but he did not suggest that the bison remains were of comparable antiquity (Williston 1902:315). In 1902 the 12 Mile Creek site was just a bison kill, and the fact that it was a deeply buried or had an associated fluted projectile had no chronological significance.

The Meserve site was discovered in 1923 by local high school students who spotted bison bones exposed some 4 to 5 ft below the surface on the south bank of the Platte River, Nebraska (Meserve and Barbour 1932:240). Returning the next week with F. G. Meserve of Grand Island College, they excavated more bones and found a projectile point beneath one scapula, a type Davis (1953:384) later designated as a Meserve point.

The initial reports on the site, however, were not published until 1932, long after the facts of Folsom were known. Indeed, when Erwin Barbour, paleontologist and Director of the Nebraska State Museum, first contacted Meserve in June 1931, it was for the purpose of obtaining some of the bison bones for display mounts. No mention was made of artifacts (Barbour to Meserve, June 4, 1931, EHB/NSM). Barbour sent C. B. Schultz to reopen the quarry in July 1931, and on that occasion a projectile point was found (Barbour and Schultz 1932). Only afterward did Barbour apparently learn that Meserve had found artifacts as well and write him asking if it was correct that "an arrow point" had been found with the bison, observing that "we are now carefully looking up all facts respecting particularly the Folsom type of arrow point" (Barbour to Meserve, November 12, 1931, EHB/NSM). Obviously, and barring the discovery of previously unseen or unpublished documents, the archaeological and paleontological significance of the Meserve site was only recognized after and because of the Folsom discovery. While Schultz (1932:274) remarked that the Meserve find leads to the conclusion that "man was contemporaneous with animals which are now extinct," that conclusion was not reached before 1927.

Lone Wolf Creek is a slightly different case. Unlike 12 Mile Creek or Meserve, it was championed prior to 1927 as providing evidence of human artifacts in the Pleistocene (Cook 1925:459). And even though Cook's announcement attracted the endorsement of the linguist/ethnographer

Pliny Goddard (Goddard 1926), few archaeologists or pale-ontologists accepted his claim. After all, the excavation had been done by that "ignorant rancher" (Boyes):

> He has had no technical training in palaeontology or geology, nor is he the least informed on archaeological subjects. Fact is, he is a fairly well-to-do rancher whom I have known for 15 years and whose interest in natural history subjects, (merely interest, not trained) has been helpful to the museum in many ways. (Figgins to Hay, December 21, 1926, OPH/SIA)

As already noted, Boyes had not recognized the signifi-cance of the artifacts associated with the bison and did not even report their presence. Cook only visited the excavations after the fact and never saw any of the points in situ, and by then there was "no matrix adhering to the artifacts and no impressions were preserved" (Cook to Hay, December 23, 1926, HAY/SOA; Figgins to Hay, December 21, 1926, OPH/SIA). The unwillingness to accept Lone Wolf Creek was rooted in the legitimate skepticism of a site excavated under circumstances that were recognized even by its promoters as far from ideal. Both Figgins and Cook's defense of Lone Wolf Creek centered on Cook's after-the-fact assessment of the geology and Boyes' honesty and ignorance, arguing that he had neither reason nor knowledge to fake the association of the points and the bison (Cook to Hay, December 23, 1926; Figgins to Hay, December 21, 1926, OPH/SIA). These were the only avenues to approach its legitimacy as an archaeological site, and they were clearly inadequate.

But perhaps the most significant difference between Lone Wolf Creek and Folsom has to do with telegrams. They were sent out from Folsom in 1927 but not from Lone Wolf Creek in 1925.

Some Are More Equal Than Others

That becomes a critical point, for on Hrdlička's advice, the evidence at Folsom—unlike Lone Wolf Creek—was wit-nessed by a group of elite scientists (Meltzer 1991b:34). The history of science shows that in times of scientific contro-versy, resolution is achieved when the elite core of the sci-entific community makes up "its collective mind on the issue" (Oldroyd 1990:345; also Rudwick 1985). That core group regard themselves, and are regarded by others, "as competent arbiters of the most fundamental matters of both theory and method within the sciences" (Rudwick 1985:420). Hrdlička was one; Kidder was another.

We know that the opinions of these individuals mattered, and not just because they themselves thought so. Cook and Figgins thought so too. In every paper they wrote, they wrote for or, rather, against, Hrdlička and Holmes (see espe-cially Figgins 1927a:229). They recognized—however much they disliked the idea, and they disliked it intensely—that Hrdlička, other skeptics, and the anthropological elite gen-erally were the group that had to be convinced of the verac-

ity of their claims for great antiquity in America. Many waited and wondered, as paleontologist Frederic A. Lucas did, whether and when "brother Hrdlička [would] express himself" with regard to Folsom (Lucas to Cook, November 3, 1927, HJC/AGFO).

Cook agreed: It was only "right and proper [that Holmes and Hrdlička] should not take without question such basic evidence as may seem necessary to establish a given fact beyond reasonable question" (Cook to Hay, December 23, 1926, OPH/SIA). On questions regarding artifacts, associa-tion, context, and age, their opinions mattered, and mat-tered very much. Thus, having a young and then relatively unknown Frank Roberts come to Folsom was important—not just because he was a representative of the Smithsonian, but because Hrdlička thought Roberts "very sensible" (Hrdlička to Hewett, December 8, 1934, AH/NAA).

Roberts proved to be just that, because after he saw what he saw, he had the good sense to telegraph Kidder at Pecos and invite him up to see Folsom as well. Kidder was just then reaching the height of his considerable powers, and was widely respected—even revered—within the archaeo-logical community. He had just a few years previously pub-lished his classic *An Introduction to the Study of Southwestern Archaeology* (1924) and was already being touted for election to the National Academy of Sciences (Boas to Fewkes, November 28, 1927, FB/APS; Kidder was elected in 1936). His Carnegie Institution appointment had even prompted an uncharacteristically effusive note from Hrdlička to the Carnegie's President:

> I have just learned, with a deep sense of thankfulness, that you have established a Department of Archaeology, to be headed by Dr. Kidder. This is a substantial step in the right direction. (Hrdlička to Merriam, June 2, 1927, AH/NAA)

That Kidder examined the Folsom site, and then within the month publicly announced that there was now evidence of a Pleistocene human presence in the Americas, carried extraordinary weight within the archaeological community. Cook's editor at *Scientific American*, Albert Ingalls, appreci-ated the significance of Kidder's testimonial and urged Cook, then preparing a paper on Folsom (Cook 1928b), to "be sure to mention Kidder and all the outfit that visited you" (Ingalls to Cook, January 31, 1928, HJC/AHC).

Once Folsom attracted the attention of Brown, Roberts, and Kidder, they in turn brought in others (Meltzer 1991b:33). Brown took the artifacts back to the American Museum to be examined by Nels Nelson, someone to whom even Kidder deferred as being "far and away the person best qualified to consider such material" (Figgins to Kidder, October 17, 1927; Kidder to Figgins, November 19, 1927, JDF/DMNS; Figgins to Nelson, August 24, 1927, DIR/DMNS). The confidence in Nelson was well placed. On his Smithsonian visit in early 1927, Figgins had been annoyed by Neil Judd's insistence that Folsom points could be found in many collections west of the Allegheny Mountains, which Figgins took to mean that Judd ascribed no particular significance or antiquity to the Folsom material (Figgins to Nelson, August 24, 1927, JDF/DMNS; Figgins to Brown, June 8, 1927, June 30, 1927, VP/AMNH; Figgins to Hay, July 1, 1927, October 13, 1927, October 22, 1927, OPH/SIA; Kidder to Figgins, November 19, 1927, JDF/DMNS). But Nelson assured Figgins that Judd, though "a young man of relatively limited experience" was right:

> If he said he could duplicate your chipped points from west of the Alleghenies he didn't go far enough. They occur also in plenty east of the Alleghenies. (Nelson to Figgins, September 13, 1927, DIR/DMNS)

Nelson, of course, knew then what is well known now: Fluted points occur across the continent in abundance (also Brown 1928a:825). He was careful to assure Figgins, however, that this fact "does not interfere with the relative antiquity of your specimens" (Nelson to Figgins, September 13, 1927, JDF/DMNS).

Roberts' enthusiastic reports on the Folsom site prompted Alexander Wetmore, then Assistant Secretary of the Smithsonian Institution, who was himself "very interested in the problem" of human antiquity in America, to throw the weight of the Smithsonian into the investigation of human antiquity in America. This took the immediately tangible form of funding Kirk Bryan's 1928 visit and field-work at Folsom (Roberts to Stirling, July 29, 1928, FHHR/NAA; Wetmore to Bryan, July 30, 1928, August 8, 1928; also Judd to Wetmore, July 30, 1928, USNM/SIA).

Kidder, soon after visiting Folsom, went to the West Coast, where he stopped in at the California Institute of Technology to meet with paleontologist Chester Stock, with whom he "discussed questions of Pleistocene and Recent and mapped out a tentative program for closer cooperation between their work and ours"—"ours," in this case, being the Carnegie Institution of Washington (Kidder to Merriam, October 7, 1927, JCM/CIW). While this had no immediate benefit, it set in motion at the Carnegie a program of interdisciplinary Paleoindian research which bore fruit for E. B. Howard at Clovis six years later, and at many other sites over the next decade.

The combined impact of these testimonials from Kidder on the site context, Nelson on the artifacts, Brown on the bison bones, and Bryan on the geology was overwhelming. The rapid acceptance of the Folsom evidence after September of 1927 attests to that.

But why hadn't the word of Cook and Figgins alone been sufficient? Why weren't they called on to testify on behalf of their sites, or explain the larger implications of what they had found? The answer, again, is a matter of relative scientific status. Just as there are few who are deemed competent to judge great questions of theory and interpretation, there are many more who are not, however much they may claim the right (Rudwick 1985:419–421). Cook and Figgins were not, and knew it. As Cook put it, he was not "on the popu-

lar, ritual-as-accepted-by-almighty-order side of this Pleistocene controversy" (Cook to Hay, October 3, 1928, OPH/SIA). Naturally, he saw it all as merely a popularity issue, not a substantive one because, of course, he believed he was on the correct side. It wasn't quite so simple. W. D. Matthew, who knew Cook best of all, was quite ready to accept Folsom, but not on Cook's word. He wanted "two or three conservative expert opinions on the stratigraphic relations. . . . And I would like to hear Nelson's opinion on the artifacts" (Matthew to Cook, August 11, 1927, HJC/AGFO).

Scientific status is, of course, partly a function of degrees and position. Neither Cook nor Figgins had an advanced degree, but in those days this was not unusual. Besides, Cook's informal education was unrivaled. He was born and raised on one of America's most spectacular fossil sites; the guests in his boyhood home included such scientific luminaries as Erwin Barbour (whose daughter would become Cook's first wife), R. S. Lull (Yale), W. Gregory, W. D. Matthew, and Osborn of the American Museum, and W. J. Sinclair (Princeton) (Cook 1968:200–203). Few of his peers had such experience and training.

Of course, Cook also had his hands full running the family ranch in western Nebraska and a geological consultancy, which consumed much of his time (Cook 1968:201; Dorothy Cook Meade, personal communication, 1993). As a result, he could take on few professional positions or research opportunities, although from 1926 to 1930—the years when he was publishing on human antiquity—he was Curator of Paleontology at the Colorado Museum of Natural History.[14] Indeed, it rankled Cook that Brown, whose position at the American Museum was comparable to Cook's and who was—at least on paper and as Cook saw it—jointly in charge of the 1928 work at Folsom, was completely ignoring Cook in his presentations on the site (Cook to Jenks, February 26, 1929; Cook to Merriam, February 25, 1929, HJC/AGFO).

Figgins, as he himself admitted, knew "nothing whatever, about archaeology and allied subjects, and but a smattering of paleontology and geology; certainly not enough to make my views of value or importance" (Figgins to Hay, December 21, 1926; also Figgins to Hay, February 23, 1927, OPH/SIA). According to Cook, Figgins "only college training" was a very short period in "some small religious college—when they tried to train him for a minister" (Cook to Colbert, June 20, 1944, VP/AMNH). Yet, despite the lack of formal training, he had achieved prominence as Director of the Colorado Museum. As an administrator, he could not devote his complete attention to scientific pursuits, but by itself that did not disqualify him from the discussion. After all, one of the most quietly influential figures in the human antiquity controversy over the first decades of the twentieth century was John Merriam, longtime President of the Carnegie Institution. Osborn, for that matter, was Figgins' administrative counterpart at the American Museum, and he too was highly regarded, despite his *Hesperopithecus* blunder.

Yet, on the face of it, and the archival records bear this out, neither Cook nor Figgins' lack of formal degrees nor their professional positions hindered them. Nor were they deemed less competent because they were not archaeologists or anthropologists: Bryan, Brown, Merriam, and Stock weren't either. Nor was Cook and Figgins' relative scientific status a consequence of *what* they found. Rather, it all came down to *what they said* about what they found.

There had long been good reason to anticipate finding evidence of a late Pleistocene human presence in the New World. Even Holmes (e.g., 1897, 1919) admitted as much. However, there had also been many claims for deep human antiquity that failed, and so there was also good reason to be skeptical. As Nelson patiently explained to Figgins, this was a volatile issue, and discretion was advised: "It is going to take quite a lot of evidence to change the prevailing view regarding the antiquity of man in America" (Nelson to Figgins, August 16, 1927, JDF/DMNS).

Yet, where angels might fear to tread, Cook and Figgins went crashing in. By 1927, they were actively campaigning for four alleged Pleistocene (and older) sites: Snake Creek, Lone Wolf Creek, Frederick, and Folsom. Worse, it was not obvious to them, as it was to almost everyone else, that Folsom was the pick of the litter. In fact, in Cook's judgment "Folsom . . . [is] the weakest and least conclusive evidence on any of our localities" (Cook to Hay, January 25, 1928, OPH/SIA). He thought Frederick provided "the most conclusive evidence of Glacial Age man yet found in America" (Cook 1927b:117; Cook to Ingalls, April 6, 1928, August 9, 1928, HJC/AHC; Cook to Ingalls, January 6, 1929, January 6, 1932; Cook to Loomis, December 30, 1927; Cook to Wissler, March 25, 1929; all HJC/AGFO).[15] Figgins wasn't so enamored of Frederick (Figgins to Cook, October 12, 1927, HJC/AGFO), but he nonetheless considered Folsom "merely confirmatory" (Figgins to Cook, September 25, 1926, HJC/AGFO).

As if to compensate, Cook and Figgins devoted much of their "breakthrough" Folsom papers in *Natural History* to Lone Wolf Creek and Frederick (Cook 1927a; Figgins 1927a). And Cook began a *Scientific American* paper on Folsom with two long paragraphs on the evidence for even deeper human antiquity at Snake Creek, Frederick, and Lone Wolf Creek (Cook 1928b). Not only were Cook and Figgins championing all four sites, but they were making daring (Hrdlička would say reckless) claims about what those sites meant. They thought such sites showed the New World had been occupied for upward of half a million years (Figgins 1928:19; also Cook to Hay, January 29, 1928, OPH/SIA)—as long as, maybe even longer than, the Old World:

> It is now obvious that mankind has lived on this continent for hundreds of thousands of years; and that even in early Pleistocene times had reached a stage of cultural development and civilization that was in some respects, at least, quite as advanced as any existing on earth at that time. (Cook 1928b:38)

Figgins (1928:19) took matters a step further, suggesting "that man had, even then [400,000 years ago], attained high skill in fashioning arrowpoints—[and] that he was a hunter. It proves that America may necessarily have been the cradle of the human race and disproves the theory that man was not retrogressively crude in the matter of culture."

Such brazen claims startled even Cook's otherwise sympathetic editor at *Scientific American*, who admonished Cook that the magazine "cannot risk [such statements] until there is more evidence" (Ingalls to Cook, March 28, 1928; Cook to Ingalls, April 6, 1928, HJC/AHC). Ingalls insisted on adding a disclaimer to Cook's paper:

> In the first two paragraphs of his most interesting article, Mr. Cook, the author, makes claims concerning the proof of the antiquity of man in America—claim which the editor regards as requiring a still larger volume of substantiation than the available evidence affords. With Mr. Cook's friendly concurrence, the present statement, in which the editor disclaims all responsibility for their inclusion, is published. (in Cook 1928b:38)

That was the reaction of a friendly editor. The archaeological community, much more familiar with the global archaeological record, was far less charitable. Nelson lectured Figgins that if everything he and Cook said were true, "We shall have to revise our entire world view regarding the origin, the development, and the spread of human culture" (Nelson to Figgins, August 16, 1927, JDF/DMNS; also Nelson 1928b:822–823). Nelson, who even after Folsom put the arrival of people in the New World at no more than 7,000 to 15,000 years ago, was not ready to do that (Nelson 1928b:823). Few were. Rather, the archaeological community almost immediately and ever after utterly ignored Snake Creek, chose to wait and see about Lone Wolf Creek, and merely dismissed Frederick as a sham (e.g., Spier 1928a, 1928b; Roberts 1940:58).

Folsom, however, was recognized straightaway as being extremely important. As Spier wrote Cook in the midst of their brawl over Frederick:

> I was quite interested in the account of the Folsom finds presented by Barnum Brown and Frank Roberts at the annual meeting of the anthropologists last Christmas. You seem to have a very likely case at Folsom. (Spier to Cook, March 22, 1928, HJC/AGFO)

But, of course, Spier believed the message about Folsom because he heard it from Brown and Roberts.

Thus, instead of convincing the archaeological and scientific community of the veracity of their evidence, Cook and Figgins' claims backfired. The fact that they couldn't critically evaluate their own evidence, had "cried wolf" so often, and considered Folsom—the importance of which was obvious to so many—the least significant of the lot, cast doubt on their competence and discredited their judgment.

The Cook and Figgins 1927 *Natural History* papers serve as a handy benchmark for the resolution of the human antiquity controversy. It is not altogether inappropriate that we cite them so—Folsom did turn out to be a late Pleistocene site. Still, these papers appeared before it was generally agreed that Folsom was, in fact, a late Pleistocene site, and in them they subsumed the discussion of Folsom in more general claims for even deeper human antiquity at Lone Wolf Creek and Frederick. These papers were not triumphant statements of consensus and resolution but, instead, opening salvos in what their authors fully expected to be a protracted dispute.

The debate over Folsom, however, was over before any real controversy got started. Folsom was a legitimate site, and because Figgins—following Hrdlička's advice—made it possible to be witnessed by a group of elite scientists, that fact rapidly became widely known and accepted.

History Repeats Itself

The events at Folsom bear a striking resemblance to those in Europe from 1859 to 1860 surrounding the establishment of human antiquity on that continent. Just as at Brixham Cave (Gruber 1965; Grayson 1983:182), the excavation at Folsom was initially for paleontological purposes. In neither case was it known or anticipated that these critical sites would prove to be archaeological.

In both cases, the sites emerged in a climate of great skepticism on the part of the scientific community toward a Pleistocene human antiquity (Grayson 1983; Meltzer 1983). In America none of the artifacts and human skeletal remains that had been found in strata with Pleistocene fauna or in Pleistocene deposits were on their face obviously or arguably ancient, as Holmes and Hrdlička had so vigorously argued. In mid-nineteenth-century Europe, Paleolithic handaxes were clearly of another age and, thus, presumptively old as Boucher de Perthes (1847, 1857) had claimed. But even though the specimens themselves gave inherent testimony of a premodern age, the road to demonstrating a Pleistocene human presence was no easier there than in America. A site still had to pass stratigraphic muster.

Brixham Cave produced Paleolithic artifacts sealed beneath a thick calcareous layer alongside the bones of extinct Pleistocene mammoth, cave bear, and cave hyena, an event that prompted a closer look at Boucher de Perthes' artifacts with extinct mammals in deep alluvial deposits (Grayson 1983:182–183). Folsom yielded artifacts between the ribs of an extinct Pleistocene bison, which could only have become embedded when that animal was alive. In both cases that evidence was ultimately judged to be sound by elite members of the scientific community. In neither case was it possible, as it had been so many times before, to argue the association of artifacts with extinct fauna was merely fortuitous.

There are further parallels. In both America and Europe there was a sharp division of scientific status between those who made the discoveries and those who were called on to judge the significance of the discovery. It's the division between Cook, Figgins, and Boucher de Perthes, on the one hand, and Brown, Kidder, Nelson, John Evans (1860), Joseph Prestwich (1860), and Charles Lyell (1860), on the other hand (Grayson 1983). In both Europe and America the discoverers were relegated to second-class status because of their propensity to make absurd arguments about their findings or cloak them in arcane theoretical contexts: Boucher de Perthes was a Noachian flood monger long after it was fashionable (Grayson 1983:128), and Charles Darwin, among others, had dismissed his work as "rubbish" (Darwin to Lyell, March 17, 1863, in F. Darwin 1898:200). Nels Nelson had to remind Figgins that everything he and Cook said about human antiquity could not possibly be true.

There are also a few significant differences between the historical cases, the most important of which was that the Folsom finds did not result in an antiquity comparable to the deep past of Europe. But by 1927 it was obvious to most archaeologists that America didn't have that kind of prehistoric time depth (Nelson 1928). Still, Folsom showed that American prehistory began at least in the latest Pleistocene or early Recent, perhaps over 10,000 years ago, which compared to what was then known of American prehistory, provided America with its own *relatively* deep past.

What Folsom Wrought

But that was enough time depth for Folsom to have far-reaching impact in American archaeology, much like the impact that followed the establishment of the Paleolithic in Europe in 1860. Having the bottom fall out of American prehistory fundamentally and in several ways changed the face of American archaeology (Kidder 1936; Kroeber 1940:474).

For one, in expanding the prehistoric timescale, Folsom raised knotty questions about the historical relationship between this ancient archaeological culture and the historically known Native American peoples. No longer would it be possible to transparently apply ethnographic data and units to the archaeological record, except perhaps in the case of the latest portion of the prehistoric record, and maybe not even there (Kroeber 1940). The common-sense use of the ethnographic and ethnohistoric record and the Direct Historical Approach, the dominant methodological tools of late nineteenth- and early twentieth-century archaeology (e.g., Thomas 1898), was never quite the same afterward.

Folsom also provided ammunition for those like Kidder who were "100% American" on the question of the origins of New World agriculture and civilization (Kidder 1936: 144). Hyperdiffusionists had claimed—much to Kidder's annoyance—that there simply wasn't enough time depth to American prehistory for native peoples to have independently developed those complex trappings, and hence they had to have had outside help from one of the Old World civilizations. Folsom soundly trumped that argument, to Kidder's "great relief," by providing a long chronological on-ramp during which New World cultural complexity could evolve and arise independently (Kidder 1936:145).

By anchoring the American prehistoric sequence firmly in the Late Pleistocene, Folsom created the daunting task of filling in that "decisive and perhaps long gap" that suddenly opened between the Late Pleistocene and the Late Prehistoric (Kroeber 1940:474). That gap left American archaeologists without a cultural historical sequence; it is not fortuitous that the first large-scale efforts at putting together time-space frameworks appear after Folsom (e.g., Ford and Willey 1941; Griffin 1946). Culture history, which dominated American archaeology during the decades from the late 1920s to the 1960s, was arguably triggered by the Folsom finds (Meltzer 1983). Of course, Folsom takes no credit for the particulars of the cultural historical approach— only for revealing to archaeologists of the 1930s just how little was known of vast spans of American prehistory. To a generation of archaeologists for whom recognized ignorance "consisted largely of the sites . . . not dug or the places and time periods . . . not investigated" (Binford 1986:459), that was revelation enough.

Finally, Folsom taught archaeologists how to find more sites like it: Look for bones of extinct animals, bison or mammoth, and then examine the spot for associated stone artifacts (Meltzer 1989a). Over the next two decades across the American West, especially as Dust Bowl drought and erosion exposed Late Glacial deposits, nearly three dozen Paleoindian sites were found in that manner—bones first (Meltzer 1989a). Most of these, of course, were kill sites, in which fluted and unfluted large lanceolate points were associated with the remains of extinct animals. As a consequence, by the late 1930s it seemed to Roberts (1939:541), among others, that Paleoindian groups were hunters "who depended entirely upon game . . . for [their] maintenance and sustenance." Where fluted points occurred it was assumed that they represented the killing of Pleistocene fauna. Of course, however true this may have been at Folsom, it was not true over vast areas of North America where there are abundant fluted points but no megafauna, nor of all Folsom sites. But believing this to be so has had a profound impact on the interpretation of Paleoindian archaeology and adaptations (Meltzer 1993).

Conclusions

As Rudwick (1985) and others (e.g., Hull 1988; Oldroyd 1990; Secord 1986) have argued, controversy in science—or at least a nontrivial one like the long-running debate over human antiquity in the Americas—is not resolved by gaining consensus across the larger scientific community. Instead, resolution is brought about by a core set within the community that occupies a particular space at the intersection of two cross-cutting dimensions: perceived competence and degree of involvement.

There is an unspoken but unmistakable hierarchy in any scientific community, based on an individual's proven ability to deliver reliable information and ideas (Rudwick 1985:419; Oldroyd 1990:345). Where an individual in American archaeology in the 1920s was situated along that "gradient of attributed competence" can be teased from published works. But a more reliable gauge is in the informal and unfiltered correspondence between members of a community (Rudwick 1985:419–421), as the brutally candid back-channel assessments of Cook and Figgins' work by Hrdlička, Matthew, and others well attest. These can provide real-time tracking of an individual's scientific "stock" as it rose or fell.

But rather than detail the relative status of each individual participant in this dispute, let me instead follow the analytical lead of Rudwick (1985) and simply place the participants in one of four broad groups along that gradient of attributed competence. Doing so runs the risk of caricaturing the various participants and their roles, and one can always quibble about the placement of individuals, but the effort is defensible on its historical merits since the broader differences in status are easily spotted, and it will serve the larger purpose of illustrating where within a scientific community controversy is resolved (fig. 2.15).

Atop the gradient are the *elite* members of the scientific community. At this moment in American archaeology the group included Hrdlička, Kidder, Nelson—and even the aged William Henry Holmes. For despite the fact that he was by then formally out of archaeology, he was still active behind the scenes—recall his actions with regard to the Lone Wolf Creek claims—and, more important, his opinion about the Folsom finds was being solicited long after it ought to have mattered (e.g., Sellards to Holmes, February 24, 1930, WHH/SIA). Not only does this testify to the inertia of a scientific reputation (good or bad), but it highlights the essential criterion for identifying the elite members of the community: Their opinions and judgments on theoretical and methodological and substantive issues matter, and matter very much.

The next group downslope—and it is important to stress that the boundaries separating one group from another are more gradual than sharp—is the *accomplished* individuals (fig. 2.15). In this case, it includes nonarchaeologists who could nonetheless contribute expert insight into chronology or vertebrate history (Brown, Bryan, Matthew) and those who were emerging as important new voices within archaeology (Roberts, Spier). Although the expertise of this group was recognized and valued, their opinions on the most fundamental matters of archaeological interpretation and theory were not binding. Roberts, the brand-new Ph.D., must have understood this at some level, for he personally saw to it that A. V. Kidder— and none of the other attendees at that first Pecos Conference—came to Folsom to examine the first point in situ that September of 1927.

Farther down still are the *practitioners*, those who have recognized but narrower competence, such as knowledge of the archaeology of a local area, abilities in fieldwork, or competence in a related subject. This group includes those whose opinions have proven to be unreliable (Cook,

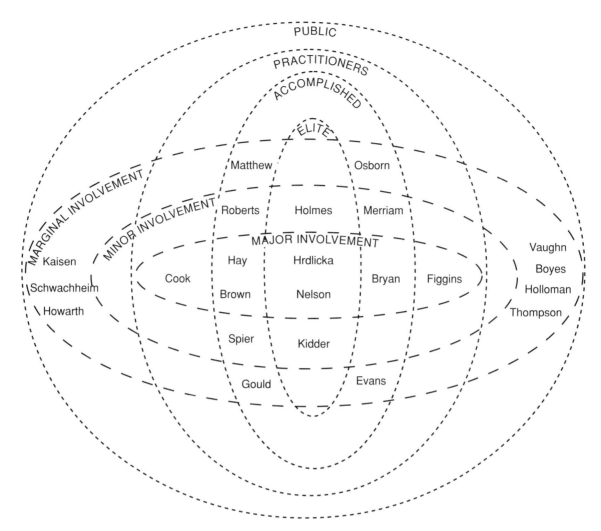

FIGURE 2.15 The intersecting orbits of the gradient of attributed competence and the degree of involvement in the human antiquity controversy in America in the 1920s. (After Rudwick 1985.)

Figins) or who simply have few opinions and implicitly understand that it is not their role to provide them (Kaisen, Thompson). And, finally, at the base of the gradient of attributed competence are the members of the nonscientific *public*, individuals who may make discoveries without fully recognizing or understanding their significance (McJunkin; Schwachheim) or whose claims about the significance of their discoveries will be received with great skepticism and will necessarily have to be corroborated (Boyes, Holloman, Vaughn) (Rudwick 1985:424).

Cross-cutting the dimension of competence is the degree of involvement. Each of those individuals participated in research on human antiquity in the Americas, but not to the same degree (fig. 2.15). There were those, like Cook, Figgins and Hay, or Hrdlička and Nelson, for whom this was the major theme of much of their research and writing during this decade. Then there were those whose involvement in the controversy was relatively minor: Kidder, Holmes, Merriam, Roberts, and Spier. Each participated at different moments; and for Kidder and Spier, at least, their testimony

at Folsom and Frederick, respectively, would essentially end their involvement in the discussion. In contrast, Holmes had long been, and Roberts soon would be, much more involved in this issue. On the outside orbit were individuals like Kaisen and Schwachheim who, though centrally involved in the fieldwork and discovery, played only a marginal part in the discussion of human antiquity.

As should now be clear, the resolution of the human antiquity controversy was the purview of those individuals who were among the elite of American archaeology in the 1920s and who had played at least a minor role in the discussion: Holmes, Hrdlička, Kidder, and Nelson. But clearly scientific status and distinction trump degree of involvement in importance: Kidder at Folsom, and, for that matter, Lyell at Brixham Cave, had influence that went well beyond the influence of those who were more actively and intimately involved in those sites and the respective disputes.

Not all scientists are created equal; some are more equal than others. It was not true, as Lucas suggested, that Hrdlička would not be satisfied with Folsom unless he "fired the arrow himself." But Hrdlička did want to know who was there when the arrow was unearthed (Lucas to Figgins, November 18, 1927, DIR/DMNS). Cook and Figgins' role was to make the finds, then get out of the way.

The inequality of science is perhaps most visible during episodes of scientific controversy, when the stakes are highest (Oldroyd 1990:345). The archaeological data provided by Cook and Figgins were extremely important, but with so much riding on the outcome of what it meant, their interpretations of these data were not. The swift dismissal of their claims, and their virtually complete exclusion from the later discussions, is evidence of that. As mere practitioners they were deemed unqualified to pronounce on the wider, theoretical implications of their evidence for the human antiquity debate. At some level, both of them understood this. Figgins admitted:

I feel this museum should avoid all expression of opinion concerning its finds and that silence is golden. The evidence must go before the final jury for ultimate opinion and for that reason our opinions are valueless. (Figgins to Gregory, December 30, 1927, DIR/AMNH)

He and Cook had an understandably difficult time following their own good advice, but then as Hull observes, "Neither humility nor egalitarianism has ever characterized scientists, and no one has ever given any good reasons why they should" (Hull 1988:31).

Ultimately, then, interpretation of the meaning of what they had found was in the hands of Brown, Bryan, Kidder, Nelson, and Roberts, whose influence over the resolution was far out of proportion to the degree of their participation in the work. Nelson hadn't even visited Folsom. But it was he and these other individuals who would go on to author the triumphant reviews of human antiquity and North American Paleoindians (Bryan 1941; Nelson 1933; Roberts 1940).

So it was that Cook and Figgins were ultimately victims of what Merton has called the *Matthew effect*: "For whosoever hath, to him shall be given . . . but whosoever hath not, from him shall be taken away even that he hath" (Merton 1968; also Oldroyd 1990:355). Hrdlička's reputation may have been roughed up by Folsom, but only slightly, and he continued to be a sought-after authority regarding human antiquity in the Americas (Hrdlička 1928, 1937, 1942). In the end, and despite their crucial role in the discovery at Folsom, neither Cook nor Figgins was invited to offer his views at the many symposia that followed or given the opportunity to stand at center stage at any of them and perhaps gain public acclaim for his contributions toward building that stage. Unfair, perhaps—but at least there were these kind, albeit private words to Figgins from the ever-gracious Alfred Kidder:

Anthropology owes you a very great deal for having handled this material so carefully and so intelligently, and I think the researches of yourself and Dr. Cook will go far towards opening a new era in the study of the question of Pleistocene man in the New World. (Kidder to Figgins, October 13, 1927, DIR/DMNS)

Epilogue: The Elephant in the Room

In 1927 research into human antiquity in Americas pivoted sharply on the Folsom discovery. As most readers of this book are likely aware, 70 years later history apparently repeated itself with the publication by Tom Dillehay in 1997 of his second and final volume on the pre-Clovis-age site of Monte Verde, the site visit that occurred almost simultaneously, and the resulting sea change in our view of human antiquity in the Americas (Adovasio and Pedlar 1997; Dillehay 1989a, 1997; Meltzer et al. 1997). I am reluctant to bring up Monte Verde here for several reasons, not least because this book is not about the peopling of the Americas. But *not* raising the issue of Monte Verde would be worse than ignoring it, precisely because it will be on reader's minds—it's that lurking elephant—and would inevitably leave the door open for invidious (or fawning) comparisons between participants then and now: Just who are today's Cooks and Kidders? Worse, it could appear blatantly self-serving, since I was a participant in the Monte Verde site visit (Meltzer et al. 1997).

Moreover, as I have argued elsewhere (e.g., Meltzer 2006), although much has been made of the Monte Verde site visit and of the parallel between it and what happened at Folsom in 1927, the cases and circumstances are not identical. The importance of site visits, the nature of the scientific community, and the role of elite scientists like Kidder have changed dramatically since the 1920s. Back then, archaeological methods and techniques were far more variable, professional training was spottier, more amateurs were in the mix working at critical sites, virtually all evidence had to be evaluated in the field, and the criteria for evaluating that evidence were less explicit or at least less universally agreed to. The testimony by a figure of towering reputation like Kidder, who had seen and evaluated the material firsthand, was therefore necessary and, understandably, influential.

Matters are very different today. Site visits are something of an anachronism, since much of the evidence emerges in post-excavation analysis of radiocarbon samples, sediment chemistry, artifact sources and residues, the isotopic composition of organic remains, and so on. Analyses in each of these areas follows well-defined and well-known protocols that can be widely evaluated across the archaeological community—we need not wait for judgment from "on high." The Monte Verde site visit did provide us the opportunity to examine the site's setting and surroundings, sediments and stratigraphy, the potential for sample contamination or mixing, and the like. However, as was emphasized in the report of that visit (Meltzer et al. 1997), our role as individuals knowledgeable about the archaeology of this time period was to recount our observations and let readers judge for themselves based on their own evaluation of our comments and, especially, of the evidence painstakingly compiled in Dillehay's two volumes on the site. Many did just that, and some reached very different conclusions than we did—clear evidence that our say-so did not carry the weight that Kidder's had 70 years earlier.

What may in the end prove to be the most historically interesting comparison between 1927 and 1997 will be what happens over the longer post–Monte Verde term, and whether in retrospect it too will have triggered an explosion of discoveries of sites of similar age and a new understanding of Pleistocene peoples in the Americas—as Folsom did.

Notes

1. Hrdlička registered his opinion across the front of his reprint of Cook's (1926) article (DOA/SI).

2. Figgins too wrote a rejoinder to what he considered Spier's "assinine [sic] comments," but his was not accepted for publication (Figgins to Hay, November 8, 1928, December 17, 1928, OPH/SIA; Figgins to Gould, October 26, 1929, JDF/DMNS).

3. Coincidentally, just a few days after writing Chamberlin, Hrdlička learned that Franz Boas had made a dramatic flanking maneuver aimed at blocking Hrdlička's election to the National Academy of Sciences. Within a month Boas published his "Scientists as Spies" letter to *The Nation*, bringing down on him the wrath and censure of the Washington anthropologists (Stocking 1968). These two episodes, although not responsible for killing the idea of a blue-ribbon panel, almost certainly distracted Hrdlička.

4. In defense of the irregularities in Cook's (1947) version, it should be kept in mind that he wrote his account 20 years after the episode and in response to a story on the Folsom discovery that appeared in the *Rocky Mountain News* (Little 1947). That story, which made no mention of Cook and gave full credit to Figgins, understandably rankled Cook. Perhaps not surprisingly, his response went too far in the other direction, downplaying the role of Figgins and McJunkin (see also Cook to/from Bailey, February 5, 12, and 21, 1947, April 11 and 20, 1947; Cook to/from Wormington, February 27, 1947, March 4 and 7, 1947, DIR/DMNS). Of course, Cook's version is no more slanted than versions sycophants would tell of Barnum Brown's role at Folsom (e.g., Barton 1941:311; for Cook's reaction to Barton's story, see Cook to Mrs. Figgins, November 30, 1944, HJC/AHC) or Figgins' accounts of the events at Folsom, later versions of which often simply omitted Cook's name altogether (e.g., Figgins 1934, 1935).

5. Barnum Brown seemed to think artifacts might be found and urged Figgins to tell his crew that if any points were found associated with the bones, they should "remain in the matrix without the relationship being disturbed" (Brown to Figgins, June 2, 1926, VP/AMNH). And he repeated that suggestion in advance of the following season, since he well appreciated that in situ evidence "no one can controvert" (Brown to Figgins, February 17, 1927, VP/AMNH).

6. As there have been incorrect assertions published about the ownership of the site (Dixon and Marlar 1997:372), it is perhaps useful here to set the record straight. In the spring of 1926, in advance of excavations, Fred Howarth (acting as Figgins' agent) drew up an agreement with J. L. Shoemaker of the nearby Crowfoot Ranch for the "sole possession of and the exclusive right and privilege of excavating and removing from his property, fossil skeletons of prehistoric animals" (agreement between J. L. Shoemaker and Colorado Museum of Natural History, May 1, 1926, DIR/DMNS). However, Howarth well knew that Shoemarker was merely the leasee. The land was owned by the State of New Mexico (Howarth to Figgins, May 3, 1926; Figgins to Howarth, May 10, 1926; DIR/DMNS). Therefore, Howarth applied to the State Land Office for permission to excavate, which was granted by the Commissioner of Public Lands (Swope to Howarth, May 5, 1926, DIR/DMNS).

7. Harold Cook would later claim that it was his carrying a message from Howarth to Jesse Figgins of Frank Figgins' misuse of Museum funds that began the deterioration of his relationship with the elder Figgins and the Museum (Cook to Jenks, January 26, 1931, HJC/AGFO; Cook to Hay, March 12, 1929, OPH/SIA). In point of fact, Howarth had already apprised Jesse Figgins that his son was spending money without reporting its purpose. Jesse Figgins promptly instructed Howarth not to let Frank "draw so freely" on the account, as he was already well supplied with food and equipment (Howarth to Figgins, July 22, 1926; Figgins to Howarth, July 24, 1926, DIR/DMNS). By Howarth's records, disbursements to Frank Figgins through mid-October amounted to $272, less than the $400 Schwachheim had by then received, but then Schwachheim had been in the field longer.

8. Cook's article needed to be revised not because it was personal or controversial, which would certainly have been in keeping with his character, but because it was "too heavy" and was not lively enough for *Natural History* (Brown to Figgins, April 4, 1927, VP/AMNH).

9. Cook was warned that Figgins was "presuming [he would] go to Washington if it is so arranged—willingly or otherwise" (Figgins to Cook, November 23, 1926, HJC/AGFO). Hay had also alerted the Chief of the Bureau of American Ethnology of the Folsom work, which prompted an exchange of letters between J. Fewkes and Figgins (Fewkes to Figgins, November 27, 1926; Figgins to Fewkes, December 1, 1926, DIR/DMNS).

10. Oliver Hay later wrote, tongue fully in cheek, that he too was worried: "The last that I saw of you, you were entering the elevator here with Dr. Hrdlička. As the minutes passed on the hours I became anxious about you and was thinking of organizing a search and relief party, but on inquiry I found that you had made a safe escape from the building" (Hay to Figgins, June 25, 1927, DIR/CMNH).

11. Figgins met both Holmes and Neil Judd and considered them "jokes, pure and simple" (Figgins to Cook, July 22, 1927, HJC/AGFO; also Figgins to Brown, June 8, 1927, VP/AMNH). Figgins gives no indication of having met Frank Roberts on this occasion, though Roberts was by then in the employ of the BAE (Judd 1967).

12. Lynch (1991:352) mistakenly implies that Brown, Kidder, and Roberts traveled great distances to examine the Folsom site, assuming, as he does, that they came from their

East Coast offices. As is clear, all of them were within a day or two of travel to the site.

13. In their exchange earlier that spring, Hrdlička had recommended to Brown, as he had to Figgins, that telegrams be sent out if anything of special interest appeared (Hrdlička to Brown, March 19, 1928). In the event, the telegrams were sent by Brown on July 23, 1928—copies of the replies are in VP/AMNH. Among the respondents were Fay Cooper Cole (University of Chicago), A. Hrdlička (USNM, Smithsonian), W. K. Moorehead (Peabody Foundation), Earl Morris (National Park Service), Frank Roberts (BAE, Smithsonian), E. H. Sellards (Bureau of Economic Geology, Texas), Alexander Wetmore (Smithsonian), and Clark Wissler (AMNH) (Schwachheim to Cook, August 10, 1928, HJC/AGFO).

14. From 1925 to 1928 Cook was listed as "Honorary Curator," but on December 4, 1928, he was promoted to "Curator," a title that he retained until he left the Museum in 1930. Biographical details can be found in various *Annual Reports* of the Colorado Museum for the years 1925 to 1930;

for Cook's version of his hiring and the circumstances of his forced resignation, see Cook to Gidley, November 26, 1930, Cook to Gould, March 5, 1929, and Cook to Jenks, February 26, 1929, and January 31, 1931, all in HJC/AGFO, and Cook to Hay, March 12, 1929, in OPH/SIA. As noted above, Cook's revelation of Frank Figgins' apparent misuse of Museum funds was said to be the root cause, but there does not seem to be evidence for this claim.

15. Cook's daughter offered an additional insight: "A further complication at the time, not then obvious to many but well-known to me, was that between 1922 and 1928, Harold Cook was divesting himself of his first wife in order to marry his second. These were the very years of the Folsom discovery, and of Snake Creek, Lone Wolf Creek and Frederick. To me, it is not unreasonable to suppose that, had his personal life been more serene at that time, his outlook on some of these matters might have been more balanced" (Dorothy Cook Meade to Meltzer, July 9, 1993).

THREE

Situating the Site and Setting
the Ecological Stage

DAVID J. MELTZER

The Folsom site is situated in northeastern New Mexico, in the Great Plains physiographic province. It is often assumed that the site is set on the High Plains portion of that province, amid the vast, flat, almost-featureless, dry, windswept grassland that reaches from Texas into Canada. Yet, although many important Folsom-age sites are in just that setting, the type site is not. Instead, and to the surprise of many first-time visitors, it is situated in the midst of an open woodland/parkland of high mesas, volcanic cones, and upland pediment surfaces, in a region dissected by small, relatively narrow streams and river valleys (fig. 3.1).

The site straddles one of those upland streams—Wild Horse Arroyo, a minor and ephemeral northwest-to-southeast trending tributary of the Dry Cimarron River. Both Wild Horse Arroyo and the Dry Cimarron River have their headwaters on nearby Johnson Mesa, one of the largest and most prominent of the regional landforms, which looms just 1600 m west of and 228 m above the Folsom site. The site's primary datum, which sits a few meters above the bonebed and half a dozen meters above the thalweg of Wild Horse Arroyo, is at an absolute elevation of ~2,109 m (~6,919 ft) (elevation data are derived from differentially corrected GPS readings).

The site itself is roughly centered within a 10-acre parcel of New Mexico State Trust Land, owned by the State of New Mexico and under the control of the Commissioner of Public Lands (New Mexico State Senate Bill 159, approved March 12, 1951, chap. 87, Laws of New Mexico, 1951; also Howarth to Figgins, May 3, 1926, DIR/DMNS; D. Reily to D. Meltzer, November 13, 1996; N. Nelson to D. Meltzer, November 15, 1996). A privately owned ranch surrounds the 10-acre state parcel.

In this chapter, I summarize the Folsom region's geological history, stone sources, soils, hydrology, and topography and its present climate, flora, fauna, and land use patterns.

While such is an almost-obligatory part of any monograph, too often the exercise is only aimed at providing a sense of place. That is important, to be sure. But I also attempt at the end of the chapter to use this information to help identify aspects of the climate, environment, and geology of potential relevance to human foragers in the area, as a foundation for identifying, and in later chapters testing, hypotheses about Late Glacial environments and adaptations.

Regional Geology and Geological History

Folsom falls within the Raton section of the Great Plains physiographic province, which forms the southern and western margin of the Great Plains. The most distinctive features of the terrain are volcanic peaks and cones, some over 2,400 m, and high basalt mesas, ranging in elevation from 1,500 to 2,400 m (Hunt 1967:220–221; Trimble 1990:6–7). The Sangre de Christo Mountains (the southern Rocky Mountains) form the western edge of the Raton Section and at their closest point are ~100 km west of the site; the High Plains form the eastern margin, ~75 km distant (Trimble 1990:23–25).

The bedrock geology and stratigraphy of the Raton section are complicated, but need only be sketched in broad outline here, save for those aspects that are directly relevant to understanding the geology of the Folsom site—notably, where potential lithic sources are located, and the local geomorphic controls. The Pleistocene and Holocene stratigraphy of the region and the site is discussed in chapter 5.

The oldest exposures in the Raton section are Triassic in age and include the Baldy Hill, Travesser, and related formations, which are roughly equivalent to what was previously designated the Dockum Group or "red beds" (Baldwin and Muehlberger 1959). The Triassic exposures nearest to the site are well east and downriver of Folsom but are of some interest to the archaeology of the site, as the lowest of

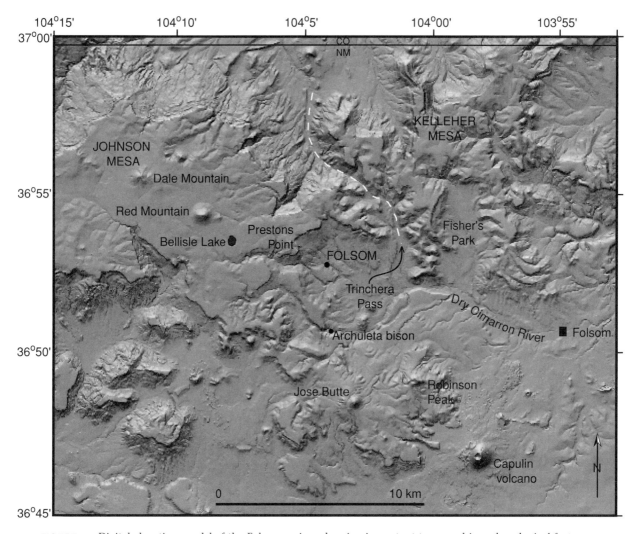

FIGURE 3.1 Digital elevation model of the Folsom region, showing important topographic and geological features.

them, the Baldy Hill Formation (Lucas, Hunt, and Hayden 1987:100), contains knappable stone. The prominent and extensive exposure occurs at the type section, a hill of the same name located ~55 km from Folsom. This formation is the lateral equivalent of the Tecovas in the Texas panhandle (Lucas, Hunt, and Hayden 1987:fig. 13) and, like the Tecovas, contains nodules of red and yellow jasper/chalcedony and some silicified wood. Despite its resemblance to both Alibates agatized dolomite and Tecovas jasper, Baldy Hill jasper can be distinguished from those sources (Banks 1990:89). At the type section, the knappable stone occurs in a <1.5-m-thick lens that forms a prominent bench near the base of Baldy Hill (Baldwin and Muehlberger 1959:37; Lucas at al. 1987:100, 115). It has been suggested that some of the artifacts at the Folsom site were made of stone procured from Baldy Hill, which, if so, would have implications for inferences about the mobility of this group (chapter 8).

Jurassic-age rocks are also relatively rare in the study area, becoming more common east of Folsom. The Jurassic

in this region includes the Morrison Formation, a massive, predominantly sandstone unit along the Dry Cimarron, with a thickness ranging from 60 to 100 m. Although no outcrops occur in the site area, the Morrison contains a distinctive and well-marked "agate bed," which occurs near the base of the formation (Baldwin and Muelberger 1959:49–50). Exposures of this unit can be found within 60 km of the Folsom site, where it occurs as a thin (<1m) stratum of intermittent chalcedony and chert nodules in silty mudstone (Neuhauser et al. 1987:158). Farther east, in the Kenton, Oklahoma, area, this unit is slightly thicker and the exposures more extensive. However, although useful as a geological marker bed, the "agate" occurs in small, relatively friable pieces, which are largely unusable for stone tool production (Banks 1990:89; also Baldwin and Muehlberger 1959:90; Meltzer, unpublished field observations).

The more extensive surface exposures of the region are Cretaceous in age, including those of the Dakota group and the Niobrara, Pierre Shale, and Raton formations (Hunt,

Lucas, and Kues 1987:fig. 2; Kues and Lucas 1987:167; Scott and Pillmore 1993; Trimble 1990). The rocks of the broadly defined Dakota Formation are primarily sandstone and shaly sandstones (Baldwin and Muehlberger 1959:56). Indurated or silica cemented quartzite varieties of this stone can be of very high quality and are commonly found in archaeological sites on the Plains (Banks 1990:89–90). A few high-quality quartzite pieces occur in the Folsom assemblage. There are extensive outcrops of the Dakota group to the east and southeast, from as close as 10 km and extending ~100 km into the Oklahoma panhandle (Barnes 1984; Scott and Pillmore 1993). However, as Banks (1990:89) notes, there are "comparatively few localities within the total extent of the outcrops [that] provide quartzite . . . of knappable quality."

The Sandy Unit of the Smoky Hill Shale Member of the Niobrara Formation (Upper Cretaceous) is the local bedrock at Folsom and the unit unconformably underlying the Pleistocene and Holocene sediments of the site. It occurs here as a gray to yellowish-gray, fine-grained, platey, well-layered calcareous sandstone, shaly sandstone, and sandy shale, which weathers to a yellowish brown to a pale yellowish brown (Scott and Pillmore 1993; also Baldwin and Muehlberger 1959:63). The shale is interspersed with occasional limestone concretions and very small pieces of carbonaceous matter, which appeared in the pollen analysis (chapter 6) and likely help explain an anomalously old radiocarbon date obtained from the Folsom bonebed (chapter 5).

It was recognized early on that the bedrock was Cretaceous in age (e.g., Cook 1927a), and in 1927 Barnum Brown initially and correctly identified it as the Niobrara Formation (Brown 1928a). However, for reasons unclear, after the 1928 season on the site he subsequently revised his opinion and referred the unit to Pierre Shale (Brown 1928b). He was not the last to do so (see also Anderson 1975:33–34; Anderson and Haynes 1979:895; Haynes, Anderson, and Frazier n.d.; Meltzer 2000). However, detailed geologic maps show that the closest Pierre Shale occurs on the west side of Johnson Mesa, some 50 km distant (Scott and Pillmore 1993). The Cretaceous bedrock in the area of the site is now mapped as the Sandy Unit of Smoky Hill Shale Member of the Niobrara Formation.

The Smoky Hill shale produces no useable stone, nor was it the surface on which the Folsom Paleoindian occupation took place, but it is relevant to the Folsom Paleoindian occupation and the archaeology of the site in several respects. First, the contours of the bedrock surface formed the landscape on which the bison were maneuvered and killed. Second, a slope wash of Smoky Hill shingle shale helped armor the top of the bonebed from weathering and erosion (chapters 5 and 7). Finally, a bedrock wall of Smoky Hill shale just northwest of the paleotributary may have helped deflect arroyo cutting and fluvial erosion, protecting and preserving the site, at least until the 1908 flood, as noted below.

The overlying Late Cretaceous and early Tertiary Raton Formation contains coal-bearing units that are 3.5 m thick and occur discontinuously from Walsenburg, Colorado to Raton, New Mexico (Flores and Cross 1991:fig. 1; McCabe 1991:475). The nearest units of the Raton formation to the Folsom site, which are not necessarily the nearest coal-bearing units, outcrop on the western escarpment of Johnson Mesa, at least 25 km west of the site (Scott and Pillmore 1993). There are no coal-bearing strata at or near the site.

Early Tertiary-age deposits in the region, primarily Paleocene and late Oligocene, include the Raton and Poison Canyon formations, the deposition of the former continuing across the Cretaceous/Tertiary boundary. These mark sediments shed from the Rocky Mountains and transported eastward, primarily by fluvial processes, during and after Late Cretaceous and early Tertiary tectonic uplift (the Laramide orogeny). In the study area the Raton and Poison Canyon occur in a broad, fan-shaped area extending west of present-day Raton, New Mexico (Scott and Pillmore 1993) (fig. 3.2).

In the later Tertiary, particularly the late Oligocene (~26 million to 22 million years ago [mya]), igneous intrusions formed the Spanish Peaks, southwest of Walsenburg, Colorado, along with an accompanying series of igneous dikes that radiate eastward almost to the Purgatoire River north of Trinidad, Colorado. These dikes help mark the northern boundary of the Raton section. The Spanish Peaks were important landmarks for Native Americans, who used them to delineate hunting areas (Jodry 1999a:139), and for early Euroamericans, serving as a navigational waypoint on the Santa Fe Trail. They appear on the earliest maps of the trail in 1834 and in subsequent guides to this and other trails (e.g., Gregg 1844; Marcy 1859:296, 299; see Hill 1992:50).

In still later Tertiary times, middle to late Miocene epeirogenic uplift (general vertical uplift not associated with mountain building) triggered a new wave of erosion along the Rocky Mountains, and a large volume of sediment was transported eastward and down slope via fluvial and aeolian processes (Gustavson and Finley 1985; Gustavson et al. 1991; Osterkamp et al. 1987; Trimble 1990). These deposits comprise the Ogallala Formation, which forms much of the High Plains surface to the east and which, at one time, likely formed a nearly continuous aggrading surface from the Rocky Mountain front east to what is now the eastern edge of the Texas Panhandle.

The Ogallala Formation has a spotty distribution in the Raton section, having been subsequently eroded or otherwise interfingered and/or capped with volcanic rocks, mostly basalts (Aubele and Crumpler 2001:72; Gustavson et al. 1991:490; Muehlberger, Baldwin, and Foster 1961:17; Scott and Pillmore 1993; Trimble 1990:24). However, a small pocket of the Ogallala Formation outcrops on the surface in the Trinchera Pass area, while another occurs just below the eastern edge of Johnson Mesa northwest of the site, both within 6 km of the site (also Anderson 1975:47). In places

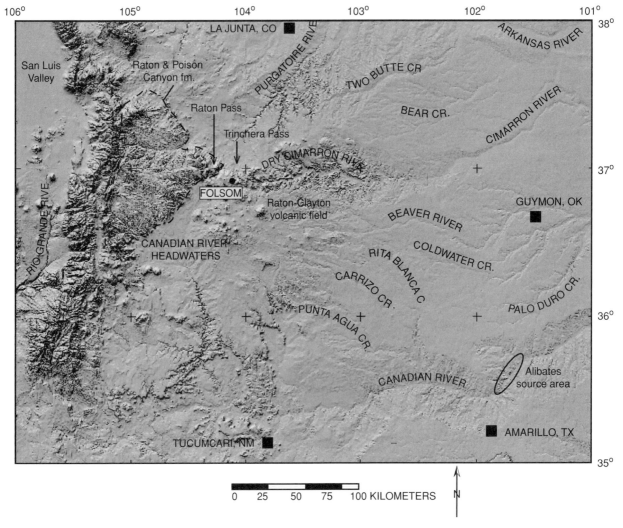

FIGURE 3.2 Digital elevation model of the southern Great Plains west to the southern Rocky Mountains, showing important topographic and geological features.

on the High Plains where this formation is more extensive, the Ogallala contains small nodules of knappable chert and quartzite, the latter usually reworked Dakota quartzite (Banks 1990:90), but the sole outcrop in the site area, currently a sand and gravel pit, showed no knappable stone (Meltzer, unpublished field observations).

The Raton Section experienced extensive volcanic activity, which began in the latest Tertiary and continued into the Quaternary. That activity occurred in two major areas: the Ocate field southwest of the Folsom site and the Raton-Clayton volcanic field. The latter covers an area of ~20,000 km² that extends across the Folsom region and represents the easternmost extent of Late Cenozoic volcanism in the United States (Aubele and Crumpler 2001; Hunt, Lucas, and Kues 1987:41). The effects of this volcanic activity on the landscape topography were considerable: Basalt flows, which form some of the major topographic features of the landscape, range in thickness from ~275 m near Raton, to ~120 m some 50 km east (Trimble

1990:24), and extend as far east as northwestern Oklahoma. The main axis of the field is aligned northwest/southeast, between Trinidad, Colorado, and Clayton, New Mexico. The field is marked by approximately 120 cones of basalt, andesite, and dacite, a number of which rise over 2,400 m (8,000 ft) in elevation (fig. 3.3). None of this volcanic activity produced obsidian, the nearest outcrops of which are ~400 km southwest of Folsom in the Jemez Mountains in the Rio Grande Valley. There are no obsidian artifacts in the Folsom assemblage.

Volcanic activity occurred in three major pulses that began nearly ~9 mya in the western part of the Raton section and 3.3 mya in the central and eastern part of the section. The most active period of eruption appears to have occurred between 4 and 1.8 mya (Dungan et al. 1989:474; see also Aubele and Crumpler 2001; Gustavson et al. 1991:490; Trimble 1990:24). These episodes produced a variety of rock units and topographically distinctive features (fig. 3.2), which in chronological order include (from Aubele and Crumpler

FIGURE 3.3 Major features of the Raton-Clayton volcanic field, as viewed from Bellisle Lake, looking southeast. (Photo by D. J. Meltzer.)

2001; Hunt, Lucas, and Kues 1987:41–51; Dungan et al. 1989:474, 479; also Baldwin and Muehlberger 1959; Calvin 1987:85; Kudo 1976; Stormer 1972).

RATON PHASE (Miocene and Pliocene; ~8.74–3.6 mya). Basalts that form and cap the high mesas of the area, such as Bartlett, Barela, Horseshoe, Kelleher, Mesa de Maya, and Black Mesa, primarily west and north of the site area, and Johnson Mesa, one of the largest mesas in the area, which lies immediately west of the Folsom site. Included as well are numerous volcanic domes, such as Red Mountain rhyodacite, which erupted through the Raton basalt on Johnson Mesa, forming Red Mountain and Towndrow Peak and, to the southwest, forming the cones of Laughlin Peak and Palo Blanco;

CLAYTON PHASE (Pliocene, ~3.6–2.0 mya). Consists of three units, which are generally similar chemically and mineralogically to the Raton basalt. The Clayton basalt flow forms a large sheet extending northwestward from the town of the same name. It includes eroded basaltic cinder cones such as Rabbit Ear Mountain and Mount Clayton (both important landmarks on the Santa Fe Trail, the latter then called Round Mound [e.g., Gregg 1844:89, 97–102; Marcy 1859:262; see Hill 1992]), Sierra Grande, a large andesitic shield volcano near the town of Des Moines, New Mexico, with a base ~15 km in diameter, which rises 600 m above the surrounding plain (fig. 3.3), and the Basanite cinder cones comprised of mafic feldspathoidal vents that formed Jose Butte and Robinson Peak.

CAPULIN PHASE (Pleistocene, 2.0 to .06 mya). Produced a variety of cinder cones, fissure flows, and basalts that cover many of the modern valley floors, including the classic cinder cones of Capulin Mountain (fig. 3.3) and the nearby, younger Baby Capulin. Capulin Volcano was recently dated at 58,000 to 62,000 years ago, based on [40]Ar:[39]Ar and cosmogenic helium, [3]He:[4]He (Sayre and Ort 1999).

Many of the basalts were erupted as sheets flowing down broad valleys. Because of the down-valley pattern of flow the basalts capping the highest mesas are usually oldest, while the lowest flows in the present-day valleys are among the youngest (Aubele and Crumpler 2001:72; Dungan et al. 1989:481). Raton basalt, which forms Johnson Mesa west of the site, and Capulin basalt, which occurs as small cinder cones and vents just a few kilometers east of the site, are the primary volcanics in the immediate Folsom site area (Muehlberger, Baldwin, and Foster 1961:31; Scott and Pillmore 1993).

Volcanic features dominate the regional landscape and topography and have been *relatively* little altered by subsequent geomorphic processes. The volcanic activity is important principally because the complex surface it formed created the particular configuration of drainages through the area, influenced the development of the local biotic communities, and, arguably, affects the local climate.

For a time, however, it was thought that the link between volcanism and human activity might be more direct. As the regional volcanic sequence was being worked out in the 1950s, Baldwin and Muehlberger (1959:127–129, 154–156) raised the possibility there were eruptions contemporary with the Paleoindian occupation at the Folsom site. Their argument was as follows:

> A single flow from Capulin Mountain entered the valley of the Dry Cimarron River. Exposures along the walls of the river . . . although not perfect, indicate that the Capulin Mountain basalt is younger than the older alluvium (containing no fragments of Capulin basalt) and older than the younger alluvium (containing talus fragments from the basalt at all levels). These alluvial fills are readily distinguishable throughout this area and can be traced upstream into the type Folsom Man locality. The older, yellowish, caliche-bearing alluvium is similar to other Folsom-bearing deposits of the Southern High Plains, which have radiocarbon ages of about 8000 B.C. Erosion of the older alluvium occurred before deposition of the dark, humus-rich, younger alluvium. Charcoal found in a fire pit close to the base of the younger alluvium near the Folsom Man locality has a radiocarbon age of 2,350 ± 250 B.C. (Baldwin and Muehlberger 1959:129)

In linking the Capulin basalt stratigraphically to alluvial units that could apparently be traced upstream to the Folsom site, Baldwin and Muehlberger bracketed the age of this particular eruption of Capulin basalts at between ~10,000 and 4,400 B.P. (Baldwin and Muehlberger 1959:127–129; see also Anderson and Haynes 1979:893, 896), raising the intriguing possibility that Folsom groups witnessed the eruption.

That possibility prompted Anderson and Haynes (1979; Anderson 1975:40–42) to reexamine the alluvial and volcanic stratigraphy to more precisely assess their chronological relationship. However, they found it impossible to correlate the deposits along the Dry Cimarron near the town of Folsom with those at the site a dozen miles distant. Nonetheless, they suspected that the deposits were not the same age: That is, there was more than one "younger" and "older" alluvium (Anderson 1975:41). Subsequent radiocarbon dating of a charred soil within one of several "older" alluvium units (their Archuleta Formation) found at the Dry Cimarron locality and apparently baked when overridden by the Capulin basalts returned an age of 22,360 (1,160 ^{14}C yr B.P. (TX-1268).[1] That age is, at best, a minimum date but clearly shows that the basalt flow well predated the Folsom Paleoindian occupation.

Additional evidence was also found by Anderson and Haynes bearing on this issue: Vesicular pyroclastic basalt pebbles and cobbles occurred in Holocene-age units overlying the Folsom bonebed. Their origin, whether primary or redeposited, was unclear to Anderson and Haynes (1979:898). However, given that these pebbles occurred in reworked channel deposits (chapter 5), coupled with the lack of evidence for post-Capulin (that is, post–approx. 60,000 B.P.) volcanic activity in the area, it seems unlikely that these scattered pebbles and cobbles represent eruptions that postdate the Folsom occupation. Paleoindians did not witness any volcanoes erupting, at least not at Folsom (Aubele and Crumpler 2001:72–73).

Glacial Activity

Folsom Paleoindians likely also did not experience glaciers in the immediate region, for it was largely, though not entirely, unglaciated. Even during the Late Glacial Maximum (LGM), glaciers at this latitude were confined to elevations >3,000 m (10,000 ft) above sea level (Richmond 1965, 1986). The nearest Pleistocene glaciers were in the Sangre de Christo Mountains (Porter, Price, and Hamilton 1983), ~100 km west and at least 1,000 m (3,000 ft) higher than the Folsom site.

By Folsom times, those glaciers were much diminished. Folsom was occupied squarely within the Younger Dryas Chronozone (YDC), a Late Glacial cold snap between 11,000 and 10,000 ^{14}C yr B.P. (12,900 to 11,600 cal yr B.P.; see Alley 2000; Alley and Clark 1999; Clark et al. 2001; Hughen et al. 2000; Peteet 1995; Severinghaus and Brook 1999; Shuman et al. 2002; Yu 2000). There is evidence that during the YDC glaciers readvanced at higher elevations in the Rocky Mountains, though not as far as their LGM limits (Armour, Fawcett, and Geissman 2002; Gosse et al. 1995; Menounos and Reasoner 1997; Reasoner and Jodry 2000; Yu and Wright 2001:349).

Although glaciers were never present in the Folsom area, and glacial activity in the distant Sangre de Christo Mountains was likely insignificant in the YDC, it seems reasonable to hypothesize, based on pollen studies north and west of the Folsom area, that temperatures in the Folsom region were nonetheless cooler and, perhaps, that regional snow lines, timberlines, and vegetation zones were lower in elevation than they are at present (e.g., Fall 1997; Reasoner and Jodry 2000; Vierling 1998). These matters are put to the test in chapter 6.

Soils and Sediments

The specific details of the sediments and soils at the site are provided in chapter 5. It need only be noted here that the sediments are derived principally from the erosion of local bedrock. In the immediate site area the soils are mapped within the Capulin series and are primarily silt loam to silty clay loam in texture, tend to be moderately calcareous, and are generally well drained. The primary depositional mechanisms at the site itself are fluvial and colluvial, although there is evidence as well for an aeolian source, at least during the Late Glacial (chapter 5). Weathering and erosion on the basalt mesas widened and deepened topographic lows, which in some cases filled with water to form perennial lakes and led to scarp erosion and the formation of extensive talus aprons and landslide deposits around the bases of the mesas.

Hydrology

Fluvial systems have repeatedly dissected the regional volcanic surface throughout the Pleistocene; basalt strath terraces are present alongside the Dry Cimarron downstream of the site. These left behind distinctive alluvial units, terraces, and pediment surfaces (Scott and Pillmore 1993). River and stream valleys in the area contain multiple cut-and-fill

FIGURE 3.4 A. Looking up Wild Horse Arroyo prior to the start of excavations, Carl Schwachheim on the North Bank, March, 1926. (Photo courtesy of Denver Museum of Nature and Science.) B. Same perspective, L. C. Todd on the North Bank, 1999. Note the erosion and retreat of both the South and North Bank faces. (Photo by D. J. Meltzer.)

sequences—and that just from Holocene times (chapter 5). The latter part of the nineteenth and the twentieth century witnessed extensive arroyo cutting (Anderson 1975:43; Anderson and Haynes 1979:893), which at Wild Horse Arroyo began or increased significantly with the 1908 flood.

There are no maps or photographs of the site prior to that flood. However, historic records suggest that the valley floor in this area was largely unbroken and undissected, and Wild Horse Arroyo, where it passes through the site, was less than 1.25 m deep (Owen 1988). Anderson (1975:43) records the local folklore that until the 1908 flood, one could walk a horse across the arroyo where the Folsom site is now exposed (also Owen 1988:27). It was apparently only after the flood that the deeply buried bison bones were exposed for McJunkin to discover. In fact, photographs taken by Cook and Figgins in March

of 1926, nearly two decades after the flood (fig. 3.4), show that even then Wild Horse Arroyo was deep, but still very narrow, and had yet to erode any significant distance into the bison bonebed, buried beneath the South Bank of the arroyo.

There is additional evidence supporting the hypothesis that Wild Horse Arroyo did not begin to incise until relatively recently, perhaps not until the early twentieth century. Currently, Wild Horse Arroyo flows east/southeast through a 4- to 5-m-deep, but very narrow (~0.5-m) and nearly vertical slot in the Smoky Hill Shale bedrock just northwest of the site (fig. 3.5). That slot opened along a joint plane in the Smoky Hill Shale, which in this area is suffused with joints, many of which are oriented almost precisely east-west. Floodwater forced through that constriction in the channel emerges on the downstream side with considerable

FIGURE 3.5 A. The upstream entrance to Firehose Canyon with exposure of Smoky Hill bedrock. B. The downstream outlet of Firehose Canyon in Wild Horse Arroyo north of the Folsom site. (Photos by D. J. Meltzer.)

velocity (Bernoulli's principle) and, perhaps, acted akin to a natural firehose.

Had that slot been open for any substantial period of time—say, much of the Holocene—by 1908 (and certainly by 1926), erosion would have considerably widened and deepened both the slot and the soft sediments of the site immediately downstream. In effect, the Folsom site would have been washed away. Obviously, it was not. The fact that the bonebed was still largely intact when excavations began in 1926, and that the slot and the channel downstream were as narrow as they were, indicates that arroyo downcutting is geologically very recent, in all likelihood occurring after and as a result of the 1908 flood.

Going farther afield from the site, the Raton section as a whole is dissected and drained by several major rivers and their tributaries (table 3.1). From north to south, the rivers are the Arkansas, the Canadian, and the Cimarron, which in New Mexico is known as the *Dry* Cimarron.[2] The Arkansas and Canadian originate in the central and southern Rocky Mountains, and several of their major tributaries originate in the Raton Section and along the New Mexico/Colorado border: The Arkansas is fed from this area by the Huerfano, Apishapa, and Purgatoire rivers, which have their headwaters just north of Johnson Mesa.[3] The Canadian is fed in this area by several tributaries, including the Cimarron (but not the Dry Cimarron) and Vermejo rivers, which come off the Sangre de Christos and join the Canadian below Springer, New Mexico, and the Beaver River (which becomes the North Canadian), Carrizo Creek, and Rita Blanca Creek, which flow out of the area around Des Moines and Clayton, New Mexico, toward Dalhart, Texas.

The Dry Cimarron, in contrast, has its origins at a lower elevation, emerging from a small lake atop Johnson Mesa, just 10 km from the eastern escarpment of the Mesa, roughly due west of Preston's Point (figs. 3.1 and 3.6).In the headwaters area today, which includes the site area, the Dry Cimarron is ephemeral (table 3.1), but as the 1908 flood showed, and as is attested in the Late Glacial and Holocene stratigraphic record (chapter 5), the river is capable of heavy flow for brief periods. Much farther downstream, where flow levels are enhanced and maintained by groundwater, the river becomes perennial (Schumm and Lichty 1963:73). Gregg (1844:197) reports that travelers on the Santa Fe Trail as far downstream as western Kansas often had to dig wells in the floor of the channel to obtain water.

Whether these rivers and streams were perennial in their headwaters in Late Glacial times is not known, although it seems reasonable to hypothesize that snow and seasonal

TABLE 3.1
USGS-Designated Watersheds and Drainages within an ~100-km Radius of the Folsom Site

USGS Watershed	USGS Unit	Upstream Watershed	Downstream Watershed	Drainage	Watershed Miles	Perennial Miles	% Reach Perennial
Purgatoire	11020010	None	Upper Arkansas	Arkansas	3,198.1	448.6	0.14
(Dry) Cimarron headwaters	11040001	None	Upper (Dry) Cimarron	Cimarron	966.1	104.6	0.11
Canadian headwaters	11080001	None	Upper Canadian	Canadian	1,309.3	250.8	0.19
Upper Canadian	11080003	Canadian headwaters	Upper Canadian	Canadian	1,881.8	290.4	0.15
Ute	11080007	None	Upper Canadian	Canadian	1,236.0	128.0	0.10
Punta de Agua	11090102	Rita Blanca	Lake Meredith	Canadian	908.8	84.9	0.09
Rita Blanca	11090103	Carrizo	Punta de Agua	Canadian	600.7	129.4	0.22
Carrizo	11090104	None	Rita Blanca	Canadian	361	15.1	0.04
Upper Beaver	11100101	None	Middle Beaver	Canadian	1,628.1	193.1	0.12

SOURCE: Data from www.epa.gov/surf3/hucs.

glacial melt from the Sangre de Christo Mountains would have maintained the flow of the Arkansas and Canadian headwater tributaries throughout this period.

Other potential sources of surface water in the area are the numerous lakes, lake basins, and wet meadows that even today occur atop Johnson Mesa (Muehlberger, Baldwin, and Foster 1961:29; Hunt, Lucas, and Kues 1987:51). Perennial lakes are today relatively uncommon in the Cimarron Headwaters watershed below Johnson Mesa—only three, covering an area of 16.3 acres, are recorded (www.epa.gov/surf3/hucs/11040001). In wetter times of the past, of course, more may have been present.

Groundwater sources are relatively rare in this area (Jorgensen et al. 1988). There are water-bearing rock units in the Raton Section, primarily in the Dakota sandstone and Ogallala Formation, but these occur far to the east and southeast of the site (Weeks and Gutentag 1988). The Ogallala aquifer, which comprises the principal and largest aquifer on the High Plains, underlying an area of about 347,000 km^2 and in places reaching a maximum thickness of 215 m, in this region occurs only in the southeastern corner of Colfax County some distance from Folsom, where it has a saturated thickness of just 9 m (Cronin 1964:2, 35; Gutentag et al. 1984; Heath 1988, Kilmer 1987). Where the Ogallala Formation occurs in the vicinity of the site, it is in small, dry, discontinuous pockets, generally isolated from water-bearing units. It may provide a conduit for groundwater drainage through other Ogallala formations, however, as springs are present in the Ogallala at the base of the basalt cliffs of Johnson Mesa (Anderson 1975:47).

Drainages, Topography, and Site Approaches

The rivers and streams in the Folsom area are important not simply because of their potential value in Late Glacial times as water sources, assuming that they were flowing, but also because of their utility as corridors across the landscape (Hofman 1999b; Stanford 1999). Such prominent features of the geography, as well as the high mesas or volcanic cones, can serve as boundaries, landmarks, or barriers, guiding or otherwise influencing the movement of people and animals across the landscape (Kelly 2003; Meltzer 2003; Zedeno and Stoffle 2003).

Most of the major river systems crossing this area are oriented east-west, with tributaries that flow northwest-southeast or southwest-northeast. Several rivers and streams have their headwaters in relatively close proximity to the Folsom site. As a result, merely traveling upriver or upstream from one of several directions would bring foragers into this area; this would be true of groups in southeast Colorado or in the Arkansas drainage moving up the Purgatoire and its tributaries, several of which head on the north side of Johnson Mesa (fig. 3.2). This would also be the case for groups traveling up several of the major tributaries of the Canadian River, such as the Punta de Agua to Rita Blanca Creek or Carrizo Creek, or the Beaver River to Corrumpa Creek, which, if followed, could have taken them into the Folsom region. Of course, groups following the Dry Cimarron River, if they traveled its full length, would have been conveyed to within just a few kilometers of the Folsom site (fig. 3.7).

This is not to say that following these river and streams systems would inevitably deposit foraging groups at the doorstep of the Folsom site. Rather, the point to be made is that coming to or through the Folsom area need not and may not have been a random movement with respect to topography, as it could be on the otherwise featureless and open High Plains to the east (though the latter could have had now-vanished game trails that pointed arrowlike toward springs or lake basins). The regional topography and drainages "guide"—and

FIGURE 3.6 The Dry Cimarron River below Johnson Mesa. The water-screening operation (described in Chapter 4) took place in this same area. (Photo by D. J. Meltzer.)

I use the term very loosely—mobile animals and hunter-gatherers into the *general* vicinity of Folsom.

Nor should one necessarily conceive of the Folsom site as a *destination* for groups traveling up the Dry Cimarron or any of the other regional drainages. In fact, there is little apparent reason to do so. There are no features of this locality that would have made it especially attractive to human foragers. There are no lithic sources and relatively few springs. As noted below, game can be abundant, but not necessarily predictable; edible plants are few. Of course, conditions could have been very different nearly 11,000 years ago, but the circumstances here in the Folsom region seemingly stand in sharp contrast to those in the San Luis valley ~125 km west, where an ecologically rich and well-watered basin supported bison and attracted considerable Paleoindian activity (Jodry 1999a:59–63). The Folsom area,

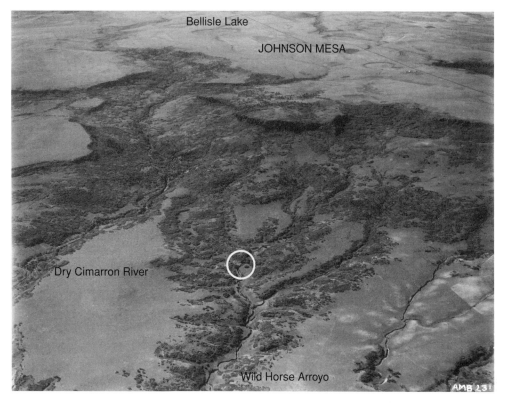

FIGURE 3.7 Aerial photograph of the Folsom site (in the white circle) and Johnson Mesa, looking west/ northwest, 1934. (Photo from Sinclair Expedition, courtesy of American Museum of Natural History.)

unlike the San Luis valley, is not a place where one would expect repeated occupations, and that is borne out in the archaeological record of Paleoindian and later groups.

Yet, though the location of the Folsom site may itself have been serendipitous, the presence of Paleoindian groups here may not have been entirely random. There is one topographic feature in the area that conceivably could have been a way point, drawn Paleoindian groups up those various drainages, and thus put them in a position to encounter and kill a herd of bison along the ancestral Wild Horse Arroyo. The Raton and Poison Canyon formations, which as noted form a large, triangular drape that extends from the eastern flank of the Sangre de Christo Mountains to an apex near Raton, New Mexico, along with the Raton-Clayton field of volcanic features and high mesas that extend east of those formations, dominate the regional landscape and topography across a 200-km stretch of the border between Colorado and New Mexico (fig. 3.2). While hardly an obstacle on the scale of the Rocky Mountains to the west, much of this is mountainous and broken terrain, and a barrier to easy north-south movement.

Along that stretch there are several mountain passes, including Raton Pass, at an elevation of 2,354 m (7,723 ft), over which the Mountain Branch of the Sante Fe Trail and the modern interstate highway cross, and, 40 km to the east, Trinchera Pass, at an elevation of 2,194 m (7,198 ft). Trinchera Pass forms a saddle between Johnson and Kelleher mesas (fig. 3.8) and is an easy 8-km walk north of the Folsom

site. Although not as well known as Raton Pass, Trinchera Pass was almost certainly used in prehistoric times to cross over this high, rough country and down onto the High Plains of southeastern Colorado. One of the earliest recorded uses was by Charles Goodnight, who, beginning in 1868 and for several years thereafter, trailed longhorn herds from Texas to Wyoming through the Pass, on what became known as the Goodnight Trail (Haley 1936:209–210).[4] Paleoindian groups headed north up the Great Plains, following the regional drainages, could well have been aiming for Trinchera Pass. All of this, of course, is a matter of speculation.

Regardless of what may have brought Paleoindian foragers into this specific area, the essential physiographic lesson is that movement into the region by Paleoindian foragers would have been relatively easy had they merely been following the regional drainages from any one of a number of places on a wide arc from well east of the region down around to areas southwest of the site. Were they coming south into the Folsom area from the Plains of southeastern Colorado, the approach to and through Trinchera Pass would have been relatively easy as well. Approaches to the Folsom area from the west, however, would have been somewhat more difficult, requiring a scramble up, and then down, the western and eastern flanks of Johnson Mesa, or else would have involved a detour south of the mesa and then east (the current New Mexico state highways do both). The talus and steep escarpments flanking Johnson Mesa were not an insur-

FIGURE 3.8 Looking north through Trinchera Pass, Johnson Mesa on the left, Kelleher Mesa on the right, and the plains of Colorado in the distance. (Photo by D. J. Meltzer.)

mountable obstacle to pedestrian traffic but were not the path of least resistance either.

Once atop Johnson Mesa, at least in summertime, foraging groups would have had a relatively easy traverse; the gently rolling landscape is flat and largely undissected. However, wintertime travel across Johnson Mesa in Late Glacial times might have been much more difficult, as it can be even today, owing to heavy snow and ice accumulations (F. Owensby and A. Reeves, personal communication, 1999).

Present Climate

The climate of the Folsom area can be described using data from the network of National Weather Service stations across the region. The closest of the stations to the site are in the town of Folsom, but these have only limited data from a few years of operation half a century ago. Some two dozen other weather stations occur within a ~30-min (~56-km/~35-mile) radius and ±150 m (±500-ft) elevation of the Folsom site. These vary in location, topographic aspect, elevation, and length, and the completeness of their records. However, half a dozen of the stations within that sampling sphere have records extending over multiple decades and are, thus, useful proxies for the climate at the Folsom site. Select summary climatic data from those stations are provided in table 3.2 (data accessed digitally from the National Climate Data Center of the National Oceanographic and Atmospheric Administration Web site; www.ncdc.noaa.gov). There are a few additional weather stations at lower elevations on the Plains to the north (Branson, CO) and southwest (Grenville and Clayton, NM) of

the site.[5] Those latter two are ~300 m (1,000 ft) and ~600 m (2,000 ft) lower than the Folsom site. Taken together, these bracket the conditions in the area of the Folsom site and provide climate records from much of the last half of the twentieth century—although, as can be seen, the individual weather stations vary in their coverage and completeness (table 3.2).

Data from these weather stations can be used to interpolate the climatic parameters for the Folsom site itself, by using the average values from these stations to create isopoll maps of each variable—e.g., mean annual temperature (MAT), mean annual precipitation (MAP), etc.—from which values for Folsom can be inferred. This was done using SURFER for all the continuous variables in table 3.2. It is important to emphasize that these are interpolations, not actual measurements. Although they seem to be reasonably accurate estimates, they do not account for the Folsom site's topographic position and down-scarp proximity to Johnson Mesa, which would influence these measures. Hence, in the discussion that follows, I primarily use the *range* of values from the surrounding weather stations to provide a sense of the general climatic patterns, rather than just these specific estimates for the Folsom site.

Overall, the Folsom region has a continental, semiarid climate. It experiences a relatively low MAP, generally between ~38 and 44.5 cm (~15–17.5 in.) at the lower-elevation stations, with significantly greater amounts at the highest-elevation station at Lake Maloya (59.5 cm [23.4 in.]) and perhaps at the Folsom site as well (table 3.2). MAP is highly variable from place to place and from year to year (Borchert 1950; Gutentag et al. 1984:3), which is not unexpected

TABLE 3.2

Summary Climatic Data from Weather Stations within a 30-Mile Radius *and* ±150 m (±500-ft) Elevation of the Folsom Site

Weather Station (Ordered by Elevation)	Elevation, m(ft)	Average No. of Days Minimum Temperature ≤0°F	Average No. of Days Minimum Temperature ≤32°F	Average No. of Days Maximum Temperature ≤32°F	Average No. of Days Maximum Temperature ≥90°F	MAT Mean Annual Temperature, °C (°F)	Mean Temperature Warmest Month, °C (°F)	Mean Temperature Coolest Month, °C (°F)	Effective Temperature (ET)[a]	Temperateness of Climate (M)[a]
Lake Maloya, NM	2,254 (7,398)	15.1	198.1	19.4	0.5	6.75 (44.15)	17.39 (63.30)	−3.5 (25.7)	12.05	50.4396
Capulin Natl. Monument, NM	2,222 (7,291)	3.14	136.2	15.71	0.93	9.11 (48.4)	19.69 (67.45)	−1.39 (29.5)	12.73	53.073
Capulin, NM	2,195 (7,203)	15.22	181.3	17.52	3.13	8.09 (46.56)	19.28 (66.7)	−3 (26.6)	12.45	50.0682
Folsom	**2,109 (6,920)**	**6**	**149**	**16**	**6**	**9 (48.2)**	**19.8 (67.64)**	**−1.4 (29.48)**	—	—
Cumico, NM	2,079 (6,822)	—	—	—	—	—	—	—	—	—
Raton, NM	2,036 (6,681)	2.74	150.9	15.2	4.43	9.67 (49.40)	20.43 (68.78)	−0.11 (31.8)	12.93	53.4567
Des Moines, NM	2,018 (6,622)	6.43	159.1	15.23	8.85	9.56 (49.2)	21.17 (70.1)	−0.94 (30.3)	12.97	51.6801
Branson, CO	1,917 (6,291)	4.35	130.9	14.34	25.04	10.94 (51.7)	22.72 (72.9)	0.83 (33.5)	13.4	52.7473
Grenville, NM	1,828 (6,000)	4	150.64	14.34	20.07	10.27 (50.49)	21.56 (70.8)	−0.33 (31.4)	13.09	52.1271
Clayton, NM	1,511 (4,958)	2.91	140.04	14.84	33.54	11.66 (52.99)	23.28 (73.9)	0.72 (33.3)	13.47	51.9425

	Last Spring Day Temperature ≤32°F (n)	First Fall Day Temperature ≤32°F (n)	Average No. of Frost-free Days (SD)	MAP Mean Annual Precipitation, cm (in.)	Maximum Annual Precipitation cm (in.)	Minimum Annual Precipitation cm (in.)	Average Annual Snow Fall, cm (in.)	JJA/SEA Seasonality of Precipitation	Period of Coverage (Not for All Variables)
Lake Maloya, NM	May 23 (19)	September 24 (18)	124 (18.7)	59.39 (23.38)	93.19 (36.69)	33.17 (13.06)	193.29 (76.1)	40.53%	1942–2000
Capulin Natl. Monument, NM	—	—	—	41.68 (16.41)	51.03 (20.09)	24.18 (9.52)	89.91 (35.4)	51.25%	1966–1979
Capulin, NM	—	—	—	39.50 (15.55)	65.63 (25.84)	19.84 (7.81)	47.24 (18.6)	53.57%	1931–1969
Folsom	**No data**	**No data**	**130**	**44.20 (17.4)**	**No data**	**No data**	**109.22 (43)**	**48%**	**None; interpolated**
Cumico, NM	—	—	—	37.82 (14.89)	66.17 (26.05)	19.91 (7.84)	46.48 (18.3)	55.12%	1940–1970
Raton, NM	May 7 (18)	October 2 (18)	149 (13.3)	44.78 (17.63)	62.05 (24.43)	24.71 (9.73)	84.81 (33.39)	47.20%	1953–2000
Des Moines, NM	May 10 (13)	September 28 (12)	141 (14.7)	43.92 (17.29)	85.83 (33.79)	29.01 (11.42)	92.71 (36.5)	47.39%	1931–1994
Branson, CO	—	—	—	41.91 (16.50)	85.52 (33.67)	19.07 (7.51)	113.03 (44.5)	41.80%	1940–1974
Grenville, NM	May 7 (19)	October 3 (18)	149 (15.4)	43.28 (17.04)	94.08 (37.04)	21.59 (8.50)	37.34 (14.70)	49.86%	1940–2000
Clayton, NM	April 23 (11)	October 8 (18)	169 (15.6)	38.55 (15.18)	65.27 (25.70)	22.68 (8.93)	56.13 (22.1)	48.84%	1947–1992

[a]Effective temperature (ET) and temperateness of climate (M) data after Bailey (1960).

NOTE: asl—above sea level; JJA—June, July, August.

SOURCE: Data from www4.ncdc-noaa.gov. Table includes several additional stations (Branson, Grenville, Clayton) to provide comparative data on nearby plains.

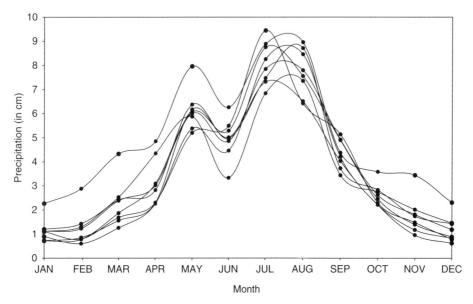

FIGURE 3.9 Monthly precipitation (in cm) at the weather stations in the Folsom area. (Data from the National Climate Data Center.)

given that much of the precipitation is associated with high-intensity summer thunderstorms that tend to be scattered and random in their behavior. Overall, this is an area of low humidity.

Most precipitation (65%–75%) falls as rain from late spring and late summer, the wettest months currently being July and/or August, with a lesser peak in May (fig. 3.9). A slight decline in June precipitation produces a bimodal pattern to the rainfall. However, precipitation during June, July, and August, the critical growing months for warm-season grasses (Paruelo and Laurenroth 1996), still comprises ~48% of the MAP. The summer precipitation peak is the result of relatively brief and sometimes heavy cloudbursts, the consequence of adiabatic cooling of warm moist subtropical air masses flowing inland from the Gulf of Mexico and, secondarily, from the southern Pacific via the Gulf of California (chapter 6). These summer convection thunderstorms tend to form in the lee of the Rocky Mountains in late afternoon and propagate eastward during the evening (Court 1974:213, 227), one of which likely led to the downcutting of Wild Horse Arroyo in August of 1908 and several of which, over the course of our fieldwork, led to uncontrolled and badly choreographed vehicle ballets on skating-rink-slick clay roads, carved into the poorly draining residuum of eroded silts and clays from the Smoky Hill Shale.

Maps of thunderstorm activity (e.g., Court 1974:fig. 23), show a peak in this area of northeastern New Mexico, though it is suspected that this higher activity may be more apparent than real, an acoustical illusion of the sound bouncing around the mesas and foothills, resulting in a greater range over which thunder can be heard (Court 1974:227).

Winter precipitation in the region is relatively low; December and January are normally the driest months of the year. However, snowfall can be heavy in the immediate area

of the Folsom site (A. Reeves, personal communication, 1999), owing to elevation, aspect, and local topographic effects, the result of the nearby scarp of Johnson Mesa, all of which produces a precipitation pattern likely closer to the Lake Maloya weather station than to, say, Clayton. Local residents report that Johnson Mesa is often buried in heavy snow during the winter. In the winter of 1997, when snowfall records were set or almost matched across the area, 620 cm (244 in.) of snow fell at Lake Maloya, just below the far western edge of the Mesa, while residents in the Trinchera Pass area just a few miles from the site reported totals of 380 cm (150 in.) of snow (F. Owensby, personal communication, 2001).

By comparison, the *maximum* snow accumulation on record at any of the other stations in the area, nearly all at lower elevations, is 217 cm (85 in.). Snowfall tends to be much lighter downslope and out on the surrounding plains (Borchert 1950:6). Snowfall, not surprisingly, does not contribute much to the annual precipitation totals across the larger region, the majority of which falls as rain (Gutentag et al. 1984:3); this is true of the Folsom site area as well.

The Gulf of Mexico is the primary source of atmospheric moisture to the region. During the summer months, particularly the latter part of the summer, low-level winds come out of the south-southeast, and Gulf moisture dominates. Under certain conditions, however, the Pacific Ocean provides significant amounts of moisture (Nativ and Riggio 1990a:157). In winter, low-level winds become more westerly and precipitation can be drawn in from the Pacific Ocean (Nativ and Riggio 1990a:160). Summer and winter precipitations have distinct isotopic signatures, with enrichment of heavy isotopes in the summer because of higher temperatures, increasing evaporation, and proximity to the Gulf sources and depletion of the heavy isotopes in winter (Nativ and Riggio 1990a:165). The implications of this for regional paleoclimates are discussed briefly below (chapter 6) but have yet to be explored (Meltzer 1999).

Because the MAP in the region is low and variable, the area is susceptible to periodic drought, especially during the spring and summer (Borchert 1950:14, figs. 9, 10; Nativ and Riggio 1990a:162). Droughts during these months are especially severe, for rainfall deficits are combined with a lack of cloud cover, higher than average temperatures, and hot, continental winds (Borchert 1950:15–16, fig. 17). Drought in this region is not necessarily associated with drought elsewhere in North America; there is or can be a sharp regional distinctiveness in these events, as has been seen historically (Borchert 1950:15; Woodhouse and Overpeck 1998:fig. 12).

The droughts that affect this area tend to be associated with abnormally strong eastward movement of dry, continental air from the Rockies during the summer (Woodhouse and Overpeck 1998:2707). In a normal year, the westerly flow dominates year-round in only a small area just east of the Rockies, and beyond that its influence is sharply limited to the few winter months of the year (Borchert 1950:fig. 18). In contrast, during drought episodes, the area dominated by westerly flow year-round is much larger (Borchert 1950:fig. 19; Woodhouse and Overpeck 1998:2707). The westerly flow brings extremely dry air masses. The stronger the westerly, the more extensive the drought area east of the Rockies (Borchert 1950:18), as Gulf of Mexico moisture is unable to penetrate into the Plains.

Strong westerlies are associated with an abnormally large south-north decrease in pressure across the midcontinent and low pressure north of the Plains, the development of large and unusually stationary high-pressure cells over the Great Basin and southern plateau, and the deflection of the moist, tropical Gulf/Atlantic air toward the southeast and away from the Plains (Barry 1983:47; Borchert 1950:18, 28–29). Those strong westerlies also bring abnormally high temperatures because there is less cloud cover and rainfall and thus more insolation than normal, and because there is less injection of cold Arctic air (Borchert 1950:19–20, 27–28). It is not known precisely how these air masses behaved in Late Glacial times, but the evidence adduced in chapters 5 and 6 provides some clues to precipitation patterns and the question of whether there was a Folsom-age drought (Holliday 2000a) in this area.

Although precipitation can vary significantly from area to area, temperature gradients are much more uniform and less abrupt and tend to grade with elevation (table 3.2). MATs across the area presently range from 6.7°C (44°F) to 11.7°C (53°F), with the warmest month, commonly July, being on average 11.22°C (20.2°F) warmer than the MAT, while the coolest month, usually December, is 9.4°C (16.9°F) colder. The difference between the coldest and the warmest months averages 21.7°C (39°F). In general, summers in the area are relatively cool, a function of the elevation. Temperatures at the higher-elevation stations rarely climb over 32°C (90°F), although on the lower, nearby surrounding plains, summer temperatures are usually much hotter (table 3.2). For example, there are on average <1 day per year of temperatures over 32°C (90°F) at Capulin National Monument, elevation 2,222 m (7,291 ft), but ~34 days over that mark at Clayton, New Mexico, elevation 1,511 m (4,958 ft). Temperatures might reach that high only half a dozen days a year in the Folsom site area.

In winter, daily lows across the area fall below 0°C (32°F) an average of 156 days of the year, with the largest average (198 days) at the highest of the stations examined here (Lake Maloya). Yet, despite these cold temperatures, daily highs in winter are commonly above 0°C (32°F) for all but an average of ~16 days of the year. Only rarely and at higher than the Folsom site itself do temperatures fall below −17.8°C (0°F) (table 3.2).

Regionwide, the months of July to August are completely frost-free.[6] The higher-elevation areas can experience freezing temperatures as late as June and as early as mid-September (Lake Maloya). Generally, however, the last spring freeze occurs before June 1, and the first freeze of the fall is not until October. The average number of freeze-free days (the growing season) varies from 124 days at Lake Maloya to as much as 169 days at Clayton, which is 760 m (2,500 ft) lower in elevation and out on the Plains. Stations at elevations closest to the Folsom site (e.g., Raton, Des Moines) have ~145 freeze-free days. The interpolated value for Folsom is 130 days (table 3.2).

The timing of the last freeze of the spring across the region tends to be spread over nearly a month's time, ranging from late April to late May. The first freeze of the fall occurs more predictably within a narrower time frame in late September to early October. In fact, the average monthly temperature curves, a composite of which is shown in figure 3.10, reveal that the fall decline in temperatures is steeper than the slope of spring warming. Fall temperatures in the Folsom site area drop quickly across the area. On that, we have the testimony of Carl Schwachheim's diary: The 1927 field season ended soon after the abrupt arrival of three days of cold, fog, and snow in late September (appendix B).

It is impossible to say, of course, precisely what the weather was on the day of a Paleoindian kill, although Wheat (1972:98) made a respectable go of it at the Olsen-Chubbuck site. Still, the evidence that the Folsom kill occurred in the fall (chapter 7), coupled with what is known of fall weather patterns today and considering the hypothesis that climates were cooler in this area during the Younger Dryas than they are at present, certainly makes the weather a potential factor in the activities and occupation of this site. Even under modern climatic patterns, the Folsom hunters could have experienced cold temperatures and driving snow, warm days and cool nights, or something in between. The weather might have been a significant factor in the amount of time spent at the site, the degree of butchering, whether the meat was dried there or packed out on limb elements, etc. (chapter 7).

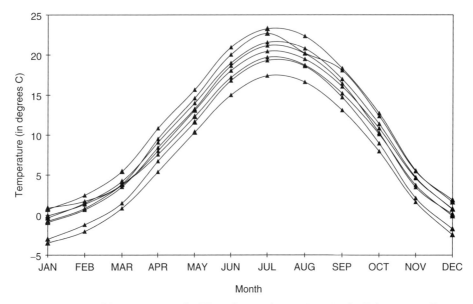

FIGURE 3.10 Monthly temperature (in °C) at the weather stations in the Folsom area. (Data from the National Climate Data Center.)

Thinking about Hunter-Gatherer Land Use

Modern climate parameters provide a useful foundation for thinking about and developing hypotheses regarding hunter-gatherer uses of the region. As Binford (1980, 2001), Kelly (1995:66–71), and others have shown, there are several environmental variables, some derived from measures of temperature and precipitation, that are generally correlated with the relative contribution of hunting and gathering in the diet of human foragers, as well as other aspects of forager adaptive strategies. The correlation is not perfect (Kelly 1995:70). Nonetheless, it is close enough that it becomes useful to employ the modern climatic and environmental parameters as a frame of reference to project what one might expect to see, all other things being equal, of forager strategies in a particular area.

Among the environmental variables of interest are effective temperature (ET), which provides a measure of the intensity and annual distribution of solar radiation, and net aboveground primary productivity (NAGP), which provides an estimate of the amount of available food (Binford 1980; Kelly 1995:66). The calculated values for these and related variables for the weather stations in the Folsom area are provided in table 3.3. The ET values, ranging from 12.05 to 13.47, confirm that this is an area of cool, seasonal climates, with a relatively short growing season (for comparison, polar regions have ET values of ~8, and tropical ET values range up to 26). In this region, the permanent standing biomass (SB) is comparatively low; there are relatively few trees, save at the highest elevations. It also experiences considerable annual turnover. In a grassland/parkland like this, of course, much of the available biomass—grass and shrubs—may not be edible to humans. However, those plant resources

are edible to animals, and secondary productivity (SP) in this area—a projection of expected ungulate prey based on empirically derived patterns of animal biomass—is relatively high (table 3.3). This ought to be a game-rich area, and in fact, it is even today.

Binford (2001) has derived a series of equations by which these and other measures of the regional climate and environment can be used to project various aspects of forager adaptation, including subsistence strategies, population density, group size and occupied area, aspects of mobility, etc. Table 3.4, which lists the projections for some of those variables of interest, is derived from Binford's (2001) equations, using the climatic and environmental data from the regional weather stations (table 3.2; additional data not listed in that table were also used). Results from two estimating methods are provided in table 3.4. The first predicts hunter-gatherer subsistence and population density (table 3.4a) using a "Terrestrial Model" that assumes the foragers have only limited knowledge of the environment and relatively simple responses to their ecological circumstances. The second, the "Hunter-Gatherer Model," from which the values in tables 3.4b and c were calculated, uses observed correlations between environmental variables and aspects of foraging strategies as measured among 339 ethnographically known foragers, to derive a set of relational equations from which one can calculate projections of forager adaptive strategies.

The estimates from both the Terrestrial and the Hunter-Gatherer Models (tables 3.4a and b) suggest that it is presently an area in which the majority of the subsistence resources would be projected to be derived from hunting, with gathering providing a lesser component of the forager diet—substantially less in the case of the Terrestrial Model. Resources derived from fishing, which includes any subsis-

TABLE 3.3
Calculated Productivity and Climatic Data

Weather Station (Ordered by Elevation)	ET Effective Temperature	Permanent Standing Biomass (SB; g/m²)	Net Aboveground Productivity (g/m²/yr; NAGP)	Biomass Accumulation Ratio (SB/NAGP)	Secondary (Ungulate) Productivity (kg/km²)
Lake Maloya, NM	12.05	13,643.16	666.32	20.48	992.92
Capulin Natl. Monument, NM	12.73	7,944.06	634.00	12.53	2,164.25
Capulin, NM	12.45	7,942.92	573.35	13.85	2,307.44
Raton, NM	12.93	8,948.48	687.26	13.02	2,791.53
Des Moines, NM	12.97	9,293.45	630.64	14.74	2,103.41
Branson, CO	13.40	7,246.61	628.20	11.54	2,226.65
Grenville, NM	13.09	7,821.96	669.09	11.69	2,586.98
Clayton, NM	13.47	6,232.35	595.50	10.47	1,557.73

SOURCE: Based on climatic data from weather stations in table 3.2 (based on formulas in Binford 2001 and A. Johnson, personal communication, 2000).

tence activity within an aquatic setting, such as collecting freshwater shellfish, crustaceans, etc., would generally contribute least to the diet. Of course, aquatic resources in terrestrial settings are often low-ranked, particularly where forager population densities are relatively low, as they are projected to be in this area even under present conditions, and there is no strong selective pressure for dietary expansion or resource intensification. In Late Glacial times, when forager population densities were assuredly even lower than those projected under present conditions, perhaps by several orders of magnitude, and higher-ranked resources were unlikely to have experienced any significant resource depression, fishing or other use of aquatic resources was likely minimal. In fact, where population density is lower, groups would be expected to focus more of their subsistence activity on hunting (Binford 2001).

Hunter-gatherers using this region should also be fairly mobile, moving a projected 17 to 30 times per year, with the projected distance moved, obviously constrained in the model by the limits on modern hunter-gatherers, still being relatively high: on the order of 340 km to 400 km per year (table 3.4; for comparative data, see Kelly 1995:128–130). Group sizes are projected to be relatively small, and the groups themselves widely dispersed. In effect, this is not an area in which one would expect to see dense human populations, engaging in intensive resource extraction.

It is important to emphasize that these projections are not an end in themselves but, rather, a starting point for thinking about how forager groups might have used an area under present-day climatic and ecological parameters, were they doing so in a fashion similar to what has been observed among ethnographic groups living on the relatively populated landscapes of the ethnographic present. Obviously, those projections cannot be applied directly to Late Glacial times, not just because climatic and environmental parameters were different, but because the social landscape was as well. These projections, taken at face value, would almost certainly overestimate some variables—population density, for example—and underestimate others, such as dependence on hunting and the frequency and distance of moves.

Regardless, there is a useful benefit to this brief exercise: Even under climatic and ecological conditions of the present, the models project forager groups in the Folsom region to be highly mobile, derive much of their subsistence resources from hunting, occupy the area at relatively low population densities, and thus be spread lightly and thinly on the landscape. As seen below, these projections match up well with inferences based on patterns of floral and faunal distribution and abundance and, also, have some measure of support in the archaeology of the late Prehistoric period in this region. Whether these general patterns are also characteristic of Folsom Paleoindian groups living during the Late Glacial is taken up in later chapters.

Modern Flora

The seasonal patterns of precipitation, length of the growing season, and summer temperatures all have a strong influence on the vegetation, both arboreal and nonarboreal, in terms of plant growth, photosynthetic rates, and health (R. Thompson et al. 1999a, 1999b, 2000). The Folsom region is marked today by a complex patchwork of biotic communities, as detailed by Anderson (1975:46–72), Dick-Peddie (1993), and Knight (1987). Dick-Peddie (1993), working at a macro scale, subsumed the area of the Folsom site within his montane scrub community, ponderosa pine/piñon pine/gamble oak series (fig. 3.11). This community, in turn was bounded by several others: montane grassland, coniferous and mixed woodland, lower montane coniferous forest, and plains-mesa grassland (see also Shelford 1963).

TABLE 3.4
Estimated Population Density, Subsistence, and Settlement Patterns

Weather Station (Ordered by Elevation)	(a) Subsistence Projections Based on Terrestrial Model			(b) Subsistence Projections Based on H-G Model			
	No. of People/ 100 km²	Hunting (%)	Gathering (%)	No. of People/ 100 km²	Hunting (%)	Gathering (%)	Fishing (%)
Lake Maloya, NM	1.96	88.0	12.0	5.29	51.5	17.3	31.2
Capulin Natl. Monument, NM	4.12	89.4	10.6	4.06	58.2	29.7	12.1
Capulin, NM	4.26	90.9	9.1	3.66	58.3	25.5	16.2
Raton, NM	5.14	88.2	11.8	4.72	55.2	32.4	12.4
Des Moines, NM	4.05	86.6	13.4	3.94	55.4	32.9	11.7
Branson, CO	4.27	86.1	13.9	3.87	53.2	38.0	8.7
Grenville, NM	4.89	87.2	12.8	3.75	57.0	33.8	9.2
Clayton, NM	3.29	81.9	18.1	3.68	53.9	41.2	4.9

(c) Projections of Group Mobility, Size, and Areal Extent, Based on H-G Model

	Projected No. of Moves per Year	Projected Distance Moved per Year (km)	Projected Minimal Group Size	Projected Area Occupied (km²)	Projected Total Population of Group	Projected % of Total Population in 100 km²
Lake Maloya, NM	29.50	406.1867	15.65	21.04	194.92	4.75
Capulin Natl. Monument, NM	19.72	359.7397	14.09	21.09	154.42	4.74
Capulin, NM	18.92	356.5180	14.05	19.49	156.33	5.13
Raton, NM	18.98	340.7063	14.17	27.67	203.62	3.61
Des Moines, NM	18.50	349.1353	14.15	28.67	174.61	3.49
Branson, CO	16.63	343.8421	13.75	26.65	171.92	3.75
Grenville, NM	18.21	355.2572	13.95	23.13	157.43	4.32
Clayton, NM	17.25	363.8371	13.30	34.43	181.58	2.90

NOTE: The Terrestrial Model does not calculate the contribution of fishing to subsistence.

SOURCE: Recent environmental data from weather stations in table 3.2. Formulas in Binford (2001) and A. Johnson (personal communication, 2000).

Anderson (1975), in contrast, divided the regional biotic communities into smaller units, tied to micro-scale variability in aspect, soil moisture availability, and topography. She distinguished eight separate biota in the region: open grassland; scrub oak, piñon, juniper forest; scrub oak, piñon, juniper parkland; moist tree oak, locust thicket; ponderosa pine forest; dense fir, pine, tree oak forest; riparian; and grassy seep.

Of course, any such classifications are somewhat arbitrary. Vegetation communities are not internally organized or clearly bounded and coherent biological entities but, instead, ways of classifying and subdividing populations of plant species that cluster historically at certain points in space and time. Save in the case of plant commensals,

which are relatively rare in this area, the distributions of those different plants need not have isomorphic boundaries. As a result, alternative classifications of vegetation distribution could be made that, depending on the analytical and geographic scales used, would either further subdivide or further aggregate biota. The effort to distinguish communities can be analytically useful in assessing the spatial patterning, co-occurrence, and density of particular resources on a landscape, as a way of helping to tease out patch choice by human foragers.

Still, there is little analytical payout to organizing the Folsom area vegetation into detailed biotic communities. This is so for several reasons; most especially, there is no reason to suppose that modern vegetation *communities* would have

FIGURE 3.11 Montane scrub community in the Folsom area: Johnson Mesa is to the left, the Folsom site is in the right, foreground. (Photo by D. J. Meltzer.)

been present in that same form in Late Glacial times, even if some of the same plant genera and species were present on the landscape—as, indeed, appears to have been the case (chapter 6). Having similar constituent elements does not ensure that those elements combined in the same fashion.

That said, there is an aspect of the modern vegetation that is of interest and that almost certainly has some bearing on and implications for the Late Glacial landscape. By virtue of its volcanic history this is an area of considerable topographic variability: The elevation drops nearly 300 m (~1,000 ft) from the lip of Johnson Mesa 2 km northwest of the site to the Dry Cimarron valley 2 km southeast of the site. If the length of that line is increased, the elevation changes even more substantially. The result is that a wide variety of plant species occur across a relatively small span of ecotonal space. This can be seen in table 3.5, which provides a listing of the flora of the Folsom region, arranged by taxon rather than by biotic community.

There are within this area recognizable patches of "grassland" and various kinds of "woodland," depending, of course, on the scale at which such are defined. But perhaps of more interest is the underlying structure, which is one of a relatively species-rich complex of vegetation. As Lisa Huckell (1998) observed in a brief vegetation census conducted in 1997, the area surrounding the Folsom site can be best described as

. . . an ecotonal mixture of grassland-scrub-woodland vegetation in which grassland is punctuated by copses or small thickets composed primarily of Gambel's oak

(Quercus gambelii), as well as New Mexican locust *(Robinia neomexicana)*, poison oak *(Rhus radicans)*, and western snowberry *(Symphoricarpos* cf. *Occidentalis)*, along with other shrubs. At slightly higher elevations upslope of the site woodland elements appear, consisting of small enclaves of Rocky Mountain juniper *(Juniperus scopulorum)*, pinyon pine *(Pinus edulis)*, mountain mahagony *(Cercocarpus* sp.), and occasional Gambel's oaks with an understory composed primarily of skunkbush *(Rhus trilobata)*, poison oak, and occasional wild roses *(Rosa* sp.). The grassland community contains a variety of grass species that includes several gramas *(Bouteloua* sp.), bromes *(Bromus* spp.), wild rye *(Elymus* spp.), and scattered ricegrass *(Oryzopsis hymenoides)*. Forbs are well-represented, with yarrow *(Achillea millefolium)*, Mexican hat *(Ratibida columnaris)*, green thread *(Thelesperma* sp.), Indian paintbrush *(Castilleja* spp.), gumplant *(Grindelia* sp.), nodding onion *(Allium cernuum)*, verbena *(Verbena macdougalii)*, elkweed *(Swertia* sp.) and flax *(Linum* sp.) among those noted. Other elements that appear to be establishing a presence in the open grassland areas include prickly pear *(Opuntia* sp.), yucca *(Yucca* sp.), snakeweed *(Gutierrezia* sp.), and sage *(Artemesia* sp.). Another biotic community can be found in a narrow band along the incised channel of Wild Horse Arroyo, in which riparian plants are growing; the narrow, shaded reach upstream of the site is particularly rich in mesic herbaceous species. Access to another biotic community, the montane conifer forest, requires traveling a short distance from the site up the escarpment of Johnson Mesa, where ponderosa pine is interspersed with dense thickets of Gambel's oak and New Mexico locust. Grassland quickly reasserts itself on the surface of Johnson Mesa.

TABLE 3.5
Flora of the Folsom Region

Grasses (by Tribe)	Season of Growth	Photosynthetic Pathway	Forage Quality for Game Species
Andropogoneae			
Andropogon gerardii (big bluestem)	Warm	C_4	(B)
Schizachyrium scoparium (little bluestem)	Warm	C_4	Fair (B)
Sorghastrum nutans (indiangrass)	Warm	C_4	
Aristideae			
Aristida longiseta (red threeawn)	Warm	C_4	Poor (B)
Aveneae			
Agrostis exarata (spikebent)	Cool	C_3	Good to excellent (E)
Koeleria pyramidata (prairie junegrass)	Cool	C_3	Good (spring and fall)
Chlorideae			
Bouteloua gracilis (blue grama)	Warm	C_4	Excellent (B)
Bouteloua sp. (grama)	Warm	C_4	(B)
Buchloe dactyloides (buffalograss)	Warm	C_4	Good to excellent (summer and winter) (B)
Hilaria jamesii (galleta)	Warm	C_4	Good (spring and summer)
Schedonnardus paniculatus (tumblegrass)	Warm	C_4	Poor
Eragrosteae			
Blepheroneuron tricholepis (pine dropseed)	Warm	C_4	
Muhlenbergia montana (mountain muhly)	Warm	C_4	Fair (D) to good (B, E)
Muhlenbergia torreyi (ringgrass)	Warm	C_4	Poor to Fair
Scleropogon brevifolius (burrograss)	Warm	$C_{4?}$	
Sprobolus airoides (sacaton)	Warm	C_4	Poor
Sprobolus sp. (dropseed)	Warm	C_4	Poor to fair (B)
Paniceae			
Panicum virgatum (switchgrass)	Warm	C_4	Fair
Setaria leucopila (plains bristlegrass)	Warm	C_4	Fair to good
Poeae			
Bromus carinatus (mountain brome)	Cool	C_3	Good (D) to excellent (B, E)
Poa fendleriana (mutton bluegrass)	Cool	C_3	Good (B, D, E)
Vulpia octoflora (six weeks fescue)	Cool	C_3	Fair (spring)
Stipeae			
Oryzopsis hymenoides (indian ricegrass)	Cool	C_3	Good (B)
Stipa comata (needle and thread)	Cool	C_3	Good (E; winter) (D; spring) (B)
Triticeae			
Agropyron sp. (western and bluebunch wheatgrass)	Cool	C_3	Fair (P)–good (D, E)– excellent (B)
Elymus canadensis (Canada wild rye)	Cool	C_3	Fair (spring)
Hordeum sp. (barley)	Cool	C_3	Poor
Sitanion hystrix (squirreltail)	Cool	C_3	

Forbs and Woody Plants (by Family)	Season of Growth	Forage Quality, All Species
Agavaceae		
Yucca glauca (yucca)	Cool	Good
Alliaceae		
Allium cernum (nodding onion)		
Alismataceae		
Sagittaria latifolia (arrowhead)	Warm	

TABLE 3.5 (*Continued*)

Forbs and Woody Plants (by Family)	Season of Growth	Forage Quality, All Species
Anacardiaceae		
Rhus aromatica (R. trilobata) (skunkbrush sumac, squawbush)	Cool	Good
Toxicodendron rydbergii (poison ivy)		
Asteraceae		
Achillea millefolium (western yarrow)	Cool	
Ambrosia psilostachya (western ragweed)	Warm	Poor
Arnica cordifolia (heartleaf arnica)	Warm	
Artemisia ludoviciana (cudweed sagewort)	Warm	Fair (D, E, P)
Artemesia sp. (sagebrush)	varies	Good (winter) (D, E) (B)
Aster sp. (aster)	varies	(P)
Chrysopsis villosa (hairy goldaster)	Warm	Poor (P)
Chrysothamnus sp. (rabbitbrush)	Warm	Poor to fair (B)
Grindelia squarrosa (curlycup gumweed)	Warm	Poor
Gutierrezia sarothrae (broom snakeweed)	Warm	Poor (B)
Helianthus annuus (sunflower)	Warm	
Ratibida columnifera (prairie coneflower)	Warm	Fair to good
Senecio douglasii (threadleaf groundsel)	Warm	
Solidago missouriensis (prairie goldenrod)	Warm	
Thelesperma sp. (greenthread)	Warm	(P)
Tragopogon dubius (salsify)	Warm	
Brassicaceae		
Descurainia pinnata (tansy mustard)	Cool	
Stanleya pinnata (desert princeplume)	Cool	Poor
Cactaceae		
Opuntia sp. (prickly pear)		
Caprifoliaceae		
Symphoricarpos albus (snowberry)	Cool	Fair (D)
Chenopodiaceae		
Atriplex sp. (saltbush)	Varies	Fair to Good (winter) (D)
Ceratoides lanata (winterfat)	Cool	Good (winter) (D, E, P)
Cupressaceae		
Juniperus monosperma (one-seeded juniper)	Evergreen	Good (D)
Juniperus scopulorum (Rocky Mountain juniper)	Evergreen	Good (D, E)
Cyperaceae		
Carex filifolia (threadleaf sedge)	Cool	Excellent (spring)
Carex geyeri (elk sedge)	Cool	Fair (D) to Good (E) (spring and fall)
Cyperus sp. (galingale)		
Eliocharis acicularis (spike-rush)		
Ericaceae		
Arctostaphylos uva-ursi (bearberry)	Evergreen	Good (D)
Fabaceae		
Daleo purpurea (purple prairieclover)	Warm	Excellent
Oxytropis sp. (loco)	Cool	Poor
Prosopis glandulosa (honey mesquite)	Warm	Poor to good (D)
Psoralea tenuiflora (slimflower scurfpea)	Warm	Poor (B)
Robinia neomexicana (New Mexican locust)	Cool	Fair to good
Fagaceae		
Quercus gambelii (Gambel's oak)	Cool	Fair to good (D)

(*continued*)

TABLE 3.5 *(Continued)*

Forbs and Woody Plants *(by Family)*	Season of Growth	Forage Quality, All Species
Gentianaceae		
Sweria sp. (gentian or elkweed)	Warm	
Iridaceae		
Iris missouriensis (wild iris)	Warm	
Lamiaceae		
Marubium vulgare (common horehound)		
Monarda pectinata (horse mint)	Warm	
Linaceae		
Linum sp. (flax)	Varies	
Loasaceae		
Mentzelia rusbyi		
(New Mexican evening star)	Warm	
Malvaceae		
Sphaeralcea coccinea (scarlet globemallow)	Warm	Excellent (D, E) (B)
Moraceae		
Humulus lupulus (hops)		
Onagraceae		
Oenothera (evening primrose)	Warm	Excellent (B)
Pinaceae		
Abies concolor (white fir)	Evergreen	
Picea pungens (blue spruce)	Evergreen	
Pinus edulis (piñon pine)	Evergreen	Good to excellent (D)
Pinus ponderosa (ponderosa pine)	Evergreen	Good to excellent (D)
Plantaginaceae		
Plantago patagonica (woolly plantain)	Cool	Poor to fair
Polygonaceae		
Eriogonum sp. (false buckwheat)	Varies	
Ranunculaceae		
Ranunculus sp. (buttercup)	Varies	
Rosaceae		
Cercocarpus sp. (mountain mahogany)	Cool/evergreen	Good to excellent (D) (winter)
Crataegus sp. (hawthorn)	Cool	Good
Fallugia paradoxa (Apache plume)	Cool	Excellent
Physocarpus monogynus (ninebark)	Warm	Fair
Prunus virginiana (chokecherry)	Cool	Good to excellent
Purshia tridentata (antelope bitterbrush)	Evergreen	Excellent
Rosa woodsii (wild rose)	Cool	Good (D, E)
Salicaceae		
Populis augustfolia (cottonwood)		Good
Salix sp. (willow)		
Saxifragaceae		
Ribes cereum (wax currant)	Cool	Fair to good
Scrophulariaceae		
Castelleja integra (Indian paintbrush)		
Penstemon barbatus (beard-tongue)		
Verbascum thapsus (mullein)	Warm	
Veronica americana (American brooklime)	Warm	
Typhaceae		
Typha latifolia (cattail)		

TABLE 3.5 (*Continued*)

Forbs and Woody Plants (by Family)	Season of Growth	Forage Quality, All Species
Urticaceae		
Parietaria pensylvanica (pellitory)		
Urtica gracilis (nettle)		
Verbenaceae		
Verbena macdougalli (verbena)	Varies	
Vitaceae		
Parthenocissus vitacea (creeper)	Warm	Fair to good

NOTE: Data on forage quality are ranges for a variety of wildlife (primarily artiodactyla) and, where no floral *species* is noted, for that genera of plant. Palatability and utility of the plant can vary by season. Where plants are especially important to particular animals, those are noted, using the following acronyms: bison (B), deer (D), elk (E), and pronghorn (P).

SOURCE: Floral data from A. B. Anderson (1975), Huckell (1997), Ivy (1995), Stubbendieck, Hatch, and Butterfield (1992), and Weber (1976); data on season of growth from Ivey (1995) and Stubbendieck, Hatch, and Butterfield (1992); data on photosynthetic pathway from Waller and Lewis (1979:table 1); data on forage quality from Elmore (1976) and Stubbendieck, Hatch, and Butterfield (1992); data on bison use from Peden et al. (1974).

The site area is quite unlike the relatively homogeneous grasslands of the High Plains to the east and north or the closed climax montane coniferous forests of the Sangre de Christo foothills to the west, though it contains individual elements of both of those communities, as well as others (table 3.5; also Anderson 1975:table 12). Because of this complex ecological structure, many plant species of different biota that are potential food resources are today more or less available to human or animal foragers within a few days' walk of the site.

Anderson (1975:50) argues that by virtue of this area's being an ecotone, it was a region rich in foods sources that human hunter-gatherers could exploit. To be sure, ecotone areas can have greater species richness than the biota that surround them (King and Graham 1981). But it is useful to qualify the point, for whether and how rich ecotones might appear, it does not necessarily follow that greater species richness entails a greater abundance of *food* resources. That obviously depends on the nature of the resources, as well as the foraging strategy. It is also important to bear in mind that as the number of species increases, the number of individuals per species decreases (MacArthur 1972; Preston 1948), which may further limit foraging options, depending on resource ranking.

Certainly, there is a variety of potential edible plants in the modern Folsom flora (table 3.5). Based on data on ethnographic usage compiled by Gilmore (1919; see also Anderson 1975; Kindscher 1987) these include yucca, nodding onion, arrowhead (which forms an edible tuber), squawbush, piñon, oak, sunflower, salsify, prickly pear, mint, chokecherry, and currant. Seeds of several of the grass species are potential food sources as well.[7] Of course, this list represents an aggregate of plant usage by a wide range of groups in different settings whose dietary demands and options, diet breadth, and historical knowledge and use of plants were different from one another's, and almost certainly from those of earlier Paleoindian groups. Thus, identifying a plant species in the Folsom area as a potential food source does not mean that it was recognized or utilized as such, even by recent groups. Grasses, for example, were generally low-ranked resources for many Plains ethnographic groups under most circumstances, and considered emergency or starvation foods only (e.g., papers in DeMaillie 2001).

Moreover, the use of these plant foods depends to a large degree on their availability and abundance and, more importantly, on the availability and abundance of higher ranked resources. Leaving the latter aside for the moment, quantitative data on the abundance of these potential plant foods are not available. However, it appears from field observations that piñon pine occurs only at a low frequency on the rocky slopes of Johnson Mesa. While the nuts seasonally produced by this tree might be a highly desirable food, as they are in many areas to many groups, harvest yields—as is well documented even in more extensive piñon stands—are often unpredictable, and periodic failure is not uncommon (Huckell 1998; Lanner 1981). Given the relatively scarcity of piñon trees in the region, and the narrow window of its harvest time, this would not be a reliable or predictable food source. Gambel's oak, in contrast, is far more abundant in the Folsom area, but it is likewise an unreliable and unpredictable food resource, with high processing costs (Reynolds, Clary, and Folliot 1970). None of the other potential plant food resources appear today in large stands. Of course, because this is an ecotone area, their abundance and distribution may be particularly sensitive to changes in climate, so their abundance today may not be representative of their abundance in the past; certain plants may have been more abundant, others less so.

Their abundance aside, relatively few of the edible plants in this area provide a significant or reliable return in fatty

TABLE 3.6
Temperature (°C) and Precipitation (mm) Ranges for Conifer and Hardwood Tree Species That Occur in the Folsom Area

Species	January lowest limiting temperature	Average annual temperature low end	Average annual temperature high end	July highest limiting temperature	Lowest annual precipitation	Highest annual precipitation
1. *Abies concolor* (white fir)	−10.7	−1.4	10.3	28	135	2,505
2. *Cercocarpus* cf. *breviflorus* (hairy mountain mahogany)	−1.0	7.3	20.7	30.8	245	640
3. *Juniperus monosperma* (one-seeded juniper)	−9.2	1.6	22	30.8	200	830
4. *Juniperus scopulorum* (Rocky Mountain juniper)	−17.9	−3.2	16.1	28.5	115	2,520
5. *Picea pungens* (blue spruce)	−12.2	−3.2	11.7	23.1	200	1,175
6. *Pinus edulis* (piñon pine)	−11.4	−0.6	17.8	30.3	135	830
7. *Pinus ponderosa* (ponderosa pine)	−14.7	−1.7	24.2	29.4	210	2,825
8. *Populis augustfolia* (mountain cottonwood)	−14.3	−3.9	18.1	28.3	145	1,175
9. *Prunus virginiana* (chokecherry)	−28.2	−5.5	18.7	27.6	155	4,370
10. *Quercus gambelii* (Gambel's oak)	−12.0	−1.5	21.7	29.3	95	1,175
11. *Robinia neomexicana* (New Mexican locust)	−6.6	3.4	18.1	28.4	150	690
Average	−11.84	−0.83	17.74	28.48	155	1,568.75
Standard deviation	3.4	2.29	4.93	2.37	43.26	889.09

SOURCE: Data from Thompson et al. (1999a, 1999b).

acids or carbohydrates for human foragers. Many can be used as herbs, to make tea, or to provide snack foods (Huckell 1998), but these are not sufficient to provide a viable subsistence backup, particularly during the critical winter and spring months when game populations are low and animals are fat-depleted (Frison 1982; Speth and Spielman 1983: 18–21; Todd 1991; Wandsnider 1997). This is not a place where one would expect human populations to overwinter (Wissler [1928] reached a similar conclusion, based on the harshness of the winter climate; see chapter 4). More generally, assuming that the relative paucity of plant resources in the modern vegetation does not reflect substantial historic modification of the landscape, such as clearing, logging, and livestock grazing, I suspect that the vegetation in this area did not and could not support substantial plant use by prehistoric hunter-gatherers.

Although this is an ecotonal area, there are underlying climatic controls on the vegetation. R. S. Thompson et al. (1999a, 1999b) plotted the geographic distribution of North American conifer and hardwood tree and shrub species and genera against a suite of climatic variables. Their study provides data on the full range—both geographic and climatic—of these taxa and, thus, illustrates the limiting factors that come into play for individual species. About a dozen of the trees from the Folsom area are so described by R.S. Thompson et al. (1999a, 1999b, 2000), and those species are listed in table 3.6, along with a few key climatic variables.

As can be seen in table 3.6, and graphically in figure 3.12, virtually all of these species' tolerance limits are comfortably within the range of temperatures and precipitation that characterize the area at present: The MAT values range from 6.7°C to 11.7°C (44°F–53°F; data from table 3.3) and most species can tolerate both lower—down to 3.4°C (38°F)—and higher—up to 24.2°C (35.5°F)—MAT values. A few species, however, are close to their limiting temperatures. For example, across its range blue spruce *(Picea pungens)* is limited to environments where mean July temperatures do not rise above 74°F (23°C); the mean July temperature in the Folsom

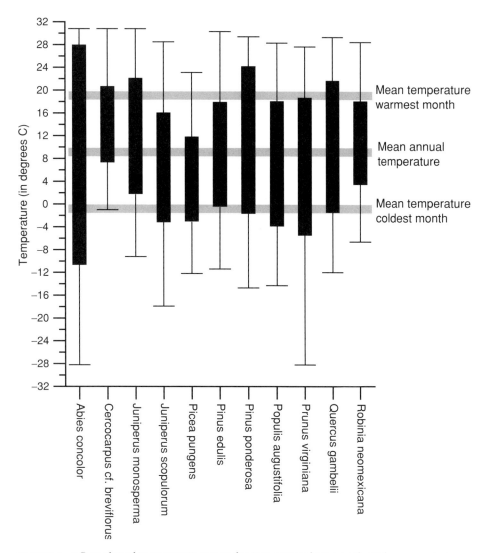

FIGURE 3.12 Box plot of temperature ranges for tree species living in the Folsom area today. (Data from Thompson et al. 1999a, 1999b, 2000.)

area hovers just below 21°C (70°F). At the other end of the temperature scale, mountain mahogany *(Cercocarpus* cf. *breviflorus)* and New Mexican locust *(Robinia neomexicana)* are close to their mean January temperature limits.

Were temperatures in the past significantly lower, as perhaps in Late Glacial times, or significantly higher, as, for example, in the Middle Holocene, records of past vegetation ought to reflect this in the disappearance or appearance of taxa able to tolerate those different climatic conditions. Looking ahead to chapter 6, if there was significant cooling in the region during the YDC, one would expect that the locust and one-seeded juniper *(Juniperus monosperma)* would not occur in the pollen or macrofossil record, but there would instead be trees that at present occupy cooler, higher-elevation settings, such as Englemann's spruce *(Picea engelmannii)* and bristlecone pine *(Pinus aristata)*.

In the Folsom region are both warm-season plants, primarily grasses, and cool-season plants, which include some grasses, but also forbs and shrubs, and evergreens, notably the woody plants and trees (table 3.5). Warm-season plants have a single cycle of growth in the summer when temperatures are highest, while cool-season forms complete most of their growth in the fall, winter, and spring (Stubbendieck, Hatch, and Butterfield [1992:2] and Bamforth [1988:32–34] discuss the growth and nutritive value of these respective forms).

As is well known, warm- and cool-season plants have distinct photosynthetic pathways (Fredlund and Tieszen 1997; Ode, Tieszen, and Lerman 1980:1304–1305; Waller and Lewis 1979:14). Cool-season grasses (e.g., genera within the Poeae, Stipeae, and Triticeae; table 3.5), most of the flowering plants, forbs, shrubs, and trees, all generate three carbon atoms as the first intermediate product of photosynthesis (the Calvin-Benson or C_3 pathway). In contrast, the warm-season forms, including the majority of xeric environment grasses, of which a number are present in the Folsom area, incorporate CO_2 into a four-carbon atom in photosynthesis (the C_4 pathway) (Chisholm et al. 1986:197; Fredlund and Tieszen 1997:table 1; Tieszen 1994:267–268; Waller and Lewis

1979:12–13). This difference is driven primarily by climate, including temperature, light, water availability, and humidity conditions, and secondarily by disturbance regime—primarily fire and herbivory (Fredlund and Tieszen 1997; Mole et al. 1994:316; Ode, Tieszen, and Lerman 1980:1304, 1310; Paruelo and Lauenroth 1996; Waller and Lewis 1979:13–14). An underlying forcing function is the amount of atmospheric CO_2; the modeled quantum yield for CO_2 uptake shows that C_4 plants are favored during times of lowered atmospheric CO_2 (Ehleringer, Cerling, and Helliker 1997; also Collatz, Berry, and Clark 1998; Cowling and Sykes 1999; Huang et al. 2001; cf. Nordt et al. 2002). This obviously has implications for Late Glacial times, a matter taken up in chapter 6.

With regard to the climatic controls, the growth of warm-season (C_4) grasses was initially thought to be correlated primarily with heat availability, particularly mean minimum July temperature. The warmer the temperature on summer nights, the more successful these species (see Teeri and Stowe 1976:4, 8–9; also Ehleringer 1978:265; Ehleringer at al. 1997; Ode and Tieszen 1980:1304–1305, 1310; Tieszen 1994:265). However, Paruelo and Lauenroth (1996) subsequently demonstrated that the MAP, as well as the proportion of precipitation falling in summer, may be a more significant influence in the occurrence of these grasses (Paruelo and Lauenroth 1996:1215–1217). Regardless, on a macro scale there is a latitudinal shift up the Great Plains in the relative percentage of C_3 and C_4 grasses, with C_3 species increasing as one moves north. The transition from C_3 to C_4 dominance occurs in the zone where daytime maximum temperature ranges are ~20°C to 28°C (68°F to 82° F), which corresponds to a present-day "crossover latitude" of 45°N (Ehleringer at al. 1997:288, 290; also, Connin, Betancourt, and Quade 1998). Recall that the Folsom site is located at ~37°N latitude.

There is a commensurate increase in the percentage of C_3 grasses with altitude (Teeri and Stowe 1976; also Steuter et al. 1995:764; Tieszen 1994:265), as a result of adiabatic cooling.[8] In turn, C_4 grasses are more abundant in the drier portion of the Plains adjacent to the Front Range and diminish in extent along the gradient of increasing precipitation toward the east (Paruelo and Lauenroth 1996:1218).

On the basis of the demonstrable relationship that exists between specific climatic variables, notably MAT, MAP, and SEA, and relative amounts of C_3 and C_4 grasses, Paruelo and Lauenroth (1996) developed a response function to provide a ballpark estimate of the percentage of C_4 grasses in a region.[9] Applying that function to climate values for Folsom (table 3.2) returns an estimate of ~52% C_4 grasses in the area today. Although merely a ballpark figure, and one subject to analytical uncertainty of the order of ±10% (see discussion in Connin, Betancourt, and Quade 1998:182), this estimate nonetheless provides a handy benchmark against which one can assess estimates of the relative amount of C_4 grasses in Late Glacial times as derived from paleoenvironmental evidence (chapter 6).

The distribution of C_3 and C_4 *sedges*, the other group in which C_4 forms are relatively common, does not appear to fol-low the clear latitudinal and altitudinal gradients to which the grasses conform, but shrubs do increase toward the west, along an approximate longitudinal gradient (Paruelo and Lauenroth 1996:fig. 7; Tieszen 1994:265–266).

The ratio of C_3:C_4 in a biotic community will also vary depending on soil types, salinity, and topography (Fredlund and Tieszen 1997; Mole et al. 1994; Tieszen 1994:267). Beyond changes across space, there are changes through time, in addition to the seasonal changes noted above. Thus, the relative composition of C_3 and C_4 plants in a biotic community will also fluctuate during warmer and cooler climatic periods.

C_3 and C_4 plants in turn produce different degrees of isotopic fractionation, with *average* $^{13}C/^{12}C$ ratios ($\delta^{13}C$) of $-26.5‰$ (parts per mil; which is the value derived as a molar ratio expressed relative to the PDB carbonate standard for C_3 plants), and $-12.5‰$ for C_4 plants. Those values will vary by species, year, season, precipitation, water stress, elevation, aspect, and topography, among other factors (Chisholm et al. 1986:197; Connin, Betancourt, and Quade 1998; Fredlund and Tieszen 1997; Mole et al. 1994:318–320; Tieszen 1994:268; Waller and Lewis 1979:13). As might be expected, within a biota like that at Folsom today in which both C_3 and C_4 plants are present, the net isotopic signature of the biomass will shift over the course of the year. The $\delta^{13}C$ values will be more negative in the spring, indicating a greater dominance by C_3 forms, will be less negative in the summer and into the early fall, reflecting the increase in the relative proportion of C_4 forms, and then will return to more negative $\delta^{13}C$ values in the later fall and winter (Ode, Tieszen, and Lerman 1980:1307–1308; Tieszen 1994:266–267). The duration and intensity of the C_4 summer "peak" can vary annually and over the longer term.

Those isotopic differences between C_3 and C_4 plants will be expressed in the bones, teeth, and tissue of herbivores that feed on them, though the absolute values will differ since the consumers further fractionate the carbon isotopes. In bison, this generally produces an upward (less negative) shift of ~5 ‰, though this can vary by skeletal or tissue element (Connin, Betancourt, and Quade 1998; Chisholm et al. 1986:197; Krueger and Sullivan 1984; Tieszen 1994: 273–275). These isotopic differences are of potential importance in the analytical effort to understand the ecology of the bison herd killed at the site and the paleoecology and paleoclimate of the Folsom region. They are discussed in more detail below and in chapter 6.

Modern Fauna

The present flora, while perhaps unable to support large hunter-gatherer populations, includes a wide range of grasses, forbs, shrubs, and woody plants attractive to a variety of birds and mammals, a summary listing of which is provided in table 3.7. This is an area presently abundant in game—a herd of several hundred elk, large flocks of wild turkeys, and families of bears, among other animals, move

TABLE 3.7
Mammals of the Folsom Region, Arranged by
Taxonomic Order

Lagomorpha
  Sylvilagus nuttali (mountain cottontail)
  Lepus californicus (blacktail jackrabbit)
Rodentia
  Eutamius minimus (least chipmunk)
  Eutamius quadrivittatus (Colorado chipmunk)
  Spermophilus tridecemlineatus (13-lined ground squirrel)
  Tamiascirus hudsonicus (red squirrel)
  Thomomys bottae (Botta's pocket gopher)
  Castor canadensis (beaver)
  Peromyscus maniculatus (deer mouse)
  Peromyscus boylii (brush mouse)
  Peromyscus difficilis (rock mouse)
  Neotoma mexicana (Mexican woodrat)
  Erethizon dorsatum (porcupine)
Carnivora
  Canis latrans (coyote)
  Canis lupus (gray wolf)
  Urocyon cinereoargentus (gray fox)
  Ursus americanus (black bear)
  Procyon lotor (raccoon)
  Taxidea taxus (badger)
  Spilogale gracilis (western spotted skunk)
  Mephitis mephitis (striped skunk)
  Lynx rufus (bobcat)
Artiodactyla
  Cervus elaphus (elk)
  Odocoileus hemionus (mule deer)
  Odocoileus virginianus (white-tailed deer)
  Antilocapra americanus (pronghorn)
  Bison bison (bison)

SOURCE: Data from Findley et al. (1975); additional information from Caire et al. (1989); Fitzgerald, Meany, and Armstrong (1994); and personal observations (1997–2000).

in the valleys and wooded mesa escarpments near the site, while antelope graze atop Johnson Mesa. The abundance of game is unquestionably driven by historical factors, principally the fragmentation and loss of habitats elsewhere; the absence of a large human population or nearby urban center; relatively low-impact land use practices and only minor modification of the local landscape; and the relative lack of human predation—except during the brief and tightly controlled hunting seasons.

Among the resident bird species are several birds, the largest of which is wild turkey *(Meleagris gallopavo),* but that also include quail, teal, and mallard. The birds feed on grass seeds, pine nuts (especially pinyon), juniper and other berries (e.g., bearberry), and currants. Seasonal avifauna include the Canadian goose.

The small mammal fauna, which forages on a wide variety of grasses and forbs, includes rabbits and rodents. Several of latter are burrowers that inhabit the immediate

site area today (e.g., *Thomomys bottae,* Botta's pocket gopher). Among the medium to large mammals, pronghorn antelope *(Antelocapra americana)* inhabit the grasslands on Johnson Mesa (Findley et al. 1975), where they feed on forbs and shrubs (including *Artemesia ludoviciana* [Cudweed sagewort], and various species of Asteraceae), juniper berries, sagebrush, and prickly pear *(Opuntia)* on occasion. Grasses constitute only a very small part of the pronghorn diet (Caire et al. 1989:366; Davis 1974; J. Jones, Armstrong, and Choate 1985:318; Stubbendieck, Hatch, and Butterfield 1992; Van Dyne et al. 1980:304–306).

Deer, primarily *Odocoileus hemionus* but also *O. virginianus,* and elk *(Cervus elaphus)* occupy the wooded and broken terrain in the site area. Deer forage on both woody plants (juniper, Gambel's oak, and piñon pine), as well as a variety of grasses and shrubs (table 3.5, also Caire et al. 1989:357; Stubbendieck, Hatch, and Butterfield 1992). Elk, which graze on grasses and forbs in summer and browse in winter, overlap somewhat with deer in their feeding strategies (table 3.5), though they also include in their diet several additional species *(e.g., Agrostis exarata* [Spikeben]) and *Carex geyeri* [Elk sedge]) and tend to browse less on woody plants, though they use them as cover (J. Jones, Armstrong, and Choate 1985:311).

Historically, the large fauna in the region also included *Bison bison* (Findley et al. 1975; Fitzgerald, Meany, and Armstrong 1994). Although no longer free-ranging, small bison herds have been pastured in recent times just below Trinchera Pass (fig. 3.13) and in the Dry Cimarron Valley downriver from the site. Bison were present in this area in prehistoric times (Anderson 1975:83, 99–100; Bailey 1931) including, of course, in the Late Glacial.

Carnivores in the region include coyote *(Canis latrans),* fox *(Vulpes* sp.), black bear *(Ursus americanus),* racoon *(Procyon lotor),* badger *(Taxidea taxus),* and bobcat *(Lynx rufus),* all of which have been observed by field crews within 10 km of the site (also Caire et al. 1989; Findley et al. 1975). These animals feed on small mammals, birds, insects, occasionally young and/or enfeebled larger mammals, and sometimes carrion (see discussion in Caire et al. 1989:278, 298, 302, 312, 325, 334). Several of these carnivores also include in their diet juniper berries, prickly pear fruit, grass seeds, acorns, and roots (Caire et al. 1989:278; 302; J. Jones, Armstrong, and Choate 1985).

Although none have been seen in the area in recent times, the gray wolf, *Canis lupus,* was almost certainly present in historic times. Historically and in Late Glacial times following the extinction of the megafaunal carnivores *(e.g., Panthera leo),* wolves would have been—along with humans—the major predators of bison (Caire et al. 1989:282–284; Guthrie 1990:287–289). When bison numbers plummeted in the late nineteenth century, cattle became the prey of choice, leading to the extirpation of wolves by ranchers.

Given the relative abundance of potential game species and the relative scarcity of high-yield plant foods, it seems

FIGURE 3.13 Trinchera Pass bison herd, Johnson Mesa in the background. (Photo by D. J. Meltzer.)

reasonable to infer that this was an area in which the primary subsistence activity was hunting, not gathering. There is archaeological evidence to support this inference. Anderson's detailed surface survey of the Dry Cimarron Valley documented some 74 sites, most of which were relatively small lithic scatters less than 1,000 years old (Anderson 1975). Only eight of those sites were classified as localities in which plant acquisition and processing took place, based on the presence of nearby stands of plant food resources and the occurrence of grinding stones—the latter otherwise being rare in the archaeological record of the region (Anderson 1975:144, 183, table 23). In addition, the total material recorded from all 74 sites was a mere 2,087 artifacts, of which 345 (16.5%) came from a single rockshelter (Anderson 1975:21, 79). Of those 2,087 objects, 108 were projectile points (Anderson 1975:appendix B). These results suggest that the prehistoric human use of this area was rather ephemeral (Anderson 1975:4, 128, 183–184, table 23), mirroring earlier observations of Schwachheim and the results of the survey conducted by Gerhard Laves for the American Museum in 1928 (chapter 4).

Assuming that those survey results do not reflect sampling biases in the archaeological record, it corresponds with the climate-based projections (Binford 2001) that forager activities consisted largely of hunting and other more limited subsistence pursuits, with only a minor contribution of gathering activities—at least in late prehistoric times. This supposition is supported in ethnohistoric accounts of the use of this region by Jicarilla Apache and bands of the southern Ute, particularly the Kapota (or Capotes) and Mowatsi (Mouache or Muhuachi), both of whom made occasional hunting forays

into the region (Simmons 2000:9; Tiller 1983b:441; see also Gunnerson 1974; Opler 1936; Tiller 1983a). There are ample reasons to suspect that a focus on hunting was true when Folsom groups were in this area as well, a matter taken up in later chapters.

Bison Diet and Its Isotopic Implications

Given the central importance of bison to the Paleoindian occupation of the Folsom site, and the paleoecological potential of an isotopic analysis of their skeletal remains, a brief excursion into bison diet is useful. The forage consumed by bison in the Folsom area in historic times is not known precisely. However, there is little doubt that bison are grazers, as was apparent in the earliest accounts of bison foraging (discussed by Larson 1940) and has since been confirmed by many field studies on the North America Plains and elsewhere, including in shrub-steppe communities at high elevations in the Great Basin. Bison consume primarily grasses, and often particular species of grasses. Less prominent in the diet are sedges, with forbs and browse generally comprising less than 10% of the diet, despite the abundance of these plants on the landscape (e.g., Guthrie 1990:175–176; Hudson and Frank 1987:72; Knapp et. al. 1999:41–43; Peden 1976; Peden et al. 1974:492–495, 497, table 4; Plumb and Dodd 1993:634, fig. 2; Steuter et al. 1995:760, fig. 3; Van Dyne et al. 1980:301, 303, 458; Van Vuren 1984:261; Vinton et al. 1993:13).

Because of their body size and substantial forage requirements, bison have evolved to become relatively efficient foragers, by adjusting their bite size and foraging time to available biomass and moving over large areas to acquire sufficient food. They have digestive systems that enable them to do well even in circumstances where forage is in limited supply (e.g., Belovsky and Slade 1986:59; Guthrie 1990: 155–156, 178–179; Hudson and Frank 1987:75).

The nature of the grasses consumed by bison changes seasonally, reflecting the seasonal cycles of grass guilds: Bison utilize cool-season (C_3) grasses from the late fall through the late spring and then warm-season (C_4) grasses from late spring increasingly through the summer and fall (Chisholm et al. 1986:200; Peden et al. 1974:table 2; Plumb and Dodd 1993:634–635; Steuter et al. 1995:761–763; Tieszen 1994:275; Vinton et al. 1993:15–16). Because of their reliance on grass, bison suffer a particularly high mortality in winter when grasses are least available and least nutritious or buried under snow (Bamforth 1988:75–76; Roe 1951:180–203).

Those seasonal changes can be readily monitored in tissue and feces of modern bison (e.g., Steuter et al. 1995:761; Tieszen 1994:fig. 11) but may not be detectable in skeletal material, modern and fossil, which tends to "average" out the seasonal contributions of C_3 and C_4 sources (Steuter et al. 1995:759; Tieszen 1994:275). The parts of the skeleton with more finite and/or seasonally specific growth periods, such as annular structures in teeth and horns (Tieszen 1994:275), or bone collagen versus bone structural carbonate (Jahren, Todd, and Amundson 1998:474), may provide the potential for detecting seasonal variance (chapter 6).

In some areas, warm-season grasses can comprise from 40% to 80% of the available forage (cf. Peden 1976; Plumb and Dodd 1993:634); recall that the estimate for the Folsom area is ~52%. The relative mix of warm- and cool-season grasses in bison diets on the Plains varies along several dimensions, including competition, fire history, forage quality, season, and patch composition and diversity (e.g., Chisholm et al. 1986:199; Knapp et al. 1999:42–43; Tieszen 1994:266, 276; Vinton et al. 1993). But on a macro scale the key dimension of variability is latitude; in this, of course, bison diets mirror the latitudinal shift in the relative percentage of C_3 and C_4 grasses on the landscape (Steuter et al. 1995:764; Tieszen 1994:277–278).

On mixed landscapes, bison are not as highly selective in their diets as other Plains mammals, such as pronghorn (Schwartz and Ellis 1981). Yet, bison can show some selectivity in their feeding under certain biotic circumstances (Larson et al. 2001; Peden et al. 1974; H. W. Reynolds, Hansen, and Peden 1978). Of the warm-season grasses utilized on the Great Plains, two species—buffalo grass (Buchloe dactyloides) and blue grama grass (Bouteloua gracilis)—when present often dominate the diet (e.g., Peden 1976:228; Steuter et al. 1995). Other grasses and forbs that are also components of bison diet on the Plains are bluestem (Andropogon gerardii), red threeawn (Aristida longiseta), and dropseed (Sporobolus cryptandrus); during the cooler months the diet includes sage (Artemesia frigida), sedge (Carex heliophila), and cool-season grasses such as wheatgrass (Agropyron sp.) and needle-and-thread grass (Stipa comata) (Peden 1976:227; Peden et al. 1974:493; Steuter et al. 1995:761; Vinton et al. 1993:16; also Caire et al. 1989:372). For bison living in the far Northern Plains or in higher-elevation areas where grasses, particularly warm-season grasses, are rare or virtually absent, sedges are often the primary component of the diet, with the available grasses having only a secondary role, even in seasons or topographic settings when or where those grass species are available (e.g., Hudson and Frank 1987; H. W. Reynolds, Hansen, and Peden 1978:586–588).

Because bison are large and mobile animals, they can also gain access over the course of a year to forage and habitats not otherwise locally available (Bamforth 1988:83–84; Chishom et al. 1986; Hanson 1984:102–103; Moodie and Ray 1976; Morgan 1980; Roe 1951:520–600). Whether that movement was systematic and under what climatic or other conditions it took place are not known, and are the point of considerable historical debate; see, for example, the long diatribe by Roe (1951:520–600). Nonetheless, there are tantalizing clues that there was a method to their mobility, most notably the possibility that some movement at an unspecified scale, and depending on latitude, was a seasonal response to temperature changes. One scenario envisions that in the late spring and summer bison foraged on the open Plains, where they had access to greening C_4 plants, and then abandoned the Plains in the fall for wooded parklands or foothills settings to the west, where vegetation cover and/or topographic features provided shelter from the elements during the coldest months. Without question, habitat selection for these large animals can be an effective form of thermoregulation (Belovsky and Slade 1986:58–59).

Yet, by most historic accounts, such movement was not an intrinsic element of bison foraging, and under different conditions, for example, during mild winters, movement may have occurred to a lesser degree or not at all (Bamforth 1988:83–84; Chisholm 1986:203; Hanson 1984:110; Moodie and Ray 1976:49–51; Morgan 1980:156; Roe 1951:194, 533). This is not surprising, since physiological and anatomical studies show that bison can cope with extremely cold temperatures (e.g., Christopherson and Hudson 1978; Christopherson, Hudson, and Richmond 1976; Telfer and Kelsall 1984). Their thick and dense coats provide high insulation, but these animals are also able to reduce their metabolic rates during especially cold periods when, say, temperatures fall below −30°C, in order to conserve energy during times of potential food scarcity (Christopherson, Hudson, and Richmond 1976:52; Christopherson and Hudson 1978).[10]

However, if strong winds are also present during periods of extremely cold temperatures (e.g., lower than −30°C), the bison hair coat is disrupted and its insulating properties are reduced, leading to a dangerous loss of insulation and body heat (Christopherson and Hudson 1978:41). Moreover bison, compared with other large North American ungulates, such as caribou, moose, Dall sheep, wapiti, bighorn sheep, whitetailed deer, and antelope, rank near the bottom of the list— just above antelope—in their ability to cope with snow.[11] This ranking was based on a series of morphological attributes, such as chest height and foot loading, and behavioral characteristics, as well as the ability to meet nutritional requirements with forage found above the snow, digging or rooting behavior, and techniques of locomotion (Telfer and Kelsall 1984:table 3; see also Guthrie 1990:200–202, Roe 1951:180–203).

Perhaps in some areas, then, the colder Northern Plains being the most obvious, there is reason to expect seasonal east-west movement of bison herds. That area is, in fact, where the majority of the historical and largely anecdotal accounts of such seasonal movement originate. However, there is also some supporting isotopic evidence in this regard (e.g., Chishom et al. 1986:201–203; Jahren et al. (1998:474).

These apparent seasonal movements of bison have been incorporated by Amick (1996, 2000) into models of Folsom Paleoindian settlement mobility on the Southern Plains. Such models must assume, of course, that human foragers kept close tabs on bison herds, which is possible but not demonstrated (chapter 1). More importantly, these models must assume a seasonally predictable movement of bison, animals whose mobility patterns are known to vary in complex ways and that are not always strictly seasonal, even on the Northern Plains, where the climatic conditions would favor such movements. Bison may not be seasonally mobile at all in relatively warmer areas like the Southern Plains (Bamforth 1988:83–84). In any case, the matter is testable, at least in principle, by examining isotopic signatures ($\delta^{13}C$) in bison bone.

Doing so raises the complicating question of whether the animals were foraging regionally or locally; that is, if their diets prove to be dominated by C_4 grasses, how might it be determined whether such were present locally or acquired at greater distances by these highly mobile animals? The answer might be had by examining isotopic signatures in other, less mobile herbivores, such as land snails, as these provide a very local record of plant species. This is a matter explored in chapter 6.

But if that issue is resolved in the positive—i.e. it can be demonstrated that bison isotopic values can reflect local vegetation—there remains a second complicating issue. That is, if bison are selective feeders, preferentially foraging on certain types of grasses in a mixed setting, say, C_4 as opposed to C_3 grasses, then their isotopic signatures may accurately reflect not the relative proportion of those grasses on the landscape, only their selectivity. It may not be possible to resolve this issue in the case of an extinct archaeological fauna. But that may not be a significant concern in this instance. As Larson et al. (2001:51–52) argue, there is good reason to assume that one of the circumstances under which bison will be less selective and more generalized in their feeding—what they dub the "bison as lawn mowers" model—would be in the case of an aggregated cow-calf herd. This is so because foraging competition attendant with aggregation will force animals to consume whatever forage is at hoof. As it happens, the animals at Folsom were part of a cow-calf herd (chapter 7), and in the absence of evidence to the contrary, I make the a priori assumption

that the isotopic signatures of these animals can serve as a useful proxy for the location vegetation (chapter 6). In any case, independent data from snails and soil carbonates can serve as a check on these data and assumptions.

Historic and Modern Land Use Patterns

The Folsom site is situated in a rural area of northeastern New Mexico. Outside of the small highway service towns and a nearby tourist destination (Capulin Volcano National Monument), this is cattle ranching country. Farming is rare, owing to the short growing season, low precipitation, and unpredictable timing of freezes. Most agricultural activity, beyond hayfields and home-produce gardens, tends to be located close in or alongside the Dry Cimarron River. Those farms are relatively small. The cattle ranches, by contrast, tend to be large, often of the order of 20,000 to 40,000 acres.

The land use practices of the area are reflected in its population numbers. This area of northern New Mexico is sparsely populated. In fact, Colfax County, in which the site is located, is one of only five counties in New Mexico to have fewer people at the end of the twentieth century than it had at the beginning of the century (data are from 26 of New Mexico's 33 counties—the remainder either were not organized in 1910 or otherwise lack data [www.unm.edu/~bber/demo/ctyshist.htm]). Two of the other counties with populations that shrank through the twentieth century are the adjoining Mora and Union counties.

The demographic history of the town of Folsom mirrors this general pattern: Its population in the 1900 census was 1,091 individuals, which had dropped to 967 individuals in 1910 (census data from Folsom 1888–1988). However, over the next three decades the population decline became precipitous—by the 1940 census there were only 360 residents. As of the 2000 census, the tally was down to just 75 individuals (www.unm.edu/~bber/demo/cityhist.htm).

These data are relevant to any discussion of the archaeological record of an area, for surface visibility and site discovery often correlate strongly with land use patterns and population (Lepper 1983, 1985; Meltzer 1987; Meltzer and Bever 1995; Shott 2002). Given the lack of farming and regular plowing of the soil, and the dearth of people walking the landscape, it comes as no particular surprise that the area has not produced the richness or density of sites seen elsewhere as, for example, on the High Plains to the east.

Still, the relative scarcity of archaeological sites in this area may not be just an artifact of poor exposure and few eyes monitoring the landscape, given the several survey efforts, some extensive, that have taken place over the years, which have yielded very few sites and small numbers of artifacts. This is so despite the fact that surface visibility is reasonably good, even in the absence of land clearing and tillage. The upland surfaces are only thinly draped with sediments. If sites were present on them, they ought to have been visible, although, of course, preservation of organic remains in such settings would necessarily be poor. The lowland areas alongside drainages can contain thick deposits of relatively recent sediments (Mann 2003, 2004), and thus there may be a bias against finding sites in that setting.

On balance, however, it seems reasonable to infer that the scarcity of archaeological sites in the area is not merely a by-product of insufficient sampling. Although prehistoric groups may have frequented the area, possibly en route elsewhere, very few apparently lingered for long periods of time in the area or engaged in the kinds of activities that produce a rich and dense archaeological signature.

What Has Been Learned from the Modern Site Setting . . .

At the outset of this chapter, I declared that a discussion of the site setting could be more than just an exercise in providing a sense of place. It's now time to pay that note. Let me do so by first reviewing some facts, and inferences drawn from facts, that have emerged from the overview of the modern setting.

1. By virtue of its Plio-Pleistocene volcanic history, this is an area of great topographic variability.
2. Because of this topographic variability, this is an ecotonal area of considerable floral diversity.
3. Yet, despite that floral diversity, there are apparently few plant species sufficiently abundant or predictable, or that could provide a critical contribution of carbohydrates and essential fats, to support long-term occupations by hunter-gatherers.
4. The flora of the region, however, can and does support large game populations, including a variety of ungulate species such as bison, deer, and elk.
5. The area has a generally dry climate but can experience heavy winter snowfalls at higher elevations.
6. Summers in the region are generally cool, but winters can be very cold—again, especially at higher elevations.
7. The preferred forage of bison, warm-season C_4 grasses, is present in the local flora.
8. The area may not be suitable as a bison overwintering ground because of the cold, snowy winters.
9. Because of the abundant game resources and the scarcity of plant resources, this is an area in which hunting, and not gathering, was likely the primary activity of human foragers.
10. This is not an area that supplies vital resources like abundant lithic raw material, rich plant foods, and copious water sources, and hence it is not an area in which one expects either repeated site occupations, or long-term occupations. With regard to the latter, if the primary activity is hunting, and there is a scarcity of plant foods, hunting pressure will lead to resource depression, which tends to trigger high residential mobility (see discussion by Kelly 1995:135–138).
11. More specifically, because of the scarcity of plant foods and the cold and snowy winters, which would potentially reduce game population numbers and/or badly deplete their fat reserves, this is also not an area in which one expects human foragers to overwinter.
12. This is, however, an area that by virtue of the macro-scale regional drainage pattern, and the local occurrence of mountain passes, may have been a potential "corridor" for hunter-gatherer groups traveling between the Southern Plains and the foothills of the Rocky Mountains.

. . . And What This Suggests of Late Glacial Environments and Adaptations

The results of the several archaeological surveys done to date support the inference that prehistoric human use of the area was ephemeral and relatively specialized, primarily focusing on hunting. Whether there was a seasonal pattern to the recent prehistoric use of the area is unclear from the evidence at hand. The more important question, of course, is: What relevance might this have for efforts to understand Late Glacial environments and adaptations?

In a very narrow, substantive sense, none at all—unless one assumes a priori that the climatic and environmental conditions of the Late Glacial are the same as those of the present. And one should not make that a priori assumption, especially in light of the evidence from nearby areas of cooling during the YDC and concomitant changes in plant communities (Fall 1997; Reasoner and Jodry 2000).

However, this review does highlight several important structural aspects of the environment: notably, its topographic and ecological diversity, which strongly influence the character of the available resources. Second, for much of the last several thousand years, and under conditions similar to the present ones, this has been an area of relatively limited use potential. Such would strongly suggest that if conditions in the deeper past were even less favorable to plant and animal species than those of the present, then the potential for human use of the area is diminished accordingly. Finally, this review helps identify some of the critical limiting factors of the regional climate and environment, the most obvious and potentially relevant being winter temperature and snowfall.

Consider, for example, what might have obtained if climates during the YDC were cooler than at present, but also assuming (though perhaps unrealistically) that all other climatic variables are at present values. Under such conditions vegetation zones would have been lowered, which would have put greater selective pressure on or regionally eliminated plant species that are currently on the margins of their ecological tolerances, and increased effective moisture by reducing evaporation. Changes such as this would have led to greater dominance of boreal elements and the growth of woodland at the expense of grassland. If some of the plant species eliminated from the local biota included the preferred forage of bison or elk, assuming that their forage preferences in Late Glacial times were as they are at present, that may have reduced the use of this habitat by these animals—and bison, in particular, almost certainly would have avoided the region in winter. The abundance of game animals in the region would have declined and their presence become less reliable and less predictable.

From that, one would further suppose that Paleoindian activity in the region would have been correspondingly reduced, since plant resources suitable for humans would have been no more abundant in Late Glacial times than they are at present and return rates on available plants and animals would have declined more rapidly. Human use of the area under these circumstances would have been highly seasonal in nature, perhaps restricted to the late spring through the early fall. In effect, there would have been even less reason for foragers to be in the area in the Late Glacial than there was in later prehistoric times, and all the more reason to consider this region merely a way station en route to someplace else. Under these circumstances, the Folsom site might have been the product of little more than a chance encounter of hunters and bison—as, perhaps, it was.

But consider an alternative scenario, which can be derived by assuming cooler temperatures overall, but this time adding the further possibility that Late Glacial climates were also drier than at present and possibly, though not necessarily, subject to periodic drought (Holliday 2000a). If that was the case, vegetation zones may still have been lower, but that shift may have been accompanied by a decrease in arboreal elements and an increase in parkland. That, in turn, may have increased faunal richness and abundance, allowing for a wider variety of game species, perhaps including lingering Pleistocene megafauna, as well as large herds of grazers in open parklands. However, because winters could have been as severe as at present, animals may still not have been available for long periods of the year, and bison would not likely overwinter in the area. Human foragers thus would have had greater foraging opportunities in summer but had even less incentive to overwinter here.

Testing and eliminating one or both of these alternatives, or any other possibilities that one can easily conjure by changing some of the critical variables, will require considerable attention to the paleoenvironmental data and evidence. And, as should be apparent from the alternative scenarios just given, it is not just a matter of whether the Folsom area was cooler in Late Glacial times, though it probably was (Fall 1997; Reasoner and Jodry 2000; Vierling 1998), or even drier, as it possibly was (Holliday 2000a). It is also a matter of whether (1) there were similar seasonal patterns in temperature or precipitation; (2) whether climates were more continental or equable—a matter that ties back to the possibility that bison were fat-depleted for shorter

periods of the year (Todd 1991); (3) whether the growing seasons were of comparable length; (4) whether the forage quality and period of availability were greater or lesser; (5) whether the ecological structure, richness, and diversity of the vegetation were the same; (6) whether the preferred forage of specific animals was available in sufficient abundance and, if not, whether those animals utilized other plant species or simply avoided the area; (7) whether the dietary preferences of modern species are even applicable to different Late Glacial species; (8) how complex the faunal community was, and whether animals were present on the landscape at an abundance similar to that in the present; and, finally, (9) how potentially different competitive relationships/niche partitioning among the fauna might have affected game distribution and predictability, to identify just some of the issues.

Not all of these matters may be resolvable given presently available data or methodologies, but they highlight the issues and questions that should be kept in mind in searching for and analyzing paleoenvironmental data and evidence, as discussed in the following chapters.

Notes

1. Anderson (1975:41) provides a slightly different age and laboratory number, and a significantly different standard deviation, for this same radiocarbon sample: 22,350 ±110 ^{14}C yr B.P. (TX-1266).

2. As that name can be a source of some confusion, let me clarify here that the *Dry* Cimarron flows through the town of Folsom, thence generally eastward out of New Mexico. When the river crosses into Oklahoma some 90 km distant, it becomes the Cimarron River, which continues east, turns north into far southeastern Colorado, then east into Kansas, ultimately finding its way south again, where it becomes a tributary of the Canadian River. The reason the headwaters section of the Cimarron River is designated the *Dry* Cimarron while still in New Mexico is to distinguish it from the much smaller Cimarron River, which flows out of the Cimarron Range of the Sangre de Christo Mountains, past the towns of Cimarron and Springer, New Mexico, where it joins the headwaters of the Canadian River. Both of the New Mexico Cimarron Rivers, then, are tributaries of the Canadian, and though they come within 100 km of each other while still in New Mexico, they join the Canadian at very different points along its traverse.

3. The Purgatoire is the modern, shortened version: the original name was El Rio de las Animas Perdidas en Purgatorio, which translates as "The River of Lost Souls in Purgatory," in honor of a renegade band of Spanish adventurers abandoned by their own priests and killed by natives at its headwaters in the late sixteenth century (Lavender 1954:9). The French transformed the name to Purgatoire, which was subsequently mangled by Americans into "Picketwire," which is how it was designated in some nineteenth-century accounts (e.g., Gregg 1844).

4. Goodnight used Trinchera Pass to save time, since Raton Pass required traveling farther west than he needed to go, and to save money, as Raton Pass was by then a toll road.

5. The data for this table were obtained for individual weather stations from the National Climate Data Center web page (www4.ncdc.noaa.gov), downloaded as digital ASCII files, and then converted into EXCEL files for descriptive and analytical purposes. The data obtained were those provided in the Summary of the Month Cooperative series (NCDC 3220). Within that series, figures on a dozen climate elements were downloaded (not all of which are used in table 3.2): DPNP (departure from normal monthly precipitation); DPNT (departure from normal monthly temperature); DT00 (number of days with minimum temperature ≤0°F); DT32 (number of days with minimum temperature ≤32°F); DT90 (number of days with maximum temperature ≥90°F); DX32 (number of days with maximum temperature ≤32°F); FRZD (freeze data with dates of various freeze thresholds); MMNT (monthly mean minimum temperature); MMXT (monthly mean maximum temperature); MNTM (monthly mean temperature); TPCP (total monthly precipitation); and TSNW (total monthly snowfall).

6. Frost-free dates were transformed into Julian dates for purposes of deriving average dates of last and first freezes.

7. A number of the plant species in table 3.5 have medicinal value (such as sagewort, gumweed, gentian, globemallow, hops, plantain, and wild rose), while others are suitable as material sources, including the pines, cottonwood, willow trees, and cattail (Kindscher 1987). But my concern here is strictly with food resources.

8. There is a cooling of ~3°F/1.66°C for every 1,000-ft (300-m) gain in elevation, which is very roughly equal to the cooling that occurs as one moved northward ~100 miles (160 km) of latitude (MacArthur 1972:8–9).

9. The formula is as follows: % C_4 grasses = −0.9827 + 0.000594MAP + 1.3528SEA + 0.2710logN MAT; where MAP is mean annual precipitation, SEA (labeled as JJA/MAP by Paruelo and Laurenroth [1996]) is the proportion of precipitation falling in summer (JJA = June, July, August) relative to the annual precipitation, and MAT is mean annual temperature. Plains-wide values put the strength of the relationship at r^2 = 0.66, F = 44, df = 3,69, $P < 0.0001$ (Paruelo and Laurenroth 1996:fig. 3).

10. In contrast, cattle increase their metabolic rate in response to cold below temperatures of −15°C. In a direct comparison, bison calves proved to be as tolerant to cold at 6 months of age as cattle were at 13 to 17 months of age (Christopherson and Hudson 1978:40–41).

11. Daubenmire (1985), in fact, uses their inability to cope with deep snow as an explanation for why bison were not known historically from areas west of the Rocky Mountains. While suitable vegetation could be found there, and in proper seasons, he argues that heavier snowfall would limit the animals' ability to survive in winter. It might be noted that ancestral forms of bison did inhabit those apparently restricted areas (e.g., MacDonald 1981: 250ff).

FOUR

Archaeological Research Designs, Methods, and Results

DAVID J. MELTZER

The investigations conducted at the Folsom site in the 1920s, and again in the 1990s, were very different in their goals, scope, design, and execution. Given the very different questions being asked across the decades separating these investigations, that is not surprising. Not should it come as a surprise that the more limited goals of the original investigations (chapter 1) were matched by a more limited research design or that there was never a published discussion of the strategies guiding that fieldwork, the methods and techniques used, and the nature and potential biases of the data that resulted. Such was the nature of paleontological fieldwork in the 1920s.

Yet, it is vitally important to understand the strategy, methods, and techniques used during those original investigations, because the decisions made in the course of that work fundamentally influenced the character of the artifact and faunal data that were recovered, established parameters for what might be gained from a reanalysis of those remains, and determined what remained of the site and where. The work done in the 1920s was a natural and necessary point of departure for the field- and laboratory work we undertook in the 1990s.

Therefore, the first part of this chapter reconstructs the largely implicit 1920s research design and how it was carried out. This reconstruction is accomplished using the correspondence between Jesse Figgins and Carl Schwachheim (as well as Fred Howarth) from 1926 to 1927 and between Barnum Brown and Peter Kaisen in 1928. Important details are also gleaned from Schachheim's diary (appendix B) and, to a lesser degree, from the spotty field notes kept by Frank Figgins (Jesse's son) in 1926 and Peter Kaisen in 1928 (none were kept in 1927). These archival materials are supplemented by the few maps—mostly sketch maps—made over the several years of excavations (e.g., Schwachheim to Figgins, September 4, 1927, September 29, 1927; F. Figgins fieldbook, July 1926, DIR/DMNS), a plan map made in 1928 by the AMNH, and photographs taken during those years and sporadically in the years and decades that followed.

The second part of this chapter addresses the research strategies and tactics used over the several seasons of our field and laboratory investigations, aimed at addressing the larger theoretical questions and analytical goals, detailed in chapter 1, that guided our reinvestigation of the site. These research questions were translated into specific strategic objectives that guided the yearly fieldwork, but the tactics were adjusted each season according to our changing understanding of the site from the prior seasons' results, as discussed below. Although the order of the discussion is chronological, this is intended not to be a detailed narrative of the work but, rather, to explain how the research design and, to a degree, the goals themselves evolved in response to our increasing understanding of the site.

A detailed discussion of the specifics of our field techniques, grid and coding systems, excavation procedures, etc., can be found in appendix A. As noted in chapter 1, such discussions can be rather dry stuff but are critical to enable readers who are so inclined to evaluate the integrity of the data and evidence and the foundations on which the inferences and conclusions are based.

The Colorado and American Museum Investigations

The 1926 and 1927 Seasons

Once the requisite permits and permissions were in hand, the 1926 excavations began in May of 1926 on the south bank of the arroyo, in what we now recognize as the western side of the bonebed. A large skull was exposed in the South Bank wall of the arroyo already, so Figgins' first instruction to Schwachheim was to dig some "prospect holes" well away from the bank, to gauge the depth and extent of the bonebed south of the arroyo. Knowing how far in fossils extended would make it possible to "clear the ground for the

FIGURE 4.1 Plaster-jacketed bison crania in camp, ca. 1926/1927. (Photo courtesy of Carl Schwachheim Collection.)

recovery of the fossils in an orderly way" (Figgins to Howarth, May 10, 1926, July 24, 1926, DIR/DMNS).

Excavation was slow going initially, hampered by a week of rain and snow and the oak trees that grew on the surface (Schwachheim to Figgins, May 24, 1926, DIR/DMNS; things were going so badly that Schwachheim admitted he'd thought of "jumping the job" [quitting]). The oaks were particularly a problem, since there seemed to be more bones beneath them than downstream in more open areas. Figgins advised Schwachheim to remove the dirt from the clear space adjoining the oaks and then excavate in and underneath them, for it would be far less trouble to undermine them at the level of the excavation (Figgins to Schwachheim, May 27, 1926, DIR/DMNS). His plan was to have Schwachheim remove the overburden down to the level of the bison bones and then "cover them with dirt and wait for Harold [Cook]" to arrive to show him how best to get the bones safely out of the ground (Figgins to Schwachheim, May 27, 1926, DIR/DMNS). Given the fragile condition of some of the bone and his own lack of experience, that suited Schwachheim (Schwachheim to Figgins, June 11, 1926, DIR/DMNS).

By late May, the weather had cleared, the site had dried out, and with the help of Lud Shoemaker from the nearby Crowfoot Ranch, who brought over his mule team, plow, and Fresno, Schwachheim had cleared "a pit about 20 × 40 [feet] from the stumps south and [had] it down about 3 ft" (Schwachheim to Figgins, May 24, 1926, DIR/DMNS). Once the excavation area was cleared by the mule team, the digging was done with pick and shovel (fig. 2.9). There is no evidence in any of the photographs or archival sources that screens were used in the excavations, which is not unexpected, given contemporary paleontological methods. Although Schwachheim was instructed to "save every scrap of bone" (Figgins to Schwachheim, May 10, 1926, DIR/DMNS), such was clearly impossible.

By early June, Schwachheim was down about 6 ft below the surface, and by mid-June Cook had arrived and the first of the bison bones were being uncovered. Ice picks were used to expose individual bones. Exposed parts of the bone were soaked with shellac cut with alcohol. The bones and surrounding sediment were taken up in blocks and encased in burlap and Plaster of Paris for shipment to Denver (fig. 4.1).

One particularly auspicious element of Figgins' strategy was the instruction to "carry . . . the dirt away from the creek, not into it" (Figgins to Howarth, July 24, 1926, DIR/DMNS). As a result, much of the overburden and backdirt from the 1926 excavations was not washed down the arroyo but still remains on-site—not all of it, however. When Cook arrived on site to help Schwachheim remove the skull, he countermanded Figgins' instructions. As Figgins complained to Howarth:

> You doubtless recall my original plan was to clear a considerable area of the overload, carrying the dirt *away* from the creek, not into it. Then when Carl had gotten it worked to near the bone layer, send someone to aid him in the removal of the bones, or so instruct him that he could proceed with the work alone. Carl worked under exceedingly difficult and adverse weather conditions, but was getting there according to instructions until Harold came and completely reversed the order of things. I had told Harold to let the north bank alone. He put Carl to cutting into it and filling up the creek. I tried to impress him with the importance of clearing away that ridge between the excavation and the creek and leave the bones largely undisturbed until we were ready. Clean up as we went along. . . . Harold could tell me nothing except how he had cut into the [North] bank to expose the skull and other bones; how soft they were; and to assure me the dirt would be taken out at the first flood. . . . In other words, I feel the necessity for saying Harold has rather messed things up. (Figgins to Howarth, July 24,1926, DIR/DMNS)

Whether or not it was part of the initial plan, within two weeks of Cook's visit, Frank Figgins was sent down from Denver to supervise the work and help with shipping the material to Denver (he came with the Colorado Museum truck—a boon to Schwachheim, who had been hauling his water over from the Crowfoot Ranch [Schwachheim to Figgins, July 9, 1926, DIR/DMNS]).

By the time Frank Figgins arrived, approximately 100 skeletal elements had already been recovered. He set about organizing and cataloging the material for shipment, individually numbering each plaster-jacketed block. A block might include several bones found together but not necessarily articulated or even from the same animal. Some provenience information was appended to the description, but it was minimal—partly because specific provenience was not being recorded, and because for the first 22 blocks F. Figgins "had to depend on Carl's memory for location." Even so, he thought "H.J.C. [Cook] can probably give additional [information] on them" (F. Figgins, field notes, DIR/DMNS). The information provided for Lot No. 20 is typical: "20. Humerus-Sternum Atlas-Lumbar Vert. Toe Bone Foot Bone, Fragments—2 Ribs, 2 Leg Bones, Lower Jaw Pelvis West Center [of the excavation area]" (F. Figgins, field notes, DIR/DMNS).

The bones recovered through early July were primarily postcranial elements, and while some of these were found close together, apparently few were articulated. After Frank Figgins arrived, they uncovered and plastered the long-exposed skull on the edge of the bank, and just a couple of feet from that an articulated segment of cervical vertebrae, which included the atlas and "five dorsal vert" (Lot No. 23A). Most important of all, "about 2 [inches] from centrum" of that vertebral section was an "arrow head" (F. Figgins, field notes, DIR/DMNS; this is specimen DMNS 1391/3; see table 8.2 for description). As Schwachheim recorded that day in his diary:

Found part of a broken spear or large arrow head near the base of the fifth spine taken out. It is about 2 inches long & is of a dark amber colored agate & of a very fine workmanship. It is broken off nearly square & we may find the rest of it. I sure hope we do. It is a question which skeleton it was in, but from the position of them it must have been in the skeleton of the smaller one & just inside the cavity of the body near the back. It was found 8$\frac{1}{2}$ feet beneath the surface with an oak tree growing directly over it 6 inches in diameter, showing it to have been there a great length of time. (appendix B: July 14, 1926; also F. Figgins, field notes, 1926, DIR/DMNS).

Unfortunately, of course, it was not found in situ. As reported in chapter 2, although Figgins sent word for Schwachheim and his son to be more careful (Figgins to Howarth, July 22, 1926, DIR/DMNS; Figgins to Brown, July 23, 1926, VP/AMNH), no points were found in place over the remainder of the field season (Figgins 1927:232; Schwachheim to Figgins, August 9, 1926, DIR/DMNS). However, near skeletal Lot No. 60 the blade of another

projectile point was found in the loose dirt (specimen DMNS 1261/1A; see table 8.2 for description), and a small wedge-shaped midsection fragment was found adjacent to a rib in that block (F. Figgins, field notes, DIR/DMNS). When the block was later cleaned at the CMNH, the point proved to refit the small fragment that had been recovered next to the rib (Figgins 1927:fig. 4).

The excavations continued to uncover bison bone, and by mid-October when the field season finally wound down, F. Figgins had tallied an additional 44 blocks of skeletal elements, including at least another 250 individual specimens, most of them postcranial elements, but also whole and fragmentary skulls from half a dozen adult and young bison.

A sketch map drawn by F. Figgins indicated that five separate clusters of bone were left in the sidewalls of the excavation. As best can be determined from his sketch maps, which have no scale, and show the "original" excavation area as being more rectangular than Schwachheim's verbal description of it as an area "20 × 40 feet" (6.09 × 12.19 m), it appears as though the 1926 excavations opened an area of approximately 3.65 × 3.65 m (12 × 12 ft) or about 13.4 m^2 (F. Figgins, field notes, DIR/DMNS). His notes also indicate that late in the season an additional prospect hole was dug 30 ft south of the excavation area, presumably to see if bones extended that far into the South Bank.

The 1927 fieldwork began in late May. Frank Figgins had recommended extending the excavations into the North Bank. However, Schwachheim—assuredly more aware than anyone else how much effort would be required to remove the deeper overburden on that side, and who certainly knew who would be doing most of the hard shoveling[1]—suggested they stay on the "same side of the ditch." There was, he assured Jesse Figgins, plenty of room to expand the excavations, and much less dirt to move (Figgins to Schwachheim, December 16, 1926; Schwachheim to Figgins, December 23, 1926, DIR/DMNS). In the end, Schwachheim's suggestion was followed: The 1927 excavations stayed on the South Bank.

The 1927 crew was comprised of Schwachheim and, for much of the season, Floyd Blair. An individual named Bob is referred to in Schwachheim's diary entries for mid- to late June (appendix B); who he was and how long he may have worked there is not known, but he appears in photographs taken in late August. Frank Figgins was not present, and as a result no field notes were kept for this season or, if kept, do not survive.

Unlike the previous season, Schwachheim and crew were on the lookout in 1927 for artifacts with the bison bone (chapter 2). Being on that watch does not appear to have influenced the excavation strategy and techniques, as these did not seem to have changed from the previous season. However, the attention required to catch artifacts in place may have slowed the pace of the work.

As before, the heavy overburden initially had to be cleared ("I have been polishing a pick handle since my arrival here in the removal of the overhead dirt" [Schwachheim to Figgins, June 4, 1927, DIR/DMNS]). But this season, only an additional area of 20 × 20 feet (~6.1 × 6.1 m) was apparently opened, perhaps because there were still portions of the block from the previous season to be excavated, and Figgins thought that would be all the exposure they would need (Figgins to Schwachheim, May 6, 1927; Schwachheim to Figgins, June 4, 1927, DIR/DMN). The 1927 excavations ultimately extended farther into the South Bank and along the arroyo both up and downstream, undercutting by nearly 2.5 m the scrub oak that covered the surface (fig. 2.9). This removal of overburden—much of which was massive clay of the McJunkin Formation (chapter 5)—and the exposure of the bonebed took place from late May through July.

Photographs taken by Schwachheim that summer provide some additional information about the excavation techniques. Specifically, and not surprisingly, horizontal units and vertical levels were not used. As a result, the surface of the excavation area was irregular, presumably contoured to the bones as they were encountered. If their excavations went down through the bonebed before stopping, which seems likely, and as the bonebed is relatively tightly constrained vertically (chapter 7), it seems reasonable to infer that little remained below the floor of their excavation. Certainly Schwachheim considered the area "worked out," once they had dug through the bone level (Schwachheim to Figgins, September 4, 1927, DIR/DMNS).

It was not until August 29 that the first in situ artifact appeared, this time between a pair of bison ribs (fig. 1.4). It was located at the upstream (western) end and corner of the excavation block, less than 2 m in from the edge of the arroyo, and underneath a large overhang of sediment and gravel bound tightly together by a thick mass of oak roots, according to photographs and a sketch map made by Schwachheim. Following Hrdlička's advice, telegrams were sent announcing the fact and inviting visitors to examine the material in place (chapter 2). But before the contingent of visitors arrived, two more points were found: one in some loose dirt apparently close to and just south of the in situ specimen, and the other from backdirt (Schwachheim to Figgins, September 4, 1927, DIR/DMNS).

After the visitors left, Schwachheim carefully removed the block with the point and bones in place and shipped it to Denver for display at the Museum (Schwachheim to Figgins, September 11, 1927, DIR/DMNS).[2] He then turned to the matter of closing up the operations at the site. Figgins directed Schwachheim to put in a prospect hole on the North Bank "down in the ditch coming in from the north" (Figgins to Schwachheim, September 22, 1927, DIR/DMNS). Based on a sketch map drawn by Schwachheim, the unit was situated ~11.6m (~38 ft) back from the edge of the arroyo. Placed in the "shallow branch east of camp," a drainage that ran north of Schwachheim's tent (appendix C), the test unit was "3 ft by 6 ft" (0.91 × 1.82 m) and reached "14 feet [~4.3 m] deep from surface of bank" (Schwachheim to Figgins, September 29, 1927, DIR/DMNS). It did not produce any bison remains or artifacts. Figgins later realized that Schwachheim had gone 20 ft too far to the north—an error for which Figgins

himself assumed responsibility, since he had trouble orienting himself at the site (Figgins to Schwachheim, October 11, 1927, DIR/DMNS). Figgins was not alone: Howarth and Schwachheim's sketch maps were routinely off by as much as ~90° in their cardinal directions (e.g., Schwachheim to Figgins, September 11, 1927, DIR/DMNS).

Following that last excavation, Schwachheim collected sediment samples from each of the major stratigraphic units they recognized. This was apparently done in order to reproduce a section through the formation (Figgins to Schwachheim, October 1, 1928, DIR/DMNS). The present whereabouts of those samples are unknown. The 1927 field season ended in late September (appendix B).

All together, the 1926–1927 excavations of the Colorado Museum removed ~34.7 m^2 of the bonebed, in an area that extended laterally along ~11 m of the south bank of the arroyo and about 3.8 m into the south bank at its eastern end.[3] There is no evidence—save the photographs and sketch maps by Frank Figgins and Schwachheim—to indicate precisely where the artifacts and bison bones had been found.

The 1928 Season

PALEONTOLOGICAL EXCAVATIONS

The 1928 field season was in principle a joint effort by the Colorado and American Museums (Brown to Taylor, February 6, 1928, DIR/DMNS; see also Cook to Brown, February 17, 1928, March 30, 1928, April 22, 1928; Brown to Cook, March 29, 1928, April 5, 1928, Brown to Howarth, March 22, 1928, April 2, 1928, VP/AMNH; Granger to Figgins, September 27, 1927, DIR/DMNS). In practice, it was mostly the American Museum's show, a matter that caused some hard feelings in later years.

The 1928 excavations began in early June and ended in early October. The goals were to expand the sample of artifacts and bison remains and resolve more precisely the age of the site. That involved excavations far more extensive than they had been the previous two seasons. Tactically, the aim was to encircle and then remove the entire bonebed. As Brown instructed Schwachheim that spring:

> . . . Make several [prospect holes] in the direction the bones seem to run—determined from previous work, so as to determine the entire area of bones. Then cut scrub brush sufficiently to permit excavation of entire area. I think we should plan to uncover all of the bone bearing layer. Next, secure the services of one or two farmers and start the excavation. I think you can use plow and scraper safely down to within two feet of the bone layer. (Brown to Schwachheim, April 9, 1928, AMNH Archives; also Brown to Cook, April 27, 1928; Howarth to Brown, May 16, 1928, VP/AMNH; also Brown 1928b).

Four such prospect holes were dug in a semicircle around the southern edge of the South Bank and labeled Pits A

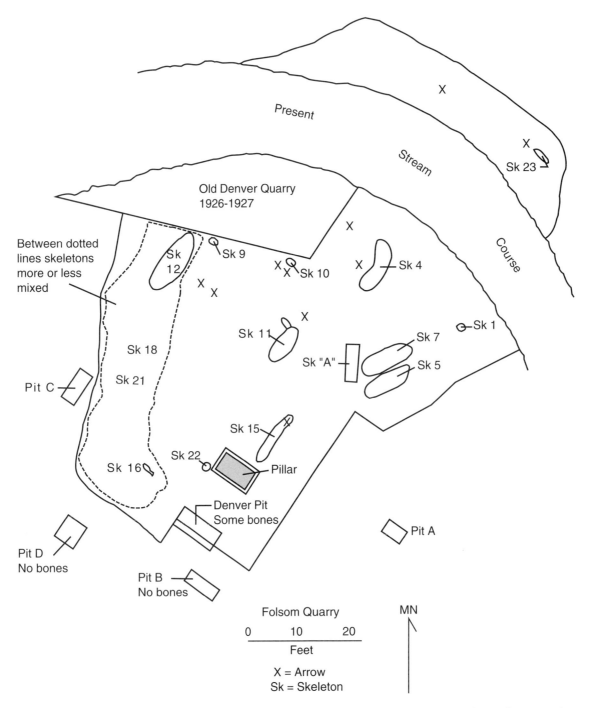

Present

Stream

Course

X

X
Sk 23

Old Denver Quarry
1926-1927

Between dotted
lines skeletons
more or less
mixed

Sk
12

Sk 9

X X
X Sk 10

X

Sk 4

X

X

X

Sk 11

X

Sk 7

Sk 1

Sk 18

Sk "A"

Sk 5

Pit C

Sk 21

Sk 15

Sk 22

Sk 16

Pillar

Denver Pit
Some bones

Pit A

Pit D
No bones

Pit B
No bones

Folsom Quarry

MN

0   10   20

Feet

X = Arrow
Sk = Skeleton

FIGURE 4.2 The American Museum of Natural History 1928 excavation plan map. (By D. J. Meltzer, after original courtesy of American Museum of Natural History.)

through D (fig. 4.2). We subsequently relocated two of those, and re-excavated one of them: AMNH Pit C. It proved to be 0.92 × 1.5 m (3 × 5 ft) in area and was dug to a depth of 3.2 m (10.5 ft) below the 1928 surface. The other pits were presumably of comparable size and depth.

The previous summers' work had "consisted on taking out the bones that were easiest to secure," so Brown was alerted at the outset it would be necessary to do "considerable team work and removal of the overlying earth before it will be possible to remove many bones" (Howarth

to Brown, March 27, 1928, VP/AMNH). Fortunately for Schwachheim, the excavation crew was larger and he did not have to shoulder as much of the excavation burden as he had in the previous years.[4] Peter Kaisen, one of the American Museum's (and Brown's) long-time excavators, was on hand to supervise the fieldwork (fig. 4.3), and they were joined by several others: Kaisen's son Ernest, Glen Streeter, and Gerhardt Laves, a University of Chicago anthropology graduate student sent to the site by Clark Wissler.

FIGURE 4.3 The 1928 Folsom field crew. From left to right: Emma Howarth (site visitor), Peter Kaisen, Glenn Streeter, Ernest Kaisen, Gerhardt Laves, Carl Schwachheim. (Photo courtesy of Carl Schwachheim Collection.)

To maintain horizontal and vertical control in the excavation area—control that was good for its time but minimal by today's standards—Brown wanted the position and depth of bison bones and artifacts measured from a fixed datum point. As there were no mapping instruments on-site, vertical measurements had to be recorded by their depth below surface. That presented a problem, however, since much of the surface was going to be removed by the excavations. Hence, Brown instructed Kaisen to leave an undisturbed pillar of earth "four or five feet in diameter" in the middle of the site as a datum (Brown to Kaisen, June 7, 1928, VP/AMNH; Brown 1928b). Kaisen was not particularly keen to have to maneuver the excavations around the pillar, and he tried resisting the idea:

> Plans work out near as outlined. Prospect holes [have] no bones. Will strip plot about sixty by sixty. *Have to leave pillar in center. If you insist will do it.* (Kaisen to Brown, June 12, 1928, VP/AMNH; emphasis mine)

Brown insisted (Brown to Kaisen, June 15, 1928, VP/AMNH). Still, Kaisen managed to circumvent, if only slightly, Brown's demand, by putting the pillar off-center and toward the southern edge of the excavations.

A possible clue as to how the measurements were taken came from the fill of AMNH Pit C, which contained a cobble

of volcanic tuff, 10.8 cm in maximum length and 4 cm in maximum thickness, on one face of which was engraved a circle 4.65 cm in diameter, with a "crosshair" incised within it (fig. 4.4). There was a dimple at the intersection of the crosshairs, suggesting that perhaps this specimen was placed on the floor of the excavation at a spot where a measurement was being taken and used to center a plumb bob being held above it.

The overburden in the excavation area was removed by "plow and scrapers down to a short distance above the bone layer," with the "final prospecting being done with small tools so that the smallest object could not escape the searchers" (Brown 1928b). It is appropriate to add, however, that the "small tools" were certainly smaller than plows and scrapers but—as in previous years—involved picks and shovels, with the expected results: "We have found to date 9 broken points. Oh! Yes, one was a fine one, but Ernie struck it with a pick breaking it ... " (Schwachheim to Brown, August 10, 1928, VP/AMNH). Fortunately, most of the artifacts could be pieced back together (chapter 8).

As before, the overburden was pushed by Shoemaker's mule teams (fig. 1.5) south and away from the arroyo onto the upland surface, ultimately forming a semicircular berm that still rings the southern portion of the site (fig. 4.5), though much diminished from its 1928 height. A small portion of the backdirt, however, appears to have been pushed

FIGURE 4.4 Possible "mapping stone" found In the fill of AMNH Pit C, 1998. (Photo by D. J. Meltzer.)

FIGURE 4.5 Looking north from the backdirt berm on the southern edge of the site, 1928. Note the pillar amidst the South Bank excavation area, the vertical face of the North Bank, and the AMNH field camp in the trees beyond. (Photo courtesy of American Museum of Natural History.)

FIGURE 4.6 American Museum crew in camp, preparing bones for shipment, 1928. Peter Kaisen is on the far left, partially obscured by the brush, Carl Schwachheim is in the center (under the black hat), and Barnum Brown Is on tha far right. (Photo courtesy of Carl Schwachheim Collection.)

to the sides of the excavation not then being worked—and possibly thrown into the arroyo bottom:

> The heavy rain we have been waiting for to wash out the creek has not come so it look [sic] like the shovel will have to do the work. (Kaisen to Brown August 19, 1928, VP/AMNH)

If this is so, there has obviously been loss to erosion of material both down the arroyo and within the perimeter of the berm.

Once the bison bones were exposed, they were shellacked and/or plaster jacketed, then hauled out of the arroyo over to the camp—which in 1928 was set in the trees on the north slope of the valley. The archival records indicate that much of the preparation for the shipment of the bones took place in camp (fig. 4.6), as confirmed by our metal detector survey (appendix C; fig. C.3), which located on the edge of that camp scores of nails and small sections of bailing wire from the shipping crate construction.

Unlike the 1926–1927 seasons, when only rough sketch maps were made, in 1928 the general provenience of skeletal remains and artifacts was plotted on a plan map of the site (fig. 4.2). However, as we subsequently discovered (see below), this map is inaccurate, possibly because (as appears from clues in the archives) it was constructed at the Museum after the field season, and apparently without field checking.

Nonetheless, it is the only formal map of the site that exists, and with it and Schwachheim's diary, it is possible to reconstruct which points were found when, where, and with which clusters of skeletal remains (chapter 8) and how the 1928 excavations moved through the site. Those excavation began in June adjacent to the arroyo (fig. 1.5) and extended south by southeast along the eastern edge of the 1926–1927 excavations. Some of the bones on this east side of the site were in poor condition (chapter 7), but the excavations yielded at least two fluted points, one found near the "No. 4 skeleton" (appendix B). However, by the first week of July, Kaisen gave up on this area:

> I am going to shift over to the west side of the quarry as we are nearly at the limits of the bone layer on the east side and some of the bones are too bad to collect. (Kaisen to Brown July 8, 1928, VP/AMNH)

During the first two weeks of July, crews were clearing the remaining overburden on the west side of the site, and no artifacts were found. But during the last two weeks of the month seven fluted points were recovered, all but one in place; the exception was found by Brown in the backdirt. Three of those points were exposed by July 23, 1928, upon which Brown sent out over a dozen telegrams inviting visitors to see the points and bone in place (chapter 2). Over the next two weeks, as visitors came in response to the broadcast, the three points were viewed in situ.[5]

The majority of the points found in 1928 were on the west side of the bonebed, close to the Colorado Museum's 1926–1927 excavation area. The higher frequency of points and bison bone from this side of the site might provide an important clue to the use of the landforms by Folsom hunters (chapters 5 and 7).

By mid-August, the "quarry work" on the South Bank was deemed complete. Excavations then shifted to the North Bank (Kaisen to Brown, August 29, 1928, VP/AMNH). The crew realized relatively quickly that faunal material in that area apparently extended "only a short distance into the bank." Even so, they excavated 3 ft beyond the last bones seen and, seeing no more, were satisfied that "the entire deposit has been exhausted" (Brown 1928b).

> One hour['s] work and all the bones in cut in north side of creek will be up. I got [a] pair of jaws, about 12 ribs, 2 humeri, femur, tibia, and perhaps 12 vert. The bones are good. On the west side stairway you know some bones was in sight. I got one jaw, one vert, and a metapodial. No more bones in sight anywhere. (Kaisen to Brown, September 1, 1928, VP/AMNH)

Brown was surprised there was so little bone on the North Bank (Brown to Kaisen September 4, 1928, VP/AMNH), but then he was not fully aware of the very different geological context of that area of the site (chapter 5).

As part of the work on the North Bank, the face was "cut straight down all along for 34 feet (10.36 m) east of stairway" (Kaisen to Brown, September 1, 1928, VP/AMNH). This might have been at the behest of Hrdlička, who had asked Brown that spring if it would "be possible to make and photograph a complete section of exposure of the wall, from say 20 feet to the left to 20 feet to the right of where the fossil bones and artifacts occur" (Hrdlička to Brown, March 15, 1928, AH/NAA). It was possible, and the bank, as Kaisen put it, "shows the strata fine," and he reported getting some "fine photos" (fig. 4.7). While some of the photographs survive (see also chapter 5), it is not known what became of the sediment samples evidently removed from the bank in copper boxes (Brown to Kaisen, August 28, 1928; Kaisen to Brown, August 29, 1928, September 1, 1928, September 8, 1928, September 15, 1928, VP/AMNH). Wissler also took sediment samples when he visited the site in August, but these cannot be located either (Wissler field diary, August 7, 1928, ANTH/AMNH).

With the anticipated completion of the North Bank excavation, there was discussion of possibly returning to the South Bank to plow the "upper end" of the bonebed, perhaps even on the far side of the backdirt to see what might occur there (Kaisen to Brown, August 29, 1928; Brown to Kaisen, September 4, 1928, VP/AMNH). Kaisen was not convinced the effort was worth it, and no further excavations were made. As he declared:

> No bones goes in anywhere in the banks of the quarry. Looks like we got around the Indians buffalo hunt. Well, I think we got a good collection with 11 arrows and all those [bones]. I am sure nothing are left worth anything when we leave. (Kaisen to Brown, August 29, 1928, VP/AMNH)

FIGURE 4.7 American Museum crew on ladders on the freshly-cut vertical face of the North Bank in 1928. Carl Schwachheim is on the ladder to the right. (Photo courtesy of American Museum of Natural History.)

As noted (chapters 1 and 2), that claim would be repeated in later months and years (Brown 1928; also Howarth to Figgins, October 12, 1928, DIR/DMNS; Brown to Figgins, February 1, 1929, DIR/DMNS; but see Cook to Jenks, March 31, 1929, HJC/AGFO).

Ultimately, the 1928 excavations cleared an area that Brown estimated in his annual report as 80 × 60 ft (24.3 × 18.2 m) or ~442 m² (Brown, American Museum fieldwork, Folsom, NM, 1928, VP/AMNH). This is a much larger excavation area than is shown on the 1928 plan map of the site. Using that map, Todd and Hofman (1991) estimated the 1928 excavation area as ~233.7 m². Obviously, the map and/or Brown's estimate is incorrect.

When the AMNH crew left in late 1928, the southern perimeter of the excavation was ringed by the semicircular berm of backdirt, which may have extended across the northern edge of the excavation area as well. The South Bank excavation area was gently sloped south-north and was relatively level east-west. The perimeter walls of the excavation were vertical for ~40 cm to 50 cm up from the bonebed floor, and then angled gently toward the upland surface above. Positioned toward the southern half of the excavation area on the South Bank was the AMNH datum pillar, which stood roughly 1 m to 1.5 m above the excavation floor, based on the photographs (figs. 2.14 and 4.5). The North Bank was left vertically faced along its entire length.

Having finished at the site itself, in mid-September Howarth led the AMNH crew to a hillside south of the town of Folsom to excavate a lava cave set into the flows associated with nearby Capulin Mountain (Kaisen to Brown, September 5, 1928, September 17, 1928, VP/AMNH). It was a large cave, extending some 30 m underground, with a vault 6 m wide and 1.2 to 2.1 m high (estimates from Laves [1928]; he identifies this locality as Cave 5). As Brown saw it, the cave was their "only hope of dating the Bison quarry," for their brief testing there in July had yielded

> . . . a Bison tibia and we should get some Bison jaws. Most of these bones and jaw appear to be recent, but there was a sloth jaw and a camel metatarsal. (Brown to Kaisen, September 4, 1928, VP/AMNH; also Brown to Kaisen, August 27, 1928, September 10, 1928, VP/AMNH)[6]

By Brown's instructions, excavations in the cave started at the entrance, worked inward, and extended down into sterile sediments below the lowest bone. They worked for over a week in the cave (appendix B). However, the material was "not prolific, and soon played out fifteen feet from the entrance" (Brown to Simpson, October 1, 1928, VP/AMNH). Nothing of archaeological interest was found (Laves, field diary, August 3, 1928, ANTH/AMNH).

While the crews were traveling from their camp to the cave via town, they hauled to the Folsom railroad depot the 35 wooden crates that had been built to ship the 8,300 lb of plastered bison bone from the site. The process of hauling the crates took several days. Once the boxes were all together at the Folsom depot, they were shipped to New York. Kaisen asked the obvious question: "Bones of course to go by freight C.O.D. What value do you want put on the shipment?" (Kaisen to Brown, September 5, 1928, VP/AMNH). What value, indeed? The field crew broke camp at the beginning of October 1928 (appendix B; Brown to Simpson, October 1, 1928, VP/AMNH).

ARCHAEOLOGICAL SURVEY AND EXCAVATIONS

The spring of 1928, Brown met with Clark Wissler of the Department of Anthropology at the American Museum of Natural History to discuss collaborative work at Folsom. Wissler had a longstanding interest in the problem of human antiquity, and both of them were keen to have an archaeological survey done "of the surrounding country for possible shelters of the ancient Folsom people" (Brown to Howarth, April 7, 1928, VP/AMNH). While in Chicago in March, Wissler had met Gerhardt Laves, a graduate student at the University of Chicago who was "specially trained in archaeology" and came highly recommended by Fay Cooper Cole (Wissler to Sherwood, April 17, 1928, ANTH/AMNH). Laves was hired to conduct the survey (Wissler to Laves, April 17, 1928; Laves to Wissler, April 23, 1928, ANTH/AMNH).[7]

Wissler detailed Laves' duties in a long memorandum. They included careful surface survey of the 6 mi^2 around the site, as well as into the surrounding country. Particular attention was to be paid to exposures along the arroyo walls, but also along the "rocky rim" of the valley, where "one may expect shelters and camp sites." At the Folsom site itself, Laves was to follow the instructions of the paleontologists in charge but be prepared to assist in the recovery of any artifacts or human remains. If the find was "merely a matter of additional chipped points found in association with a bison skeleton," he need not do any more than examine and photograph the find.

The goal of the summer's work was not to excavate but rather to explore, with the aim of locating as many sites as may exist that might warrant subsequent excavation (Wissler to Laves, April 17, 1928, ANTH/AMNH). Wissler well appreciated, however, that once on the ground, the nature of the task would become clearer, and Laves might need—and should feel free—to make "radical changes in the program" (Wissler to Laves, May 7, 1928, ANTH/AMNH).

In the end, Laves mostly followed Wissler's plan. The arroyos, however, were not in his view worth the investment of time and effort, since the sediments and few bones he came across appeared to be recent, and any older material would not be in situ. On Brown's recommendation, he devoted more of his energy to searching for shelters and overhangs on the mesa rim, in the hopes of finding older materials (Laves to Wissler, July 3, 1928; Laves, field diary, July 1, 1928, ANTH/AMNH).

Unfortunately, there were few areas with caves "worthy of shelter" (Laves to Wissler, July 15, 1928, ANTH/AMNH; also Laves 1928). Ultimately, Laves found only three in the general vicinity of the site: one just off the eastern edge of Johnson Mesa (Cave 1) along the Folsom to Raton road (now New Mexico Highway 72); another along Trinchera Creek north of Trinchera Pass (Cave 2); the third in a small gully on a mesa behind the Owen house (now the Hereford Park Ranch) (Cave 3). Excavations in Caves 1 and 2 yielded just a few artifacts and some human skeletal remains. Cave 1 had "charcoal beds of some extent." But all three were otherwise "devoid of significant specimens" (Laves 1928) or, at least, ones that might be tied to the occupation at the Folsom site.

Casting his net farther afield, again on Brown's recommendation (Laves to Wissler, July 3, 1928, ANTH/AMNH), Laves also spent a few days exploring Oak Canyon, some 15 km east and south of the Folsom site. While pictographs were located, and a small shelter test was excavated, he concluded that work in this area was also "not likely to yield the remains of any great antiquity, except by chance" (Laves 1928; Laves, field diary, August 14, 1928, ANTH/AMNH).[8]

Brown had recommended that Laves talk to the "natives" and examine their artifact collections, which he did when the branding season brought many of them together. But even with the offer to pay for specimens, he came away with few good leads to sites and only two projectile points, neither of which were of the Folsom type (Laves to Wissler, July 15, 1928,

ANTH/AMNH). In fact, as he reported, Carl Schwachheim—who knew the region well and had been collecting there for some time—"declares that this basin is very barren of Indian remains despite its natural attractiveness" (Laves to Wissler, July 3, 1928, ANTH/AMNH; also Laves 1928).

At the end of Laves' three-month survey there was, as Wissler himself admitted, little to show for the effort. Still, Wissler thought the work worthwhile for demonstrating the "general scientific result, namely, that the region was never permanently inhabited" (Wissler to Laves, September 13, 1928, ANTH/AMNH). The probable cause for that lack of habitation, he supposed, was that "the winter climate is inhospitable." Even so, it was a good place to hunt at other times of the year, since the constant rainfall on the Mesa "would have made fine summer pasture for bison" (Wissler 1928).

There were tentative plans to expand the archaeological survey the following summer off to the east and north (Wissler to Laves, December 4, 1928, ANTH/AMNH), but these were never realized. However, Wissler was not going to give up the search for ancient human remains. Instead, he was going to broaden it. As he explained to Figgins the following spring:

> . . . My intention is not simply to hunt for points of the Folsom type, but to make a thorough archaeological survey of the whole strip from Laramie to southern Texas. Of course this will take several years to complete. . . . (Wissler to Figgins, April 4, 1929, ANTH/AMNH)

Wissler was an optimist. No such survey was completed then, or since.

Fieldwork at Folsom, 1929–1996

What the Folsom site had not yielded was "the habitation site of the Folsom bison hunters" (Jenks to Cook, April 29,1929, HJC/AGFO). Puzzled over where it might be, Jenks inquired in 1929 about the possibility of doing more work at the site. As a professional courtesy, he asked for and received permission to do so by both the Colorado and the American museums but, in the end, chose not to go to Folsom (Jenks to Figgins, February 18, 1929, DIR/DMNS; also Brown to Figgins, February 1, 1929, February 19, 1929; Figgins to Brown, January 26, 1929, February 11, 1929, DIR/DMNS).[9]

It is unclear what influenced Jenks' decision. Brown's confident assertion that his crews had exhausted the bonebed may have played a role, but it is also true that both Brown indirectly and Cook directly assured Jenks that there might still be material at the site, and where to find it. As Brown explained:

> By all means grant Dr. Jenks permission to work in the Folsom quarry—we will not undertake any further work there. There is just a possibility that Dr. Jenks might secure a few more bones and, of course, there is always a possibility of additional arrowpoints on the west side of the quarry. We thought we had exhausted the quarry, but there were a few bones close to where work stopped on the west side. (Brown to Figgins, February 1, 1929, VP/AMNH)

The point was echoed by Cook:

> In regard to the quarry itself. While it is a gamble, I am confident that more material can be had there. On the upstream [west] side where work stopped, bones were numerous almost to the edge of the present cut, and there is nothing to indicate that the floor cuts off at that point. On the north, deep side, the difficulty, I should say, is principally one of moving a heavy overburden, and, so, expensive. I would not plan on doing an enormous amount of work there in one fell swoop. (Cook to Jenks, March 31, 1929, HJC/AGFO)

While Jenks would ultimately choose not to follow their suggestions, they would prove valuable clues in guiding our own excavation efforts seven decades later (below).

Over the years of excavation, there had been talk of the importance of mapping the site area.[10] But it was only in the summer of 1931, three years after the excavations were over, that a detailed topographic map was made of the site and its surroundings by surveyor Robert Merrill, at Clark Wissler's request and with Wissler's help in the field (Merrill to Wissler, October 26, 1931, ANTH/AMNH). Produced with a transit-stadia, and done at a scale of 1 in = 20 ft, with contour intervals of 1 ft, the map is reasonably accurate. Merrill used the 1928 pillar to represent the mean ground surface, which he assigned an elevation of 7,100 ft above sea level. Unfortunately, because of the erosion of the site in later decades and the lack of permanent landmarks in the area (or shown on Merrill's map), it is difficult—though not impossible—to overlay his map on current maps of the landscape (the effort is made below).

For his part, Howarth occasionally monitored the site and lamented the fact that "there is an occasional bone exposed but it soon goes as every month or two someone passing thru [sic] visits this place and of course takes anything in sight or destroys it so that no one else can get it" (Howarth to Cook, July 25, 1933, HJC/AGFO; on other ad hoc collections, see Cook to Sellards, November 10, 1931, HJC/AGFO; Howard to Brown, January 7, 1937, VP/AMNH; Howarth to Cook, November 6, 1931, February 26, 1932, July 25, 1933, HJC/AGFO). Indeed, over the decades, the site was visited by an untold numbers of individuals, some of whom are known to have recovered projectile points and/or bison bone from the site. In 1934 E. B. Howard found the base of one fluted point that he believed joined with a tip found seven years earlier; it did (Howard 1943:228; chap. 8). In the 1950s, the then-Caretaker of Capulin Volcano National Monument collected several fluted points from the site (chapter 8). In 1994, nearly 70 years after the original excavations were completed, a member of a tour group from the Denver Museum recovered a point from a corner of the site (Dixon and Marlar 1997). All together, nearly 30 fluted projectile points have come from the site, although the current whereabouts of some are unknown, and a few points from the site may yet lurk in private collections not yet recorded.

FIGURE 4.8 Bison skeletal remains on the floor of the Trinidad State College excavations on the North Bank, 1972. (Photo courtesy of Louden-Henritze Archaeology Museum.)

In late November 1936, a large bison skull fragment (Maxwell Museum 60-12) and a "side-scraper"—actually, a large flake knife—were found eroding at the site by Joseph H. Toulouse and Ele Baker (Reed 1940:4). Photographs taken during that visit show Ele Baker pointing to the spot where the artifact (or skull) was found (fig. 8.2), on the North Bank of the arroyo. Given the erosion since 1936, it is almost certain the sediments in which these remains were found have long since washed away. The artifact was kept in the Baker family, and available for study. Such is not the case for several other flake tools recovered from the site over the years; although these are mentioned by Roberts (1940:59), and at least one photograph exists (fig. 8.3), their current whereabouts are unknown (chapter 8).

On one of his own visits to Folsom in the 1930s, Howarth collected a sample of charcoal, which produced the first radiocarbon date from the area—though not, as initially supposed, from the Paleoindian occupation at the site (chapters 1 and 5).

Also in the 1930s there was an unconfirmed report of a discovery "by a Mr. Jim Macey of Clayton, New Mexico, who has the specimens in his barber-shop in that town, of a cache, apparently, of ten or a dozen Folsom points on the rimrock [Johnson Mesa] west or north of the [Folsom] bison-quarry" (Reed 1940:4–5). That a Folsom cache might have been found in the general vicinity of the site is intriguing, especially given the general scarcity of any archaeological material—let alone Paleoindian remains—in the area and the virtual absence of any other known Folsom caches—here or elsewhere on the Plains.

The Folsom Ecology Project under Anderson's direction did not expressly focus on Folsom (chapter 1) but did include excavation of <10 small test pits and backhoe trenches in and around the site, mostly on the South Bank. In addition, the arroyo walls were cleaned and pollen profiles were obtained from two nearby sections. Radiocarbon samples were collected from the arroyo walls; these included bone fragments and a small amount of charcoal, which occurred as a series of flecks dispersed throughout the sediment in what appeared to be a secondary context (Haynes et al. 1992:87). Finally, Anderson conducted an extensive surface survey of the surrounding area of the Upper Dry Cimarron Valley, which yielded some 74 additional sites and 192 isolated artifact occurrences (Anderson 1975:14, 80), though only a very small fraction of those was Paleoindian in age. These results mostly served to show that human use of this area in prehistory was ephemeral and consisted largely of hunting (Anderson 1975:4, table 23)—much as Schwachheim and Laves had discovered 50 years earlier.

Aspects of Anderson's work at the Folsom site are touched on in her dissertation (Anderson 1975), though the attention there is primarily on the results of the survey and land use patterns of all the prehistoric material in the area. The Folsom site results were slated for a separate report (Anderson 1975:82), which was never published. The results of the radiocarbon dating were summarized by Haynes et al. (1992).

At about the same time that Anderson was conducting her survey of the area, a group from Trinidad State College, with help from Willard Louden and other local avocational archaeologists, excavated bison remains from the North Bank (chapter 1). These included a cranium, ribs and at least one thoracic vertebra, according to photographs taken at the time (fig. 4.8). The cranium and photographs are curated at the Louden-Henritze Archaeology Museum at Trinidad State

FIGURE 4.9 A. Looking northwest at the freshly-cut vertical face of the North Bank, 1928. (Photo courtesy of American Museum of Natural History.) B. Same perspective, with C. Vance Haynes, 1990. (Photo by D. J. Meltzer.)

Junior College. The whereabouts of the remaining bones and any field notes are not known.

The SMU/QUEST Investigations

At the time our investigations began it was evident the Folsom site was relatively small, the prior excavations had been highly localized, much of the bonebed had been removed, and there had been substantial erosion in the decades since the 1920s. As our fieldwork was unavoidably bound by these conditions, let me turn first to the appearance of the site when we arrived in 1997, so that the subsequent discussion of our field strategies and tactics can be put in context.

The Folsom Site as It Appeared Prior to Our Investigations

The Folsom site has changed dramatically from when it was first photographed in 1922 and, for that matter, from how it appeared before the start of excavations in 1926. Wild Horse Arroyo, which used to be no more than a few meters wide from bank to bank, was by 1997 approximately 10 m to 15 m wide, bank to bank. Based on early photographs, the floor of the arroyo is now much wider and deeper—by at least several meters—than it was in 1928 (fig. 3.4).

Most of the changes are a result of the 1928 excavations or, more precisely, erosion that took place in the decades afterward. The erosion was set in motion when, at the end of that

FIGURE 4.10 A. Looking west across the South Bank excavation area, 1928. (Photo courtesy of American Museum of Natural History.) B. Same perspective, 1997. Individuals left to right, Pel-Lin Yu, David J. Meltzer, and Robert L. Kelly. (Photo by L. C. Todd.)

field season, the American Museum crew left without backfilling or stabilizing the site. This was not uncommon for its time, nor wantonly irresponsible—after all, they assumed there was little, if anything, left of the site and, thus, little that needed to be preserved. The consequences of their actions are nonetheless apparent when comparing photographs taken of the site in late 1928 with views taken over subsequent years and with its appearance in 1997 (fig. 4.9) when we began work.

The backdirt berm had by 1997 been stabilized by grass and small brush, although it appeared to have diminished since 1928. Its "wings" were less pronounced, and any trace of its presence on the northern edge of the South Bank had vanished all together. Where it remained on the uphill or south

side, in places it was breached by small erosional rills. By 1997 the South Bank excavation area had become a large bowl, with a much steeper south-to-north slope than it had in 1928 (fig. 4.10). A deep Y-shaped gully through the bowl had also developed, which had cut down to the level of the thalweg of Wild Horse Arroyo. Smaller tributary gullies extended from that central gully up into the South Bank basin.

Not surprisingly, the 1928 earthen datum pillar had disappeared by 1997. Left standing by itself, the pillar was vulnerable to weathering, mechanical erosion, and damage by site visitors, who were naturally drawn toward it to see what might be eroding out of it. Game and livestock leaning or scratching themselves against it likely took a toll as well.

FIGURE 4.11 Using the Colorado Museum truck to haul a plaster-jacketed bison skull up the North Bank wall, 1927. (Photo courtesy of Denver Museum of Nature and Science.)

When Merrill mapped the site in 1931, the pillar was still standing (Merrill to Wissler, October 26, 1931, View 9, ANTH/AMNH). Photographs taken of the site area by Ele Baker on his visit in 1936 show that the pillar had by then diminished in size (Maxwell Museum, photograph 88.27.3). In those photos, it is probably no more than 1 m² at its top. By 1949, there was at best only a modest bump where the pillar once stood—judging by photographs taken by C.V. Haynes (Haynes collection, 1949).

Substantial erosion had also taken place on the North Bank. During the original investigations, it was possible for the crew to back their trucks up to the lip of the North Bank in order to hoist heavy, plaster-jacketed bones out of the arroyo and over to their camp (fig. 4.11). Such would have been possible in 1949 as well. Since that time, however, the North Bank slope has sought its angle of repose;

what was a nearly vertical face in 1928 is now at a slope of ~40°, although the profile-topping McJunkin stratum (chapter 5) was able to maintain more vertical integrity and has an average slope closer to ~75° (figs. 3.4 and 4.12).

Moreover, by 1997 an ~15-m-long segment of the North Bank that had been faced vertically in 1928 had begun to slump, perhaps as a result of a game and livestock trail cut into and behind it (possibly enhanced by the erosion of stair steps cut into the wall in 1928). In 1997 the highest point on the lip of the North Bank—which in 1928 was level with the upland surface—was now >50 cm below that surface and was separated from it by a deepening swale.

All of this meant there would be no straightforward means of tying our site grid to the datum used during the original investigations. Furthermore, elevation and depth data from the 1920s would not be directly comparable to

FIGURE 4.12 Looking north/northeast across the South Bank, with the North Bank in the middle distance, 1997. Note the copse of oak trees on the right, corresponding to that in Figure 4.5. individuals on the South Bank, Virginia Hatfield, Luis Alvarado, Michael Bever; on the North Bank, Douglas Anderson and Pei-Lin Yu. (Photo by L. C. Todd.)

such measures today, given the removal by excavation, erosion, or slumping of the surface from which such depths were taken in the 1920s. Any excavations we would conduct at Folsom on the South Bank would necessarily be confined to the outer edges of the 1920s excavations. Finally, work on the North Bank would have to move in from the eroding slope or excavate several meters down from the upland surface. These were conditions that framed our efforts.

Field Strategies, Tactics, and Guiding Results, 1997–1999

THE 1997 FIELD SEASON

Our fieldwork in 1997 had several strategic objectives: (1) to determine whether intact portions of the bonebed were still present and whether there was somewhere nearby a Folsom-age camp or habitation area; (2) to extend our knowledge of the stratigraphy, geochronology, and paleoenvironment of the site, especially beyond the bonebed area; and (3) to assess the possibility that volcanic activity—known to have predated the Folsom-age occupation at site (Anderson 1975:34–37, 40–42, Anderson and Haynes 1978)—could also have been contemporaneous with or postdated that occupation, as hinted at by Muehlberger, Baldwin, and Foster (1961:10–11).

Critical to resolving the first two objectives was gaining a clearer sense of precisely where the original excavations had taken place. Doing so not only would identify where intact Folsom-age deposits—bonebed or associated camp areas— might occur, but also would have the incidental potential of spatially situating the artifacts and bones shown on the 1928 plan map. The third objective resolved itself in the negative soon after the first profiles were cleaned to examine the stratigraphy.

The 1997 fieldwork was conducted under an Archaeological Excavation Easement permit from the State of New Mexico (Permit Reference AE-74) over a total of 16 working days in the summer and early fall of 1997. During that time a range of activities took place on-site, including the following.

- Construction of a detailed, close-interval contour map of the site (fig. 4.12) using a total station, to replace the transit-stadia map made by Merrill in 1931.
- Excavation of eight shallow slit trenches totaling 22.62 m (average dimensions, 2.72 m long × 0.75 m wide × 0.19 m deep). These slit trenches were cut in order to relocate the perimeter "prospect holes" or perhaps remnants of the pillar, or the walls and corners from the 1926–1928 excavations, and thus hopefully render useable the historic map and excavation notes on the site (fig. 4.13).
- Excavation and careful dry screening through 3.175-mm (0.125-in.) mesh of three units (two 2 × 2-m units and one 1 × 1-m unit) in the 1928 backdirt berm, in order to gauge the amount of archaeological remains missed in the original excavations and, thus, provide insight into the nature of the data previously recovered at the site.
- Clearing and sampling of several profiles along the North Bank of the arroyo, along with hand augering (22 holes) and Giddings machine coring (15 holes) within and extending on a radius ~50 m to 100 m

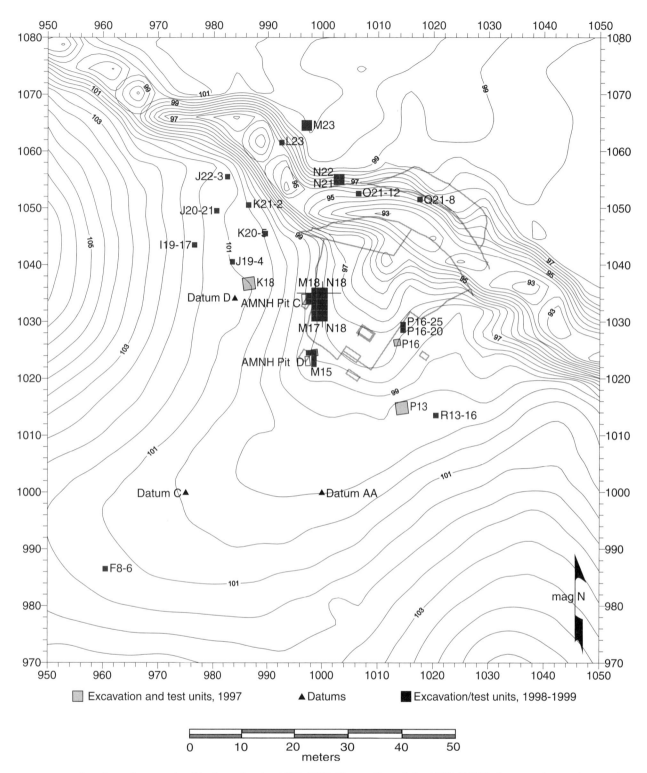

FIGURE 4.13 Location of excavation blocks and units, 1997–1999. The outline of the 1928 AMNH plan map is superimposed in gray, and was scaled to fit this map by matching it to the actual locations of AMNH Pits C and D. The fit is imprecise.

away from the 1926–1928 excavation area, to map the subsurface topography and the extent of the key stratigraphic units.

- Electrical resistivity and seismic refraction surveys, to complement and enhance the coring and augering data on subsurface bedrock topography.

- Excavation and dry screening of two test units (one 2 × 2-m unit and one 1 × 1-m unit) on the western and southern margins of the 1928 excavations, to determine if the archival hints (noted above) were correct and that intact portions of the bonebed or possibly a camp might still exist in this area.

- Archaeological surface survey and reconnaissance of the site and the surrounding area, to assess possible camp loci away from the bonebed and the possible presence of other archaeological components Paleoindian or otherwise.
- Survey of the modern vegetation for plant resources that might have been used by recent hunter-gatherers, to serve as a baseline that, coupled with planned paleo-environmental studies, would help assess the potential for plant foods at the time of the Folsom occupation.
- Sampling of excavated test units, exposed sections (along Wild Horse Arroyo), and a nearby lake bed on Johnson Mesa for sediments, charcoal, archaeological bone, gastropods, phytoliths, and pollen.
- A metal detector survey in the areas of the 1926–1927 and 1928 camps, in order to assess the nature of the historic components at the site, the original field camps having, by then, become themselves part of the archaeological record (appendix C).

The 1997 fieldwork, though brief, nonetheless demonstrated a number of important facts about the site. For one, considerable erosion had taken place since the close of excavations in 1928, but most of the erosion—save that on the North Bank—had taken place in already-excavated areas of the site and had not significantly impacted previously unexcavated areas of the South Bank.

In addition, it was apparent that those unexcavated areas might be much more extensive than the archival records indicated. This became apparent when AMNH Pit C was found. A slit trench was subsequently dug on a perpendicular from AMNH Pit C toward the center of the 1928 excavation area, in order to pick up the outer edge of that excavation. That edge ought to have been just 50 cm east of AMNH Pit C, based on the 1928 plan map. However, the trench had to be extended ~4 m before the western edge of the 1928 excavations was finally encountered. While exposing and cleaning the western wall of the 1928 excavations to examine the stratigraphy, in situ bison bone was found including several small fragments, a 27.3-cm section of rib, and a virtually complete sacrum of a calf—all in a relatively small area. On that evidence, there were excellent prospects for finding more intact portions of the bonebed in the intervening and unexpected 4 m of deposit between the 1928 excavation edge and AMNH Pit C.

That discrepancy between the 1928 plan map and excavation reality naturally called into question the accuracy of that map. Two other excavation corners were relocated some 10 m southeast of AMNH Pit C, but it proved impossible to ascertain whether these marked the corners of AMNH Pit B or the "Denver Pit" (fig. 4.13). Although both of these perimeter pits were ostensibly southeast of AMNH Pit C, in neither case did their distance and bearing from AMNH Pit C as shown on the 1928 plan map match up with the actual measurements between AMNH Pit C and those two relocated corners.

The effort to relocate the 1928 pillar yielded what we suspected might have been its remnant base. That feature was no more than a subtle 38-cm rise on an irregular surface; it was so subtle, in fact, that it might simply have been a random bump on the surface and wholly unrelated to the pillar. That caveat notwithstanding, aligning Merrill's 1931 map with ours using site contours (which cannot be done with great precision owing to erosion since 1931) puts the pillar at ~N1030.5 E1009.5 on our grid. The possible pillar base we relocated was at ~N1029 E1007. Further, Merrill's map shows the top of the pillar to be ~6.1 m (20 ft) above the thalweg of Wild Horse Arroyo (Merrill to Wissler, October 26, 1931, ANTH/AMNH). The elevation of the possible pillar we relocated is just 4.66 m above the thalweg; however, if one assumes that the pillar once stood as much as 1.5 m higher (as noted above), that would put the top of this feature at ~6.2 m above the thalweg. These are reasonably close horizontal and vertical matches under the circumstances, and we may indeed have relocated the base of the pillar. Of course, the end game of this exercise was to match up not with Merrill's 1931 map but, instead, with the 1928 plan map, for it is only the latter that shows the location of artifacts and bison remains. Unfortunately, there are again problems: There is a significant discrepancy in the distance between the pillar and AMNH Pit C as shown on the 1928 plan map (~9.1 m) and the measured distance between those relocated features on the ground (~11.2 m). Thus, although there is a good possibility the feature we found was indeed the base of the 1928 datum pillar, that fact is ultimately of limited utility.

Surface surveys in the area, particularly on the upland interfluves between Wild Horse Arroyo and the Dry Cimarron, and between Wild Horse Arroyo and Cherry Creek, failed to yield any trace of Folsom-age artifacts, seemingly eliminating those areas as possible locations for Paleoindian habitation. Equally significant, the surveys here and elsewhere in the area yielded virtually no archaeological material of any age. This result corresponded with Anderson's earlier (1975) and far more extensive survey results, as well as with observations made in the 1920s, which described the area as "barren" of artifacts (Laves to Wissler, July 3, 1928, ANTH/AMNH). The scarcity of archaeological material in the region, Paleoindian or otherwise, was puzzling, especially in light of the rich game resources known to have been in the area in historic—and presumably in late prehistoric—times.

One useful clue to explain that fact emerged from the modern vegetation survey: Subsistence plant foods are very rare in the area today. Were that the case in the past, it would certainly have had a bearing on land use patterns in Paleoindian times, but in 1997 the nature of the paleoenvironments at Folsom was not known.

Another result of this initial fieldwork was a better understanding of the geological setting and geomorphic history of the site. The coring and augering revealed that the Folsom bonebed was not situated within a paleovalley of Wild Horse Arroyo but, instead, appeared to extend from what would

FIGURE 4.14 General view of excavations in the M17 block, 1993. (Photo by D. J. Meltzer.)

have been the southern edge of that channel up into a paleo-tributary, which headed south/southwest of the paleovalley (and Wild Horse Arroyo). This work also indicated that the western valley wall of the paleotributary was at least 2 m above the bonebed floor. Whether this valley wall influenced or otherwise guided the tactics of the kill (e.g., helping trap or maneuver the bison) remained unknown.

THE 1998 FIELD SEASON

Several questions emerged from the 1997 field season that guided the following season's work. The first pertained to the bison bone on the western edge of the site: Were the elements found in the side walls scattered fragments from the outer edge of the bonebed, or did a portion of it continue in that direction and, if so, how dense was it, and how far did it extend? More intriguing, what was to be made of the annotation on the 1928 AMNH map that bison remains in this area were "more or less mixed," while remains from the other areas of the site apparently occurred as complete skeletons? All of this raised the interesting possibility that different activities had taken place across the bonebed, and that perhaps a bison processing area was present but unrecognized in this part of the site, and might still remain. As a result, a key objective of the 1998 field season was to extend excavations into this area.

Then too there was Brown's observation, conceivably related, that the western side of the bonebed yielded a higher density of bison remains than other parts of the site (Brown 1928b). If this was not a function of processing activities, might it reflect where the animals were killed? Were they trapped and dispatched against what appeared to be a high valley wall on the western side of the paleotribu-

tary? Was there a possibility that the animals were stampeded off that valley wall into the paleotributary? Resolving such questions required more detailed data on the geometry of the paleotributary, in order to reconstruct the form of the landscape in Paleoindian times.

A further objective for 1998 was to increase the understanding of the location, depth, and extent of the Folsom-age deposits beyond the immediate area of the bonebed, in order to identify potential surfaces where additional faunal remains or associated habitation debris might occur.

Work was also initiated on the North Bank since it was there in 1936 that Eli Baker found one of the few documented tools from the site along with a cranial fragment. This was also where, in 1972, apparently in situ bison remains, including a skull, were recovered by Trinidad State Junior College. Precisely where those remains came from relative to the main portion of the bonebed and, indeed, how the deposits on the North Bank articulated stratigraphically with those on the South Bank were of considerable interest.

Finally, the paleoenvironmental context of the site still was largely unknown. As a result of the 1997 field season it was evident that certain paleoecological indicators were present, potentially in great abundance, notably gastropods and charcoal. These needed to be sampled in a systematic fashion.

With these as guiding themes, fieldwork was conducted in 1998 under Archaeological Excavation Easement permit AE-78, this time extending over three 10-day sessions in June and July. The 1998 season was focused on (1) intensive excavation on the western side of the site (fig. 4.14) to assess the nature, extent, and archaeological preservation and potential of the intact bonebed deposits discovered in 1997; (2) continued mapping of the subsurface by hand augering and machine coring using a trailer-mounted Giddings rig, in order to enhance the understanding of the site's topographic and geological setting and history; and (3) intensive collection of sediment samples in an attempt to recover charcoal, gastropods, phytoliths, pollen, beetles, and other material from excavation units, exposed sections along the arroyo, core and auger holes, etc., in order to gain a better understanding of the environment at the time of the Paleoindian occupation. The specific field activities included the following.

- Completion of the detailed topographic map of the site. Most of the site area had been mapped in 1997 at an ~0.5-m contour interval; however, because of time constraints the mapping did not extend any farther up Wild Horse Arroyo than the bedrock constriction immediately west of the northwest corner of the site. Wild Horse Arroyo and the surrounding uplands 100 m upstream of the site were thus mapped. This effort included the placing of multiple concrete datums on the site (there are now a total of four on the South Bank and one on the North Bank (appendix A; fig. 4.13).
- Creation of a hierarchical grid and block system, in which all units were referable to a uniquely numbered 1 × 1-m square within a 5 × 5-m alphanumeric block, as described and shown schematically in appendix A, figures A.1 and A.2.
- Intensive excavation of twenty-one 1m x 1m and partial units covering a total area of ~15 m² in two separate blocks, the M15 and M17 blocks on the west side of the South Bank—both of which had yielded bison bone in the 1997 season. These excavations were aimed at determining the density, condition, and richness of remaining portions of the bonebed and recovering any associated archaeological remains. They involved detailed piece-plotting and, where needed, plaster-jacketing of large and/or fragile bison bones and systematic sampling within all excavation units and 5-cm levels for paleoenvironmental remains, as well as ad hoc sampling as items of potential paleoecological value were encountered, such as gastropods (appendix A).
- Water-screening of all excavated sediments through nested 3.175-mm (0.125-in.) and 1.587-mm (0.0625-in.) mesh (which took place along the Dry Cimarron, in the area shown in fig. 3.6).
- Hand augering (21 holes) and Giddings machine coring (18 holes) within and extending on a radius ~50 m to 100 m away from the 1926–1928 excavation area, in order to continue the mapping of the paleotopography of the site, including the depth and lateral extent of the Folsom-age deposits, the depth and variation of the local Cretaceous bedrock, and the configuration of the valley walls of the paleotributary and the paleovalley, and to "ground-truth" the results of the 1997 seismic and resistivity surveys.
- Wall clearing and shallow trenching on the North Bank to relocate the walls and corners from the 1972 Trinidad State Junior College excavations, which led

FIGURE 4.15 Looking east across the South Bank, 1998. The M17 excavation block is in the immediate foreground. Jason LaBelle and Todd Surovell are standing on the eastern wing of the backdirt berm. (Photo by D. J. Meltzer.)

to the discovery farther upstream of newly exposed additional skeletal elements.
- Detailed mapping of the North Bank area, in order to gather baseline information to track the slumping of the bank.

The 1998 fieldwork confirmed that a portion of the Paleoindian bonebed remained along the western side of the South Bank. It was also readily apparent that the most abundant faunal elements being recovered in this area were those that do not yield much meat, and were generally parts discarded in the course of butchering (Cannon 2003; Emerson 1993, Todd 1987a; Wheat 1972). This pattern contrasted with the bison remains ostensibly recovered during the original investigations, which reportedly were more complete skeletons (Kaisen, unpublished field notes, VP/AMNH). We considered two possible explanations for this difference. The first was that our excavations had come down onto a well-defined and spatially distinct activity zone—such as a bone discard area—not encountered during the 1920s excavations. On its face, this possibility seemed unlikely given how closely the excavation areas were situated to one another (fig. 4.15). The second

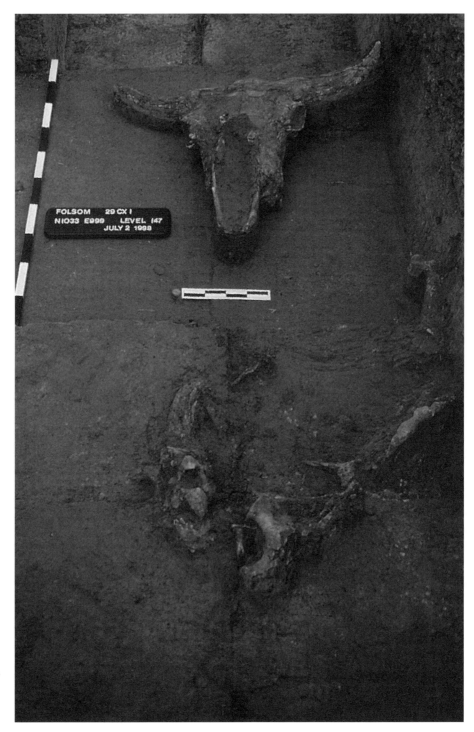

FIGURE 4.16 Two of the complete bison crania recovered in the M17 excavation block, 1998. Only the occipital condyies and a portion of the horn cores are visible in the cranium in the foreground. (Photo by D. J. Meltzer.)

possibility was that the contrast in recovery was more apparent than real, with the difference attributable to the tendency in the 1920s to focus on skeletons, rather than on individual skeletal parts, and refer to clusters of bones as "skeletons." Given the close proximity of the respective excavations, it seemed reasonable to assume the null hypotheses that the faunal assemblages recovered in the 1920s and 1990s were similar—unless further analysis proved otherwise (chapter 7).

The 1998 fieldwork also demonstrated that the bonebed remains were extremely well preserved. Among the elements recovered that season were two large bison crania (fig. 4.16) with incisives and horn cores intact. Their excellent preservation opened the possibility of examining bone elements

FIGURE 4.17 Isolated bison humerus on the North Bank. Note the missing proximal end, its place taken by a carbonate nodule. (Photo by D. J. Meltzer.)

and surfaces for traces of butchering patterns and taphonomic history—in both the elements we recovered and those found during the 1920s and curated in the Denver and American museums—assuming, that is, that the faunal material collected in the 1920s was in comparable condition and that their recovery methods did not unduly damage the bone. In general, both assumptions proved true (chapter 7).

Why the bone was so well preserved became apparent in the examination of the stratigraphy of the site, which that season revealed key aspects of the geomorphic history of the site. First, the butchered bison carcasses and bone were not exposed on the surface for very long after the kill but, instead, were buried relatively quickly and virtually completely by fine-grained silts—possibly aeolian in origin. Second, the sediment that buried the bones was, in turn, relatively quickly buried by a slope wash of shingle shale, which apparently originated on the valley wall immediately above the bonebed, and which effectively armored and protected the underlying bonebed (chapters 5 and 7).

A better sense of the extent of the site emerged during the 1998 field season. Following up on the recovery of a few bone fragments in one of the 1997 test units, excavations were opened in the M15 block 5 m south of the M17 block. Only a very few bone fragments were recovered in the excavations in the M15 block, indicating that the bonebed does not extend into this area. Though few bones were present in this block, AMNH Pit D proved to be. This pit was roughly the same size as AMNH Pit C (0.92 × 1.5 m [3 × 5 ft]; we did not remove the fill to check its depth) and proved to be surprisingly close to where it ought to be in distance and bearing from AMNH Pit C, based on the 1928 plan map. Unfortunately, the two excavation corners found the previous season still defied assignment to any features shown on the 1928 plan map—

though with the 1928 map now better oriented with the location of AMNH Pits C and D, they were closer to where the "Denver Pit" ought to be, rather than AMNH Pit B.

While the M15 block excavations helped delimit the extent of the bonebed, the discovery of an isolated bison humerus eroding out of the North Bank (fig. 4.17) ~26 m northwest of the bonebed substantially increased the known extent of the site. Of course, much of the intervening space includes the present channel of Wild Horse Arroyo, so the gain in area that might potentially contain intact deposits was not as substantial as it might appear. Assessments of that potential were further tempered by the observation that portions of the paleovalley had also been incised by a substantial cut-and-fill episode, which we surmised might be Middle Holocene in age—a supposition that radiocarbon dating later proved to be correct (chapter 5).

Nonetheless, this observation suggested not only that Paleoindian activities took place in the paleotributary where most of the excavation to date had occurred, but that portions of the kill and/or processing area were also present under the North Bank in the paleovalley itself. The possibility that the skeletal elements in the paleovalley had simply washed in from the paleotributary was easily eliminated by the fact that the skeletal elements found in the paleovalley were ~14 m upstream of the junction of the paleotributary and the paleovalley (chapter 5). All things considered, there appeared to be potential for additional Folsom-age deposits in the paleovalley.

Finally, coring and augering clarified aspects of the site's geological and paleoenvironmental history, notably the realization that the fine-grained sediments in which the bones were found were possibly aeolian in origin, suggesting the working hypothesis that the environment at the time of the occupation was relatively dry and treeless. To help resolve this matter, charcoal and gastropods as well as sediment samples were collected individually and in bulk samples to allow for detailed analyses (chapters 5 and 6).

THE 1999 FIELD SEASON

Several results from the 1998 field season guided work during the 1999 field season. First, the prior seasons had clearly demonstrated that there were parts of the bonebed remaining and strongly indicated that what remained of the bonebed in the paleotributary had spatial structure and integrity. There was no evidence to indicate that the bison remains there had been subject to any significant postdepositional movement or modification save, perhaps, for the skeletal elements on the eastern side of the paleotributary excavated by the AMNH in 1928. Kaisen observed that the bone from that side of the site was more heavily weathered than in the rest of the site, but he had no explanation to account for the disparity (e.g., Kaisen to Brown, July 8, 1928, VP/AMNH; Kaisen to Cook, both July 8, 1928, HJC/AGFO). In addition, all evidence indicated that the bone in the paleotributary was in primary context, the carcass processing apparently having taken place close to where the animals were dropped. It therefore seemed appropriate during the 1999 season to expand the excavations out from our initial excavations in order to open a larger, contiguous horizontal block, so as to gain a better understanding of the distribution and patterning of the skeletal remains.

Second, although it was apparent that the site extended out into the paleovalley, how extensive the remains might be in the paleovalley was uncertain. What was certain was that it would not be easy to resolve the matter, given the several meters of overburden in that part of the site. Nonetheless, it was thought that test units on the slope of the North Bank might provide stratigraphic and archaeological control on the depth and extent of the Folsom age deposits, reveal how far the bonebed might extend to the north and west and at what depth, and possibly indicate the density of skeletal material in that area of the site. All of this would clarify the overall extent of the kill and, perhaps, even reveal the location of an associated camp.

The paleoenvironmental setting of the occupation still remained a critical unknown. Large samples of gastropods were recovered during the 1998 field season, but these represented only one line of environmental evidence. On its first run through the laboratory, pollen did not appear to be preserved in the sediments at the site (but see chapter 6). However, it was hoped that pollen might be preserved in sediments from lakes atop nearby Johnson Mesa, which may have served over time as traps for pollen, charcoal, macrofossils, diatoms, ostracods, and other potential ecological indicators.

Our previous two seasons had provided valuable data on the paleotopography and the extent of the Folsom-age surface and the depth and variation of the local bedrock. However, a few areas remained unmapped or poorly known, including the upper reaches of the paleotributary. Also of interest was the deeply incised, apparent Middle Holocene arroyo channel discovered on the North Bank in 1998. It was evident that this cut/fill episode must have truncated at least a portion of the site, and mapping that channel might help resolve the effects of this geomorphic process on the bonebed.

Finally, our work the first two seasons intensified the interest in finding an associated camp or habitation area. Following the 1998 field season, the number of identified specimen (NISP values) recovered during the CMNH, AMNH, and our own (SMU/QUEST) excavations were plotted and statistically compared, and this analysis made it clear that the several assemblages are, in fact, quite similar. All were dominated by low-utility skeletal elements, the high-utility elements having apparently been removed (chapter 7). The place to which they were removed was not known, though there are at least two obvious possibilities—on-site or off-site (chapter 7). But in either case there presumably was a habitation area somewhere nearby.

Work over previous seasons, and that includes the surveys by Anderson in the 1970s and by the CMNH and AMNH crews in the 1920s, had shown that if a Folsom-age camp was present, it was not located in any of the obvious places. Our own surveys had largely eliminated the possibility that a camp was preserved on the interfluve areas above the site—areas that were, in any case, several hundred meters from the kill and from freshwater. Careful attention was paid to the material from the coring and augering, but in no case was archaeological debris recovered.

The discovery of bison bone on the North Bank suggested two possible locations for a Folsom camp. Assuming that there were bison carcasses being worked by the group in both the paleovalley and the paleotributary, the most convenient place to camp closest to each area would either be on the floor of the valley where the two drainages intersected or atop the valley wall above that intersection. The former would afford protection from the wind but might be unsuitable if there was water in the drainage, while the latter would provide a dry surface underfoot and better exposure to the sun. Of course, the former was also under several meters of overburden and showed signs of cut-and-fill episodes in which the likelihood of finding remains in primary context was commensurately reduced.

With all of that in mind, fieldwork was conducted at the site over 25 working days, mostly in July, under Archaeological Excavation Easement permit AE-83. The work had three major components. (1) Excavations were conducted in three areas: the southern half of the M17 and N17 blocks, in order to complete units begun in 1998 and open a larger contiguous block; hand augering and then test excavations on the knoll immediately west of the M17 block, to test for a camp area; and systematic clearing and testing on the slope of the North Bank, to assess the density of material in that area of the site. (2) Continued mapping of the subsurface was done by augering, as well as careful

FIGURE 4.18 Excavation on the North Bank, N21 and N22 blocks, 1999. This is the area in which Howarth likely found "deer" remains in 1926, and where the antilocaprid specimen discussed in Chapters 5 and 7 was recovered in 1999. Individuals are Nicole Waguespack and Jason Meininger. (Photo by D. J. Meltzer.)

measurement of the North Bank area using mapping points set in during the 1998 season to determine if there had been additional slumping, which might endanger that area of the site. (3) A lake on Johnson Mesa was cored to obtain a pollen record. The specific field activities included the following.

- Intensive excavation of 11 1 × 1-m units in the M17 block, in order to complete units unfinished from the 1998 field season and extend excavations to the west to link up with AMNH Pit C, thereby filling in the gap between this "prospect hole" and the western wall of their excavations in the bonebed and gaining a larger, contiguous horizontal excavation block (fig. 4.13). As before, these excavations involved detailed piece-plotting of all faunal material, systematic sampling in all excavation units for sediments and paleoenvironmental remains, and—as in 1998—fine-mesh water-screening of all excavated sediments.

- Test excavation on the eastern side of the 1928 excavation area, in order to attempt to understand the striking differences evident in bone preservation. It was thought that the shale shingle gravel that armored the bonebed on the west side might have been absent on the east side, but augering on the east side in 1998 suggested that could not be correct, as gravel lenses were present over the *f2* sediments. The most expedient way of resolving the matter was to directly examine the context of bone elements on the east side, hence two 1 × 1-m test units were placed in this area.

- Hand augering (21 holes) across the paleotributary and perpendicular to its side walls, to map the subsurface topography.

- Excavation of seven 1 × 1-m test units across the site, primarily on the upland knoll immediately west of the M17 block, but also on the eastern uplands of the site and in the northern reaches of the paleotributary, ~40 m south of the bonebed/main excavation area. Unlike the excavations in the bonebed, these units were dug using slightly different procedures (appendix A). Unfortunately, none of these units yielded any archaeological remains.

- Testing and intensive excavation on the North Bank of the site, which included the excavation of a large (2.5 × 2.5-m) and deep (~3-m) unit on the North Bank (M23 block) near where the bison bone had been recovered in 1998, aimed at opening a window into any bonebed deposits in this area of the site; the excavation of seven whole and partial 1 × 1-m units in the slopes of the bank to examine the position and context of bone, especially in the area of the 1972 excavations (fig. 4.18); and the clearing of the face of the North Bank to examine the stratigraphy.

- Sediment coring for a pollen record at Bellisle Lake, a small (~45-acre) and relatively permanent lake atop Johnson Mesa, some 5.5 km distant and at an elevation 261 m above the Folsom site (fig. 3.1).

- Continued excavation of shallow slit trenches, this time in an attempt to locate the remaining AMNH perimeter pits that had so far eluded us.

The most important fact to emerge from the 1999 field season was the demonstration of the substantial differences in the stratigraphic histories of the paleotributary and the paleovalley (chapter 5), which had a pronounced effect on the taphonomic history of the bison bone deposited in those different settings (chapters 5 and 7). In brief, it became clear during our work that the faunal remains in the paleovalley do not occur—as they do in the paleotributary—as a discrete archaeological horizon or even a recognizable "bonebed" in primary context. Instead, most of the elements are in a secondary context, having been reworked by fluvial action. This observation proved to have important implications for the interpretation of data recovered from the North Bank in the 1920s and other studies of the geological history and radiocarbon chronology of the site (chapter 5).

The 1999 fieldwork also further delimited the southern edge of the bonebed—at least along the west side of the site. The density of bison bone dropped off significantly in the units on the southern margin of the M17 block. That observation, coupled with the virtual absence of bone in the M15 block 5 m to the south, suggests that the edge of the bonebed is in this vicinity.

On the eastern side of the South Bank, the test excavations showed considerable shingle gravel atop the Folsom age sediment in this part of the site as well, eliminating the absence of gravel as an explanation for why bone preservation is poor in this area. In none of the deposits, however, were any bones or artifacts encountered, though all excavated sediment from this unit was water-screened.

We continued to be unsuccessful at relocating AMNH Pit A, despite trying as many possibilities as conceivable based on our knowledge of where the two known (and one suspected) other perimeter pits were located and on the distances and angles between them as shown on the 1928 map. Unfortunately, a large backdirt pile prevented testing one area where Pit A might occur. We cannot rule out the possibility that traces of this perimeter pit were obliterated during the 1928 excavations.

In the end, the testing for an associated Folsom age camp was unsuccessful, the only benefit to the work being that it made clear where such a camp did *not* occur. That was only small recompense. Left unanswered was the question of whether a camp had once been there and either was still undetected or had since eroded away, or whether there had never been an associated camp nearby. Subsequent radiocarbon dating of charcoal from one of the test units atop the valley wall overlooking the intersection of the paleotributary and paleovalley shed light on this matter (chapter 5).

Finally, our efforts to core Bellisle Lake were only partly successful, as it proved exceedingly difficult from a floating platform to get sufficient purchase to penetrate and then extract the sediment corer from the heavy clays in the bottom of the lake. However, enough material was recovered to assess whether pollen was present in these sediments and obtain a Late Pleistocene radiocarbon age near the bottom

of the section, which provided sufficient warrant to try again in winter (chapter 6).

EXCAVATION SUMMARY, 1997–1999

Overall, just under 100 m² at the Folsom site was excavated from 1997 to 1999, including ~28 m² of shallow slit trenches, 32.25 m² of test units, and 37 m² within bison bone-bearing deposits.[11] The latter included 19 m² of intensive excavation within the main portion of the bonebed in the paleotributary between the western edge of the 1928 excavations and AMNH Pit C. This included units in the M17, N17, and M18 blocks (fig. 4.13). Not all of these were whole units: The two units in the M18 block were "windows" less than 1 × 1 m in size, cut at the level of the bonebed to more easily remove mandibles and long bone elements that extended from the units immediately to the south.

As shown in figure 4.19, the eastern edge of AMNH Pit C cuts through the western half of units N1033 E997 and N1034 E997. Ironically, skeletal elements—including bison crania—were found within centimeters of the edge of that "prospect hole." The 1928 map (fig. 4.2) indicates that "no bones" were found in either AMNH Pit B or AMNH Pit D, but no such annotation is appended to AMNH Pit C. The close proximity of bison bone to Pit C suggests that they might have found bone in it—the map does not exclude the possibility. One might hazard the guess that the skeletal elements that were found were judged to be of insufficient size or interest to warrant extending the excavations into this area, bearing in mind that in 1928 one of the primary goals of the excavation was to acquire well-preserved museum-quality specimens. This might also reinforce Hofman's (1999a) observation that the bison carcasses were widely dispersed, with large gaps between perhaps filled by a few scraps of bone. Obviously, this raises the potential for additional remains to be found on-site. So too does the fact that bison remains are now known to extend into the paleovalley. That all of the latter were in a secondary context suggests the more remote possibility that farther upstream of these finds bison bone may yet occur in primary context within protected bends in the paleovalley.

Collections Research

Faunal remains from the earlier investigations are curated at the American Museum of Natural History, the Denver Museum of Nature and Science, and the Louden-Henritze Archaeology Museum at Trinidad State Junior College, Trinidad, Colorado. These collections represent a substantial proportion of the material recovered from the site and can shed varying amounts of light on matters such as the season of the kill, the age and sex composition of the herd, the size of the herd, its foraging history, evidence of environmental or dietary stress, the butchering, processing, and utilization of the carcasses by the hunters, and the tapho-

FIGURE 4.19 Outline of AMNH Pit C (dark, rectangular area on the left), on the western side of the M17 block, 1999. Note the calf cranium on the floor, and the articulated vertebrae. See also Figure 7.2. (Photo by D. J. Meltzer.)

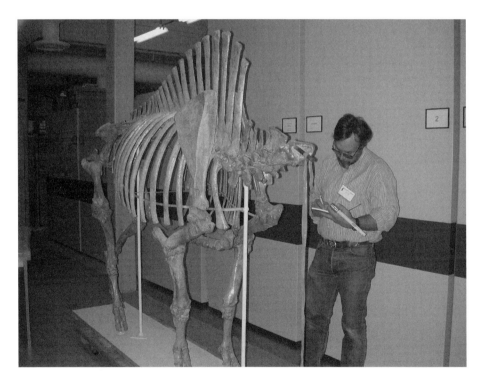

FIGURE 4.20 Lawrence Todd recording data on the original mounted bison skeleton from Folsom at the Denver Museum of Nature and Science, 2000. (Photo by D. J. Meltzer.)

nomic history of the bonebed. Yet, these materials had not previously been fully analyzed. Accordingly, each of these collections was inventoried and examined by Meltzer and Todd and the resulting data form the basis of the discussions in chapter 7 (fig. 4.20). This effort involved

• An inventory of all identifiable elements, their portions, segments, and sides, in all collections, as well as counts of unidentifiable fragments;
• Examination of surface condition and recoding (where feasible) of the weathering stage, the presence/abun-

dance and location of any root etching, carnivore modifications, rodent gnawing, burning, green bone breaks, impact damage, cut marks, and excavator damage;

- Analysis of tooth eruption and wear on all mandibles and patterns of bone fusion on relevant elements;
- Collection of measurements on long-bone elements well suited to distinguishing bulls from cows;
- Examination of teeth and bones for pathologies or age-excess wear; and
- A search of collections for species other than bison— that is, deer and small mammals—that reportedly were recovered from the site.

Insofar as possible, we also sought to reconstruct the archaeological provenience of the skeletal remains. It was often possible, especially with the AMNH faunal collection, to reunite skeletal elements that had been recovered in close proximity to one another. But the state of the records is such that it was often difficult to reliably locate where that cluster of elements had been found. Nonetheless, there was sufficient information available to allow us to identify elements recovered in the 1920s from the eastern side of the South Bank and from the North Bank, and these—along with the data we gathered—proved vital to helping us understand the taphonomic history of the site.

Projectile points and other artifacts previously recovered from the site are housed at the Denver and American museums, the University Museum of the University of Pennsylvania, Capulin Volcano National Monument (New Mexico) and in several private collections in New Mexico and Colorado. Detailed metric and nonmetric data were compiled by Meltzer on these points, to examine the stylistic and technological variability among them, evidence of use-wear, their degree of resharpening and reworking, evidence of tool use, and the source of the stone used in manufacture (chapter 8).

The examination of these collections was mostly conducted after the 1998 field season and, thus, had a limited role in shaping field strategies and tactics.

2000–2004 Field Activities

Brief visits have been made to the Folsom site every summer since the close of excavations. During these visits we mapped and collected any bison bone that had been exposed on the surface, continued to monitor the slumping on the North Bank by taking a series of record photographs and mapping topographic control points, and undertook efforts at erosion control on the South Bank. In addition, a more successful coring attempt was made at Bellisle Lake in early 2001 (chapter 6). Finally, in 2002 Daniel Mann initiated a study of the geomorphology of the Upper Dry Cimarron, in part to provide insight into the region's Quaternary climatic and environmental history and, also, to determine whether sediments of comparable age to those

of Folsom occur in places beyond the site itself. This study is ongoing, but some preliminary results are available (Mann 2003, 2004), parts of which are incorporated here (chapter 5). Mann's study did not include detailed work at the Folsom site itself.

THE ARCHULETA BISON

In the course of a geological reconnaissance in 2002, Mann discovered the remains of a bison along Archuleta Creek, a tributary of the Dry Cimarron River, at a spot ~4 km southeast of the Folsom site. The bones were exposed ~4 m below the surface in a deeply undercut bank and consisted of a series of ordered vertebrae lying flat with their dorsal surfaces protruding out, suggesting that the skeleton was on its side, oriented roughly parallel to the present drainage. The remains were lying along the upper, undulating surface of what appeared to be Pleistocene-age gravel and largely contained within and overlain by fine, overbank sediments.

On subsequent field examination by Mann and Meltzer, it was apparent that the bones were from a very large bison, perhaps an animal within the size range of *Bison antiquus*. Its suspected age and taxonomic identity, and proximity to the Folsom site, raised the question of whether it was a paleontological or an archaeological occurrence and, if the latter, whether the animal had escaped from the kill at Folsom and died on the floor of this nearby drainage.

The layout of the exposed vertebrae suggested that much of the skeleton was still contained within the bank, but removing it would require deeply undercutting an already undercut profile, endangering both the skeletal remains and the crew. It was decided to excavate only the visible and most vulnerable skeletal elements and examine the remains for associated artifacts.

Ultimately, 14 bones were exposed, but no artifacts were found with them. All bones are part of the vertebral column and include the atlas and axis, thoracic spines, lumbar vertebrae including the sacrum, and scattered rib fragments. The bones extend along a horizontal distance of ~2 m, but over a narrow vertical span (<15 cm). The atlas and axis were removed for study; the remaining elements extended too deeply into the wall for safe removal. The latter were recorded, plaster-jacketed, covered, and left in situ. A rock diversion wall was then built to deflect the stream's energy away from the section.

In general the bone is in excellent condition, having had only minimal subaerial exposure and surface weathering and no apparent carnivore modification. With a few exceptions, the bones lay flat or nearly so—modal inclination values were <5°, which, along with the lack of patterned orientation and the absence of scratch marks on the bone surface, suggests that fluvial reworking or animal trampling was minimal. There was, however, some slight postmortem but predisarticulation contortion: Both the neck and the sacrum twisted outward toward the stream bed. The cranium would thus have been one of the first elements exposed and lost when the bank began to erode.

The absence of a cranium hindered ready taxonomic identification. However, metric data on the atlas and axis from the Archuleta bison were compared with those of *B. antiquus* and *B. bison* from a variety of sites (Meltzer, Mann, and LaBelle). The comparison places the Archuleta bison within the range of *B. antiquus* and indicates that it was a large bull.

A crucial question is whether the Archuleta specimen is of the same antiquity as the Folsom site bison. The axis of the Archuleta bison yielded an age of 10,190 ± 30 ^{14}C yr B.P. (CAMS-96033), younger by some 300 radiocarbon years than the remains at Folsom (chapter 5). Although it seems unlikely that the Achuleta and Folsom bison were part of the same herd, these ages do fall squarely within the Younger Dryas Chronozone, with its radiocarbon-distorting plateaus, and may ultimately prove to have more significant temporal overlap. Whether this animal did escape from the melee at Folsom will require a more thorough investigation of the Archuleta bison for evidence of artifacts.

Isotopic data from the Archuleta bison are used for comparative purposes in chapter 6.

Notes

1. When the AMNH crews got to the North Bank the following season, the excavating proved as difficult as Schwachheim anticipated it would be. As Kaisen described it, "it was hard work to pick some of that stuff. It is just like rubber. We was all glad the day we struck the bone layer" (Kaisen to Brown, August 29, 1928, VP/AMNH).

2. As of 2004, the specimen was still on display.

3. Calculations of the excavated area done by Todd and Hofman (1991), who digitized the 1928 excavation plan of the site. That map is of suspect precision, as discussed below, so the figures should be taken as rough estimates. It is almost certainly of the correct order of magnitude, however imprecise in detail.

4. This was intentional on Brown's part, as he'd learned from Figgins that Schwachheim "was a faithful worker, but somewhat lacking in initiative" (Brown to Howarth, April 7, 1928, VP/AMNH). Kaisen, on the other hand, had "plenty of initiative" (Brown to Howarth, April 27, 1928, VP/AMNH).

5. Scientists were not the only visitors Kaisen and the crew had to contend with. As Kaisen reported to Brown, " . . . last Sunday here was about 40 people. The postmaster told me the New Mexico papers was full of things from here. Now a lot of people come here to see the boys and the wonderful lot of Bisons that was found. . . . I have to get up early Sunday morn-

ing to get ready for the visitors but let them come (Kaisen to Brown, August 19, 1928, VP/AMNH).

6. There had earlier been some question as to whether the specimen was camel (Wissler, field diary, August 4, 1928, ANTH/AMNH). But Brown had fellow curator George Gaylord Simpson take a quick look at the specimen before he left on vacation, and he confirmed the identification (Brown to Kaisen, August 27, 1928, VP/AMNH).

7. As usual, Harold Cook had his own unique slant on Laves. Cook claimed it was his idea that the American Museum send out someone from their archaeology department to investigate some of the caves and shelters Howarth and he had found. When Laves appeared, however, Cook was disappointed: "the 'great man' . . . turned out to be a delightful, well-mannered boy, a product of fine schools and an obviously cloistered life, from a mid western school, whom Barnum Brown had secured to 'determine' for him the archaeological evidence of the Folsom deposits. Fred Howarth was even more shocked than we were. Fred told us that the boy had asked him what geological formation the arroyo at the Folsom quarry was in. Fred said, 'Well, the old cut or wash was in the Pierre Shale.' The boy said, 'Well, then, as I understand the problem, I am to search for and determine the cultural evidences to be found in the Pierre Shale.' Inasmuch as the Pierre is a purely *marine* deposit, recognized to be well over 60 million years old, our shock at the background, training, and fitness of this otherwise very nice boy, for the job, is understandable. . . . The boy spent little time at the quarry, but any day we could hear his portable phonograph playing, somewhere not too far from camp, in the hills. The evidence of early man in the Pierre Shale remained unsolved" (Cook 1948).

8. From 2002 to 2003, Dan Mann located deposits of Late Glacial age in Oak Canyon, but no archaeological remains (Mann 2003, 2004).

9. Jenks had been closely following the work at Folsom since 1927 and had been among those receiving a telegram that August to come visit the site. Unfortunately, he was in Taos, New Mexico (only a few hours from Folsom), when the telegram was sent and did not receive it until he returned to Minneapolis two weeks later—much to his regret (Jenks to Figgins, September 15, 1927, October 3, 1927, DIR/DMNS).

10. At one point late in the 1928 season, Brown asked Alexander Wetmore, Smithsonian Institution Secretary, if he would ask the War Department to send a plane to the area to create a "photographic map of the entire valley and rim" (Brown to Wetmore, August 3, 1928, USNM/SIA). Wetmore explained to Brown that this method of mapping was not nearly as easy, inexpensive, or efficient as the press made it out to be and that he was much better having a map made by a surveyor—which he did not offer to pay for (Wetmore to Brown, August 17, 1928, USNM/SIA).

11. For comparison, and as noted earlier, the total area excavated by the CMNH was ~34.7 m^2, while the AMNH excavated ~233.7 m^2.

FIVE

Geology, Paleotopography, Stratigraphy, and Geochronology

DAVID J. MELTZER, WITH VANCE T. HOLLIDAY

Many preserved kill sites are indeed "geological oddities," as Albanese (1978:61) put it, and Folsom is no exception. To understand why and how this site was preserved where it is, and to use that information to gain insight into the form of the ancient landscape, we need to explore Folsom's geological context and stratigraphic history. Although important insights into these topics emerged from earlier work here, especially that of Anderson and Haynes in the early 1970s, there were significant gaps in our knowledge of these matters when our fieldwork began, as detailed in chapter 1.

To fill those gaps and complement the prior research at the site, we undertook our investigations with several goals in mind. First, we attempted to reconstruct the topography of the site as it was in Late Glacial times, in order to explore how the landscape appeared and may have been used by Folsom hunters to exploit their prey. It was in the course of this effort that it became apparent that a large portion of the site was situated within a paleotributary, while the remainder extended into an adjoining paleovalley (chapter 4). As a result of that discovery, a second goal of our work was to understand the depositional and erosional histories of the paleotributary and paleovalley and how these may have differed—with an eye on understanding, among other things, differences in the taphonomic history and preservation of faunal remains in these two settings (a matter also explored from the vantage of the bones in chapter 7). Finally, we sought to develop a more precise radiocarbon chronology for the site and the Paleoindian occupation than previously available.

Initial Efforts to Resolve Folsom's Age

The original investigators at Folsom had a broad sense of the site's stratigraphic history. As Figgins explained it to Oliver Hay, in reference to the now-famous photograph taken September 4, 1927, of Schwachheim and Brown next to the first in situ Folsom point (fig. 2.12):

This photo also illustrates the three strata overlying the bone layer. The lower [light-colored] member is composed of clays which Mr. Brown declares were deposited in an old stream channel, cut into Cretaceous deposits. The bones lie almost directly upon the Cretaceous. The stratum varies in thickness, as do the overlying members, but at the point shown in the photograph it is 36 inches (to the left this increases to 60 inches). The next [dark-colored] stratum is composed of clays and silt, and carries a high percentage of carbonaceous matter, together with occasional freshwater shells. At the point shown in the photograph, it is 22 inches, but increases to the left to a depth of 4 feet or more. The top [lighter-colored] layer is composed of Cretaceous clays and pieces of shale up to four or five inches in length. Also granite bowlders [sic] of equal size. You will appreciate the density of the deposit when I tell you the upper layer has been undercut fully 6 feet, the pick in the foreground being within a vertical line of the overhanging deposit. It is, of course, partially supported by roots. (Figgins to Hay, September 15, 1927, OPH/SIA)

As Brown himself observed, on the north side of the arroyo there was much more than 4 ft of the upper clay:

North side of quarry section shows 9 feet of black clay (& thin stripes) of yellowish in local areas which merges into brownish yellow at base indistinguishable in character from that below. This dark color is probably due to the oxidization of clays by plant life, for under every large clump of scrub oak the masses of rootlets reach to about this point. Bone layer extends from 10'–12' in deepest part of deposit which was center of stream course or water holes although most of skeletons rested on sloping bank of original N × S stream course 1/2 mile in length above quarry. A few inches above, with, and several inches below bone layer there are irregular lime concretions, with veinlets of pure gypsum. . . . [?] of ground waters through the eroded Pierre [sic] shales which throughout most of quarry have been dissolved into clays. (Brown, ca. September 4, 1927, Field Notes, VP/AMNH)

While these and other descriptions (see also Cook 1927a, 1928b; Hay and Cook 1930) lack detail, all seem to agree that there were at least three main deposits sitting atop the Cretaceous bedrock at the site (also Cook to Hay, January 25, 1928, OPH/SIA). Embedded within their descriptions was some hint of the paleotopography and the sloping surface on which the bone was resting (Brown 1928b) as well as of the stratigraphic variability across the site—or at least of the difference in depositional histories evident in the profiles of the South Bank and North Bank, though such differences were scarcely remarked on or understood at the time.

There was, however, a marked difference of opinion with regard to the depositional context of the sediments in which the bison bone was found. Cook, on the one hand, believed that

the fine, mucky character of the matrix (mostly clay-silts) . . . and, its freedom from coarse materials such as a creek or stream would commonly carry, and especially in a region where steep gradients were the rule, as here, makes it obvious that this must have been some sort of muck-hole, marsh, or slough at that period. (Cook to Hay, January 25, 1928, OPH/SIA)

This interpretation was supported by the apparent presence of freshwater invertebrates in the deposits (also Cook 1927a:244; Cook 1928b:39; Cook to Hay, December 23, 1926, OPH/SIA). For his part, Bryan (1937:141–142) saw the deposits as a "clayey alluvium," of a floodplain deposit. In later years, it was generally supposed that the kill had taken place in "an old bog or waterhole" in which the animals were trapped in the mud (e.g., Roberts 1940:59).

On his first visit to the site in September 1927, Brown supposed that the bonebed was situated in "a stream course or water hole." By the following spring, the water hole had become a lake, formed behind a lava dam (Brown 1928a:826; also Wissler, field diary, August 2–10, 1928, ANTH/AMNH). However, six months later, and after his own season (1928) of excavations, Brown revised that opinion, identifying the gastropods in the bonebed as "pulmonate land shells." He further observed that throughout the bone-bearing stratum there were irregular "limestone nodules" (i.e., calcium carbonate nodules), secondary gypsum crystals, and fine gravel matrix, but "only one stone was encountered during the three years' work, a piece of lava the size of one's head" (Brown 1928b). Brown reasoned, as Cook (1928b:39) had, that if these were sediments of fluvial origin, boulders of all sizes would have been found.

Quite unlike Cook, however, Brown (1928b) then concluded that the deposit was "of aeolian origin accumulated during a long period of little or no rainfall." Beyond these general observations, there is no surviving record of any detailed sediment or stratigraphic analysis, although Brown's AMNH crew was obviously interested in the matter, given the labor they invested in clearing that 3-ft profile along the North Bank (fig. 5.1; chapter 4).

The initial estimate of the age of the bonebed was by Cook (1927a:244), who put it at "certainly thousands of years" and of later Pleistocene age. This was his public position; privately, he guessed that it might date to "some interglacial stage of the Pleistocene (Cook to Hay, January 25, 1928, OPH/SIA). Yet, and as discussed earlier (chapter 2), he never thought it "had any such antiquity as the Frederick, Oklahoma and Colorado, Texas, evidence" (Cook to Ingalls, January 6, 1929, HJC/AHC; Cook to Wissler, March 25, 1929, HJC/AGFO; Cook 1928b:39).

Cook was reluctant to express an opinion of the site's absolute age in years, but A. V. Kidder was not (chapter 2). He thought it indicated an antiquity of the order of 15,000 to 20,000 years ago (Kidder 1927, 1936:144). Similarly, Brown estimated, based on observations he made on his first visit to the site, that it would take at least 24,000 years for the sediments lying atop the bison bone to accumulate (Brown 1928a:828). Following the 1928 season on the site, he would only suggest that the extinct species of bison and the overlying sediments of "highly restratified earth" indicated great antiquity, perhaps dating to the close of the Pleistocene (Brown 1929:128). A few years later, he again provided numerical estimates, echoing Kidder's more conservative supposition that Folsom groups were on the landscape "15,000 to 20,000 years ago" (Brown 1932:82; also Brown to Howard, May 16, 1935, VP/AMNH). In those preradiocarbon years, such estimates were necessarily based on general assumptions about sediment accumulation rates or the observation that if the extinct fauna was Late Pleistocene in age, then it would therefore be in that temporal range, based on the antiquity then assigned to this period in geological history (e.g., Antevs 1925, 1931).

Still, the co-occurrence of artifacts and bison, on which the relative age estimates were based, was necessary but not sufficient evidence of the site's antiquity, since the taxonomy and age of the extinct bison were still in question (e.g., Antevs 1935:303; Bryan 1941:507; Roberts 1937:155; Romer 1933:70). Brown was confident this was a late Pleistocene species of bison; Hay thought it earlier (Brown to Frick, August 28, 1928, VP/AMNH; Romer to Cook, May 26, 1931, HJC/AGFO). For others, the possibility was still open that this species survived into the Holocene (Cook to Hay, January 25, 1928, OPH/SIA; Simpson to Brown, July 25, 1928, VP/AMNH). As Kirk Bryan lamented, "We know so little about the Pleistocene faunas" (Bryan to Wetmore, August 16, 1928, USNM/SIA).

Therefore, during his Smithsonian-sponsored fieldwork at Folsom in 1928 (fig. 2.14; chapter 4), Bryan sought

Stratum designations

w

m2

m1

f3

f2

FIGURE 5.1 Peter Kaisen in front of the North Bank profile, 1928, with the stratigraphic formations used in this analysis demarcated. (Photo courtesy of American Museum of Natural History.)

independent geological evidence of the site's antiquity. Using remnant terraces and benches in the valley (see Bryan 1929, 1937), Bryan reconstructed four main stages in the region's alluvial history—the last of which was represented by the present valley of Wild Horse Arroyo. In each successive stage, rivers and streams cut below the level of older and broader valleys, leaving behind a stair step of valley-floor remnants. The development of these "four successive stages doubtless required all of Pleistocene time" (Bryan 1937:143).

The floodplain deposits in the most recent of these valleys were dominantly comprised of younger alluvium, little different from floodplain deposits in other New Mexico streams that contained "relics of the Pueblo Culture." But these deposits also contained a rare pocket of older allu-

vium, within which was the Folsom bison bonebed (Bryan 1937:142–143). If the younger alluvium was of the order of 1,000 years old, then the older alluvium "on the ordinary criteria used by geologists . . . would be considered Early Recent or very Late Pleistocene" (Bryan 1937:143; also Bryan 1929:129). Several years later, Bryan (1941:511) would round up the age of the Folsom culture (not site) to "25,000 ± years ago," based on his work at Lindenmeier and correlating terraces and glacial deposits of the southern Rockies with continental glaciers of the Middle West (for a historical discussion of Bryan's work, see Haynes 1990, 2003; Holliday 2000b).

Many who worked at or visited the site made note of the fact that the bone-bearing deposits contained a great deal of charcoal. Yet, none of it, so far as Clark Wissler could tell, appeared to be cultural in origin: "It is found scattered and generally, but not suggestive of a camp fire" (Wissler, field diary, August 2–10, 1928, ANTH/AMNH; also Wissler to Howarth, August 26, 1931, VP/AMNH). Although radiocarbon dating was still two decades into the future, the charcoal was nonetheless of interest. Wissler collected some while he was there in 1931 overseeing Merrill's mapping of the site (Wissler to Howarth, August 26, 1931, VP/AMNH).

In early 1933, Cook wondered if charcoal from the site might be dated using dendrochronology. He asked Howarth if, on his next trip to the Folsom site, he would look in the arroyo a short distance downstream

> to see if there is still any charcoal exposed in the bottom of it,—where that was, in those "fire pits (?)" in contact with the undisturbed Pierre,—just about the head of that little narrow wash. I'm sure you know the spot. If you find any bits big enough to show any number of annual rings,—please collect them, and send them up. I want to examine them; and will forward them to a man who is specializing on this,—Douglass,—to see if he can date them. I doubt it,—but it is interesting to try. I will give you full credit for doing it. . . . (Cook to Howarth, February 17, 1933, HJC/AGFO)

Howarth knew the spot Cook was directing him to and was sure the deposits would be "near the same age as the finds in the Folsom pit" (Howarth to Cook, February 28, 1933, HJC/AGFO). In July, 1933, Howarth finally got out there, a few days after a heavy rain had fallen and exposed charcoal in the arroyo walls—and two projectile points at the site (chapter 8). Howarth collected the points and the charcoal and sent the latter to Cook (Howarth to Cook, July 25, 1933, HJC/AGFO).

The charcoal Howarth collected must have held little dendrochronological promise, for Cook merely saved the sample. Given the limited temporal range of dendrochronology in those days, he may have realized submission of the sample was pointless, even if annual rings were visible. But soon after radiocarbon dating was invented, and at the suggestion of Frank Roberts (Roberts to Cook, October 10, 1949, HJC/AGFO), Cook submitted the sample to Willard Libby at the University of Chicago, with the following explanation:

> The sample of old charcoal I collected in July, 1933, from below the Folsom bison and artifact level, in the arroyo of the type site of that cultural group. The arroyo which had cut a narrow, steep channel, down through this bison bone and artifact level, as I first saw it before any excavation work was done there, had exposed the edge of what appeared to be an old "fire pit," just a little below, and downstream from the horizon in which the bones occurred. When I was there, at the time I collected this charcoal, a recent heavy rain had better exposed this "fire-pit. (Cook to Libby, December 7, 1949, HJC/AHC; portions in Arnold and Libby 1950:10)

That Cook reneged on his promise to credit Howarth for collecting the sample is troubling but hardly surprising. More unfortunate is the fact that he submitted a sample assuming, but not actually knowing, where it was obtained. That would prove to be a mistake.

Libby was sure the sample would provide enough carbon for dating (Libby to Cook, December 15, 1949, HJC/AHC) and, within the year, had results. Using the original solid carbon method, two ages were obtained: $4,575 \pm 300$ and $3,923 \pm 400$ [14]C yr B.P. These were averaged by Libby to $4,283 \pm 250$ [14]C yr B.P. (C-377).[1] Libby offered the laconic observation that the age was "surprisingly young" (Arnold and Libby 1950:10).

This first radiocarbon date on a Paleoindian site "caused considerable comment when [it] was released" (Roberts 1951:20; also Evans to Cook, January 20, 1951, HJC/AHC). While the charcoal sample was believed to have come from *below* the bonebed, the age "was entirely too low in the opinion of many archaeologists and geologists and was completely out of line with dates for other materials known to be later stratigraphically" (Roberts 1951:20). For that matter, it was far too young for either Early Holocene or Late Pleistocene.

Cook privately admitted to Glen Evans that he'd condensed "too much in the statement of where I got the charcoal," although he never admitted that it was not so much condensed as it was simply untrue. He claimed that during the original excavations he had been so immersed in the work in the bone quarry that he had not spend enough time on the stratigraphic details, which proved to be more complex than he had originally perceived (Cook to Evans, February 7, 1951. HJC/AHC). In order to clear up this "widespread misunderstanding," in June 1950, Cook

> revisited the Folsom site (where I had charge of the original excavations),[2] for the purpose of re-examining the site. Here a condition I had suspected was clearly seen; namely, that the fire pit from which the dated charcoal came, while old, was definitely much younger than the deposit from which the original Folsom *Bison* and artifacts were recovered. Erosion of the past eighteen years has better exposed these beds. . . . (Cook, in Roberts 1951:20).

The stratigraphic details may well have been hard to see in 1933, and only clearer as a result of subsequent erosion. Still, much of the misunderstanding was a result of Cook's dissembling: He reported that the charcoal came from "a narrow valley" that cut through the "original [Folsom age] deposit." But he never specified where that fire pit was relative to the bonebed, even in his clarification note. In one paragraph, Cook put it "a few yards downstream" of the bonebed, while just a few paragraphs below he had it "some hundred feet [30 m] plus or minus," east of the bonebed (Cook, in Roberts 1951:20). In a letter to Libby written at the same time, Cook admitted he couldn't be certain about the stratigraphic position of the charcoal until he could examine early photographs "to be sure just where I got that charcoal, in relation to present erosion,—I know off-hand within a very few feet,—from memory" (Cook to Libby, September 30, 1950, HJC/AHC; also Roberts to Cook, October 5, 1950, HJC/AGFO). That was a convenient excuse, of course, since he could not have had either memory or photographs of collecting the charcoal, since he had not actually collected the charcoal—Howarth had.

There is a deeply incised ravine matching Cook's instructions to Howarth (Cook to Howarth, February 17, 1933, HJC/AGFO) that enters Wild Horse Arroyo a few hundred meters downstream of the Folsom site, and high in the stratigraphic section lenses of apparently burned earth. Samples from this ravine, and from deposits upstream of the site, were subsequently investigated and sampled by Anderson and Haynes in 1970 (Haynes, unpublished field notes). Anderson (1975:39–40) suggested that the features sampled were not hearths at all but, instead, burned sediment lenses from natural forest or range fires.

Still, not knowing precisely where or from what context Howarth obtained the sample, it is impossible to say whether or not he sampled hearths. Cook himself (in Roberts 1951:20) claimed that "in size and shape the charcoal lens looks like" a hearth, but by 1950 erosion had removed "all traces" of the spot from which Howarth had collected the charcoal nearly 20 years earlier. However, there are two points worth noting: first, Cook's description of the deposit in which the charcoal was obtained—a sediment "much darker and readily distinguished" from the *f2* sediments of the bonebed, and which filled a valley cut "as deep or deeper than the current arroyo bottom" and thus was below the level of the bonebed (Cook, in Roberts 1951)—easily fits what would later be designated the McJunkin Formation. The ages of the McJunkin Formation are roughly similar to the C-377 determination, as discussed below. It seems likely the sample Cook submitted came from this unit (Anderson and Haynes 1979:897). Second, there may have been anthropogenic charcoal in that stratum; one of our dated samples (CAMS-57518) from the base of the McJunkin Formation on the South Bank was obtained from a lens of charcoal that may have been a hearth, which is discussed more below. Thus, the first age on the Folsom site, although irrelevant to the Folsom Paleoindian occupation, is nonetheless a usable age for marking an episode in the Holocene history of the area.

Establishing a Stratigraphic Framework: The Folsom Ecology Project

Subsequent geological and geoarchaeological fieldwork (Anderson and Haynes 1979; Haynes, Anderson, and Frazier, 1976; Haynes et al. 1992), which included a series of backhoe trenches and profiling of exposed arroyo walls, was conducted in the 1970s, primarily on the North Bank. On the basis of these investigations, Haynes named and described three major Pleistocene/Holocene formations overlying the Cretaceous shale: From top to bottom these were the Wildhorse, McJunkin, and Folsom formations, and they are essentially equivalent to the Upper, Middle, and Lower units described by Figgins, Brown, and others (above).

Haynes has since further subdivided some of those stratigraphic units, and/or refined those subdivisions, but the broadly defined formations remain (Haynes, personal communication, 1997). The units are shown (as best we can determine their boundaries) on the North Bank behind Kaisen in 1928 (fig. 5.1) and are listed in table 5.1, which also provides the ages from radiocarbon samples obtained during the Folsom Ecology Project, as well as from subsequent site visits—all of which comprise the radiocarbon data available at the time we began our work.

By the mid-1990s the stratigraphic history and geochronology at the site were broadly understood: The eroded Cretaceous surface was overlain unconformably by the *f1* sediments that began accumulating in latest Pleistocene times. The more massive overlying stratum *f2*, a pale brown clayey silt, contained the Folsom bison bone and artifacts in its upper part—with those remains partially coated with secondary calcium carbonate. The depositional origins of the *f2* were unknown or at least unspecified. There was no "distinct surface of occupation" on which the archaeological material was resting (Haynes et al. 1992:87).

There were indications, notably in the "inclination of the cultural zone," suggesting that slope wash had contributed to the in-filled sediments (Haynes et al. 1992:87). Just how much slope wash may have occurred and, more significantly, how much of the material in the bonebed had been reworked would only become apparent following our more extensive exposure of faunal remains on the South and North Bank. Because only limited work had been done on the South Bank in the 1970s, the geological and paleotopographic context of the main portion of the bonebed was apparently unknown.

Haynes, Anderson, and Frazier (1976) observed that there was a knickpoint in the floor of Wild Horse Arroyo a few tens of meters upstream of the site, and inferred that it might have been an obstacle of sufficient magnitude to block the passage of bison long enough to enable them to be dispatched by the Folsom hunters. To be sure, along the present channel there is a constriction and an abrupt rise in the arroyo floor (fig. 3.5), but we suspect this channel is relatively recent (chapter 3). We do not know if there was a similar knickpoint in the bedrock within the Late Glacial paleovalley.

The top of the *f2* was observed by Haynes and others to be crosscut by the coarsely laminated *f3* sediments, indicating "a very shallow pond or a low gradient discharge and shallow water table during deposition" (Haynes, Anderson, and Frazier, 1976). Such may have accounted for the iron stains in the *f1*, the lowest part of the Folsom Formation, which implied saturation by groundwater some time after

TABLE 5.1
Haynes's Stratigraphic Unit Descriptions and Radiocarbon Ages for the Folsom Site

Stratigraphic Unit	Radiocarbon Age (^{14}C yr B.P.)
Wildhorse Formation *(w)*	
Sandy silt: dark gray, interbedded light and dark gray layers (10–20 cm thick) of clayey sandy silt with thin layers of charcoal or decayed plants near top of unit and lenses of pebble to cobble gravel and shingle. Unconformably overlies *m2*.	
McJunkin Formation 2 *(m2)*	4,470 ± 90 (TX 1272)
Clay: very dark gray to black organic clay with rootlet molds and moderate, medium, prismatic soil structure breaking to angular. Unconformably overlies *m1*.	4,850 ± 120 (TX 1270)
McJunkin Formation 1 *(m1)*	6,060 ± 500 (TX-1452)
Clayey silt: yellowish-brown clayey silt with alternating light and dark layers and thin layers of carbonized plants. Separated from *m2* by a weak erosional contact that truncates a black soil on *m1*. Unconformably overlies *f3*.	6,910 ± 110 (TX-1271)
Folsom Formation 3 *(f3)*	10,630 ± 80 (AA-7089; humates)
Silty clay: interbedded brown and dark grayish-brown silty clay with caliche nodules and strong, medium to coarse prismatic soil structure breaking to fine blocky. Dispersed charcoal in lower 25 cm. Unconformably overlies *f2*.	11,100 ± 130 (AA-7090; carbon residue)
Folsom Formation 2 *(f2)*	10,260 ± 110 (SMU-179)
Clayey silt: pale brown clayey silt with coarse caliche nodules, rootlet molds, and iron stains. Interfingers with shingle colluvium adjacent to shale bedrock. Contains Folsom artifacts and scattered charcoal. Conformably overlies *f1*.	10,760 ± 140 (AA-1709)
	10,780 ± 100 (AA-1213)
	10,850 ± 190 (AA-1711)
	10,890 ± 150 (AA-1710)
	10,910 ± 100 (AA-1712)
	11,060 ± 100 (AA-1708)
Folsom Formation 1 *(f1)*	12,355 ± 210 (AA-7090; carbon residue)
Silty clay: gray, iron-stained clay. Unconformably overlies bedrock Pierre Shale (Smoky Hill Shale) in observed exposures.	12,395 ± 90 (AA-7091; humates)
Pierre Shale (Smoky Hill Shale)	

SOURCE: After Anderson and Haynes (1979:table 1).

deposition. The top of the *f3* itself was in turn eroded and subsequently buried by a younger alluvium, the McJunkin units of Middle Holocene age, and then by the alluvial Wildhorse Formation.

Samples for radiocarbon dating were collected by Haynes and Anderson from the site's exposed North Bank. These samples included bison bone fragments and charcoal from various strata, including the *f2*. Samples were also collected from the McJunkin Formation as exposed in Wild Horse Arroyo upstream and downstream of the site and from cut banks along the Dry Cimarron—the latter as part of an effort to assess the age of volcanic activity at Capulin (Anderson and Haynes 1979; Haynes, unpublished field notes, July 1970).

The bone fragments from the site were analyzed as part of an experiment in the efficacy of bone dating (Hassan 1975). The laboratory extraction produced good bone collagen and

the first radiocarbon age for the Folsom level at the type site: 10,260 ± 110 ^{14}C yr B.P. (SMU-179; Hassan 1975:table 19; Haynes et al. 1992). This age was in close agreement with then-available ages from other Folsom sites (Anderson and Haynes 1979:896–897). Still, bone dates from the 1970s were suspect, given their potential for contamination and the state of extraction and laboratory protocols (R. R. Taylor 1980).

Haynes had also collected charcoal samples from the *f2*, but these were too small for dating using conventional techniques then available (Anderson and Haynes 1979:896). When radiocarbon dating by accelerator mass spectrometry (AMS dating) became available a decade or so later, these samples were submitted for analysis. Five of them were individual charcoal flecks, while the sixth was a composite of the others. The dates ranged from 10,760 ± 140 ^{14}C yr B.P. (AA-1709) to 11,060 ± 100 ^{14}C yr B.P. (AA-1708) and produced a

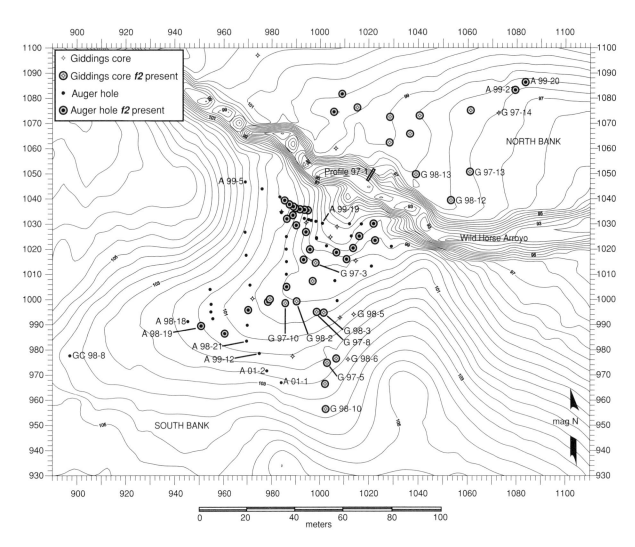

FIGURE 5.2 Location of Giddings soil core and bucket auger holes, 1997–2001.

mean age of 10,890 ± 50 ^{14}C yr B.P. (Haynes et al. 1992:84). Haynes supposed that this average age on charcoal was more accurate than the radiocarbon age derived from the bison bone collagen (Haynes et al. 1992:87). Yet, that supposition rested on the assumption that the charcoal was anthropogenic, derived from a Folsom-age hearth. Haynes admitted he could not preclude the possibility the charcoal had its origin in natural fires. Given the wide discrepancy—630 radiocarbon years—between the charcoal and the bone collagen radiocarbon ages from within the *f2*, ostensibly from the same event, more work on the age of the Folsom occupation was obviously necessary.

Recent Investigations into the Geology of the Folsom Site

As noted at the outset of this chapter, the goals of our investigation of the geology of the Folsom site were, broadly, threefold: (1) to gain a sense of the configuration of the paleotopography, in order to identify features of the land-

scape that may have been used by the hunters to reduce the risks of the hunt; (2) to understand the stratigraphic histories of the paleotributary and paleovalley, so as to gain insight into the depositional and erosional processes (and the climatic and environmental conditions behind them) in those different settings and derive a first approximation of the stratigraphic context and taphonomic history of the bonebed (as a prologue to more detailed analyses in chapter 7); and (3) to develop better chronological control over the site's stratigraphic history and, of course, narrow down when the Paleoindian bison kill occurred. By gathering evidence along these several lines, we also would be in a better position to identify areas of the site where previously undiscovered intact Late Glacial deposits might occur (or ascertain such were missing) and, therefore, better assess the likelihood of finding traces of associated habitation areas. The remainder of this chapter is divided generally along those three lines, in each case first presenting the basic geological evidence, then in a separate subsection describing what that evidence implies for our understanding the archaeology of the site. Before that discussion, however, a brief summary is provided of the geological and geophysical field methods used in these investigations.

Geological and Geophysical Methods

The methods in use included machine coring, hand augering, and geophysical remote sensing techniques, as well as mapping of exposed sections along the North Bank of Wild Horse Arroyo and in our excavation areas. The latter were somewhat limited in the information they provided, insofar as they occurred within the area of the bonebed, where the upper 2 m to 3 m of deposits had already been removed by the 1920s excavations and by our clearing of remaining overburden prior to excavation. Sections along the North Bank of the arroyo were relatively complete and several profiles were described in detail, although these are relevant primarily to the stratigraphic history of the paleovalley.

Therefore, the more complete picture of the stratigraphy on both the South and the North banks was obtained from machine coring and hand augering. The coring was done using a trailer-mounted Giddings soil probe during the 1997 and 1998 seasons of fieldwork. Overall, 33 soil cores were obtained, 21 of which were on the South Bank. The cores varied in depth but on average were 3.83 m (12.56 ft) deep. In general, cores were placed within a 100-m radius of the bonebed. The placement of cores did not follow a specific plan but, rather, was generally aimed at mapping the extent and occurrence of Folsom-age bone-bearing (*f2*) sediments, variation in the stratigraphic sequence across the site, and the subsurface depth and configuration of the bedrock in both the paleotributary and the paleovalley and, occasionally, to aid in positioning excavation units. As cores were completed and the sediment examined, the observations made would often guide the placement of subsequent cores.

Hand augering was done over all three seasons. All together, 64 auger holes were put in at an average depth of 2.17 m (7.12 ft), of which the great majority (60) was on the South Bank in the area in and around the bonebed. Because augering does not produce intact sediment segments, and often cannot reach the depths that a machine corer can, the augering was mostly done in areas where it was not possible to maneuver the Giddings rig or was aimed at otherwise filling in gaps in the stratigraphic coverage. Furthermore, because it is difficult to detect fine-scale stratigraphic changes in the sediment churned up in a bucket auger, the aim of the effort was often restricted to recording the presence of particular stratigraphic units or the depth to bedrock.

All sediments brought up in core drives and augers were examined for archaeological debris. Once pulled, each core was split longitudinally and described using standard methods of field soil description (e.g., recording of horizons, depth, texture, structure, color, boundary conditions, and effervescence). More limited descriptions were obtained from sediment in the augers. Distinctive or potentially diagnostic sediments were often sampled from cores and,

FIGURE 5.3 Douglas Wiens firing signal shotgun during seismic refraction surveys of the site area, 1997. (Photo by D. J. Meltzer.)

to a lesser extent, augers. Where relevant or of interest, charcoal from cores was collected for radiocarbon dating. The depth of the sample was determined to ±5 cm, as vertical measurements are not as precise in coring as they can be in controlled excavations. The locations of all core and auger holes were mapped by EDM/Total Station and are shown in figure 5.2.

Geophysical remote sensing work was conducted at Folsom in the fall of 1997, by advanced students from Washington University's Environmental and Exploration Geophysics course, under the direction of Drs. Roger Phillips and Douglas Wiens. Two methods—seismic refraction (fig. 5.3) and electrical resistivity—were used.

A total of six seismic lines were run, four on the South Bank and two on the North Bank of Wild Horse Arroyo (fig. 5.4). These ranged from 60 to 120 m in length; the combined length of these lines was 550 m. Geophone and shot spacing varied on each line; geophone intervals varied from 1 to 2 m; shot intervals, from 5 to 20 m. The refracted waves were analyzed, and the time delay method used to calculate the depth to bedrock for each line. There were four electrical resistivity lines run (RL1–RL4), all on the South Bank and primarily on the western side of the 1928 excavation area, and these extended over a total of 128 m. Wenner arrays were used on lines 1, 3a, and 3b; dipole-dipole arrays were used on lines 2, 3, and 4. Cathode and electrode spacing varied on the lines.

The remote sensing was done prior to the completion of the bulk of the coring, augering, and excavations on site and, thus, could not benefit from foreknowledge of the subsurface topography. Although in retrospect the lines were not ideally positioned, the methods were nonetheless well suited to mapping the interface of the bedrock and overlying unconsolidated sediment and identifying general trends in the bedrock morphology, which could subsequently be ground-truthed with coring and augering.

Resolution of the remote sensing was somewhat constrained by the lack of a sharp boundary between the Smoky

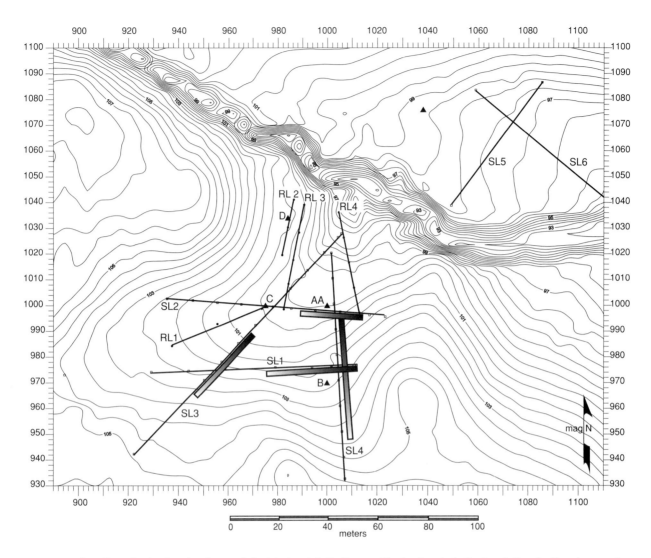

FIGURE 5.4 Location of seismic refraction and electrical resistivity lines, with shading to indicate relative depth, where such could be reliably inferred.

Hill Shale and the overlying sediment. The upper surface of the shale, as is evident in exposures along the present valley wall, weathers into a gradual rubble of "shingle shale" that diffuses the stratigraphic boundary (Anderson and Haynes 1979). Hence, the modeled contact is that of the underlying consolidated bedrock. Use of the core/auger data to map the bedrock surface is also limited by the diffuse nature of the boundary between the Smoky Hill Shale and the overlying sediments, insofar as auger bucket and core barrel can be obstructed by the shingle shale rubble. However, this is likely more of a problem nearer the margins of the bonebed, where there is an added component of shingle shale from slope wash.

Moreover, where core and auger holes are close to the remote sensing lines, there proved to be good agreement, generally within 1 m, between the elevation of the top of the Smoky Hill Shale as measured in the core and auger holes and the elevation as modeled by seismic refraction, suggesting a reliability to the methods. There was, however, a much greater discrepancy—upward of 4 m—between the

core and auger data and the depths modeled by electrical resistivity, likely a function of groundwater effects and the relatively coarser resolution of this technique (Roger Phillips, personal communication, 1998).

Mapping Bedrock and Reconstructing Paleotopography

It was known from the prior work of Anderson and Haynes (1979) that a thick sequence of Quaternary sediments rests unconformably on the Cretaceous-age Smoky Hill Shale bedrock, and that the Smoky Hill Shale (fig. 3.5) is the primary control on the topography of the site area today—as it presumably was in Late Glacial times. Although those sediments obscure the contours of the Late Glacial surface (also Bryan 1929), the present topography at the Folsom nonetheless provides a few clues to the form of that buried landscape.

The most obvious of those is the fact that immediately upstream and downstream of the site a high bedrock wall flanks the southern edge of Wild Horse Arroyo. However, at the site itself the bedrock wall is absent, the gap extending

South Bank

North Bank

Late Glacial
debris flow fan

recently eroded
shingle shale

A

Late Glacial
debris flow fan

consolidated
bedrock

B

FIGURE 5.5 A. View west up Wild Horse Arroyo, showing recent (2001) debris flow off the South Bank and the Late Glacial debris flow fan preserved at the base of the North Bank, 2001. B. Looking south at the South Bank, showing the Late Glacial debris flow fan overlying consolidated shingle shale bedrock, 2001. Ethan Meltzer. (Photos by D. J. Meltzer.)

over a linear reach of >30 m along the arroyo. In effect, and this must have been apparent during the 1920s investigations, the site and bonebed on the South Bank are located within a low, sediment-filled area between two bedrock uplands.

Fronting that gap is an ~32-m-wide, fan-shaped deposit of shingle shale that rests atop the Smoky Hill Shale bedrock (fig. 5.5). The shingle shale comprising this unit is not in primary context, evidenced by high inclinations of individ-ual pieces of shingle shale. Where exposed in profile, this debris flow fan is of varying thickness, reaching a maximum of 1.70 m along the South Bank. The gap in the bedrock and the debris flow fan appear to be related features, and repre-sent the mouth of a paleotributary that opened into the paleovalley of Wild Horse Arroyo.

A further and final surface hint of the paleotopography: There is on the surface today a shallow and subtle drainage extending away from that debris flow fan and bedrock gap. Although it is difficult to see in the immediate site area owing to the modifications wrought by the 1920s excavations and subsequent erosion, that drainage extends south/southwest from Wild Horse Arroyo for some 30 m, then (at ~N1020 E1010 on our site grid) curves west/southwest and ultimately disappears ~100 m from the site.

On the supposition that these several features are the expressions of a buried paleotributary, we ran one series of seismic lines along and across this area to map the contours and configuration of the underlying bedrock (fig. 5.4). Seismic line 3 (hereafter SL3), run southwest to northeast, partly overlays that modern drainage; SL4 is on a converging bearing to SL3, though it originates >50 m to the east and runs almost due north. On both lines, seismic data reveal that the bedrock slopes from south to north as it approaches the arroyo. There is good agreement between these two lines on the elevations of that slope: The northern and deeper end of SL3 bottom out at a modeled elevation of ~94.7 m, while the northern end of SL4 bottoms out at modeled depth of ~95.2 m (Phillips et al. 1998). The overall drop of the bedrock in SL4 is greater, however, owing to the shallower depth of the bedrock at the starting point of this line.[3]

Even more suggestive, along SL3 there appeared to be a steep drop in the bedrock at a point ~50 m up from its north end (Phillips et al. 1998)—that is, slightly north of where SL3 and SL1 intersect (fig. 5.4) and, more importantly, north of the bonebed. In effect, there appears to be a bedrock sill or headcut in the paleotributary, as it extended up from the paleovalley.

Equally intriguing evidence came from SL1 and SL2, which revealed a gradual west-to-east slope across the site but then, toward the eastern ends of each line, a sharp plunge in the bedrock, which in SL1 is modeled as an ~3-m drop in just 7 m of horizontal distance, to reach a depth of ~6 m below the present surface. An even greater dropoff was modeled for SL2 (a drop of ~4 m across a comparable horizontal distance). SL2 also revealed a subsequent rise in the bedrock east of where it had plunged. SL4, which intersected at nearly right angles to both SL1 and SL2, showed a corresponding drop in the bedrock that corresponded closely to the position and depth of the drop off in the bedrock modeled on both these east-west lines (fig. 5.4).

These depth estimates are somewhat problematic, however, since they are based on data from the ends of the seismic line, where depth is estimated from only a few or just one datum point and not, as in the central portion of the line, from a series of points based on several time delay calculations (Phillips et al. 1998). Nonetheless, together these data seemingly confirmed that the paleotributary entered from the southwest, turned northeast, and emptied into the arroyo at the bedrock gap—much like the present surface drainage. They also revealed that there was possibly a second prong of the paleotributary

(that plunges at the eastern ends of SL1 and SL2), coming in more or less from due south, but without any visible surface expression.

In order to ground-truth the geophysical modeled depths and test the hypothesized two-prong configuration of the paleotributary, a series of core and auger holes was strategically placed across the site (fig. 5.2). Those confirmed that there are indeed two prongs to the paleotributary. The wider and longer of the two prongs enters the site area from west of GC 98-8, then trends east/northeast toward AH 98-19 and AH 98-21, from which it continues to GC 97-10 and GC 98-2 and then north to the bedrock gap at the arroyo junction. This portion of the paleotributary is apparent from the drainage visible on the modern surface, which must have in-filled it. There is, as noted, an abrupt headcut along this segment of the paleotributary.

The head of the second prong of the paleotributary is near GC 98-10. From there, the channel runs almost due north, where it joins the other prong in the vicinity of GC 97-10 and GC 98-2. The auger and core data provided, as the seismic data from SL2 data did not, a more precise measure of the rise in the bedrock on the east side of this prong. As figure 5.6 shows, the bedrock surface rises ~5 m, some 12 m west of where it bottomed out. In fact, the coring indicated that there is also a comparable rise in the bedrock close to the end of SL1 (5.25 m in the 4.5 m separating GC 97-5 and GC 98-6)—too close to the end, apparently, to have been be detected by the seismic surveys. This second, narrower prong of the tributary is largely invisible on the present surface, although one of the trenches dug during the Folsom Paleoecology Project may have encountered it (Haynes, Anderson, and Frazier, unpublished field notes, Trench 5 profile). The two prongs of the paleotributary were separated at their upper reaches by a high bedrock peninsula (encountered in AH 01-1 and AH 01-2) and came together ~20 m north of there and ~30 m south of the bonebed (fig. 5.6).

The paleotributary drained into a paleovalley that was configured somewhat differently than Wild Horse Arroyo is today, which, of course, is a recent feature of the landscape and merely the latest in a series of Holocene cut-and-fill episodes (of which, more below; also Mann 2003, 2004). The thick sediments of the North Bank almost completely bury the paleovalley, but evidence of its depth and configuration emerges in SL5 and SL6 (fig. 5.4). The modeled depth-to-bedrock along the paleovalley axis is as much as 4 m to 6 m below the modern surface (Phillips et al. 1998). Coring and augering on the North Bank confirm this estimate: The deepest core reached 6.15 m below surface (GC 97-13), for a basal elevation of 91.41 m. The thalweg of the present Wild Horse Arroyo is at about the same depth.

Also apparent in SL5, as well as several cores (GC 98-12, GC 97-13, GC 97-14) and augers (AH 99-21, AH 99-20) that were put in along that same line, is that the bedrock surface rises nearly 4 m as it approaches the northern valley

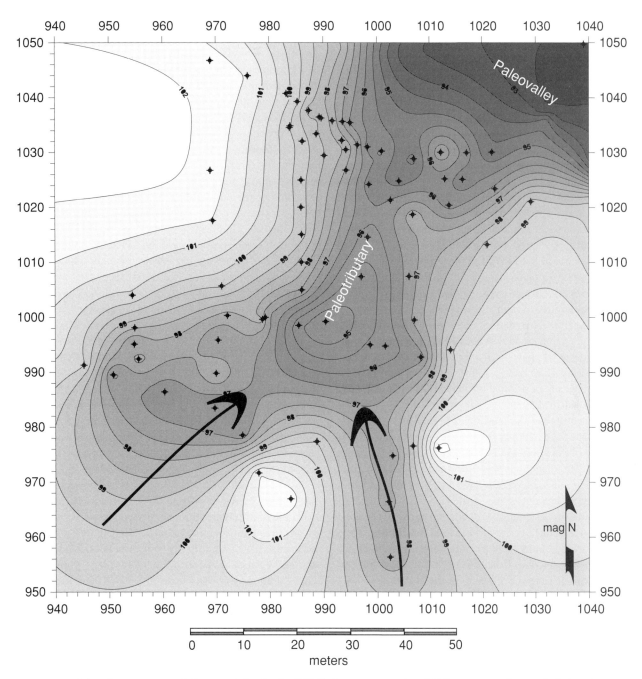

FIGURE 5.6 Subsurface contours of the top of the Smoky Hill Shale, as measured at Giddings core and auger hole locations. Darker areas of the map are deeper.

wall. Whether that rise in slope in Paleoindian times was gradual or more abrupt depends on how much sediment was draped over the valley margin. We suspect, based on weathering patterns observable in the Smoky Hill Shale today, that these were abrupt plunges—a vertical falling-away of the paleovalley wall—rather than steeply angled slopes.

The axis of the paleovalley thus appears to run roughly parallel to Wild Horse Arroyo but is ~30 m farther to the north. The portion of the paleovalley presently exposed by Wild Horse Arroyo along the North Bank is therefore the

margin of the paleovalley, with a stratigraphic sequence differing in details from that in the central portion of the channel, largely as a result of its proximity to inflow from the paleotributary. Consequently, the exposures on the North Bank are a complicated record of geomorphic processes occurring close to the intersection of both the paleotributary and paleovalley—the intersection likely spanned the area now occupied by Wild Horse Arroyo. The deposits in the South Bank, along with the main area of the bison bonebed, are set squarely within the paleotributary. However, bison bone and artifacts, as noted, are also found on the North Bank within the sediments of the paleovalley (chapter 7).

Maneuvering for the Bison Kill

This brings the discussion to what the seismic and stratigraphic data reveal of how the landscape may have appeared and possibly been used by the Folsom hunters. Although the majority of the bison remains that were recovered were in the paleotributary, bison remains were also found in the paleovalley. Thus, in thinking about how Folsom hunters may have used the topography to advantage, one must consider both settings.

We assume for the sake of discussion that the bison herd was initially confronted in the paleovalley. If the floor of the paleovalley was as flat as Wild Horse Arroyo is today (which drops at a rate of just 2 m vertically over a 100-m horizontal distance), and if there were exposed bedrock walls flanking the paleovalley as high and as abrupt as presently occur along Wild Horse Arroyo (fig. 3.5), which in the immediate site area are as much as 8 m above the arroyo floor, those lateral walls would have been a formidable barrier to a herd of bison looking to escape from a group of hunters who were strategically positioned to cut off the herds' exit up or down the paleovalley.

The paleotributary, sloping down toward the paleovalley, would have presented a possible escape outlet for the herd, and some of the bison clearly took that route or were maneuvered into this landform and killed there by the hunters. There are limited data on how steep a climb the bison would have faced in scrambling out of the paleovalley up into the paleotributary. There is a 3.61-m difference in elevation between the highest point at which we encountered bison remains on the South Bank and the lowest point at which such were recovered on the North Bank (elevations = 98.07 and 94.64, respectively, across a horizontal distance of 27 m). A similar vertical difference was apparent in 1928: The highest bone on the South Bank was 1.5 m below the surface; bone on the North Bank was ~3.6 m below the surface (chapter 7). Connecting those as end points of the same surface yields a relatively steep slope from the paleovalley up into the paleotributary, which would be over three times steeper than modern arroyo gradients in the area. However, since the North Bank bones are likely not in primary context (chapter 7), their depths may not reflect the base level of the paleovalley in Folsom times but, instead, depths reached during subsequent downcutting and later emplacement of those bones with the arroyo fill of the paleovalley.

Although we cannot say how easy or difficult it would have been for a herd of bison to scramble out the paleovalley into the paleotributary, we do have data on what obstructions may have blocked their escape once in the paleotributary—leaving aside, of course, strategic positions the hunters may have taken. As noted, several of the seismic lines (especially SL1 and SL2) show that the walls of the paleotributary were high and steep, a matter subsequently confirmed by

core and auger lines put in at right angles to the paleotributary. The data are shown as profiles in figure 5.7. All profiles are at the same scale; there is a 10× vertical exaggeration.

These profiles are not necessarily a precise rendering of the walls of the paleotributary as they may have appeared in Paleoindian times. After all, the upper surface of the bedrock was eroded, and portions of it were mantled in sediment by the time of the Paleoindian occupation, though the depth of that mantle, and the degree to which it modified the topography, is unknown. This makes it difficult to assess just how steep the walls of the paleotributary were at the time of the kill—and thus how difficult it might have been for the bison herd to have maneuvered in or escaped from the paleotributary.

These caveats notwithstanding, what is striking about figure 5.7 is that with the exception of Profile 3, which runs down the center of the eastern prong of the paleotributary, the side walls of the paleotributary were high and steep, with vertical dropoffs of the order of 3 m to 4 m in the span of a few horizontal meters. Profiles 1 and 2 are especially interesting in this regard, since both are close to the terminus of the larger, western prong of the paleotributary. These indicate the presence of a knickpoint at the upper end of the paleotributary, which could have been an obstacle to bison seeking escape via that route. Obviously, were hunters positioned atop the 3- to 4-m-high valley walls ringing the paleotributary, they would have been able to attack the bison while staying safely out of danger.

That the majority of the bison recovered were apparently killed in the paleotributary (chapter 7) is consistent with a hypothesis that these landscape features were used to advantage by the hunters in the kill. Of course, the herd could also have been initially confronted in this area of the site and escaped down into the paleovalley, reversing the topographic scenario discussed above. Or the hunters may have caught the animals in both areas simultaneously. Unfortunately, we have no data on how this process played out.

A final note with regard to the paleotopography and its consequences for the archaeology: The high bedrock walls on the west and northwest margin of the paleotributary effectively shielded the sediments within the paleotributary from fluvial action in the paleovalley and, thus, helped ensure the preservation of the site. Of course, those valley walls would not have protected the bonebed from erosion within the paleotributary, but there is little evidence that this occurred, and if it did, it was not on the scale of the erosion in the paleovalley. The absence of arroyo cutting within the paleotributary may also indirectly testify to the effects of the shingle shale debris flow fan (fig. 5.5) at the junction of the paleotributary and paleovalley, for by raising the base level of the paleovalley it would trigger aggradation, rather than incision, within the paleotributary. There are still other geological processes that came into play in the paleotributary that also served

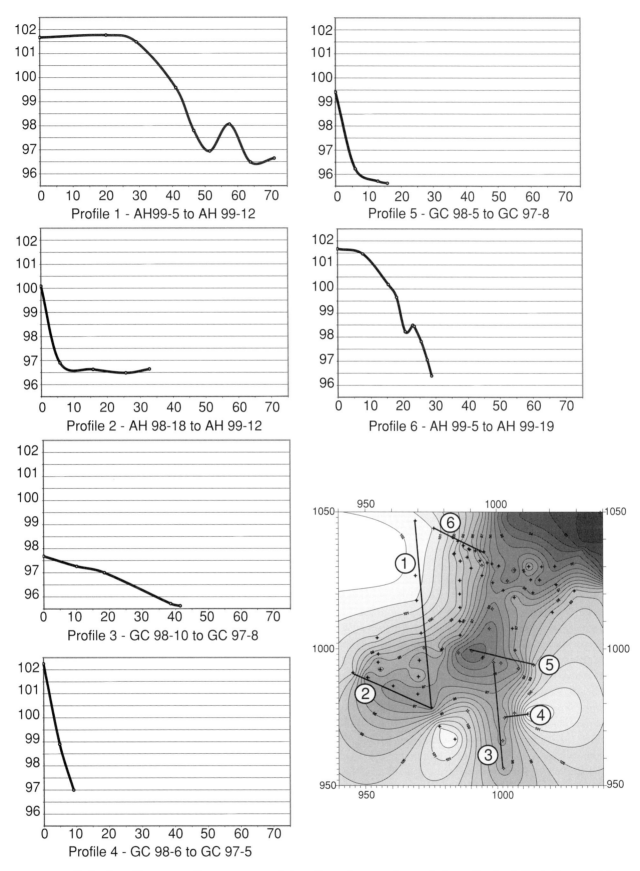

FIGURE 5.7 Bedrock profile cross-sections, as measured along lines of core and auger holes on the South Bank. Lower right map, from Figure 5.6, shows location of the cross-sections lines superimposed on the top of the Smoky Hill Shale.

to protect the bonebed, discussed in more detail below (also chapter 7).

Site Stratigraphy and the Geological Context of the Bison Bonebed

As noted, the depositional environments are very different in the paleotributary and paleovalley, with the sediments in the paleovalley possibly derived from multiple processes, including alluvial, colluvial, and eolian deposition, as well as being more susceptible to cut-and-fill episodes and fluvial reworking. The sediments in the better-protected paleotributary were less subject to (and show less evidence of) fluvial deposition and erosion. Bison bones and artifacts in the two areas will, as a result, likely have very different taphonomic histories. In this section we discuss the details of the stratigraphic sequences and sediments of the paleotributary and the paleovalley, then the geomorphic processes affecting these areas, followed by a discussion of the stratigraphic position and depositional context of bison remains within them. The latter addresses in part the taphonomic history of the bonebed (explored in more detail in chapter 7).

The lithostratigraphic subdivisions and terminology for the late Quaternary fill in Wild Horse Arroyo we use generally follow the tripartite scheme developed by Haynes (fig. 5.1 and table 5.1; Anderson and Haynes 1979; Haynes, Anderson, and Frazier, 1976) though with modifications. The modifications are necessary because the sequence was developed primarily on profiles and trenches from the North Bank and thus reflects the paleovalley sequence, which as noted differs from that of the paleotributary—although there are broad similarities, reflecting their response to common climatic and geomorphic triggers.

Two additional caveats. First, as just noted the North Bank only roughly approximates the paleovalley sequence, as this wall forms the paleovalley's southern margin, not its axis. Second, and more critical, stratigraphic interpretation has been complicated by disturbances due to subsequent geomorphic processes and archaeological activities. The present arroyo has cut down to the Cretaceous bedrock, and the 1926–1928 excavations removed much of the Holocene sediment from the South Bank, which triggered erosion on both banks. The net effect has been the removal of a critical portion of the stratigraphic record, namely, the top of the section on the South Bank, and of the junction of the paleotributary and the paleovalley. We include here (table 5.2) the stratigraphic descriptions of a representative section from the South Bank from Giddings Core 97-3, which was placed on the backdirt berm immediately south of the bonebed and thus in an undisturbed area of the site, and of Profile 97-1, which describes the stratigraphic sequence of the North Bank. Although it is difficult to connect the strata between the paleotributary and the paleovalley, we can surmise some of their stratigraphic relationships based on

observations made during the original work at the site and in the course of our fieldwork.

THE FOLSOM FORMATION

Stratum *f* rests unconformably on the eroded Smoky Hill Shale bedrock. The sediments are predominantly silt with layers of redeposited, angular shale fragments—shingle shale derived from the bedrock—more common at the base. Altogether the Folsom Formation is up to 290 cm thick, but the thickness varies significantly depending on (1) the topography of the underlying bedrock—the formation is thinner where the bedrock is higher, and (2) the amount of erosion of the Folsom Formation prior to its burial by the McJunkin Formation. The Folsom Formation is divisible into three subunits (hereafter, *f1*, *f2*, and *f3*). Paleoindian bison remains in the paleotributary occur solely within the *f2*; in the paleovalley, those remains occur in both the *f2* and the *f3*.

In most cores and North Bank exposures the basal unit of the Folsom Formation—stratum *f1*—consists of 50 cm to 170 cm of shingle shale lenses interbedded with layers of silty deposits (silty clay loam and clay loam) modified by iron oxide mottling. The shingle is light olive brown (2.5Y 5/4 dry) to light yellowish brown (2.5Y 6/4 dry) and derived from Smoky Hill Shale bedrock. The silty interbeds typically are light yellowish brown (2.5Y 6/4 dry). Stratum *f1* comprises the debris flow fan and represents episodic accumulation of angular shale fragments, perhaps in rapid succession, delivered as outwash from the paleotributary as well as from erosion of the high bedrock walls of the paleovalley. There is some sorting and imbrication of the shingle shale in this unit. Between periods of shingle accumulation the valley filled with layers of silty clay. Some time after burial, these basal layers of shingle and silt were subjected to a fluctuating water table that produced the distinctive iron-oxide (FeOx) mottles.

Stratum *f2* overlies *f1* in most cores and sections. Stratum *f2* is a massive deposit, ranging in texture from silt loam to silty clay to silty clay loam. Sedimentological data on *f2* sediments, primarily from the bonebed in the paleotributary, are provided in table 5.3; the textural data on sediments from unit *f2*, as well as other units for comparison, are illustrated in figure 5.8. Stratum *f2* is over 2 m thick in protected areas of the paleotributary (e.g., GC 98-3) but only ~1 m thick on the north wall of Wild Horse Arroyo. There is no distinctive *f1/f2* contact; instead, *f1* grades into the *f2*. Stratum *f2* has a yellow to light-brown hue, typically light yellowish brown (2.5Y 6/4 dry) to light brownish gray (2.5Y 6/2 dry). Mottles of more neutral grayish colors (grayish brown, 2.5Y 5/2 dry, to light gray 2.5Y 7/2) occur locally.

Stratum *f2* is calcareous (table 5.3) and commonly exhibits secondary, probably pedogenic deposits of calcium carbonate as threads, films, and fine tubules and nodules. The presence of pedogenic carbonate in *f2* was noted by the original investigators (Brown 1928b; Bryan 1937:142; Cook 1928b:39), who saw it as evidence of the great age of this deposit (Anderson and Haynes 1979:897). It is unclear what duration

TABLE 5.2
Stratigraphic Descriptions, North and South Banks

(A) Profile 97-1, North Bank

Unit	Soil Horizonation	Depth Below Surface (cm) and Description of Color, Texture, Structure, Boundary	% Sand	% Silt	% Clay	USDA Texture
w	A–C	0–40 cm: light yellowish-brown (2.5Y 6/4 dry) to olive brown (2.5Y 4/4 moist) silty clay; A = 0–10 cm, v. weak subangular blocky, C = 10–20 cm, weak subangular blocky; 20–40 cm, dense coarse gravel; abrupt lower boundary	13	47	40	SiC
m2		40–74 cm: light olive brown (2.5Y 5/4 dry) to olive brown (2.5Y 4/4 moist) clay with some mottling and few faint carbonate threads; moderate angular blocky; clear lower boundary	10	59	31	SiCl
	ABb1	74–100 cm: grayish-brown (2.5Y 5/2 dry) to dark grayish-brown (2.5Y 4/2 moist) clay, prismatic to angular blocky, faint olive mottling; gradual lower boundary	5	69	25	SiL
	Btgb1	100–136 cm: grayish-brown (2.5Y 5/2 dry) to dark grayish-brown (2.5Y 4/2 moist) clay; subangular blocky; v. dark-gray clay films (organs?) on ped faces; abrupt lower boundary	7	69	24	SiL
m1	ABb2	136–175 cm: dark grayish-brown (2.5Y 4/2 m) to v. dark grayish-brown (2.5Y 3/2 moist) clay with common films and threads of carbonate; ; prismatic and angular blocky common clay films (probably pressure faces?) on ped faces	4	74	22	SiL
		175–235 cm: laminated zone with grayish-brown (2.5Y 5/2 dry) to dark grayish-brown (2.5Y 4/2 moist) silt loam; subangular blocky; and dark grayish-brown (2.5Y 4/2 moist) to v. dark grayish-brown (2.5Y 3/2 moist) clay; angular blocky; the blocky clay has continuous clay films (probably pressure faces); abrupt lower boundary	7	93	23	SiL
f3	Bkb3	235–250 cm: light brownish-gray (2.5Y 6/2 dry) to dark grayish-brown (2.5Y 4/2 moist) clay with some lighter olive mottles; common threads of carbonate; common flecks of charcoal; common fragments of rock, especially in lower 10 cm; clear lower boundary	4	66	30	SiCL
	Bk1b4	250–300 cm: dark grayish-brown (2.5Y 4/2 moist) to v. dark grayish-brown (2.5Y 3/2 moist) clay, strong angular blocky with common threads and films of carbonate; clear lower boundary	11	62	27	SiCL
f2	Bk2b4	300–330 cm: grayish-brown (2.5Y 5/2 dry) to dark grayish-brown (2.5Y 4/2 moist) silt loam with rock fragments common; subangular blocky; common threads and films of carbonate; common flecks of charcoal; clear lower boundary	6	67	27	SiL
		330–360 cm: light brownish-gray (2.5Y 6/2 dry) to dark grayish-brown (2.5Y 4/2 moist) silt loam; weak olive-gray mottles; strong subangular blocky; clear lower boundary	5	59	36	SiCL

(continued)

TABLE 5.2 (*Continued*)

Unit	Soil Horizonation	Depth Below Surface (cm) and Description of Color, Texture, Structure, Boundary	% Sand	% Silt	% Clay	USDA Texture
f1		360–440 cm: light yellowish-brown (2.5Y 6/4 dry) to 2.5Y 5/4 moist) silt loam; heavily mottled; common Fe-ox stains, especially near bottom; strong angular blocky; few threads especially and films of carbonate; common shale fragments, in lower half	9	60	31	SiCL
		440–540 cm: covered				
Bedrock		540+ cm: shale				

(B) Giddings Core 97-3, South Bank

Unit	Soil Horizonation	Depth Below Surface (cm) and Description of Color, Texture, Structure, Boundary	% Sand	% Silt	% Clay	USDA Texture
1920s fill		0–20 cm: back dirt from berm				
	A	20–27 cm: dark gray (10YR 4/1 dry) to v. dark gray (10YR 3/1 moist) silt loam; weak angular blocky to granular; clear boundary	5	69	26	SiL
	Bt1	27–50 cm: v. dark gray (10YR 3/1 dry) to black (10YR 2/1 moist) silt loam; weak prismatic and strong subangular blocky; thin, continuous clay films on ped faces; clear boundary	6	70	24	SiL
	Bt2	50–80 cm: dark gray (10YR 4/1 dry) to v. dark gray (10YR 3/1 moist) silty clay loam; strong prismatic and strong angular blocky; thin, continuous clay films on ped faces; clear boundary	8	62	30	SiL
	Bt3	80–125 cm: dark gray (10YR 4/2 moist) silty clay; thin continuous clay films on ped faces; strong prismatic and strong subangular blocky; clear boundary	4	48	48	SiC
f3		125–157 cm: light yellowish-brown (2.5Y 6/4 dry) to light olive brown (2.5Y 5/4 moist) silty clay loam; moderate subangular blocky; carbonate and shale "shingle" at 153157cm; clear boundary				SiCL
		157–203 cm: light yellowish-brown (2.5Y 6/4 dry) to light olive brown (2.5Y 5/4 moist) silty clay loam; moderate subangular blocky; clear boundary				SiCL
		203–285 cm: light brownish-gray (2.5Y 6/2 dry) to dark grayish-brown (2.5Y 4/2 moist) silty clay loam; moderate subangular blocky; black clay lens at 203–204 cm; clear boundary	9	59	32	SiCL
		285–343 cm: light yellowish-brown (2.5Y 6/4 dry) to dark grayish-brown (2.5Y 4/2 moist) silty clay loam; moderate subangular blocky; common faint Fe-ox mottles; distinct Fe-ox lenses 298–305 cm; clear boundary	9	59	30	SiCL
f1		343–370 cm: light gray (2.5Y 7/2 dry) to light olive brown (2.5Y 5/4 moist) silty clay loam; moderate subangular blocky; common distinct Fe-ox mottles; distinct Fe-ox concentration 365–370 cm; abrupt boundary	12	54	34	SiCL
Bedrock		370 cm: shale				

NOTE: See figure 5.2 for location. Soil horizon notation provided only where pedogenic modifications present. We subdivided the section of Core 97-3 from 125 to 203 cm after sampling and laboratory analysis and, therefore, do not have separate textural data for the two units within that section. Sand, silt, and clay percentages rounded to the nearest whole number. v., very.

TABLE 5.3

Sedimentological Data on Samples from Unit f2

Sample No.	Very Coarse Sand	Coarse Sand	Medium Sand	Fine Sand	Very Fine Sand	% Sand	% Silt	% Clay	USDA Texture	% Organic Carbon	% Organic Matter	% CaCO$_3$
L23-8-12	<1	<1	<1	1	6	9	62	29	Silty clay loam	0.61	1.05	1.25
M15-24-46	1	Trace	Trace	Trace	3	4	51	45	Silty clay	0.45	0.78	16.47
M17-9-63	0	0	Trace	Trace	2	3	48	49	Silty clay	0.49	0.84	20.35
M17-21-16	0	0	Trace	<1	7	9	45	46	Silty clay	0.60	1.03	17.74
M17-23-90	0	0	0	Trace	2	2	52	46	Silty clay	0.48	0.83	16.64
M17-24-95	0	0	Trace	Trace	6	6	55	39	Silty clay loam	0.42	0.72	13.35
M18-4-16	0	0	0	Trace	4	5	53	42	Silty clay	0.48	0.83	15.39

NOTE: Particle size percentages rounded to the nearest whole number. Trace amounts are <0.5.

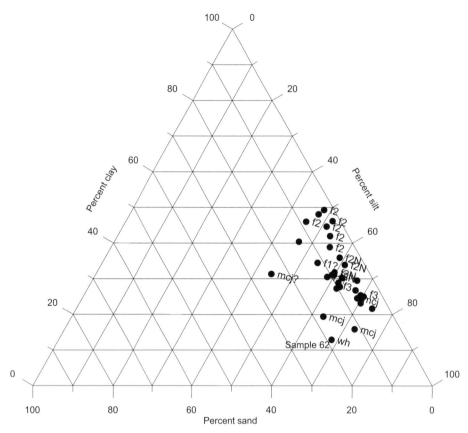

FIGURE 5.8 Textural triangle of Folsom site sediments from Folsom formation 2 (*f2*), Folsom formation 3 (*f3*), McJunkin (mcj), and Wildhorse *(wh)*.

of surface stability is indicated by these features of weak pedogenesis: Observations of modern Entisols developing in overbank settings along streams in the area suggest that under the present climate the accumulation of diffuse secondary carbonates, clay illuviation, and the development of prismatic structure occur rapidly, perhaps in a matter of centuries or less (Mann 2004). Other pedogenic modifications of *f2* include formation of subangular and prismatic soil structure with coats of illuvial clay on ped faces. Distinct Bk, Bt, and Btk horizons were observed in some but not all cores and exposures, making them difficult to trace through the site. Some finely divided gypsum is also present in *f2*, as noted by Brown (1928b; unpublished field notes, September 4, 1927, VP/AMNH) and in thin section (Goldberg and Arpin 1999).

Since the bison bonebed in the paleotributary occurs in the upper portion of stratum *f2* and is covered thinly by more *f2* sediments, the origin of this unit and what it may reveal of environmental conditions at the time of deposition are of considerable interest. We previously (Meltzer, Todd, and Holliday 2002) called attention to the similarity stratum *f2* displays to the physical characteristics of loess. Loess is not widely reported for the region, but as noted, Brown (1928b) also inferred an aeolian origin for this unit. Loess was observed in a study of volcanic rocks of the area: Collins (1949:1023) remarks that on some of the basalt

mesas "quaternary loess has been added to the decomposition products" but does not elaborate. Allen (1959), in examining soils formed on the basalt uplands, observed that loess (and volcanic ash) is an important component of the parent material of some soils. Similarly, the soils of Johnson Mesa, mapped as the Barela Series, and the Capulin series down in the valleys, "formed in . . . residuum derived from basalt and other volcanic debris that were modified by mixed eolian material" (G. G. Anderson et al. 1982:15). Loess would not be out of place in the region given the site's proximity to glacial and periglacial processes in the southern Rocky Mountains during the Late Glacial. The late glacial timing of *f2* deposition would also be appropriate for loess accumulation. Of course, as recent work has shown, not all North American loess is glaciogenic (Aleinikoff et al. 1999; Busacca et al. 2004; Mason 2001).

Alternatively, Mann (2004; personal communication, 2003) points out that the Smoky Hill Shale underlying the site is a source of abundant silt through physical and chemical weathering, and that fine sand, silt, and clay are the predominant sediments in valley fills. He observed deposits of silts and clays that closely resemble stratum *f2* elsewhere in the Upper Dry Cimarron area, dated to the early and middle Holocene, when regional-scale loess deposition seems less likely than during the Late Glacial. He notes, further, that although these are similar deposits, their distribution is patchy. Loess ought to form widespread "blankets" of sediment, since it is carried so high in the air column. In the absence of other indicators of aeolian activity—e.g., sand sheet sediments—and given the distance of Folsom from retreating ice sheets, outwash rivers, or other obvious sources, Mann argues that the *f2* was derived not from airborne loess, but primarily by colluvial deposition or slope wash from a local bedrock source.

To be sure, the Smoky Hill Shale is a ready source of fine-grained sediment, but evidence against the supposition that Stratum *f2* was largely derived from colluvium or slopewash includes the following.

- In places within the paleotributary the *f2* is several meters thick and largely devoid of any particles coarser than fine sand. Were the *f2* sediments derived from the Smoky Hill Shale, we should see lenses of shingle shale and a greater contribution of coarse particles throughout—as we do in the *f3* (below). The coring data further indicate that there was massive silt deposition inset against bedrock; the only coarse material is immediately adjacent to bedrock, even where there was 3 m to 4 m of relief over a distance of just a couple of meters—that is, even where steep slopes were present.
- Although weathering of the shale would produce fine sediments similar to *f2*, it would also produce more clay than is indicated by the textural analysis (table 5.3 and fig. 5.8). Also, in order to derive sufficient silt from the shale there would have to have been very rapid weathering, exceeding the weathering of the *f2*—and we see no soils or other weathering zones buried in *f2*.
- Silt deposits have been detected on the uplands and in the lowlands by us and others (above) and an airborne source accommodates this. Folsom is downwind from the Sangre de Cristos, which, while not subjected to massive glaciation, were and still are subject to periglacial processes that could produce silt.
- Those sources are not likely to produce massive quantities of silt, and though such silt would initially cover upland and lowland areas, given the regional topographic relief it would probably get quickly redeposited in lowland areas.

Examination of thin sections from stratum *f2* sheds some light on the matter. Goldberg and Arpin (1999) found that the *f2* consists of domains of quartz silt in which the fine fraction appears to have been elutriated, interspersed with finer, calcareous clay bands. In places the sediments are finely bedded, but they are commonly disrupted. There are occasional, coarse clay coatings, some of which are very thick. Calcium carbonate is present in four forms: (1) within the clay, (2) in precipitated hypocoatings around voids, (3) as grains in the sand and silt fraction, and (4) as shell frag-

ments. The carbonate is therefore both primary and secondary. A thin section across the boundary at the top of the Bk horizon just below the bonebed (discussed below) shows that it is much finer grained than other *f2* samples and is more porous. The sample also contains gypsum precipitated in large chambers and voids at its base. Gypsum formation disrupted the structure of the sediments resulting in the higher porosity.

What these thin sections indicate is that the *f2* may not be primarily airfall loess, however, but rather remobilized or redeposited silts derived from airborne loess. The thin sections display fine bedding with stringers of clay, indicating syndepositional reworking (Goldberg and Arpin 1999). Nevertheless, the absence of coarse clastics or pronounced bedding in the *f2*—except for several widely separated lenses of shale gravel—suggests that the loess was not extensively reworked by fluvial/colluvial runoff or sheetflow before it began to accumulate in the paleotributary.

Evidence for pedogenesis within the *f2* unit further complicates interpretations and indicates that the sediments accumulated there episodically. Anderson and Haynes (1979:897) note evidence for pedogenesis in *f2* in the form of "calcareous root molds and carbonate coating on ped surfaces" but propose that these pedogenic features suggest "aggradation at a rate rapid enough to prevent clear differentiation of soil horizons." The presence of distinct buried soils in *f2* in some cores indicates, however, that aggradation ceased or slowed significantly for substantial amounts of time. Further, because distinct buried soils seem to be discontinuous through *f2*, this unit may have been subjected to several cycles of erosion as it aggraded.

If the *f2* was ultimately aeolian in origin, it must have once been much more extensive than it is at present. Examination of the exposed walls along Wild Horse Arroyo, as well as along that portion of the Dry Cimarron valley from the base of Johnson Mesa to its downstream junction with Wild Horse Arroyo, reveals that deposits like those of the *f2* are exceedingly rare (also Bryan 1937:142–143). Sections of fine-grained silts are visible in the walls of Wild Horse Arroyo in a few areas upstream and downstream of the site and in a few spots along the Upper Dry Cimarron (Mann 2004). Unlike the situation at the site itself, these silts contain a higher concentration of platy and angular fragments of Smoky Hill Shale, indicating deposition in higher-energy settings.

At the Folsom site, *f2* sediments occur in both prongs as well as throughout the main portion of the paleotributary; essentially, all of the 1920s excavation areas, as well as our excavation blocks (fig. 5.2). The unit also occurs in most the core and auger placed on the North Bank. Assuming that the *f2* at one time more or less draped the region, its present scarcity on the landscape is testimony to subsequent widespread erosion.

Within the paleotributary, *f2* is capped by unit *f3*, which consists of angular, platy fragments of Smoky Hill Shale, generally <5 cm in maximum length. The shingle shale

FIGURE 5.9 CaCO$_3$ horizon below the level of the bonebed in the M17 block. (Photo by D. J. Meltzer.)

tends to be imbricated, poorly sorted, angular, and primarily from a single source—downslope movement of the Smoky Hill shale off the bedrock walls flanking the paleotributary. The shingle shale for the most part flowed across the top of the f2, and in the area of the bonebed it formed a lens—sometimes sets of lenses—between 10 and 30 cm in thickness. In the upper reaches of the paleotributary the shingle thins and altogether disappears. The distribution of shingle shale in the M17 and M15 excavation blocks is discussed in more detail in chapter 7, as it pertains to the taphonomic history of the bonebed. Here we note that in just a few places in the paleotributary the f3 came to rest directly on bison bone; overall the f3 deposit played an important role in the preservation of the site, for it effectively armored and protected the underlying bonebed from subsequent disturbance (e.g., erosion, rodent burrowing).

In contrast to the relatively homogeneous shingle shale capping the bonebed in the paleotributary, the clasts comprising stratum f3 in the paleovalley tend to be more rounded (i.e., gravel), show more size sorting, and occur in multiple, complex lenses of gravels, which include secondary carbonate nodules that may have been transported as well (as also observed by Anderson and Haynes 1979). These clasts are embedded in sediments that appear to represent continued deposition of fine-grained silt; while these silts bear a strong resemblance to those of the f2, and presumably result from the same depositional mechanism, they are stratigraphically defined as the fine component of the f3.

The f3 deposits in the paleovalley are of variable thickness, ranging from 70 to 232 cm, the variation a consequence of the slope of the channel both downstream and

from the margins to the axis of the channel, the irregular surface it filled, and the erosion that subsequently took place across the f2/f3 contact. The f3 here also has more complex layering and fine laminations than in the paleotributary. The laminated interbeds of silty clay contain scattered fragments of shale, all of which mark repeated episodes of low gradient fluvial erosion and redeposition. The size of the gravels, packets of which range from coarse to very fine (i.e., <8 mm), suggests that water velocity and turbidity here on the edges of the valley were irregular, as was stream competence. There is also, overall, a fining upward through this unit, from the complex lenses of sand, gravel, and occasional faunal materials lower down into laminated brown and dark grayish-brown silty clay loam and silt loam lenses higher up.

Although the f3 looks different in the paleotributary and the paleovalley, and the depositional mechanisms in those two areas are likely also different (colluvial vs. fluvial), we nonetheless suspect that the deposition of f3 across the site was essentially penecontemporaneous and that depositional processes in both areas were responding to the same underlying geomorphic/climatic trigger.

THE STRATIGRAPHIC CONTEXT OF THE FOLSOM BISON. In both the paleotributary and the paleovalley the bison bone occurs near the top of stratum f2, but not atop the f2.[4] In the paleotributary, where the stratigraphic context of the bone is more straightforward, there is no disconformity or other stratigraphic indicator of the surface on which the bonebed rests; the Folsom site may be rather unusual in this regard (Frison 1991; Hofman 1989, 1996; Holliday 1997).

Although there is no well-defined stratigraphic surface on which the bonebed is resting, the bones in the paleotributary were apparently deposited on a relatively level and well-defined surface, as they are distributed across a very narrow vertical span (chapter 7). There is a subtle change in soil texture and chemistry just below the bonebed: There is an increase in carbonate content (table 5.3) and a pronounced Bk horizon (fig. 5.9), which coincides with a change in texture and porosity of *f2*. The sediments below the bonebed are finer-grained and more porous than those associated with the bone. Bk horizons can be formed by such textural changes (Birkeland 1999:17; Gile 1975) and, more importantly, may be indicative of activity surrounding the kill and butchering. The actions of humans and their large prey could result in the introduction of coarser sediment from the valley walls and uplands and also reduce porosity through trampling.

The absence of a stratigraphic break marking an occupation surface, and the apparent rapidity with which the skeletal material in the paleotributary was buried (based on the very slight weathering of the bone surfaces; chapter 7), implies that *f2* deposition continued essentially uninterrupted through Folsom times, almost completely blanketing the bonebed soon after the carcasses were deposited. This further substantiates the inference that airfall loess alone might not have been the sole mechanism of *f2* deposition, unless one assumes that there was a substantial amount of dust falling out of the atmosphere and a very large and nearby donor source, which we do not. However, there is no evidence to suggest that the bison carcasses within the paleotributary, once deposited, were moved appreciably by fluvial action associated with either stratum *f2* or stratum *f3*. Indeed, there are several lines of evidence, detailed in chapter 7, to indicate that the faunal remains here are in primary context.

In sharp contrast, some of the bison bones recovered in the paleovalley were resting at very high angles (up to 79°). These bones are situated in and among multiple lenses of gravel in the *f3*, while other bone elements (or portions thereof) are found within small packets of fine-grained sediment (fig. 5.10). Some of those packets may, in fact, be noneroded "islands" of stratum *f2* surrounded by *f3*. Other packets, however, are clearly part of the *f3* fine component; in one particular instance, two ribs were found in proximity to one another, lying at high angles (37° and 49°) within fine-grained sediments sandwiched between lenses of fine gravels. Sitting directly atop the upper fine gravel lens was a large cervical vertebra, coated with calcium carbonate, amid fine-grained sediments. This evidence, in turn, hints that the deposition of *f3* took place over a relatively long period of time or that there were separate episodes in which bison bone was plucked from primary context upstream and then transported and deposited downstream.

The bone found in clean and gravel-free sediments in the paleovalley, that is, within those *f2* islands, or *f3* fine sediments, tended on the whole to be in better condition than the bone found within the gravel lenses. None of the bone in the paleovalley was articulated, and much of it from the *f3*

FIGURE 5.10 Bison rib fragments in unit *f3* on the North Bank, lying at high angles amidst gravel lenses. (Photo by D. J. Meltzer.)

experienced relatively greater amounts of breakage on the articular ends and surface attrition in comparison to the bone in the paleotributary, presumably from being rolled by water.

As a result, faunal remains in the paleovalley do not form—as they do in the paleotributary—a discrete archaeological horizon or even a recognizable bonebed. Furthermore, they are not solely within *f2* sediments or protected by an overlying shale shingle armor. Instead, they tend to occur as isolated elements in secondary context at high angles indicative of fluvial transport, sometimes jutting upward in the lowest levels of the *f3* (anchored in *f2* islands) or within the *f3*. The *f2* in the paleovalley has an irregular, erosional upper contact.

Despite these differences in stratigraphic context, we infer that the bison bones in both the paleotributary and the paleovalley are from the same event. The evidence supporting that inference is straightforward.

1. There is no stratigraphic evidence that more than a single kill took place in any area at this site.
2. The bison dental age cohorts are tightly grouped (chapter 7).

3. Radiocarbon ages on bison bone from both the paleotributary and the paleovalley are virtually identical, as discussed in more detail under Radiocarbon Dating and Geochronology, below.

All of this suggests that the postdepositional history of the faunal remains in the paleovalley was very different from the history of those in the paleotributary, and that divergence began soon after the original deposition of the bison carcasses. In both settings, the bones were deposited on top of, and in turn were buried by, *f2* silt. Slope wash in the paleotributary relatively soon thereafter carried in shingle shale that sealed the faunal remains in place, while fluvial action in the paleovalley dispersed and redeposited skeletal elements. In the paleovalley, erosion and redeposition likely continued at intervals throughout the Holocene.

Although the bison bones in the paleovalley were obviously transported and in a secondary context, and the remains in the paleotributary were in primary context, it is important to stress that it is not the case that the bison remains in the paleovalley were necessarily washed out from the paleotributary. Indeed, it may well be that some, perhaps even a majority, of the bison remains in the paleovalley were originally deposited there, as suggested by the following evidence.

1. Bison remains in the paleovalley have been found at least ~35 m *upstream* of the mouth of the paleotributary and, thus, could not have been redeposited from that source.
2. In 1928 projectile points were found in association with bison bone, and in 1972 an intact and well-preserved bison cranium was recovered from within the paleovalley. Either of these finds could have been in a secondary context; unfortunately, no records exist to indicate their precise depositional contexts, angles of orientation and inclination, etc.[5] Yet, their condition opens the possibility that they are in a primary context. That inference is not incompatible with the observation that reworking has been extensive in the paleovalley. Fluvial action need not have affected all of the carcasses deposited in this area.
3. Finally, although *f2* sediments were eroded by flow down the paleotributary in Early and Middle Holocene times (as discussed below), there is little corresponding evidence that bison elements in the paleotributary were moved or otherwise influenced by fluvial action (chapter 7). If this area was a significant feeder source of the bone in the paleovalley, there ought to be more evidence of movement in the paleotributary than is otherwise present. Admittedly, however, the absence of faunal disturbance is known only from our excavations on the west side of the paleotributary; we have no data on the east side of the paleotributary.

The possibility that bison were killed in both the paleotributary and the paleovalley raises several intriguing but likely unanswerable questions: Were the bison attacked in the paleovalley and sought the paleotributary to escape the hunters? Or was the kill in the paleotributary, and the animals tried to flee down the paleovalley? Or were bison dispatched—and processed—in both areas? If the main activities of the kill took place in the paleovalley, that might account for why the density of carcasses in the paleotributary is lower than that seen in other Folsom sites (e.g., Bement 1999b; Hofman 1999a). And if the bison remains in the paleovalley were deposited as partial carcasses, that raises the question of whether processing and habitation areas were once located in the paleovalley and might still be preserved under the thick overburden of the North Bank.

THE MCJUNKIN FORMATION (*m1–m2*) AND HOLOCENE CUTTING AND FILLING

Following the deposition of stratum *f3*, there was a series of cut-and-fill episodes, producing stratum *m*, the McJunkin Formation. We place the boundary between the Folsom and the McJunkin formations at the appearance of the laminated and bedded dark-gray and dark-brown layers above the *f3* shingle. The McJunkin Formation deposits filled and ultimately obscured the paleovalley and the paleotributary.

Stratum *m* covers stratum *f* throughout the South Bank and the North Bank and in all observed exposures along Wild Horse Arroyo. Stratum *m* is largely fine-grained silts and clays and is typically 200 cm to 300 cm thick (and up to 350 cm thick in places). Layers of shingle and gravel are common. Visually, stratum *m* is quite distinct from stratum *f*, because *m* is stratified and is generally much darker in color. The very low chromas and values and the relatively high organic carbon content of stratum *m* suggest deposition of fine sediments from dust, or slopewash in a heavily vegetated or wet environment, the latter inferred from the presence of gleying and mottling. Analogous slackwater deposits are common in the area today in tributary arroyos when the Dry Cimarron backfills its tributaries (Mann, personal communication, 2003).

The most distinctive and obvious stratification in stratum *m* is in its lower half on the North Bank, a zone identified as stratum *m1*. This subunit consists of layers of silt loam, silty clay loam, and silty clay, each a few centimeters to a few decimeters thick, with a few sandy interbeds and thin shingle and gravel layers. Dry colors of the layers are variable and include dark grayish brown (2.5Y 4/2), grayish brown (2.5Y 5/2), light olive brown (2.5Y 5/4), light brownish gray (2.5Y 6/2), light yellowish brown (2.5Y 6/4), and dark gray (10YR 4/1). The laminated zone of *m1*, though distinctive along the North Bank exposure (Profile 97-1), does not extend north of the arroyo to any significant degree, based on our coring. The presence of the laminations only in exposures opposite the paleotributary raises the possibility that these lighter-brown and yellowish-brown silts may be redeposited *f2* or *f3* flushed out into the paleovalley. The absence of the laminated zone along the axis of the paleovalley may be due to subsequent downcutting, or simply because these sediments were never deposited that far out from the mouth of the paleotributary.

Stratum *m2*, like *m1*, is stratified with layers of silt loam, silty clay loam, and silty clay. In *m2*, however, the individual strata are thicker (decimeters) and coarse clastics (i.e., shingle) are less common except in proximity to the bedrock valley wall. Stratum *m2* also is generally darker and duller in color than *m1*; the layers are black (10YR 2/1), very dark gray (10YR 3/2), dark gray (10YR 4/1), grayish brown (2.5Y 5/2), and dark grayish brown (2.5Y 4/2) (all dry colors).

The silty and clayey layers comprising stratum *m* contain lenses of shingle and gravel, typically along the valley margins. Sections toward the valley axis have little or no shingle or gravel. This may be a simple facies change—shingle and gravel are more common lithologies in proximity to the bedrock valley walls—but this characteristic may also indicate that fine-grained valley-axis units are cut into the bedded deposits. Cores from the center of the North Bank contain thick deposits of *m* over a deeply eroded surface of *f2*. For example, the top of the *f2* is at an elevation of 94.138 in core 98-13 along the valley axis but at an elevation of 98.084 in Profile 3 on the valley margin. Moreover, the laminated deposits of stratum *m1* (discussed above) are not found in the valley axis, suggesting that they may have been cut out.

Some of the silty layers within unit *m* exhibit sharp, irregular upper boundaries and several of these are capped by shingle or gravel, which is suggestive of cut-and-fill episodes. Earlier investigators also observed cut-and-fill sequences along the vertical walls of Wild Horse arroyo below the site (Anderson and Haynes 1979; Cook, in Roberts 1951). The most obvious stratigraphic evidence for McJunkin-age channel cutting was exposed during excavation of the L23 block, where a well-defined channel ~2 m deep was discovered. The incision of this channel completely removed the *m1*, *f3*, and *f2* units and eroded the surface of the *f1*, which was then buried in *m2* fill. This channel was visible only in cross section, so its upstream and downstream course is uncertain. It appears to have come in from the northwest, where a part of it may be visible ~40 m to 50 m upstream, pivoted against the Smoky Hill Shale bedrock on the South Bank, and then turned east/northeast back toward the valley axis. There is no evidence that the bonebed in the paleotributary experienced any such significant Middle or Late Holocene erosion; it was buried and well armored by the time these processes took place.

Evidence for postdepositional and probably some postburial alterations of stratum *m* is common (Anderson and Haynes 1979:897 [including their *f3*, now part of our *m1*]). Pedogenesis is marked by the accumulation of organic matter in some zones within *m* (indicated by the very dark colors and relatively high organic carbon content; table 5.2),

the development of prismatic structure, the slight reddening (10YR hues, in contrast to dominantly 2.5Y hues in the Folsom Formation strata), and the presence of films and threads of carbonate on ped faces in some horizons (e.g., Profile 97-1: AB-Btg horizonation in *m2* at 74–136 cm and AB horizon in *m1* at 136–175 cm; Bk horizonation in *f2* at 250–330 cm; and in Giddings Core 97-3, A-Bt horizonation at 20–125 cm; table 5.2). Some horizons also were subjected to gleying (e.g., Btg horizon in Profile 97-1 at 100–136 cm; table 5.2). The evidence for soil formation within unit *m* in turn indicates episodes of landscape stability during the evolution of the deposit.

THE WILDHORSE FORMATION (*w*)

Stratum *w* is the youngest layer in the late Quaternary stratigraphic sequence at the Folsom site. The distinctive characteristics of *w*, according to Anderson and Haynes (1979:897, table 1), and in contrast to *m* and *f*, are that the unit consists of "sandy silt" in layers ranging in color from dark gray (10YR 4/1) to grayish brown (2.5Y 5/3) interbedded with cobbles, gravel, or crystalline rock and shingle shale. Limited laboratory data plus field textures from cores indicate that the fines are lithologically similar to the silty clays of *m* and *f*. Weak soil development is apparent in some layers within the Wildhorse Formation. The principal difference between stratum *w* and stratum *m*, therefore, are the multiple layers of shingle and gravel in stratum *w*. In the absence of the coarse clastics, therefore, *w* is very similar to some facies of *m*.

Stratum *w* is best expressed in the North Bank arroyo exposure opposite the paleotributary. This hints that perhaps some of the gravel in *w* came down the paleotributary. The profile for trench 5 of Haynes, Anderson, and Frazier (1976) is relevant here: It shows a very thin and gravel-free stratum *w* resting unconformably on *m*, further hinting that *w* is related to the paleotributary. That said, other source areas are clearly indicated, given that there are gravels in *w* that in places are much larger than the clasts in the paleotributary and, in fact, occur upstream of the mouth of the paleotributary.

Stratum *m2* is at the surface in many parts of the site, suggesting that *w* may be a channel fill inset against *m*, and it is so indicated by Haynes et al. (1992:fig. 3.1). Such crosscutting relationships were not observed. The only relevant data we have come from an ~3-m section exposed in an excavation unit in the M23 block on the North Bank, where a 30- to 40-cm-thick lens of cobbles and coarse gravel, the base of a channel cut, was exposed ~1.5 m below the surface and subsequently topped with very dark-gray silty clay (10YR 3/1). While this zone of large clasts was deposited, it was not necessarily associated with an erosional event, and it may well be a pulse of gravel interrupting otherwise quiet and ongoing deposition of mud (McJunkin). In effect, stratum *w* may simply be a facies of *m* that, within the main valley, contains large clasts.

In any case, accounting for the relatively common presence of crystalline gravel is somewhat problematic. The source is likely gravel from high Pleistocene strath terraces of Wild Horse Arroyo and the Upper Dry Cimarron, which is dominated by basalt but also includes gravel from the Ogallala Formation, which outcrop out around the area (chapter 3). Why the gravel is more common in *w* than in any of the older units is still unclear, however. It should have been available for redeposition in stratum *f* or *m*. Perhaps deposits of gravel were only available for erosion and redeposition following a unique combination of climatic, floral, and geomorphic processes.[6]

Radiocarbon Dating and Geochronology

All together, there are 49 radiocarbon ages available from the Folsom site and nearby sampled sections along Wild Horse Arroyo. These vary in their precision and accuracy in providing age control over the stratigraphic sequence and the age of the Paleoindian bonebed. A comprehensive list of all radiocarbon dates obtained from the site, from 1950 to the present, is provided in table 5.4, along with provenience, context, and other information.

Ten of those ages are separately dated fractions from the same sample of charcoal (e.g., AA-7088 and 7089, AA-7090 and 7091, CAMS-57513 and 57514, CAMS-57518 and 57519, CAMS-57520 and 57521). Where charcoal and humic acid fractions from a single charcoal sample were separately dated (the CAMS pairs), the charcoal fraction is used, on the assumption that it is less likely to be contaminated. Where carbon residue and humate fractions from a single sample were separately dated (the AA pairs), the carbon residue fraction is used, again on the same assumption. We further assumed that radiocarbon ages from charcoal as opposed to bison bone represent maximum-limiting ages for sediments in which it is contained.

The four radiocarbon ages listed at the end of table 5.4 are unacceptable based on stratigraphic and/or other evidence and are not considered further.[7] Thus, a total of 40 radiocarbon ages form the basis of the discussion that follows.

Radiocarbon ages from the laboratories at the University of Arizona (AA), Southern Methodist University (SMU), and The University of Texas Radiocarbon Facility (TX) were obtained by C. V. Haynes (Haynes et al. 1992). All but one of the samples with a CAMS designation were prepared by T. Stafford of Stafford Research Laboratories; the other CAMS sample (CAMS 96034) was prepared by P. Matheus at the Alaska Quaternary Center, University of Alaska.

METHODS

Charcoal samples received standard acid/base pretreatment using hydrochloric acid (HCl) decalcification and potassium hydroxide (KOH) cycles to removed humates (Stafford 1998). The SMU sample of bison bone was prepared by A. Hassan,

using then-experimental techniques, as described in detail by Hassan (1975). The laboratory protocols for all other bison bone samples followed the procedure developed by Stafford (see Stafford et al. 1987:fig. 1, 1988, 1991), which is also used by Matheus (1997:appendix 2). In general, that procedure involves a series of steps to separate the collagen and mineral (bioapatite) portions of the bone, which is accomplished by repeated soaking of the sample in weak HCl at low temperatures until the sample is demineralized. Once the sample is rinsed to neutralize the HCl, it is soaked in KOH to partially remove humic and fulvic acids. The collagen is then gelatinized to separate it from other proteins and organic compounds in the demineralized bone and given a final "filtering" via liquid chromatography using nonionic, hydrophobic, styrene (XAD) resins to remove fulvic acids (Matheus 1997; Stafford, Brendel, and Duhamel 1988). All CAMS sample targets were run at the Lawrence Livermore National Laboratory Center for Accelerator Mass Spectrometry (CAMS).

Calibration of radiocarbon ages was done using both OxCal Version 3.9 (Ramsey 2003) and CALIB Version 4.4 (Stuiver, Reimer, and Reimer 2000). The OxCal results are used in table 5.4 and the figures because they provide more illustrative graphical results. We note that OxCal and CALIB returned slightly different calibrated ages for the same ^{14}C age, but the discrepancy does not appear to be significant. Assessments of statistical contemporaneity and averaging of ages are based on chi-square analysis using an SPSS algorithm developed by H. Hietala (1989). Ages are rounded to the nearest decade.

The 1σ errors on half of these radiocarbon ages—which includes all of the *f2* ages—fall between 13,200 and 11,260 cal yr B.P., thus overlapping the Younger Dryas Chronozone, with its multiple radiocarbon plateaus and reversals (Beck at al. 2001; Delaygue et al. 2003; Hajdas et al. 1998; Hughen et al. 2000; Kitigawa and van der Plicht 1998; Leuenberger, Siegenthaler, and Langway, 1992; Monnin et al. 2001; Taylor, Stuiver, and Reimer 1996). Hence, the calibration and precision of radiocarbon ages from this period are hardly straightforward, since a given radiocarbon age can correspond to an unusually wide range of calibrated calendar ages (table 5.4). The ages are discussed in stratigraphic order.

AGE OF THE *f1*

There is but a single age available on unit *f1*: 12,355 ± 210 ^{14}C yr B.P. (AA-7090; the humate split from this sample [AA-7091] is statistically indistinguishable). The precise provenience of this sample is uncertain. It is reported only as having come from "a 0.5 cm thick carbonaceous band ca. 40 cm below 1 Folsom 90" (Haynes, unpublished field notes, July 18, 1990). It is unclear whether the chronological gap between this sample and the oldest of the ages from the *f2* (~11,500 ^{14}C yr B.P.) is indicative of a depositional hiatus or stratigraphic unconformity that has been otherwise overlooked or whether the sample is minutely contaminated.

TABLE 5.4

Radiocarbon and Calibrated Radiocarbon Ages from the Folsom Site

(a) Accepted Radiocarbon Ages

Stratum	Provenience	Material Dated	Laboratory No.	^{14}C Age, yr B.P.	Cal Age, yr B.P.: 68.2% Probability	Cal Age, yr B.P.: 95.4% Probability	Comments
w	North Bank, collected July 1999, from M23 profile (M23-NAP-29; 99-29) (N1063.792 E997.491 Z99.561)	Oak charcoal	CAMS-74651	700 ± 40	560–590 (19.1%) 640–680 (49.1%)	550–610 (30.3%) 620–710 (65.1%)	
m2?	North Bank (?),collected by F. Howarth, July, 1933, "some hundred feet, plus or minus," east of the site (Cook, in Roberts 1951:20)	Charcoal	C-377	4,575 ± 300 3,923 ± 400 average = 4,280 ± 250	4,450–5,300 (68.2%)	4,150–5,650 (95.4%)	From possible hearth in-fill in a secondary channel; analyzed using solid carbon method; see discussion in text
m2	North Bank, collected July 1998, (98-479) (N1065.567 E998.177 Z94.732)	Charcoal	CAMS-57517	4,460 ± 50	4,970–5,080 (25.9%) 5,100–5,130 (6.6%) 5,160–5,280 (35.6%)	4,870–4,940 (7.3%) 4,960–5,300 (88.1%)	From gravels at the base of the McJunkin arroyo; dates the onset of filling in the arroyo
m2	North Bank, collected by C. V. Haynes July 1970, from "charred log and reddened silt" in "Loc. upstream from type site" (CS 70-9)	Charcoal	TX-1272	4,470 ± 90	4,970–5,290 (68.2%)	4,850–5,350 (95.4%)	Haynes, unpublished field notes, July 13, 1970
m2	North Bank, collected July 1997, from Giddings Core 97-14. (97-49F) (N1074.134 E1072.558 Z 93.649)	Pine charcoal	CAMS-74644	4,640 ± 60	5,300–5,470 (68.2%)	5,050–5,200 (7.1%) 5,250–5,600 (88.3%)	Core taken close to north valley wall, on presumed edge of paleovalley. Sample from 3.92–3.93 m below surface. Date indicates m2 fills this portion of paleovalley
m2	North Bank, collected by C. V. Haynes July 1970, from "charcoal with red band, downstream from confluence (car park)" (CS 70-14)	Charcoal	TX-1270	4,850 ± 120	5,480–5,510 (14.3%) 5,580–5,660 (53.9%)	5,460–5,710 (95.4%)	Haynes, unpublished field notes, July 13, 1970; same location as TX-1271 and TX-1452

(continued)

TABLE 5.4 (*Continued*)

Stratum	Provenience	Material Dated	Laboratory No.	^{14}C Age, yr B.P.	Cal Age, yr B.P.: 68.2% Probability	Cal Age, yr B.P.: 95.4% Probability	Comments
m2	South Bank, collected July 1998, Feature 1, upper half of m2 (M17-3-1; 98-304) (N1030.301 E997.608 Z98.242)	Pine charcoal	CAMS-57518 (from same sample as CAMS-57519)	4,910 ± 50	5,590–5,670 (61.4%); 5,690–5,710 (6.8%)	5,490–5,510 (1.1%); 5,580–5,750 (94.3%)	Average of CAMS-57518 and -57519 = 4,880 ± 35 ^{14}C yr B.P.
m2	South Bank, collected July 1998, Feature 1, upper half of m2 (M17-3-1; 98-304) (N1030.301 E997.608 Z98.242)	Pine charcoal—humic acid fraction	CAMS-57519 (from same sample as CAMS-57518)	4,850 ± 50	5,480–5,510 (14.3%); 5,580–5,660 (53.9%)	5,460–5,710 (95.4%)	Average of CAMS-57518 and -57519 = 4,880 ± 35 ^{14}C yr B.P.
m1	North Bank, collected by C.V. Haynes July 1970, from a "meter down in this unit, where exposed 300 m downstream from the site" (Haynes and Anderson, unpublished); (CS 70-13)	Charcoal	TX-1452	6,060 ± 500	6,350–7,450 (68.2%)	5,750–7,950 (95.4%)	Haynes, unpublished field notes, July 13, 1970; same location as TX-1270 and TX-1271
m1	North Bank, collected by C.V. Haynes July 1970, from "charcoal in fire pit in [stratum] Qy in main draw" "downstream from type site and forks" (CS 70-16)	Charcoal	TX-1271	6,910 ± 110	7,610–7,640 (4.1%); 7,650–7,850 (62.7%); 7,900–7,920 (1.4%)	7,570–7,950 (95.4%)	Haynes, unpublished field notes, July 13, 1970; same location as TX-1270 and TX-1452
m2?	North Bank, collected June 1997, from Profile 97-1 (97-7F) (N1051.445 E1021.947 Z96.937)	Charcoal—humic acid fraction	CAMS-57511	7,500 ± 40	8,200–8,270 (27.2%); 8,290–8,370 (41.0%)	8,190–8,390 (95.4%)	Sample from lower portion of m2, ~20 cm above m1/m2 contact. Date seems too old for m2 but would fit with m1.
f3	North Bank, collected July 1998, *Bison* carpal. Bone from stratigraphically above f2/f3 contact and just above *Bison* humerus (element L23-8-10; see below) (L23-8-9) (N1061.742 E992.593 Z96.558)	Bone (XAD-gelatin & KOH collagen)	CAMS-74654	9,220 ± 50	10,260–10,350 (28.9%); 10,360–10,430 (27.7%); 10,440–10,480 (11.6%)	10,230–10,510 (94.0%); 10,520–10,550 (1.4%)	Bone not from Paleoindian bison kill

Facies	Context	Material	Lab no.	Age (14C yr B.P.)	Calibrated age ranges		Comments
f3	North Bank, collected July 1999, antilocaprid femur (N22-3-1) (N1055.797 E1002.889 Z95.857)	Bone (XAD-gelatin & KOH collagen)	CAMS-74934	9,270 ± 50	10,280–10,300 (3.9%) 10,310–10,320 (2.4%) 10,330–10,340 (2.2%) 10,390–10,510 (48.2%) 10,520–10,560 (11.6%)	10,240–10,580 (95.4%)	In same area of the North Bank where Howarth reported finding "deer" bones, which he believed came from Folsom bison bonebed. If this is the same individual, it indicates the animal is not from Paleoindian bison kill.
f3	North Bank, collected July 1998, in f3 immediately (5 cm) above Bison carpal (see L23-8-9 above) (L23-8-21; 98-489) (N1061.802 E992.633 Z96.608)	Pine charcoal—humic acid fraction	CAMS-57515	9,340 ± 50	10,420–10,450 (5.0%) 10,490–10,600 (49.6%) 10,610–10,640 (11.8%) 10,600–10,670 (1.9%)	10,390–10,700 (95.4%)	
f3	North Bank, collected July 1999, from auger hole 1.55 m below floor of M23 unit; 4.75 m below present surface. Sample from f3 ~5 cm above f2. (M23-NAP-45; 99-45) (N1064.47 E997.944 Z95.434)	Pine charcoal	CAMS-74653	9,440 ± 50	10,570–10,750 (68.2%)	10,500–10,850 (82.0%) 10,900–11,100 (13.5%)	
f3	North Bank, collected July 1998, in f3 immediately (4.7 cm) below Bison carpal (see L23-8-9 above) (L23-8-20; 98-488) (N1061.752 E992.543 Z96.511)	Charcoal—humic acid fraction	CAMS-57516	9,780 ± 40	11,170–11,205 (61.1%) 11,215–11,225 (7.1%)	11,140–11,235 (95.4%)	
f3 facies?	South Bank, collected July 1999, from bedrock surface atop bedrock wall immediately west of the bonebed (J19-4-10; 99-58) (N1040.479 E983.502 Z299.971)	Pine/conifer charcoal	CAMS-74647	9,820 ± 40	11,190–11,230 (68.2%)	11,160–11,260 (94.4%) 11,280–11,300 (1.0%)	Indicates that Folsom Paleoindian-age sediments not preserved on this surface

(continued)

TABLE 5.4 (Continued)

Stratum	Provenience	Material Dated	Laboratory No.	${}^{14}C$ Age, yr B.P.	Cal Age, yr B.P.: 68.2% Probability	Cal Age, yr B.P.: 95.4% Probability	Comments
f3	North Bank, collected June 1997, from basal portion of f3, 8 cm above f2/f3 contact, Profile 97-1. (97-12F) (N1051.445 E1021.947 Z95.155)	Charcoal—humic acid fraction	CAMS-57512	10,370 ± 50	11,950–12,050 (4.5%) 12,100–12,400 (40.4%) 12,450–12,650 (23.4%)	11,750–12,850 (95.4%)	Marks period of reworking of f2 sediments on North bank
f3	North bank, collected July 1999, from amid gravels in paleovalley (Q21-8-16; 99-61) (N1051.534 E1017.803 Z94.623)	Charcoal	CAMS-74648	10,420 ± 140	11,950–12,050 (3.3%) 12,100–12,650 (61.5%) 12,700–12,800 (3.4%)	11,650–12,950 (95.4%)	
f2	South bank, collected July 1998, from above bone bed (N17-21-18; 98-170) (N1034.158 E1000.160 Z97.620)	Conifer charcoal	CAMS-74645	10,010 ± 50	11,260–11,280 (2.4%) 11,290–11,450 (37.7%) 11,460–11,570 (26.2%) 11,610–11,630 (1.9%)	11,200–11,700 (92.4%) 11,850–11,950 (3.0%)	Minimum limiting age for f2 deposition
f2	North Bank, Bison bone (radius) collected by C. V. Haynes July 1970) (CS-70-11)	Bone—collagen	SMU-179	10,260 ± 110	11,650–12,350 (68.2%)	11,350–12,850 (95.4%)	Haynes considered this age less secure than charcoal from f2 (Haynes et al. 1992:87).
f2	South Bank, collected July 1998 from M15 block. (M15-24-43) (N1024.051 E998.593 Z97.733)	Charcoal	CAMS-57520 (from same sample as CAMS-57521)	10,380 ± 50	11,950–12,050 (2.6%) 12,100–12,400 (38.7%) 12,450–12,650 (26.9%)	11,900–12,800 (95.4%)	Bonebed virtually absent in this area; date indicates sediments of Folsom Paleoindian age are present. Separately dated fractions are not contemporaneous.

	Description	Material	Lab no.	¹⁴C age B.P.	Calibrated age ranges (1σ)	Calibrated age ranges (2σ)	Comments
f2	South Bank, collected July 1998 from M15 block. (M15-24-43) (N1024.051 E998.593 Z97.733)	Charcoal—humic acid fraction	CAMS-57521 (from same sample as CAMS-57520)	10,600 ± 40	12,380–12,480 (24.3%) 12,610–12,670 (10.4%) 12,710–12,860 (33.6%)	12,300–12,550 (32.3%) 12,600–12,950 (63.1%)	Same as CAMS-57520 (above)
f2	South Bank, collected July 1998, *Bison* (right) humerus (M17-19-106) (N1033.091 E998.899 Z97.627)	Bone (XAD-gelatin & KOH collagen)	CAMS-74656	10,450 ± 50	12,150–12,250 (4.5%) 12,300–12,650 (60.1%) 12,750–12,800 (3.6%)	11,950–12,850 (95.4%)	
f2	South Bank, collected July 1999, *Bison* (right) tibia (M17-24-310) (N1034.312 E998.220 Z97.538)	Bone (XAD-gelatin & KOH collagen)	CAMS-74658	10,450 ± 50	12,150–12,250 (4.5%) 12,300–12,650 (60.1%) 12,750–12,800 (3.6%)	11,950–12,850 (95.4%)	
f2	North Bank, collected June 2002, *Bison* thoracic vertebra (R20-14-5) (N1047.760 E1023.113)	Bone (XAD-gelatin & KOH collagen)	CAMS-96034	10,475 ± 30	12,320–12,550 (44.9%) 12,560–12,650 (15.5%) 12,740–12,800 (7.8%)	12,050–12,850 (95.4%)	
f2	South Bank, collected July 1998, *Bison* (right) mandibular molar (M17-24-141) (N1034.765 E997.992 Z97.567)	Bone (XAD-gelatin & KOH collagen)	CAMS-74657	10,500 ± 40	12,330–12,530 (43.1%) 12,580–12,650 (13.2%) 12,730–12,810 (12.0%)	12,050–12,900 (95.4%)	
f2	South Bank, collected July 1999, from test unit in upper portion of tributary headcut, ~65 m south of the southern edge of the bonebed (F8-6-20; 99-68) (N986.460 E960.999 Z99.789)	Pine charcoal	CAMS-74649	10,510 ± 50	12,330–12,530 (41.4%) 12,580–12,660 (13.2%)	12,100–12,900 (95.4%)	Indicates sediments of Folsom Paleoindian age present in this area of the site, although no archaeological traces were found

(continued)

TABLE 5.4 (*Continued*)

Stratum	Provenience	Material Dated	Laboratory No.	¹⁴C Age, yr B.P.	Cal Age, yr B.P.: 68.2% Probability	Cal Age, yr B.P.: 95.4% Probability	Comments
f2	North Bank, collected July 1998, *Bison* (right) humerus from below f3/f2 contact (L23-8-10) (N1061.474 E992.465 Z96.289)	Bone (XAD-gelatin & KOH collagen)	CAMS-74655	10,520 ± 50	12,730–12,820 (13.7%) 12,330–12,520 (40.6%) 12,580–12,660 (12.4%) 12,730–12,820 (15.2%)	12,100–12,250 (6.6%) 12,300–12,900 (88.8%)	
f2/f3	North Bank, collected July 1999, *Bison* (left) tibia in reworked sediments (Q21-8-52) (N1051.620 E1017.792 Z94.587)	Bone (XAD-gelatin & KOH collagen)	CAMS-74659	10,520 ± 50	12,330–12,520 (40.6%) 12,580–12,660 (12.4%) 12,730–12,820 (15.2%)	12,100–12,250 (6.6%) 12,300–12,900 (88.8%)	
f2	South Bank, collected July 1998, associated with *Bison* metatarsal. (M17-25-102) (N1034.518 E999.652 Z97.425)	Charcoal—humic acid fraction	CAMS-57525	10,670 ± 50	12,630–12,760 (31.2%) 12,770–12,900 (37.0%)	12,350–12,500 (13.8%) 12,600–13,000 (81.6%)	As age predates bone dates, indicates not all charcoal in f2 sediment is anthropogenic
f2	North Bank, collected by C. V. Haynes July 1970; part of a sample of "single lumps or flecks widely dispersed throughout the [f2] zone, 10–30 cm thick, as it was exposed in scraping the arroyo wall (CS-70-7.3)	Pine charcoal	AA-1709	10,760 ± 140	12,400–12,500 (1.4%) 12,600–13,000 (66.8%)	12,300–13,150 (95.4%)	Haynes et al. (1992:87) average AA-1708 to AA-1712 plus AA-1213 together. As age predates bone dates, indicates not all charcoal in f2 sediment is anthropogenic
f2	North Bank, collected by C. V. Haynes July 1970, as above AA-1709 (CS-70-7.0)	Charcoal	AA-1213	10,780 ± 100	12,650–12,740 (17.5%) 12,800–12,990 (50.7%)	12,400–12,500 (3.9%) 12,600–13,150 (91.5%)	This sample is a composite of fragments from AA-1708 to AA-1712.

				14C age			
f2	North Bank, collected by C.V. Haynes July 1970, as above AA-1709 (CS-70-7.4)	Juniper charcoal	AA-1711	10,850 ± 190	12,650–12,740 (11.5%) 12,800–13,140 (56.7%)	12,150–13,450 (95.4%)	As above, AA-1709
f2	North Bank, collected by C.V. Haynes July 1970, as above AA-1709 (CS-70-7.3)	Juniper bark or hardwood charcoal	AA-1710	10,890 ± 150	12,820–13,140 (68.2%)	12,350–12,500 (3.0%) 12,600–13,200 (92.4%)	As above, AA-1709
f2	North Bank, collected by C.V. Haynes July 1970, as above AA-1709 (CS-70-7.5)	Juniper bark or hardwood charcoal	AA-1712	10,910 ± 100	12,860–13,050 (55.0%) 13,060–13,130 (13.2%)	12,600–12,750 (10.8%) 12,800–13,200 (84.6%)	As above, AA-1709
f2	South Bank collected July 1998, from M15 block, which marks southern margins of bonebed (M15-24-55) (N1024.495 E998.236 Z97.682)	Charcoal—humic acid fraction	CAMS-57523	10,970 ± 50	12,910–13,050 (50.6%) 13,070–13,140 (17.6%)	12,660–12,720 (4.8%) 12,840–13,160 (90.6%)	Although older than the age of the kill, the lack of any overlying stratigraphic unconformity or truncation of f2 in this area indicates sediments of Folsom Paleoindian age likely present here; absence of bone is evidence of absence.
f2	North Bank, collected by C.V. Haynes July 1970, as above AA-1709 (CS-70-7.1)	Pine charcoal	AA-1708	11,060 ± 100	12,950–13,160 (68.2%)	12,650–12,750 (3.1%) 12,800–13,400 (92.3%)	As above, AA-1709
f2	South Bank, collected July 1998, in close proximity to Bison mandible (M18-4-22; 98-319) (N1035.117 E998.391 Z97.447)	Conifer charcoal—humic acid fraction	CAMS-57524	11,070 ± 50	12,980–13,150 (68.2%)	12,650–12,750 (1.9%) 12,850–13,200 (93.5%)	Bison mandible is M17-24-146
f2	North Bank, collected by C.V. Haynes July, 1990, precise provenience unknown. "Collected at base of dark unit (f?) in buff (yellow) loess?" (1 Folsom 90A)	Charcoal—carbon residue	AA-7088 (from same sample as AA-7088)	11,100 ± 130	12,900–13,200 (68.2%)	12,650–12,750 (2.3%) 12,850–13,450 (93.1%)	Sample was plotted by Haynes as occurring at the base of f3; field description matches f2 sediments.
f2	North Bank, collected by C.V. Haynes July, 1990, precise provenience unknown. "Collected at base of dark unit (f?) in buff (yellow) loess?" (1 Folsom 90A)	Charcoal—humates	AA-7089 (from same Sample as AA-7088)	10,630 ± 80	12,390–12,480 (17.8%) 12,620–12,680 (11.5%) 12,700–12,880 (39.0%)	12,150–12,250 (1.1%) 12,300–13,000 (94.3%)	Sample plotted by Haynes as occurring at the base of f3; field description matches f2 sediments

(continued)

TABLE 5.4 (*Continued*)

Stratum	Provenience	Material Dated	Laboratory No.	^{14}C Age, yr B.P.	Cal Age, yr B.P.: 68.2% Probability	Cal Age, yr B.P.: 95.4% Probability	Comments
f2	North Bank Profile 1, collected June 1997, Profile 97-1. Sample from ~28 cm below f3/f2 contact and below level of bonebed (97-2F) (N1051.445 E1021.947 Z94.8)	Charcoal	CAMS-57513 (from same sample as CAMS-57514)	11,370 ± 150	13,140–13,490 (68.2%)	12,950–13,850 (95.4%)	Average of CAMS-57513 and -57514 = 11,500 ± 40 ^{14}C yr B.P.
f2	North Bank Profile 1, collected June 1997, Profile 97-1. Sample from ~28 cm below f3/f2 contact and below level of bonebed (97-2F) (N1051.445 E1021.947 Z94.8)	Charcoal— humic acids	CAMS-57514 (from same sample as CAMS-57513)	11,510 ± 40	13,200–13,250 (1.5%) 13,300–13,550 (49.0%) 13,700–13,800 (17.6%)	13,150–13,550 (66.6%) 13,600–13,850 (28.8%)	Average of CAMS-57513 and -57514 = 11,500 ± 40 ^{14}C yr B.P.
f1	North Bank, collected by C. V. Haynes, July, 1990; precise sample position and elevation unknown. From "0.5 cm thick carbonaceous band ca. 40 cm below 1 Folsom 90" [AA-7088/AA-7089] (2 Folsom 90A)	Charcoal— carbon Residue	AA-7090 (from same sample as AA-7091)	12,355 ± 210	14,050–15,050 (68.2%)	13,750–15,550 (95.4%)	Average of AA-7090 and -7091 = 12,390 ± 80 ^{14}C yr B.P.
f1	North Bank, collected by C. V. Haynes, July, 1990; precise sample position and elevation unknown. From "0.5 cm thick carbonaceous band ca. 40 cm below 1 Folsom 90" [AA-7088/AA-7089] (2 Folsom 90A)	Charcoal— humates	AA-7091 (from same sample as AA-7090)	12,395 ± 90	14,100–14,500 (35.5%) 14,550–15,050 (32.7%)	14,050–15,450 (95.4%)	Average of AA-7090 and -7091 = 12,390 ± 80 ^{14}C yr B.P.

(b) Rejected Radiocarbon Ages

Stratum	Provenience	Material Dated	Laboratory No.	^{14}C age, yr B.P.	Comments
m1	North Bank, collected July 1999, from M23 profile (M23-NAP-24; 99-24) (N1064.198 E997.093 Z97.983)	Oak charcoal	CAMS-74650	840 ± 50	Date younger than expected, based on stratigraphic position
f3	North Bank, collected July 1999, from M23 profile (M23-NAP-44; 99-44) N1063.887 E998.617 Z97.526)	Pine charcoal	CAMS-74652	820 ± 40	Date younger than expected, based on stratigraphic position, and because it was thought to come from f3 (compare age from sample 99-45)
f2	South Bank, collected July 1999, in close proximity to *Bison* crania (M17-23-69; 99-11) (N1034.269 E997.429 Z97.602)	Conifer charcoal	CAMS-74646	14,800 ± 2,500	Sample was only 5 μg and was evidently contaminated
f2	South Bank, collected July 1998, in close proximity to *Bison* rib, upper part of bonebed (M17-25-50; 98-24) (N1034.348 E999.389 Z97.709)	Charcoal	CAMS-57522	55,500 ±	Sample was not especially small but may have been minutely contaminated by carbonaceous material eroding out of Smoky Hill Shale

NOTE: Arranged by stratum and chronologically within stratum, except in the case of paired fractions from a single sample, which are listed together, with the age that is used listed first. See text for discussion. Calibrations were done using OxCal version 3.9 (Ramsey 2003)

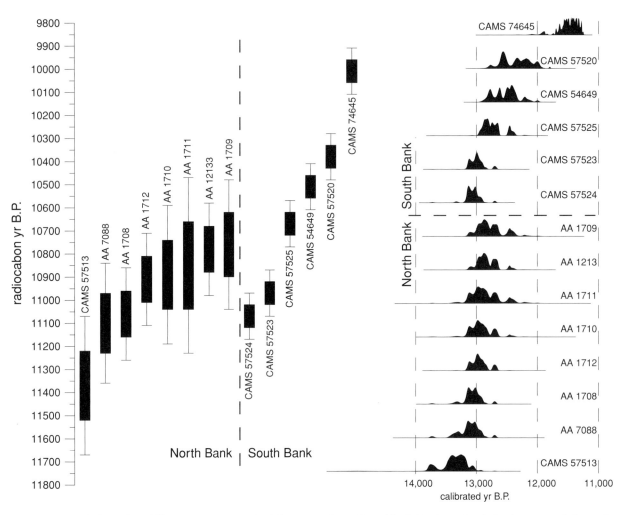

FIGURE 5.11 Left, box plot of ^{14}C ages to two standard deviations from *f2* on North Bank vs. South Bank, arranged chronologically within each area; right, those same ages calibrated using OxCal, and plotted chronologically within each area.

AGE OF THE *f2*

Leaving aside for the moment the half-dozen radiocarbon ages obtained directly from bison bone, radiocarbon ages derived from charcoal samples from the *f2* on both the South and the North banks range from 10,010 ^{14}C yr B.P. (CAMS-74645) to 11,370 ^{14}C yr B.P. (CAMS-57513) (fig. 5.11). The combined 1σ errors of these ages span the interval from 9,960 to 11,520 ^{14}C yr B.P. (11,260 to 13,490 cal yr B.P.). Taken as a group, these ages suggest that the deposition of this unit took place over a relatively long period of time, some 1,560 radiocarbon years (2,230 cal years). Combined with the absence of stratigraphic unconformities and the evidence for pedogenesis and buried soils (above), these ages suggest that *f2* accumulation was uninterrupted through much of the Late Glacial.

That general picture becomes more complicated, however, when radiocarbon ages from the paleotributary and paleovalley are separately examined. The ages on the *f2* in these two areas overlap for a span of 500 years (both ^{14}C and cal), but the maximum and minimum ages of the *f2* from the

paleovalley are older than the corresponding ages from the paleotributary. The difference in the mean ^{14}C ages between the two areas is significant, as measured by *t*-test ($t = 2.267$, $P = 0.043$). In effect, it appears as though *f2* deposition began earlier and ended earlier in the paleovalley. That the ages of the paleotributary and paleovalley are out of phase, however, should not be taken at face value, as the offset is clearly a by-product of both sampling and erosion. In terms of the former, we did not excavate significantly below the bonebed in the paleotributary, so *f2* ages from the South Bank do not date the onset of *f2* deposition. In contrast, deeper and older portions of stratum *f2* were visible in the North Bank profiles and sampled for radiocarbon. Moreover, on the North Bank erosion within the paleovalley may have stripped away the upper portion of the *f2* and, with it, any of the younger charcoal that might have represented the final period of deposition of that unit in that area. In the higher, more protected paleotributary, the upper portion of the *f2* remains intact and, not surprisingly, yields younger ages.

While these ages bracket the *f2* deposition, they do not pertain directly to Paleoindian activities at the site. The charcoal that produced these ages commonly occurs as small fragments throughout stratum *f2* and, to a lesser degree, in the lower reaches of the *f3*. No hearths, burned areas, or other discrete anthropogenic features were found within the *f2* in the paleotributary or in the paleovalley. The ubiquity of the charcoal and the absence of any cultural features lead us to conclude that charcoal in the *f2*—and the *f3*, for that matter—is a result of natural fires in the area (also Bryan 1937:142; Wissler, field diary, August 1928, ANTH/AMNH). The wooded slopes of Johnson Mesa today produce abundant charcoal in slope wash due to frequent forest fires. Thus, the ages from charcoal collected sitewide from the *f2* can be used only to bracket, not pinpoint, the age of the Paleoindian occupation. From that, and the erosion of *f2* sediments in the paleovalley, it also follows that the previously reported ages from charcoal sampled from the North Bank provide, at best, only a ballpark figure for the Folsom kill (cf. Haynes et al. 1992).

Two of the *f2* charcoal ages are nonetheless of particular interest: CAMS-57520, which yielded an age of 10,380 ± 50 ^{14}C yr B.P. (CAMS-57520), came from the M15 excavation block. Bison bone was virtually absent from this block, prompting the question of whether this part of the site was simply beyond the margins of the kill, or whether bone had once been present but had degraded in place, or whether bone-bearing sediments of the proper age had simply been removed. This radiocarbon date indicates that sediments of the appropriate age are present. The question of whether the bone disappeared as a result of erosion or weathering, or was simply not deposited here in Paleoindian times, is taken up in chapter 7.

CAMS-74569 came from a charcoal sample in *f2* sediments 1.5 m below the surface in a test pit (N986 E960) dug in the upper reaches of the western prong of the paleotributary, and yielded an age of 10,510 ± 50 ^{14}C yr B.P. (CAMS-74649). This unit was excavated to assess whether traces of a habitation area associated with the Paleoindian kill could be detected—the spot being some 65 m south of the bonebed—and as a check to assess whether sediments of the right age are present in this area. No trace of habitation debris was found, save for an enigmatic rhyolite cobble (described in chapter 8). However, this radiocarbon age also indicates that sediments of the right age are present.

THE AGE OF THE FOLSOM BISON KILL. As Haynes et al. (1992:87) anticipated, the best solution for determining the age of the bison kill would be bone dating using "technological improvements allowing the isolation of specific amino acids for dating." We have now done this: Six samples from various bison bone elements (long bones, primarily)—three from the paleotributary and three from the paleovalley—were radiocarbon dated.

The resulting ages are statistically indistinguishable and provide a mean age of 10,490 ± 20 ^{14}C yr B.P. (CAMS-74655 through CAMS-74659, and CAMS-96034). The two right humeri that were dated, one from the paleotributary and the other from the paleovalley, were obviously from different individuals. We cannot rule out the possibility that the remainder of the specimens were from the same two animals, but the likelihood that five bison bones scattered over such a large area and separated by topographic barriers that preclude transport between the paleotributary and the paleovalley areas makes that possibility remote. Of course, if the three North Bank specimens are from the same individual, this provides additional support for the stratigraphic evidence of reworking, for these specimens were found 27 m apart, with one coming from *f2* sediments only a few centimeters below the *f3/f2* contact, while the other two came from amid reworked gravels in the *f3/f2*.

This mean age of these six ages is older than the single age obtained by Hassan on bison bone collagen several decades ago (10,260 ± 110 ^{14}C yr B.P.). Although this may attest to the unreliability of that bone collagen age, as Haynes et al. (1992) anticipated, it must also be observed that the 1σ calibrated age for this sample (11,650–12,350 cal yr B.P.) has at least a 200–calendar year overlap with the 1σ calibrated age range for the other dated bone (12,150–12,820 cal yr B.P.). Its δ^{13}C value also corresponds with values obtained on the other bison bone (chapter 6), suggesting that the then-experimental efforts of Hassan (1975) to extract collagen were reasonably reliable.

The mean age on the bison bone is also several hundred years younger (fig. 5.12) than the age of the Paleoindian occupation previously inferred from charcoal dates, which produced a mean age of 10,890 ± 50 ^{14}C yr B.P. (Haynes et al. 1992). This substantiates the suspicion that the charcoal came from natural fires, but also the fact the charcoal represents a maximum limiting age of a deposit—it can be older than the sediments in which it was embedded. Thus, we put the age of the Paleoindian bison kill at the Folsom at ~10,500 radiocarbon years ago, based on the direct dating of the bison bone.

That *f2* deposition continued after the kill, as observed stratigraphically, is confirmed by several younger radiocarbon ages on charcoal. Precisely how long deposition continued is unclear, at least in the paleovalley, where erosion removed the upper portion of the *f2*. The latest age on the *f2* here, if one excludes the bone dates, is only 10,760 ± 140 ^{14}C yr B.P. (12,400–13,000 cal yr B.P.). The latest age for *f2* deposition on-site is likely that from the paleotributary, where the *f2* yielded an age of 10,010 ± 50 ^{14}C yr B.P. (11,260–11,630 cal yr B.P.).

AGE OF THE *f3*

The complications of erosion within the paleovalley predictably carry over into the assessment of the age of stratum *f3*. Further complicating matters, there are no radiocarbon ages on the *f3* in the paleotributary. The majority of the ages on the *f3* postdate 10,000 ^{14}C yr B.P. Yet, several of the oldest overlap chronologically with (or are

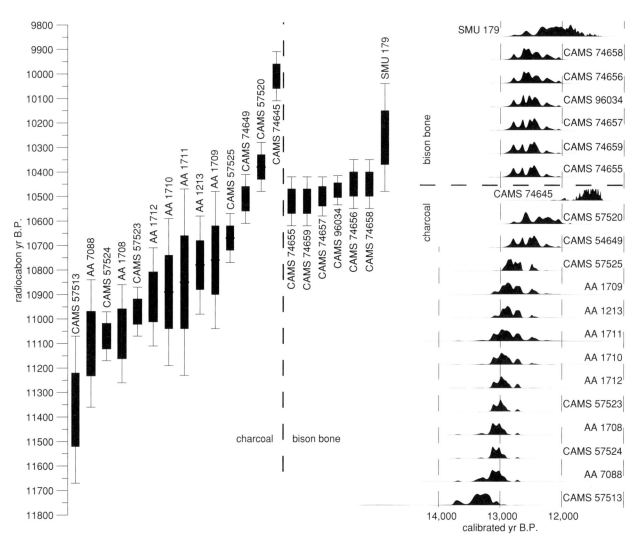

FIGURE 5.12 Left, box plot of ^{14}C ages to two standard deviations from *f2* on charcoal vs. bison bone, arranged chronologically; right, those same ages calibrated using OxCal, and plotted chronologically.

older than) the youngest of the *f2* ages (fig. 5.13). The oldest of these, 11,100 ± 130 ^{14}C yr B.P. (AA-7088), was from a sample Haynes collected from the base of a "dark unit(f?) in buff (yellow) loess." Although no further provenience information is provided, we suspect that the sample may come from the base of the dark band we observed at 250 cm to 300 cm in Profile 97-1 (table 5.2), which would put the sample immediately above the *f2/f3* boundary at 300 cm. Because the age of this sample clearly overlaps that of the *f2*, there is a strong possibility that this particular sample was reworked from older sediments and came to rest on the stratigraphic contact. The other two older ages, 10,420 ± 40 ^{14}C yr B.P. (CAMS-74648) and 10,370 ± 50 ^{14}C yr B.P. (CAMS-57512), come from fine gravels in the *f3* and from silts 8 cm above the *f2/f3* contact, respectively. Also in the *f3* on the North Bank, of course, are radiocarbon dates on redeposited bison bone that dated to ~10,500 ^{14}C yr B.P. All of this, of course, is in keeping with the observation that stratum *f3* includes older reworked material.

Excluding those radiocarbon ages >10,000 yr B.P., the oldest of the ages on *f3* in the paleovalley is 9,780 ± 40 ^{14}C yr B.P. (CAMS-57516), this from a charcoal sample immediately above the *f2/f3* contact in a topographically higher portion of the *f3* in the paleovalley. Perhaps not coincidentally, this date is statistically identical to one of 9,820 ± 40 ^{14}C yr B.P. (CAMS-74647) obtained ~23 m away on charcoal resting directly atop the Smoky Hill Shale bedrock (in block J19-4). The latter sample was not from *f3* sediments per se, and hence is not included in figure 5.13, but instead was on the Smoky Hill Shale bedrock surface on top of the valley wall immediately west of the paleotributary. This was the source area from which the shingle shale was washed out over the bonebed, and thus the date on this fragment of charcoal, deposited on an exposed bedrock surface, then blanketed in sediment, marks the maximum age at which sediment deposition began on this surface, after it served as a shingle shale source.

It is significant that two radiocarbon dates from widely separated areas of the site and from very different strati-

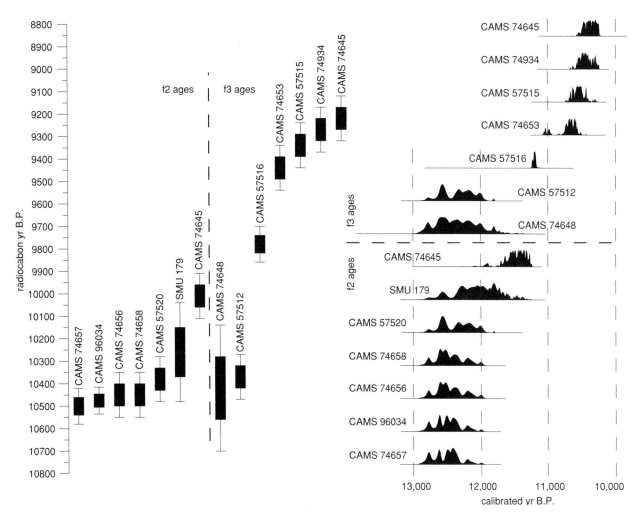

FIGURE 5.13 Left, box plot of ¹⁴C ages to two standard deviations from *f2* and from *f3*, arranged chronologically; right, those same ages calibrated using OxCal, and plotted chronologically.

graphic and topographic contexts yield statistically identical ages for the onset of *f3* deposition. If we are correct in the inference that *f2* deposition ceased ~10,010 ¹⁴C yr B.P., and that following a brief episode of incision a period of deposition and greater landscape stability began again around 9,800 ¹⁴C yr B.P., then clearly the most active period of shingle shale movement in the paleotributary, and erosion of the *f2* within the paleovalley, took place during the intervening ~200 radiocarbon years. It was likely at this time that bison bone in the paleovalley, first deposited half a millennium earlier, was reexposed, transported, and reburied. There is emerging evidence that a similarly timed cycle of Late Glacial and Early Holocene deposition-erosion-deposition occurred in other drainages in the Upper Dry Cimarron (Mann 2004), at Bellisle Lake (chapter 6), and elsewhere on the Great Plains and Rocky Mountains (Mayer et al. 2005), suggesting panregional geomorphic and climatic controls on arroyo filling around the Pleistocene/ Holocene boundary.

The age obtained from the charcoal sample resting on the Smoky Hill Shale bedrock surface above the bonebed is sig-nificant for another reason. It provides an important clue as to why no associated Folsom camp was found in this area of the site. This was the source area for the *f3* shingle shale that washed out across the bonebed. If Paleoindian activities had taken place on that surface, their traces would have been removed along with that shingle shale, for it appears that surface was largely swept clean of clasts between 10,000 and 9,800 ¹⁴C yr B.P. For the record, no Paleoindian artifacts have been found mixed in with the *f3* shingle shale armor-ing the bonebed, but it should be added that this unit was not systematically searched or even screened.

The youngest age from the *f3*, a date of 9,220 ± 50 ¹⁴C yr B.P. (CAMS-74654) was obtained from a bison carpal above the *f2/f3* contact. It is well bracketed by charcoal dates from just above and below the specimen (CAMS-57515 and CAMS-57516), and younger by 1,300 years than a bison humerus resting 27 cm below (CAMS-74655). Thus, this bison carpal is not a reworked element from the Paleoindian kill but is, instead, from an unrelated animal. This finding raises the cautionary flag that not all bison remains in the paleovalley, and even near the *f2/f3* contact, can be assumed to be reworked elements from the Folsom occupation.

Similarly unrelated to the Paleoindian kill, and of particular significance for Folsom subsistence, is the radiocarbon age of 9,270 ± 50 ^{14}C yr B.P. (CAMS-74934) obtained on a femur of an artiodactyl found on the North Bank. Its significance derives from the fact that Howarth found in 1926 what he identified as deer bones at the site (see Howarth to Cook, May 18, 1928, HCP/AFNM; see also Brown 1928b; Hay and Cook 1930:30; see the discussion in chapter 7). Howarth's taxonomic identification cannot be confirmed and the specimen cannot be relocated (chapter 7), but this recovery of a deer (or whatever species it might be) has long held open the possibility that other prey species may have been part of the assemblage at Folsom. Yet, because the precise stratigraphic position of the material Howarth found was unknown, that possibility could not be resolved. Howarth, did, however, leave a very rough sketch map showing where the remains were found, placing them on the North Bank just upstream of a large overhanging mass of oak roots—still present at the time of our excavations—that marks the upstream end of the 1920s clearing (see Howarth to Cook, May 18, 1928, HCP/AFNM).

Test excavations in this same area in 1999 yielded the proximal shaft of a femur from an unidentified species, but one that more closely resembles an antilocaprid than a cervid (chapter 7). Importantly, this element was found not in the *f2* sediments, but in the overlying *f3* stratum. Assuming, for the sake of discussion, that this artiodactyl specimen comes from the same animal Howarth found in this same area, its Early Holocene radiocarbon age negates the claim that deer were exploited at the site by Folsom hunters.

AGE OF THE HOLOCENE STRATA

There is a series of dates now available on the McJunkin units, including the one obtained by Harold Cook in 1950. With one exception (CAMS-57518), all of the samples are from deposits on the North Bank, but not necessarily from the site.

Ages on the *m1* range from ~6,000 to ~7,500 ^{14}C yr B.P.; two of those (TX-1271 and TX-1452) are not from the site proper but from some distance downstream. There is a 2,000-year gap in the chronology above the *m1*, which roughly correlates with the Middle Holocene/Altithermal period. This, as well as the stratigraphic unconformity seen in Profile 1, suggests erosion or lack of deposition during this period; however, as that unconformity is seen only in Profile 1, this evidence is not conclusive.

The *m2* dates to the later Middle Holocene: The ages from this unit fall within a relatively brief and well-defined interval from ~4,900 to ~4,200 ^{14}C yr B.P. The oldest age on the *m2*—4,910 ± 50 ^{14}C yr B.P. (CAMS-57518)—came from a small concentration of charcoal, possibly the remnants of a hearth, found in the sidewall of the M17 block on the South Bank.[8] This is not where one would expect the oldest age for stratum *m2*: Initially the *m2* in-filling would have taken

place in the deepest part of the paleovalley, and then the deposits would have reached up into the paleotributary. And yet charcoal from gravels at the base of the McJunkin channel within the paleovalley, 3.5 m lower than the possible hearth, was dated at 4,460 ± 50 (CAMS-57517). There is a 420–radiocarbon year difference in age between these samples and they do not overlap, even at 2σ. These samples are nonetheless from the same stratigraphic unit, which is multiaged and time-transgressive and represents sediments from discretely different alluvial events. The onset of deposition of *m2* clearly begins sometime around the older of the two ages. That the charcoal collected from gravels at the base of the McJunkin channel is younger than the charcoal higher up in the stratum may mark minor episodes of erosion within the paleovalley or vagaries of sampling.

There is only one radiocarbon date available on the uppermost part of the stratigraphic sequence at the site: An age of 700 ± 40 (CAMS-74651) was obtained on charcoal from the Wildhorse Formation on the North Bank. The charcoal came from sediments ~80 cm above the thick lens of cobbles and coarse gravel in the M23 block, just 50 cm below the present surface. It provides an approximate age for the last major pulse of deposition in the valley.

Obviously, there is a large gap in the chronological sequence between stratum *m2* (late Middle Holocene) and stratum *w* (Late Holocene). This gap may in part reflect a lack of sampling, rather than a depositional hiatus. Therefore, we do not know the age for the onset of deposition of the Wildhorse Formation and, thus, have little chronological help in addressing the question of whether the Wildhorse Formation was simply a facies of the McJunkin Formation. More work is required to resolve this matter.

Summary: The Quaternary Geology of the Folsom Site

The Folsom bonebed is within Late Glacial sediments that fill the lower portions of a small, two-pronged paleotributary and the adjoining paleovalley—the ancestral Wild Horse Arroyo—into which it drained. Both the paleovalley and the paleotributary are incised into Smoky Hill Shale, and in each area this Cretaceous unit is overlain unconformably by late Quaternary sediments. It is not known precisely when these bedrock valleys were incised or when the landscape began to take on the topographic appearance it had when Paleoindian groups entered the area. Given the erosion of latest Tertiary and early Quaternary basalt flows, as well as the very spotty distribution of Ogallala formation gravels throughout the region, it appears that a considerable amount of erosion took place in post-Tertiary times, perhaps beginning in the early Quaternary and continuing through much of that period (see also Bryan 1937).

In-filling of stratum *f1* sediments onto the Cretaceous surface began after 12,400 ^{14}C yr B.P. By the time of the Folsom occupation, ~1 m to 2 m of fluvial and colluvial deposits had filled the lower reaches of these channels. Beginning some time after 11,500 ^{14}C yr B.P., the paleotributary and the paleovalley began to fill with sediments of stratum *f2*, fine-grained, calcareous, light yellowish-brown silts, similar in physical characteristics to loess and deposited through a combination of airfall and redeposition of silts derived from airborne loess. Evidence from thin sections shows fine bedding with stringers of clay, indicating syndepositional reworking. Still, and as noted earlier, the absence of coarse clastics or bedding suggests that the loess was not extensively reworked by fluvial/colluvial runoff or sheetflow.

Deposition of the *f2* took place during a geologically brief period of time. The slight degree of pedogenesis in the sediments of the *f2* and the absence of any visible erosional breaks are consistent with a single episode of aggradation in which the surface never stabilized for more than several years to several decades. That deposition appears to have been continuous through the time of the Folsom kill. The top of the bison bone within the paleotributary is covered, thinly in spots, by *f2* sediments.

Around 10,500 ^{14}C yr B.P. Paleoindian hunters killed a herd of at least 32 bison, dropping the animals in both the paleotributary and the adjoining paleovalley. The kill was made on a ground surface that was essentially dry underfoot, at least within the paleotributary. The *f2* within the adjoining, topographically lower paleovalley may have had more moisture, but unfortunately we cannot be certain, as we have virtually no evidence that speaks to this matter. The iron-oxide mottling of the *f1* clearly indicates that water flowed in the paleovalley after that unit was deposited, but we do not know how long afterward or whether that rise in the water coincides with the time of the Paleoindian occupation. In neither area of the site do the bison bones occur on a distinct stratigraphic surface or unconformity; the ground on which the animals were killed was not long exposed. However, a backplot of the bones in the paleotributary shows that there is a well-defined archaeological surface on which they are resting (chapter 7), and there are subtle differences in soil texture and chemistry just below the bonebed.

There is no stratigraphic or paleontological evidence for more than a single Paleoindian kill taking place at this site, though the bones on the North Bank suggest that the kill was spread out over a large area and may have taken place within the paleotributary and the paleovalley. The statistically identical radiocarbon ages on bison bone from these different parts of the site substantiate this inference, within the limits of radiocarbon dating. We believe, all evidence considered, that these bison were killed at the same moment in real time.

The most precise age for that event comes from the dates run directly on the organic fractions of bison bone, which yield a mean age of 10,490 ± 20 ^{14}C yr B.P. The majority of the charcoal ages from stratum *f2* are as a group slightly—but consistently—older than the ages from the bison bone in that same unit (fig. 5.12), and the bison bone ages are more tightly constrained in age. However, the calibrated ages of both groups overlap at 1σ, suggesting geological contemporaneity, granting the inaccuracies of the radiocarbon timescale during the age plateaus of the Younger Dryas Chronozone and the fact that there is a considerable amount of natural charcoal here.

Although the *f2* deposits may at one time have blanketed much of the area, they are today relatively rare, being limited to the immediate area of the site and a few spots up- and downstream of it. The scarcity of *f2* deposits, particularly within the paleovalley, could be the result of one or more of the subsequent cut-and-fill episodes of the Holocene, the last following the 1908 flood that exposed the bison bone for George McJunkin to later discover. That a sizable portion of the bonebed was protected from these later erosional episodes is partly a consequence of many animals having been killed within the confines of the paleotributary, where their skeletal remains were protected from fluvial processes in the paleovalley by the flanking Smoky Hill Shale bedrock walls.

The geomorphic process that had the most impact on the archaeological deposits was the erosional and depositional cycle that began soon after the kill. Erosion started sometime after 10,100 ^{14}C yr B.P. and was followed by deposition of the *f3*, which began before ~9,800 ^{14}C yr B.P. and lasted until ~9,200 ^{14}C yr B.P. The effects of this cycle played out in different ways between the paleotributary and the paleovalley. Soon after the deposition of the bison carcasses, shingle shale washed off the nearby bedrock uplands and paleotributary walls. The shingle shale was deposited onto the bonebed in a thick lens—sometimes sets of lenses (chapter 7)—that rode across the top of the *f2* and effectively armored and protected the underlying bonebed. It is not known how much of the *f2* was removed in the process. This slope wash of shingle shale testifies to a scarcity of vegetation on the landscape: If the bedrock walls had been anchored by a vegetative groundcover, it would have been unlikely that the shingle shale could have moved so readily and en masse downslope.

In the paleovalley, the geomorphic processes that triggered the deposition of stratum *f3* had a detrimental impact on the bison remains. The bison bones in this area show evidence of having been dispersed by fluvial action, occur at high angles and largely as isolated elements, and are found jumbled amid channel gravels of varying size, all of which is indicative of fluvial transport and a secondary depositional context. These bison bone elements are not capped by a protective shingle armor, nor do they constitute an integrated and distinctive bonebed. Although much of the bone that we examined in the paleovalley was subject to transport and redeposition, not all of the bone from this area was similarly impacted, as we infer based on meager records of an earlier excavation of a bison cranium in the 1970s.

Importantly, the bison bone found in the paleovalley could not have simply washed out from the paleotributary; the bone was recovered well upstream of the intersection of the two. Thus, the kill must have extended over a large area, and was not restricted to either the paleotributary or the adjoining valley. How many animals may have been dropped in either area is unknown, as is the larger question of whether the hunters' strategic objective was to maneuver and kill the bison in the paleovalley, the paleotributary, or perhaps both. Although the greatest concentration of bison bone was recovered from the paleotributary, this could be an artifact of preservation, and not necessarily the main locus of the kill. Indeed, it is conceivable that the kill was centered in the paleovalley and that a comparable or greater number of animals were dropped there, and their traces were either moved or removed by subsequent erosion or still remain buried and archaeologically invisible beneath the deep Holocene fill of the paleovalley. A kill extending over this large an area would not be off-scale for hunter-gatherers (O'Connell, Hawkes, and Blurton-Jones, 1992).

In either setting, the hunters could have maneuvered, trapped, or otherwise disadvantaged the animals, by using the high bedrock walls of the paleovalley and the paleotributary and, perhaps, a knickpoint within the paleotributary, to reduce the risks of the hunt. Subsequent sedimentation in the area, including colluvial movement of shingle shale off the valley walls, has obscured the precise configuration of the land surface at the time of the kill.

The *f3* erosion and reworking of older *f2* sediments further complicate efforts to tease apart the precise chronological relationship between these depositional units, given the admixture of charcoal throughout, the removal of an unknown amount of *f2* sediment from the paleotributary and the paleovalley during the erosional episode, and the overlap in radiocarbon ages between the top of stratum *f2*, which ranges in age from ~11,500 to as recent as ~10,010 ^{14}C yr B.P., and the base of stratum *f3*, at ~10,420 ^{14}C yr B.P. (fig. 5.13).

Deposition of stratum *f3* continued into the early Holocene. The stratigraphy of the early Holocene and the onset of the middle Holocene is somewhat obscure, admittedly because less investigative effort focused on this portion of the sequence. Stratum *f3* itself was eroded sometime after ~9,200 ^{14}C yr B.P., Filling of the valley appears to have been episodic after *f3* deposition; there was deposition of bedded muds and silts (*m1*) from ~7,500 to ~6,000 ^{14}C yr B.P. The gap in the chronological sequence from 9,200 to 7,500 ^{14}C yr B.P. may overestimate the period of erosion, as far fewer radiocarbon dates are available for this part of the sequence.

Stratum *m1* was, in turn, eroded sometime between ~6,000 and ~4,900 ^{14}C yr B.P. and was followed by deposition of organic rich muds (*m2*) from ~4,900 to ~4,200 ^{14}C yr B.P. Again, there appears to be a chronological gap—this time from ~6,000 to ~4,900 ^{14}C yr B.P.—between stratum *m1* and stratum *m2*. Unlike the earlier gap, this one is better expressed stratigraphically, and better controlled chronologically, since the onset of *m2* deposition is securely dated. The ages from the Folsom site of this stratigraphic

and chronological hiatus, and subsequent deposition, roughly correlate with the Altithermal erosional-depositional sequence on the Great Plains (Holliday 1995; Meltzer 1999). While we suspect that this marks a local expression of Altithermal climatic changes, we recognize that more work and more radiocarbon ages would be necessary to resolve the matter.

The youngest sediments of the valley are those of stratum *w*, with a single Late Holocene age of ~700 ^{14}C yr B.P. Stratum *w* is characterized by muds and interbedded gravel lenses. We were unable to determine whether this deposit was inset into *m2* or just a late Holocene facies thereof. The gravel that characterizes stratum *w* contrasts with the fine-grained sediment devoid of gravel in the underlying McJunkin strata. This textural difference may simply reflect whether the exposed sections mark channel axis as opposed to channel margin deposits.

In the American Southwest arroyo cutting in the late Holocene is interpreted as evidence for cycles of drought, although the environmental significance of erosion and deposition in arid and semiarid environments has long been debated (Bryan 1922, 1925, 1940; Bull 1991; Butzer 1980; Knox 1983; Miller 1958). That late Holocene cycles of erosion and deposition in the Folsom region may be related to cycles of drought is supported by data from the High Plains, where reactivation of dunes due to aridity after ~1,000 ^{14}C yr B.P. is widely reported (Forman and Maat 1990; Forman, Oglesby, and Webb, 2001; Holliday 2001; Madole 1995; Muhs et al. 1996, 1997).

Notes

1. The two ages are not statistically distinct as measured by chi-square test (Hietala 1989), so averaging is appropriate. A more precise average of the two would be 4,340 ± 240 yr B.P., but to avoid undue confusion, the original Libby average is used.

2. Like so many of Cook's claims about the work at Folsom, this one aggrandizes, if not flat-out misrepresents, his role. Figgins was nominally in charge of the excavations from 1926 to 1927; Schwachheim was the principal excavator. Cook visited the site sporadically over those two seasons and, on occasion, gave instructions that had to be countermanded by Figgins (chapter 4).

3. The modeled depths in the different seismic lines did not always match each other, even at points at which the lines intersected. There are likely several reasons for this, not least that each line was analyzed by a different individual using different analytical parameters. The differences are not significant, however, and in the end absolute depths matter less than the more reliable relative trends in the depth and elevation of the bedrock.

4. Few specific details of the depositional context of the bones as they appeared at the time of the original excavations survive; however, the stratigraphic exposures we had from 1997 to 1999 are likely representative.

5. Even though we found no records of that 1972 excavation, we did find two corners of their excavation unit. The elevation at what appears to have been the floor of that unit is

96.15, which puts it ~1.5 m below the average level of the bone in the M17 block some 21 m distant.

6. There are pre–Late Glacial episodes of massive reworking of these large cobbles and gravels in the form of terrace deposits along the Dry Cimarron River.

7. Some comments on the rejected radiocarbon ages follow.

a. The dated charcoal samples that yielded ages of 840 ± 50 yr B.P. (CAMS-74650) and 820 ± 40 yr B.P. (CAMS-74652) were from strata *m1* and *f3*, respectively. Thus, not only are they far younger than they ought to be (ages on these units ought to be Early to Middle Holocene, not latest Holocene), but they are not even in proper stratigraphic order—despite being vertically separated by 45 cm of deposits. Moreover, despite being recovered from very different strata at different depths, the two ages are statistically identical. It appears, therefore, that younger, penecontemporaneous charcoal worked its way down into older deposits.

b. It was suspected even during pretreatment that sample 99-11 (M17-23-69) might not yield a suitable age, since the amount of available carbon was extraordinarily small, even for AMS dating, and it was therefore highly susceptible to contamination (T. Stafford, personal communication). That it ultimately produced an age of 14,800 ± 2,500 yr B.P. (CAMS-74646) confirmed those suspicions.

c. The still older age, 55,000 ± yr B.P. (CAMS-57522), run on charcoal from the bonebed came as a complete surprise. The sample was of suitable mass for dating and came from a secure context relatively high up within the Folsom bonebed on the South Bank (that is, not from any stratigraphic context that would be even close to that antiquity). There are several factors that might account for this "pre-Clovis" date from the Folsom site, most obviously the minute inclusion of geologically ancient carbon, which has since been identified within the Cretaceous-age Smoky Hill Shale (chapter 6). Contamination may have occurred via particulate matter or, perhaps, via groundwater percolating through the shingle shale that caps the bonebed. Such contamination, fortunately, was not a chronic problem at the site, given the otherwise good agreement between ages run on charcoal and on bone from the same stratigraphic unit.

8. This feature was spotted in the western wall of the M17 excavation block, and was photographed and profiled, but could not be excavated to examine the top or nearby surface; it remains in the ground.

SIX

Late Glacial Climate and Ecology

DAVID J. MELTZER, WITH CONTRIBUTIONS BY
MEENA BALAKRISHNAN, LINDA SCOTT-CUMMINGS,
BONNIE F. JACOBS, AND JAMES L. THELER

The Folsom area presently experiences a continental, semi-arid climate of cold winters and relatively warm summers (chapter 3). Summer rainfall is dominant (fig. 3.9), but there is the potential for heavy winter snowfall. It is an area of significant topographic variability and, therefore, considerable biotic diversity, the vegetation ranging from open oak and locust parkland in the immediate vicinity of the site, to pine and spruce galleries and rolling grassland on the slopes of and atop (respectively) nearby Johnson Mesa (figs. 3.6, 3.8, 3.11). The area today has few plants capable of providing the carbohydrates and essential fats necessary to support long-term occupations by hunter-gatherers. However, the vegetation does support a rich vertebrate fauna, including large mammals, such as elk, deer, and antelope today, and free-ranging bison in historic and prehistoric times (fig. 3.13).

The climate and environment of the Folsom site area in Late Glacial times are less well known. The topic was not the focus of attention during earlier investigations, and the few observations made along these lines were not always consistent. There have been a number of studies of Late Glacial climate and environment in the Great Plains and Rocky Mountains, especially in recent decades; however, most of these took place at localities hundreds of kilometers distant and/or hundreds of meters higher or lower in elevation than Folsom (e.g., Allen and Anderson 2000; Amundson et al. 1996; Armour, Fawcett, and Geissman 2002; Barnosky, Anderson, and Bartlein 1987; Betancourt 1990; Connin, Betancourt, and Quade 1998; Elias 1996; Fall 1997; Gosse et al. 1995; Holliday 2000a; Lovvorn, Frison, and Tieszen 2001; Markgraf and Scott 1981; Nordt et al. 2002; Polyak, Rasmussen, and Asmerom 2004; Reasoner and Jodry 2000; R.R. Thompson et al. 1993; Vierling 1998; Yu and Wright 2001). Although useful and suggestive, these studies provide only indirect evidence of the paleoclimate and paleoenvironment at Folsom. Similarly, modeling of climatic conditions during the Younger Dryas Chronozone (YDC), within

which the Folsom site was occupied, mostly serves to emphasize the variability in climate and environment in this region and elsewhere during this rapidly changing millennium (e.g., Bartlein et al. 1998; Kutzbach 1987; Kutzbach and Ruddiman 1993; Kutzbach et al. 1993; Shuman et al. 2002; R. Thompson et al. 1993; Whitlock and Grigg 1999; Williams, Shuman, and Webb 2001; Yu and Wright 2001).

Thus, in order to understand the environmental and climatic conditions specific to the Folsom area during the Late Glacial, new data were required. Guiding their acquisition were the questions laid out in chapters 1 and 3. To recap briefly: Was the vegetation community in Folsom times open or closed? Were the apparent richness and diversity of the vegetation the same as they are at present? Were tree lines lowered? Were there MAST trees in the area to provide forage for people? Was the preferred forage of other animals—notably bison—present in sufficient abundance to support local herds?

Is there evidence here of a Folsom-age drought, as occurs on the nearby plains (Holliday 2000a)? More generally, what was the climate before, during, and immediately after the occupation? There is certainly a priori reason to expect that the average annual temperature was colder during the YDC than at present, but was it? Were climates more continental or more equable? And were conditions drier or wetter than at present? How widely available was surface water? Were there seasonal differences in precipitation—for example, was winter snowfall more pronounced and, perhaps, a critical limiting factor in occupation of the area? Is there any reason to believe, contrary to expectations derived in chapter 3, that this is an area in which Paleoindian foragers might be expected to overwinter?

To address these questions—and there were no illusions that we would be able to answer all of them—we acquired several kinds of data, including pollen, macrofossils, land snails, and faunal remains, from the site as well as from a lake core on Johnson Mesa. The data and analyses are detailed below, after a summary of the work on site paleoenvironments done during earlier investigations on-site.

Previous Views of Past Environments

There are likely several reasons for the relative lack of attention on the part of the original investigators to the climate and environment at the time of the bison kill. For one, this was not a topic of particular interest to archaeologists or paleontologists of the 1920s (Holliday 1997; Meltzer 2006); analytic expertise in such matters was still in its infancy (pollen analysis was just then being imported from Europe to America), and of course, much of the attention was riveted by the new species of bison (chapter 7) and the previously untyped fluted projectile point (chapter 8) that had been discovered. Nonetheless, passing remarks were made of sediments and snails from the site that show some interest in the site's paleoenvironmental context.

As discussed in chapter 5, the site was initially supposed to have been wet underfoot. In his opening report on Folsom's geology, Cook (1927a:244) reported that "shells of characteristic freshwater invertebrates occur, but have not been studied," and also that the bison bones showed evidence of having been trampled on by other animals, while lying buried in the mud" (chapter 7).

Cook never published anything further on those freshwater snail taxa, although in a letter to Oliver Hay, he tallied the taxa he had identified:

> Considering the invertebrates, one genus, *Planorbis*, is a species not over one half an inch across, is the dominant form I have observed, occurring in the bed with the *Bison* bones. However, I also saw tiny specimens that I could not distinguish from *Physa* but only about one eighth of an inch long. In the bone level I also found two broken, freshwater *Unios* of a species probably an inch and a half long judging by what was preserved of them. A bit lower down I also found two specimens of *Lymnea* of small size but not in the bonebed. Unfortunately no importance or interest was attached to such evidence by those in charge of the work and no care has been taken to collect and save such evidence, save in a few convenient instances . . . the presence of such life in these waters, and the presence of *Bison*, is, to my mind, good evidence of non-glacial conditions at that time. Undoubtedly a specialist in fresh water invertebrates can deduce far more than this from that evidence. (Cook to Hay, January 25, 1928, OPH/SIA)

As best can be determined, those specimens apparently sent by Figgins to Washington were never analyzed—or, if analyzed, were never reported. Nonetheless, Cook correctly identifies the habitats of the species listed; all are indeed shallow-water air breathers or pulmonates. Still, questions Cook himself raises about the context in which these specimens were recovered make their assignment to the Folsom bonebed open to question.

But not to Cook. Aquatic snails and a mud-hole seemed to him clear evidence that "swampy, marshy conditions existed for some time in a considerable area in the valley bottom" (Cook 1927a:244). That it was stagnant body of water rather than a flowing stream was supported by the observation that "the matrix is free from coarse materials such as sand and gravel, which characterize stream beds in such a location as was this one" (Cook 1928b:39; also Cook to Hay, January 25, 1928, OPH/SIA; Cook to Hay, December 23, 1926, OPH/SIA). Moreover, such a setting made sense to Cook (1928b:40) in terms of bison behavior: "Buffalo like to wallow and roll in mud-holes, as is well known; and wounded, feverish animals would naturally seek cool waters to lie down in when suffering from such wounds."

The idea that the bonebed was set within a small marsh or lake would be repeated in later years (e.g., Brown 1928a: 826; Roberts 1936:14, 1938:534, 1940:59). However, by late 1928 Brown had a different interpretation of the origins of the deposit, which conjured a distinct view of the climate. If, as he suspected, the sediments were of aeolian origin and accumulated during "a long period of little or no rainfall" (Brown 1928b:5), this implied "a climatic condition different from that now existing in the same area..." (Brown 1936).

Fortifying Brown's view of the climate was a very different suite of snails collected by his crew during the 1928 season. The complete (and only) description of that material is as follows:

> In the bone layer near the center of the quarry we found many shells which have been identified by W. [Wendell] C. Mansfield as *Succina avara*, *Oreohelix strigosa*, and *Pyramidula cockerelli*. These are all pulmonate land shells "that dwell in moist places, but are not indicative of a pond as at first thought." (Brown 1928b:4)

The question of whether the snails in the Folsom bonebed are aquatic or terrestrial can be readily resolved based on the results of our excavations. Those excavations involved intensive recovery efforts that yielded a detailed stratigraphic record of >2,300 specimens, not including juveniles and eggshells, of nearly two dozen taxa of snails. There is not a single aquatic form among them, as discussed in detail below.

Cook may well have been correct in his species identifications, but in all likelihood the aquatic specimens he found were not associated with the bonebed. That, in turn, raises two possibilities: Either those snails came from higher or lower stratigraphic levels—recall that the thick pond sediments of the McJunkin Formation overlay the bonebed in the paleotributary—or the shells were recovered from the North Bank in the deposits of the paleovalley. Unfortunately, at this historical distance it is not possible to determine which of these possibilities is correct, and given the nature of 1920s excavation techniques, it could be either. All that can be reasonably assumed, given the more representative and stratigraphically controlled sample we recovered, is that Cook's snails did not come from the bonebed proper.

In contrast to Cook's snails, the ones recovered by Brown's crews in 1928 match up reasonably well with those recovered during our excavations. However, the match is not perfect. Part of the problem is simply taxonomic: Ambiguities in the type specimen and description of *Succina avara*, for example, have prompted some authors to drop it altogether as a taxonomic entity (Metcalf and Smartt 1997: 49–50), or synonymize it with *Catinella vermeta* (Hubricht 1983:16,18). But complicating the matter still further is the fact that identifying genera and species within the Succineidae Family is "notoriously difficult" (Metcalf and Smartt 1997:47). Thus, our excavations yielded members of the Succineidae—but *which* Succineidae and whether it was the same as the ones recovered in 1928 are not clear. Because of such taxonomic ambiguity, there is little that can be said about the habitat of these snails, save that several species of succineads (including *C. vermeta*) are known to occur at present in this montane region of New Mexico (Metcalf and Smartt 1997:40–41, 50).

The *Pyramidula cockerelli* recovered in 1928 is now synonymized with *Discus shimekii*. *D. shimekii* presently occurs at high elevations in the northern mountains of New Mexico (Metcalf and Smartt 1997:40). In fact, it occurs at altitudes upward of 2,560 to 3,535 m (8,400 to 11,600 ft) and may well be a Pleistocene relic in this portion of the Rocky Mountains (Dillon and Metcalf 1997:113–114; Metcalf and Smartt 1997:41). Below that lower elevation, the more common *Discus* is *D. whitneyi*. It is, of course, quite possible that the elevation range of *D. shimekii* shifted downward during the Pleistocene and that its occurrence at Folsom at ~2,109 m (~6,919 ft) reflects that (*D. shimekii* also occurs in Midwestern Pleistocene loess deposits). However, given that only *D. whitneyi* was found in our excavations, and the difficulty of separating *D. shimekii* from *D. whitneyi* at lower elevations where both occur and may even hybridize, it remains an open possibility that the specimens recovered in 1928 were *D. whitneyi*.

The *Oreohelix* recovered during the 1928 excavations was the same species as the one recovered during our subsequent excavations, *O. strigosa*, the Rocky mountainsnail. *O. strigosa* is also a form that inhabits higher-elevation regions, lack of moisture being an important limiting factor in its lower range of elevation, which in modern transects does not extend below about 2,560 m (8,400 ft) (Dillon and Metcalf 197:115; Metcalf and Smartt 1997:41).

The land snails recovered in 1928 thus present a picture of a high-elevation gastropod fauna in the Folsom bonebed, in which at least two of the species occur below their present ranges, suggesting a downward shift in temperature during late Pleistocene times.

In the 1970s, Anderson and Haynes collected pollen samples from "all alluvial deposits in the valley" but were only able to process six of those. Of those six, only four were

TABLE 6.1

Pollen Types in Folsom Site Samples Recovered During the Folsom Ecology Project

Stratum	Taxon
Qo2 (=*m1*)	*Pinus edulis*
	Quercus
	Cheno-Am
	Compositae
	Artemesia
	Graminae
Qo2 (=*f3*)	*Pinus edulis*
	Quercus
	Cheno-Am
	Compositae
	Graminae
Qo1b (=*f2*) (top of bonebed)	Compositae
	Cheno-Am
	Artemesia
	Graminae
	Quercus
	Juniperus
	Pinus edulis
Qo1b (=*f2*) (base of bonebed)	*Artemesia*
	Cheno-Am
	Compositae
	Graminae
	Quercus

SOURCE: Data from A. Anderson (1975:fig. 6, and pages 75–76). Order is apparently by frequency of occurrence.

from archaeological deposits. Two were from the "bottom and the top of the Folsom-bearing deposit, Qo1b" (=*f2*); the third was from the base of the "overlying deposit, Qo2" (=*f3*) (A. Anderson 1975:75–76). The fourth was obtained 300 m downstream of the site in what was initially identified as the top of unit Qo2 (A. Anderson 1975: fig. 6) but was later assigned to unit *m1* based on an associated ^{14}C age of 6,060 ± 500 (TX-1452) (Haynes, Anderson, and Frazier 1976). A very limited amount of pollen was present in the samples; only 50 grain counts were obtained (A. Anderson 1975:76). The pollen types identified in the analysis are listed in table 6.1.

Anderson interpreted these results as indicating that the pollen from the top and bottom of the *f2* were similar, "except for a frequency replacement of oak . . . Pinyon pine . . . and Juniper." These two samples differed in turn from the pollen in the overlying unit, which yielded "a high percentage of Pinyon pine pollen and no sage *(Artemesia)*" (A. Anderson 1975:76). She cautioned, however, that the limited nature of the sampling constrains the reliability of these results. That said, those results are generally consistent with ours, though they differ in several respects, as discussed below.

Reconstructing Folsom Paleoenvironments

The data discussed here were acquired from three sources: a sediment core from Bellisle Lake on Johnson Mesa, from which pollen was extracted; the systematically collected sediment samples (chapter 4) from the excavation units within the bonebed and the *f2*, from which we sought to extract beetles, charcoal, gastropods, phytoliths, plant macrofossils, pollen, and other potential environmental indicators; and, the bison bone from the site. Obviously, with the exception of the pollen core from Bellisle Lake, the samples discussed here were obtained directly from the Folsom site.

Analysis of these data variously involved identification of taxa, their habitats, and their ecological tolerances; studies of stable isotope geochemistry—specifically carbon (δ^{13}C), nitrogen (δ^{15}N), and oxygen (δ^{18}O); analysis of ground paleotemperature; and studies of ancient (bison) DNA. Some of these analyses are ongoing and will be reported elsewhere. Efforts to recover beetle remains from the sediments proved unsuccessful; there was simply insufficient organic detritus preserved in the bonebed deposits (Scott Elias, personal communication, 1998).

The discussion begins with plant remains, pollen, and macrofossils, then ranges up the "ladder of nature" to invertebrates (snails) and then on to vertebrates (bison).

The Pollen Core from Bellisle Lake

With Bonnie F. Jacobs

In order to gain an understanding of the vegetation in Late Glacial times, we obtained a sediment core from Bellisle Lake, as well as sampled for pollen and macrofossils in the sediments at the site itself. These data inform at different spatial scales: The lake core provides a record of the regional pollen rain and vegetation, while the site sediments, particularly the charcoal, help identify plants growing in the vicinity of the site. Each data source has its own taphonomic history, which may be more or less complicated, depending on the nature of the depositional environments and processes, vagaries in preservation, the catchment area(s) being sampled, the degree of long distance transport, and the like. For these reasons, the Bellisle Lake and Folsom site data are discussed separately, beginning with the more regional record from Bellisle Lake.

A number of basalt-rimmed lakes occur atop Johnson Mesa; these may be erosional and weathering features (Lucas, Hunt, and Hayden 1987:51), topographic low spots in the basalt, or perhaps maars. Many of those lakes are ephemeral; others appear to be relatively permanent, at least on the historic timescale (at least one of the ephemeral lakes has archaeological sites adjacent to it [A. Anderson 1975]).[1]

Bellisle Lake (36°53'40"N and 104°07'50"W), one of the larger and presumably more permanent lakes on Johnson Mesa (fig. 3.1), is a shallow basin at 2,364-m elevation. The lake is circular, with a diameter of ~500 m; at the time of the initial coring, the water was ~1.8 m deep across this flat-bottomed basin. The vegetation around the lake is short-grass steppe and midgrass prairie, but on the distant hillsides and particularly to the west toward Raton, are conifer woodlands dominated by *Pinus ponderosa* (ponderosa pine). From Bellisle Lake, Rocky Mountain upland communities are more continuous to the northwest, and Great Plains vegetation becomes more widespread and continuous to the north and northeast (chapter 3; also Dick-Peddie 1993; B. Thompson et al. 1996).

The majority of the precipitation—31.5 cm (53% of the MAP)—falls between May and August (fig. 3.9). The MAT over the same period is 6.6° C with peak minima and maxima occurring in June, July, and August (fig. 3.10). Thus, summer rains arrive at the hottest time of year and are most critical for plant growth (chapter 3). These temperature and precipitation parameters, here and elsewhere across the Southwest, are influenced by topographic features and the movement of moist air masses inland from the Pacific Ocean, the Gulf of California, and, especially, the Gulf of Mexico. Temperature changes through time would have been approximately equal at equivalent elevations across the Southwest (New Mexico, Colorado, Arizona, Utah) because lapse rates, the inverse relationship between temperature and elevation, are essentially uniform across the region (Meyer 1992). Precipitation parameters, on the other hand, cannot be expected to increase and decrease uniformly across the Southwest, even at equivalent elevations, because they are determined by the source of moist air masses, which have varied from region to region since the Pleistocene.

Today, the proportion of annual precipitation that falls in the winter increases toward the west across the Southwest and summer rain becomes more important toward the east. Winter rain originates primarily with Pacific air masses, while summer monsoon rain comes primarily from the Gulf of Mexico, with a secondary contribution from the Gulf of California and tropical eastern Pacific (chapter 3; Adams and Comrie 1997; Bowen 1996). Variability in moisture from the Gulf of California and Gulf of Mexico has ecological ramifications in eastern New Mexico, parts of Arizona, and Colorado, where summer rainfall is most important. Wet summers are usually associated with a more northerly position of the subtropical ridge of high pressure, the Bermuda High, which moves toward Mexico and the Southwest and helps to direct moist air inland from the Gulf of Mexico and Gulf of California (Adams and Comrie 1997). Moist air from the Gulf of Mexico is especially important as a source of rain for the uplands of New Mexico, the Colorado Plateau, and the southern Rocky Mountains, because it moves inland at a higher altitude than air masses from the Gulf of California and reaches the eastern part of the Southwest first (Adams and Comrie 1997; Bowen 1996).

Johnson Mesa and Bellisle Lake are situated at a high enough elevation and far enough east to intercept moist air from the Gulf of Mexico and, therefore, are in a strategic position for detecting temporal changes in moist air influx from the Gulf of Mexico through time. Changes in forest

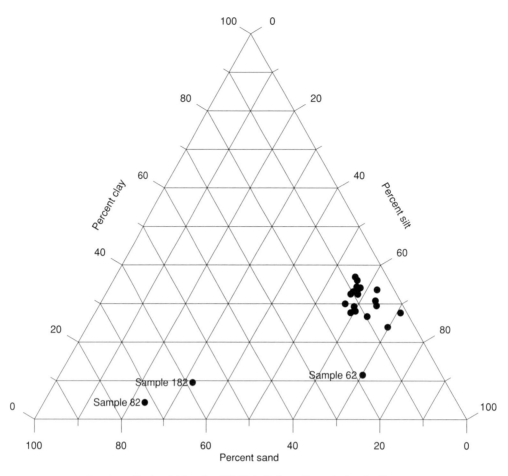

FIGURE 6.1 Textural triangle of Bellisle Lake sediments, by depth.

cover caused by these changes in moisture were recorded by pollen assemblages in the Bellisle sediment core.

METHODS

Bellisle Lake required two coring efforts. The first took place during the summer of 1999 with collaborators from Arizona State University, using a coring platform floated onto the lake and anchored. Unfortunately, the lake clays are extraordinarily dense and compact, and a floating platform offers no resistance and thus no leverage. Hence, it proved extremely difficult to extract the coring apparatus once it had been driven into the lake floor.[2] By using various devices—Livingstone and Dachnowsky corers and, as a last resort, a bucket auger—a broken and discontinuous set of samples was recovered to a depth of 212.5 cm. Such were insufficient for detailed pollen work. However, the sediment from the deepest part of the core yielded pollen, along with organic material that produced a radiocarbon age of 10,310 ± 60 ^{14}C yr B.P. (Beta-137087).

That evidence justified a second attempt to retrieve a core, and accordingly, the work was conducted with collaborators from the U.S. Geological Survey in February of 2001, when the lake was frozen and offered a stable working surface. Coring was done using a modified Livingstone device;

a hand-operated weight attached to the apex of a large tripod was used to hammer the 5-cm-diameter piston corer into the sediment. Extraction of the core barrel was aided by the use of a winch attached to the tripod apex. A total of 217 cm of sediment was recovered from the lake in five drives. The core bottomed in loose sands.

In the laboratory, core sections were scraped clean and split lengthwise. Pollen samples of 1-cm thickness were taken at 10-cm intervals, after accounting for compaction. Samples were processed at the Northern Arizona University Laboratory of Paleoecology, Flagstaff, using standard preparation techniques, adding a known quantity of *Lycopodium* spores to each sample for later calculation of pollen concentrations. Aquatic taxa were counted outside the pollen sum.

SEDIMENTS AND RADIOCARBON CHRONOLOGY

The sediments comprising the core were nearly uniformly fine-grained throughout, averaging about 60% silt and 30% clay and falling primarily into the silty clay loam and silt loam textural classes (fig. 6.1). Thus, the bulk of the core represents the relatively gentle accumulation of mud on the floor of the lake. Two samples, however—one from a depth of 81 cm to 82 cm, the other at 181 cm to 182 cm—were comprised of 72% and 58% sand, respectively, and fall into

TABLE 6.2
^{14}C Radiocarbon Ages from Bellisle Lake Core

Core Drive	Depth (cm)	Laboratory No.	^{14}C Age (Yr B.P.)	$\delta\ ^{13}C$	Cal ^{14}C Age (Yr B.P.; 1σ Probability)
1	10–11	Beta-155335	800 ± 40	−24.7	676–732 (1.0)
1	50–51	Beta-155336	570 ± 40	−24.3	535–566 (0.438)
					597–634 (0.562)
1	53–54	Beta-159158	640 ± 40	−24.4	560–600 (0.618)
					628–652 (0.382)
2	80–81	Beta-155337	3,800 ± 40	−28.1	4,093–4,118 (0.178)
					4,145–4,241 (0.822)
2	100–101	Beta-155338	5,230 ± 40	−27.3	5,926–5,994 (0.878)
					6,083–6,085 (0.015)
					6,152–6,165 (0.106)
3	120–121	Beta-155339	5,680 ± 40	−27.5	6,409–6,493 (1.0)
3	138–139	Beta-159159	8,240 ± 40	−26.2	9,092–9,098 (0.030)
					9,129–9,221 (0.594)
					9,233–9,282 (0.290)
					9,292–9,294 (0.013)
					9,361–9,367 (0.030)
					9,388–9,396 (0.043)
3	140–141	Beta-155340	7,920 ± 40	−25.5	8,638–8,777 (0.724)
					8,831–8,854 (0.114)
					8,883–8,897 (0.060)
					8,918–8,931 (0.062)
					8,942–8,947 (0.015)
					8,969–8,975 (0.024)
4	150–151	Beta-155341	3,000 ± 40	−25.9	3,080–3,089 (0.064)
					3,111–3,126 (0.098)
					3,140–3,150 (0.070)
					3,158–3,260 (0.745)
					3,312–3,316 (0.023)
4	155–156	Beta-159160	5,530 ± 50	−26.3	6,285–6,321 (0.447)
					6,326–6,349 (0.225)
					6,369–6,397 (0.328)
4	160–161	Beta-155342	6,890 ± 40	−25.8	7,674–7,747 (1.0)
4	164–165	Beta-159161	8,130 ± 40	−25.7	9,011–9,034 (0.230)
					9,037–9,089 (0.535)
					9,102–9,127 (0.221)
					9,223–9,226 (0.014)
4	170–171	Beta-155343	8,060 ± 40	−25.9	8,786–8,800 (0.068)
					8,807–8,827 (0.146)
					8,868–8,878 (0.056)
					8,901–8,911 (0.062)
					8,980–9,029 (0.664)
					9,054–9,055 (0.004)
4	180–181	Beta-155344	7,790 ± 40	−26.0	8,462–8,466 (0.027)
					8,479–8,494 (0.130)

(continued)

TABLE 6.2 (*Continued*)

Core Drive	Depth (cm)	Laboratory No.	^{14}C Age (Yr B.P.)	$\delta\,^{13}C$	Cal ^{14}C Age (Yr B.P.; 1σ Probability)
4	183–184	Beta-159162	11,480 ± 50	−25.0	8,517–8,530 (0.104)
					8,537–8,599 (0.740)
					13,191–13,247 (0.173)
					13,309–13,499 (0.741)
					13,744–13,778 (0.086)
4	190–191	Beta-155345	8,300 ± 40	−25.4	9,157–9,166 (0.045)
					9,267–9,328 (0.416)
					9,338–9,401 (0.477)
					9,410–9,421 (0.062)
5	200–201	Beta-155346	11,240 ± 40	−25.1	13,042–13,070 (0.082)
					13,134–13,217 (0.373)
					13,241–13,383 (0.544)
5	210–211	Beta-155347	11,170 ± 40	−25.1	13,004–13,193 (0.976)
					13,301–13,311 (0.014)
5	215–216	Beta-159163	11,040 ± 40	−25.6	12,977–13,141 (1.0)
5	216–217	Beta-155348	10,260 ± 40	−25.2	11,765–11,809 (0.169)
					11,920–12,146 (0.624)
					12,227–12,304 (0.004)
Auger II, Drive 7	182–200	Beta-137087	10,310 ± 60	−23.7	11,777–11,793 (0.037)
					11,941–12,336 (0.963)

NOTE: All dates are on bulk carbon. Organic pieces were too small for AMS analyses.

the sandy loam textural class. These sand spikes represent a significant change in the depositional history of the lake and are likely indicative of drying and above normal erosion, the coarse material left behind during deflation. Alternatively, it could represent changes in sediment load coming into the lake, but given the flat bottom of the lake, much of that sand would likely accumulate in sediment deltas on the lake margins. These textural anomalies aside, we did not see any other significant depositional changes or stratigraphic unconformities in the core.

Initially, 14 samples of sediment from the core were sent to Beta Analytic Inc. for radiocarbon dating. They returned a sequence of dates that ranged from 570 ± 40 ^{14}C yr B.P. at 53 cm (Beta-155336) to 11,240 ± 40 ^{14}C yr B.P. at 200 cm (Beta-155347) (table 6.2). However, five stratigraphic reversals occurred at intervals along the core. We considered several possible hypotheses that might account for these reversals, including (1) complexities in the sedimentation history of the lake, such as hiatuses, deflation, and episodes of redeposition; (2) errors in sampling the core for radiocarbon; (3) sediment mixing during collection; and (4) influxes of dead carbon. We felt reasonably comfortable eliminating the last possibility, since the substrate and local bedrock are basalt. Because each core segment was extruded intact and apparently unmixed, we considered mixing during collection to

be unlikely but could not exclude the possibility altogether. Mostly, we suspected sampling errors and changes in depositional history were keys, the latter because this shallow lake is sensitive to below-average precipitation and can dry out completely, as happened in the summer of 2002 (also fig. 3.7).

In an attempt to resolve which factors were in play and to clarify the chronology, we submitted an additional six samples for radiocarbon dating, five of which were from the lowermost 75 cm of the core (table 6.2; Beta-159158 through Beta-159163). These were selected because we were especially interested in clarifying the Late Pleistocene and Early Holocene chronology and, also, to ensure that no radiocarbon sample was collected more than 10 cm from the adjacent samples in this portion of the core—most, in fact, are within 5 cm or less of their neighbor. This close-interval sampling would help eliminate the possibility that vagaries in the radiocarbon column were due to sampling error, since it would make it very unlikely that we simply missed collecting a sample from the time period in question.

All 20 radiocarbon ages are plotted by depth in figure 6.2. There remain reversals in the sequence, but when it is viewed as a whole, there is a strong linear relationship between age and core depth ($r^2 = 0.824$). Calculating residual scores identifies which ages are significant outliers along that regression line; there are four, and they are identified in

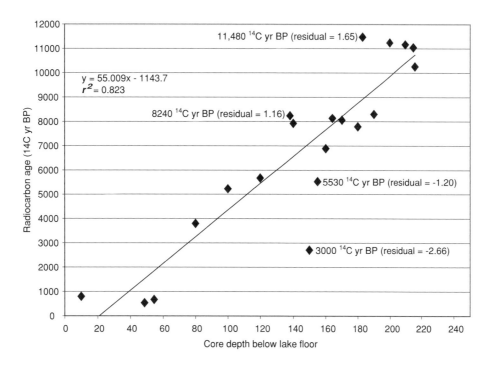

FIGURE 6.2 Plot of radiocarbon ages by depth, Bellisle Lake core, with outliers as noted.

figure 6.2. Of those four, two are younger than they ought to be, given their position in the core, and two are older.

There are various mechanisms by which younger organic remains can be incorporated into older sediments, but two are most likely: the down-core movement of material via mud cracks or redeposition when the lake dried, which would then be welded into the deposit when the lake was recharged; and/or the down-core movement of material during the coring process itself, when younger material falls into the core hole as the device is pulled out or reinserted with each drive. In the event of the former, anomalous ages ought to occur during periods for which there is independent evidence of drying; the latter might appear as anomalously young ages at the top of a drive segment as perhaps, for example, the 3,000 ± 40 ^{14}C yr B.P. age (Beta-155341) from the top of Drive 4. With regard to the reverse situation—older organic remains incorporated into younger sediments—the obvious mechanism in this setting is drying-out of the lake. The exposed floor will not evenly erode, and older material could easily be scattered across younger surfaces, then covered when the lake is recharged.

When arranged chronologically (fig. 6.3) the radiocarbon ages fall into a number of groups, separated by millennia-long temporal gaps. The earliest cluster of radiocarbon dates ($n = 5$) are Late Glacial in age, between 11,480 ± 50 ^{14}C yr B.P. (Beta-159162) and 10,260 ± 40 ^{14}C yr B.P. (Beta-155348) (fig. 6.3). The oldest of those ages was not, as it ought to have been, at the base of the core but, instead, was a significant outlier at a depth of 183 cm in a portion of the core that was otherwise dated to ~8,000 ^{14}C yr B.P. We infer that older material from some other part of the lake floor had

been eroded at a later date and was redeposited at this spot in the core. In support of that, we note that the anomalous date comes from the portion of the core containing one of the two sand spikes (at 181–182 cm), clearly indicating the possibility of significant disturbance. With the exception of the youngest age (e.g., 10,260 ^{14}C yr B.P.), the radiocarbon ranges in this first group overlap at 1σ (calibrations using OxCal 3.9), indicating a relatively tight cluster and continuous sedimentation onto the lake floor.

This first group is followed by a ~2,000–radiocarbon year gap that covers the initial millennia of the Early Holocene. Given the density of samples from this portion of the core, we are confident this gap is not a result of inadequate sampling. Rather, it is more likely a consequence of deflation, a lack of sedimentation, or both.

The second group of ages ($n = 6$) clusters in the latter part of the Early Holocene, beginning at 8,300 ± 40 ^{14}C yr B.P. (Beta-155345) and extending to 7,790 ± 40 ^{14}C yr B.P. (Beta-155344). This group spans ~1,000 calibrated years, with ranges that do not altogether overlap, even at 2σ. It also includes several small stratigraphic reversals. Four samples came from Drive 4 at a depth of 164 cm to 190 cm; the other two were from Drive 3 at a depth of 138 and 140 cm. All of this suggests rather steady deposition through this millennium, followed by later reworking. As to when that reworking took place, we suspect the Middle Holocene, since this segment of the core (corresponding to Drives 3 and 4) is a mix of both older and younger materials, and two of the ages within this group (7,920 ± 40 ^{14}C yr B.P. and 8,240 ± 40 ^{14}C yr B.P. [Beta-155340 and Beta-159159]) were in the midst of a range of Middle Holocene dates.

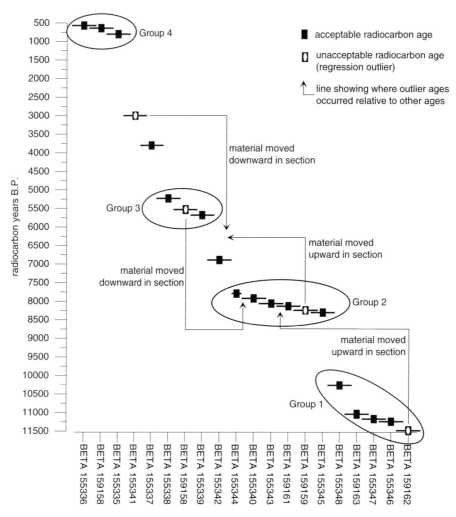

FIGURE 6.3 Box plot of Bellisle Lake ^{14}C ages to one standard deviation. Acceptable radiocarbon ages shown as solid blocks; unacceptable radiocarbon ages shown as hollow blocks. Arrow tips mark where these unacceptable ages occurred relative to other dated samples.

There is a millennium-long gap that follows the second group—the next dated sample is 6,890 ± 40 ^{14}C yr B.P. (Beta-155342) at a depth of 160 cm. That single age, in turn, is followed by another millennium-long gap to 5,680 ± 40 ^{14}C yr B.P. (Beta-155339). These two gaps in the temporal sequence fall squarely within the Altithermal and, not coincidentally, bracket the portion of the core in Drives 3 and 4 in which several significant outliers occur; that is, those between 138- and 155-cm depth. All of this suggests that the depositional history during this period was complicated by deflation, and mixing. Sediments of this age may have once been present but were lost to erosion either during or after this period. That samples dated to 3,000 ± 40 and 5,530 ± 50 ^{14}C yr B.P. (Beta-159160) came out of that part of the core that is otherwise early Middle Holocene in age suggests the possibility that desiccation and drying of lake opened mud cracks, down which younger material subsequently moved.

The third group (n = 3) has radiocarbon ages between 5,680 ± 40 and 5,230 ± 40 ^{14}C yr B.P. (Beta-155338). It appears—based on the 20 cm of sediment that accumulated on the floor of Bellisle Lake between the end-members of this group, which are in proper stratigraphic order—that there was rapid sediment buildup during the later Middle Holocene.

Following the third group is another long gap in the temporal sequence, between 5,230 ± 40 and 3,800 ± 40 ^{14}C yr B.P. (Beta-155337). This temporal gap coincides with 20 cm of core depth and could reflect either a slowdown in sediment deposition or a depositional hiatus. The latter seems likely, given the sand spike that also occurs within this portion of the core (81- to 82-cm depth).

The presence of material dated to 3,000 ± 40 ^{14}C yr B.P., although out of stratigraphic sequence, nonetheless indicates that some deposition was taking place between 3,800 and 3,000 ^{14}C yr B.P. After that, however, there is a 2,200-year gap in the radiocarbon sequence to 800 ± 40 ^{14}C yr B.P. (Beta-155335). As this gap corresponds to a 25-cm segment of core (55- to 80-cm depth) that was not sampled for dating, the gap may be more apparent than real. The fourth and final group of radiocarbon ages (n = 3) is all younger than 800 ^{14}C yr B.P., and the calibrated ages are all within 2σ of one another.

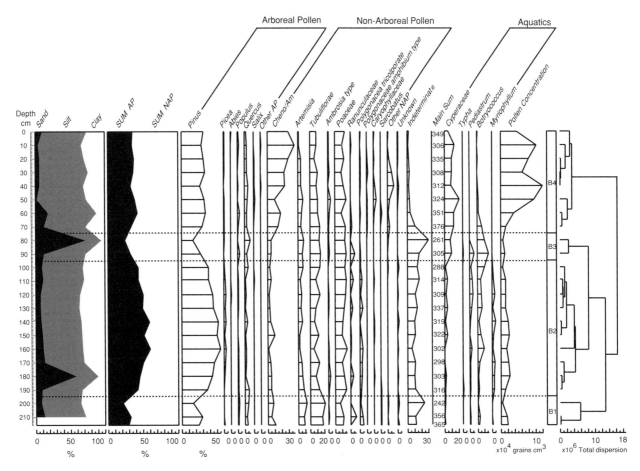

Depth
cm

Sand Silt Clay SUM AP SUM NAP

Arboreal Pollen

Pinus Picea Abies Populus Quercus Salix Other AP Cheno/Am

Non-Arboreal Pollen

Artemisia Tubuliflorae Ambrosia type Poaceae Ranunculaceae Polygonaceae tricolporate Polygonaceae amphibium type Caryophyllaceae Sarcobatus Other NAP Unknown Indeterminate

Aquatics

Main Sum Cyperaceae Typha Pediastrum Botryococcus Myriophyllum Pollen Concentration

0
10
20
30
40
50
60
70
80
90
100
110
120
130
140
150
160
170
180
190
200
210

349
306
335
308
312
324
351
376
261
305
288
314
309
337
319
322
302
298
303
316
242
356
365

B4

B3

B2

B1

0   50   100 0   50   100 0   50 0000 000 30 0 0 2000 0 0 0000 0 0 30   0 2000 0 0 0   0   10   0   10   18
%   %   %   x10⁴ grains cm³ x10⁶ Total dispersion

FIGURE 6.4 Pollen diagram, Bellisle Lake. (Prepared by B. Jacobs.)

Leaving the outliers aside, the radiocarbon ages from the core fall throughout the Holocene, though they do not do so evenly, which we attribute to changing periods of deposition and erosion at Bellisle Lake and changes in sedimentation rates on the lake floor. In general, however, the ages increase with depth and, thus, provide some chronological control on the core. Importantly, they also show that sediments of Late Glacial age are present. Although that portion of the record is of greatest relevance to the matter at hand, we nonetheless provide here the data on the full pollen sequence from Bellisle Lake, to help put the environment of this period in the larger Pleistocene/Holocene context.

POLLEN, VEGETATION, AND CLIMATE HISTORY

The Bellisle pollen diagram (fig. 6.4) is divided into four biozones using the CONISS program (Grimm 1987) and available as an option in the PSIMPOLL plotting program used here (version 4.10 Bennett 2002). CONISS provides objective pollen zonation by constraining cluster analyses to samples in successive stratigraphic positions based on relative pollen percentages. All pollen taxa were used in the analysis, but nonpollen categories such as sediment texture were omitted.

Zone B-1, which represents the Late Glacial, and initial sedimentation in the lake, is characterized by relatively low ratios of arboreal-to-nonarboreal pollen (NAP). NAP comprises between 60% and 65% of pollen counts in this zone and is dominated by Asteraceae (primarily *Artemisia* and other Tubuliflorae), Poaceae, Chenopodiinae (Chenopodiaceae plus *Amaranthus*), and other herbaceous taxa present at amounts of 5% or less.

Pine pollen comprises the bulk of the arboreal pollen in Zone B-1 at Bellisle Lake, but that constitutes only 20% to 30% of the total pollen assemblage. This contrasts with the 35% to 40% pine pollen in modern pollen spectra at Bellisle, which is typical of open settings in the Southwest and San Juan and Rocky mountains (Fall 1992; Hall 1994; Maher 1963). Today, pine trees are located on hillsides within ~2 km of Bellisle Lake (e.g., fig. 3.8). This suggests that the Late Glacial landscape would have been open as it is today, but with conifer communities located at greater distances from the lake and with a greater presence of sage *(Artemisia)* and other members of the Asteraceae in the immediate area. Together, this evidence indicates that during Late Glacial times summer rainfall was lower than at present, and insufficient to support tree growth on this part of Johnson Mesa.

The lake itself was probably shallow, periodically supporting the green algae *Botryococcus* and *Pediastrum*, as well as *Myriophyllum*, an aquatic plant. When they are present, the pollen of aquatic plants occurs in amounts of 3% or less in this zone. Their presence, albeit in small amounts, indicates permanent standing water, which, given lower-than-modern summer rainfall, would have been possible only if cooler conditions made Late Glacial warm-season evaporation less than it is today. Therefore, this period was likely relatively dry and cool in comparison with today.

It is not possible to resolve the amount of winter precipitation relative to today from the pollen assemblages at Bellisle Lake, although abundant spring and summer snow melt could have contributed to maintenance of the lake even if it was not sufficient to support tree coverage at modern levels.

A significant hiatus in deposition occurs at the end of the Late Glacial, at the boundary between Zone B-1 and Zone B-2 (about 195 cm), where the 2,000-year period from 10,260 to 8,300 ^{14}C yr B.P. is not represented by sediment. The older date provides only the maximum age for the beginning of the hiatus, because some younger sediments could have been lost to deflation. This hiatus overlaps the erosional episode that occurred between the deposition of the *f2* and that of the *f3* unit at the Folsom site (chapter 5) and in other localities in the area (Mann 2004), indicating that this was not a lake-specific phenomenon but a response to changes in regional climate.

When water flow to the lake resumed, deposition onto an uneven surface must have caused erosion of local highs and redeposition of older sediments in reverse order, as they are between 217 and 195 cm in depth.

Zone B-2 extends from ~8,300 to about 5,000 ^{14}C yr B.P. (fig. 6.4) and is characterized by a relatively high ratio of arboreal pollen to NAP and *Pinus* percentages of 36% to 55%. Other arboreal taxa, such as *Picea*, *Quercus*, and *Populus*, are present, but at percentages that together do not exceed 6%. Pollen of Asteraceae (including *Artemisia*) comprises about 15% of the pollen count, much less than in Zone B-1, where the Asteraceae total is at least twice that amount. There is a slight increase in pollen of Poaceae (grass), from about 10% in Zone B-1 to about 15% in Zone B-2.

The freshwater algae, *Pediastrum* and *Botryococcus*, and the aquatic plant, *Myriophyllum*, are more abundant here than in Zone B-1, ranging from 5% to 10%. These indicate that standing water was probably deeper, or at least more continuously present, than it was during the deposition of Zone B-1. Cyperaceae (sedge), which occurs today in moist or wet areas and would have grown at or close to the lake margin, is present at amounts of 5% or less.

Modern pollen studies document pine pollen present at relative amounts of between 37% and 80% in central and southern Colorado in surface pollen samples from montane conifer forest (Fall 1992; Maher 1963). High pine pollen percentages also occur in the alpine zone due to upward transport of pollen into an area where local pollen productivity is low. However, montane and alpine environments can be distinguished by the presence of at least 20% *Picea* pollen (Fall 1992). At Bellisle Lake, *Picea* pollen does not exceed 5% in any of the samples. Thus, the vegetation around Bellisle Lake during Zone B-2 was likely montane as opposed to alpine forest, dominated by *Pinus* spp., with an understory of grass or grass-dominated small clearings.

Pollen assemblages in Zone B-2 document a wetter climate at Bellisle Lake, at least episodically, than during any other time period. The presence of coniferous forest close to or surrounding the lake would have required ample summer rain to support tree growth during the warm growing season, and the relative abundance of aquatic pollen indicates a permanent body of standing water during most of this time period. However, the occurrence of occasional peaks in both Cheno/Am and Poaceae, the multiple radiocarbon reversals and gaps (fig. 6.3), and the high sand spike at 181- to 182-cm depth indicate that there was episodic warming and drying, which complicated depositional history throughout this time interval.

Zone B-3, from 95 to 75 cm in depth, is bracketed by radiocarbon ages of 5,230 and 3,800 ^{14}C yr B.P. (fig. 6.4), but as noted no samples were submitted for dating in the 20 cm of core that separates these two ages. Within that span, pollen assemblages are characterized by a substantial increase in indeterminate pollen to a maximum of 30% at 80-cm depth, a decline in pine pollen to a minimum of ~25%, and a slight increase in oak pollen, to 10%. NAP increases to between 60% and 70%, the bulk of which is Asteraceae, Poaceae, and Ranunculaceae (buttercup family) pollen. Among the aquatic taxa, *Botryococcus* and Cyperaceae increase, but both *Myriophyllum* and *Pediastrum* disappear at the top of this zone. Thus, the lake not only was quite shallow, but may have been marshy at times when not completely dried out.

The sediment texture curve, as noted, documents a sand peak (72%) at 81- to 82-cm depth, indicative of drying and erosion. This coincides with the peak in indeterminate pollen, which is accompanied as well by relatively low pollen concentrations. These factors indicate that pollen preservation was poor in Zone B-3 even during times when there was enough moisture to support sediment deposition in the lake. Zone B-3 represents a transition between the late-middle and late Holocene during which time there was a hiatus or hiatuses in deposition that may have been accompanied by deflation.

The poor condition of the pollen indirectly testifies to the climate at the time, as the likely cause of the damage was oxidizing conditions resulting from repeated wetting and drying of sediment. Consistent with this interpretation is the strongly mottled sediment from 75 to 85 cm, indicative of numerous wet-dry cycles, and the sand spike at 81 cm to 82 cm, which indicates a short-lived relatively high-energy event. The most likely explanation for Zone B-3 environments is that warm and dry conditions prevailed during this transitional period from the late middle Holocene to the late Holocene.

Zone B-4, from 75 cm to the surface, represents the last ~800 years (fig. 6.4). This zone is characterized by 55% NAP in nearly all samples, up from approximately 45% in Zone B-2. A prominent feature of Zone B-4 is an increasing amount of

Chenopodiineae pollen toward the surface, reaching a maximum of 35% at 10-cm depth. *Pinus* pollen once again comprises the bulk of arboreal pollen and is present consistently at or near 35%. There is a greater diversity and amount of pollen from families of primarily herbaceous plants, but this may be due to the high pollen concentration and excellent preservation, as indicated by only 5% or less indeterminate pollen. Pollen concentration increases in this zone to a maximum of 120,400 and 102,340 grains/cm^3 at 40 and 10 cm, respectively, from ≤25,000 grains/cm^3 in Zones B-1 to B-3. Aquatic taxa are represented almost entirely by Cyperaceae, with *Botryococcus* present at amounts of <5%.

Along with these conspicuous changes in pollen composition in Zone B-4, there is a significant sedimentological change at the boundary between Zone B-3 and Zone B-4. The sediment is characterized by strong brown mottling (7.5YR 5/8), most common between 75 and 85 cm, and becomes dark reddish gray to reddish black above the B-3/B-4 boundary, indicating a large increase in organic content. The environment at Bellisle Lake during this period was essentially as it is today. Pine trees grew nearby, but conditions were open around the lake. Chenopodiineae are usually indicative of disturbance in settings other than saltpans or deserts, and the increasing percentage of plants represented by this superfamily is consistent with relatively recent land use by humans or their cattle. The presence of Cyperaceae at amounts of 10% to 20% indicates marshy conditions at or near the margins of the lake. Altogether, climate conditions are essentially modern.

DISCUSSION

At Bellisle Lake the Late Glacial appears to have been drier and cooler than today. Because this part of New Mexico receives most of its moisture from summer monsoon rains, a weaker monsoon during the time period represented by Zone B-1 is consistent with fewer trees on the landscape due to drier conditions. Late Glacial weakening of monsoon circulation patterns around the world relative to today is supported by climate models (Kutzbach et al. 1993), which also indicate a concomitant increase in winter precipitation in the North American West due to a southerly displacement of the jet stream originating in the Pacific. Many proxy records from the American West document winter precipitation greater than that today, especially west of the Continental Divide, and/or decreased summer rainfall during the Late Glacial, when temperatures were also depressed relative to modern ones but warmer than during the full glacial (Barnosky, Anderson, and Bartlein 1987; Betancourt 1990; Fall 1997; Holliday 1997; Nordt et al. 2002; R. Thompson et al. 1993; Vierling 1998). Fall suggests that jet stream westerlies associated with winter precipitation were at a maximum at 11,000 ^{14}C yr B.P. in central Colorado, as their position receded northward behind diminishing continental glaciers (Fall 1997; also Kutzbach 1987; R. Thompson et al. 1993). The more southerly location of Johnson Mesa and the

Folsom site relative to the track of winter Pacific air masses may account for the lack of sufficient winter precipitation to compensate for diminished summer rains.

The depositional hiatus from about 10,260 to 8,300 ^{14}C yr B.P. at Bellisle Lake is interpreted as a period during which sediment either was not deposited, was eroded away, or both. Other proxy records in the Southwest from this time period document warm conditions, in some cases warmer than the present (Fall 1997), and with greater rainfall in the southern Rocky Mountains and Colorado Plateau, perhaps a result of the strengthening of the summer monsoon (Betancourt 1990; Fall 1997; Markgraf and Scott 1981; Vierling 1998). However, at Bellisle Lake precipitation was still apparently insufficient to exceed evaporation to the degree necessary for maintenance of permanent water and a depositional record. Summer temperatures may have been warmer than today: Orbital calculations for 9,000 ^{14}C yr B.P. suggest that the Northern Hemisphere experienced 8% greater than modern July insolation (Kutzbach and Ruddiman 1993). Thus, the climate during all, or at least the latter part, of this hiatus would have been warmer than during Zone B-1 time but not wet enough to overcome the effects of warm-season evaporation.

The time period represented by Zone B-2, approximately 8,300 to 5,000 ^{14}C yr B.P., was generally wetter than the preceding period, suggesting that precipitation exceed evaporation, because summers were either cooler or wetter or both. That said, we also note that portions of the sediment record during this period are mixed or missing, aquatics are relatively low, and there are minor spikes in NAP, including grass. Clearly, the record at Bellisle Lake is complex and bears a resemblance both to higher-elevation sites in central Colorado and the Colorado Plateau (Betancourt 1990; Fall 1997; Markgraf and Scott 1981), which document continued maximum expansion of both upper and lower treelines in response to warmer and wetter conditions from 10,000 ^{14}C yr B.P. to as late as 4,000 ^{14}C yr B.P., and to the lower-elevation southern High Plains to the east, where drought conditions intensified from the Early through the Middle Holocene (Holliday 1997; Meltzer 1999).

The duration of Zone B-3, from approximately 5,000 to 3,800 ^{14}C yr B.P., is represented by oxidized and mottled sediment as the result of continual wetting and drying. Consequently, pollen is poorly preserved in Zone B-3 and sediment is likely to be missing due to hiatuses in deposition, or erosion, or both. Late Holocene climate is not as well documented by pollen, packrat middens, and sedimentological data as earlier periods; however, pollen data from central Colorado are interpreted as indicating cooler conditions than the middle Holocene from 4,000 to 2,000 ^{14}C yr B.P. (Fall 1997) or modern climate conditions by 4,000 ^{14}C yr B.P. (Markgraf and Scott 1981).

Pollen Zone B-4, spanning at least the last 800 years, is marked by rich dark sediment, excellent pollen preservation, and consistent pollen assemblages representing the modern environment. The sedimentary change between Zones B-3

TABLE 6.3
Pollen and Spores from Folsom Bonebed

Sample	Taxon (Common Name)	Taxon Unique to This Sample (Common Name)
Unit: N1034 E1000 Level: 147 Number: N17-21-16 Context: above bonebed, in an area where bone is scarce Mass sample: 408 g	*Juniperus* (juniper) *Picea* (spruce) *Pinus* (pine) *Quercus* (oak) *Artemesia* (sagebrush) Low-spine asteraceae (includes ragweed, cocklebur, etc.) High-spine asteraceae (includes aster, rabbitbrush, snakeweed, sunflower, etc.) Cheno-Am (includes amaranth and pigweed families) Poaceae (grass)	*Salix* (willow) *Tsuga* (hemlock) *Euphorbia* (spurge) Ranunculaceae (buttercup) *Typha angustifolia* (cattail) Monolete (fern spore) *Selaginella densa* (little clubmoss spore)
Unit: N1034 E997 Level: 149 Number: M17-23-89; M17-23-90 Context: lower part of bonebed, adjacent to bison cranium Mass sample: 362 g Note: two samples from same quad/level, collected in small pieces, so greater surface area exposed to possible contamination	*Juniperus* (juniper) *Picea* (spruce) *Pinus* (pine) *Artemesia* (sagebrush) Low-spine asteraceae (includes ragweed, cocklebur, etc.) High-spine asteraceae (includes aster, rabbitbrush, snakeweed, sunflower, etc.) Cheno-Am (includes amaranth and pigweed families) Cyperaceae (sedge) Poaceae (grass)	Convolvulaceae (convolvulus) *Ephedra nevadensis* (Mormon tea) *Polygala* (milk-wort) Trilete (fern spore)
Unit: N1031 E999 Level: 151 Number: M17-9-63; M17-9-64 Context: below bonebed Mass sample: 401 g Note: two samples from same quad/level	*Juniperus* (juniper) *Picea* (spruce) *Pinus* (pine) *Quercus* (oak) *Artemesia* (sagebrush) High-spine asteraceae (includes aster, rabbitbrush, snakeweed, sunflower, etc.) Cheno-Am (includes amaranth and pigweed families) Cyperaceae (sedge) Poaceae (grass)	*Robinia* (locus) Anacardiaceae (sumac family) Rhamnaceae (buckthorn) Rosaceae (rose) Rosaceae striate

SOURCE: Analysis and identification by L. Scott-Cummings; all samples from *f2* sediment from South Bank excavations.

and B-4 indicates a change in depositional mode, which may correspond to a change to the modern climatic regime.

Pollen and Charcoal from the Folsom Bonebed

With Linda Scott-Cummings

In order to examine the vegetation closer in to the site, 3 pollen and 17 charcoal samples from the Folsom bonebed were analyzed. The pollen was extracted from sediment samples taken from immediately below, within, and immediately above the bonebed (in excavation Levels 151, 149, and 147, respectively). All samples are from unit *f2*, in the M17 excavation block. The samples all weighed 350 g to 410 g and were from portions of the excavation no more than ~10 × 10 cm in area (table 6.3).

The charcoal samples were recovered from multiple areas on-site, as (1) scattered occurrences throughout the *f2* sediments on the South Bank, all of which appear to be natural in origin; (2) Feature 1, a concentration of charcoal—perhaps representing a hearth—which was exposed in the sidewall of the M17 block in m2 sediments and, therefore, postdates the Paleoindian occupation (chapter 5); (3) the stratigraphic profile from the M23 block on the North Bank; and (4) various buried channel deposits, including the Holocene-age McJunkin Arroyo. Most of these charcoal samples were subsequently radiocarbon dated (chapter 5).

METHODS AND CAVEATS

A chemical extraction technique based on flotation is the standard preparation technique used for the removal of the pollen grains from a large volume of sand, silt, and clay with which they are mixed. This particular process was developed by Cummings at the Paleo Research Laboratory for extraction of pollen from sediments where preservation has been less than ideal and pollen density is low.

Hydrochloric acid (10%) was used to remove calcium carbonates present in the sediment, after which the samples were screened through 150-μm mesh. The samples were rinsed until neutral by adding water. After the samples stood for 2 hr, the supernatant was poured off. A small quantity of sodium hexametaphosphate was added to each sample, then the beaker was again filled with water and allowed to stand for 2 hr. The samples were again rinsed until neutral, the beakers filled only with water. This step was added to remove clay prior to heavy liquid separation. The samples were then dried and pulverized. The samples were mixed with sodium polytungstate (density, 2.1) and centrifuged at 2,000 rpm for 5 min to separate organic from inorganic remains. The supernatant containing pollen and organic remains was decanted. Sodium polytungstate was again added to the inorganic fraction to repeat the separation process. The supernatant was decanted into the same tube as the supernatant from the first separation. This supernatant was then centrifuged at 2,000 rpm for 5 min to allow any silica remaining to be separated from the organics. Following this, the supernatant was decanted into a 50-ml conical tube and diluted with distilled water. These samples were centrifuged at 3,000 rpm to concentrate the organic fraction in the bottom of the tube. After rinsing the pollen-rich organic fraction obtained by this separation, all samples received a short (10- to 15-min) treatment in hot hydrofluoric acid to remove any remaining inorganic particles. The samples were then acetylated for 3 min to remove any extraneous organic matter.

A light microscope was used to count the pollen grains at a magnification of 600×. Pollen preservation in these samples varied from moderate to poor. Comparative reference material collected at the Intermountain Herbarium at Utah State University and the University of Colorado Herbarium was used to identify the pollen to the family, genus, and species level, where possible.

Pollen aggregates were recorded in one instance. Aggregates are clumps of a single taxon and may be interpreted to represent pollen dispersed over short distances or to represent the introduction of portions of the plant into a deposit. Aggregates were included in the pollen counts as single grains, as is customary, but are identified by an "A" next to the frequency on the pollen diagram (fig. 6.5). Pollen diagrams are produced using TILIA, developed by E. Grimm of the Illinois State Museum. Pollen concentrations are calculated in TILIA using the quantity of sample processed, the quantity of exotics (spores) added to the sample, the quantity of exotics counted, and the total pollen counted. Indeterminate pollen includes pollen grains that are folded, mutilated, and otherwise distorted beyond recognition. These grains are included in the total pollen count, as they are part of the pollen record.

Complicating the Folsom site pollen analysis was the presence of ancient, redeposited pollen, likely released from the local, Smoky Hill Shale Cretaceous bedrock. The analytical distinction between this redeposited Cretaceous pollen and the Late Glacial pollen was twofold. First, and most reliably, a pollen grain was considered redeposited if it was of a plant type that no longer grows in North America, or at least has not grown in North American in Quaternary times. Second, and based on the observation that pollen grains identified by the first criterion were routinely flattened, were distinctly two-dimensional, and took up the laboratory staining in a distinctive manner, any pollen grains displaying that morphology and color but not otherwise identifiable were placed in the redeposited category. Pollen grains that did not meet either of those criteria were considered part of the Late Glacial assemblage. The most difficult task when working with these samples was to separate the Late Glacial pollen from the redeposited Cretaceous pollen on a consistent basis.

As it happens, the redeposited pollen and spores occur at frequencies approximately four to eight times higher than does the Late Glacial pollen. The abundance of this ancient organic matter, occasional flecks of which were observed coming out of shingle shale from auger holes, likely also helps explain the errant radiocarbon date of 55,000 ^{14}C yr B.P. from the site (chapter 5). Absolute concentrations of the Late Glacial pollen, once it was separated from the more abundant Cretaceous pollen, proved to vary from a low of ~60 to a high of ~500 pollen grains per cubic centimeter of sediment. These were sufficient quantities for analysis and interpretation.

Processing the charcoal samples was complicated in several cases by the presence of hard clays; these samples (97–49, 99–24, and 99–44) were initially floated in the laboratory. The samples were added to 500 ml of hot water, then stirred until a strong vortex formed. The disaggregated materials were then poured through a 150-μm-mesh sieve. Additional water was added and the process repeated until no sediments remained in the beaker. The floated portions were allowed to dry on clear, sterile plastic sheets.

Because charcoal and possibly other botanic remains were to be sent for radiocarbon dating, clean laboratory procedures were implemented during flotation and identification

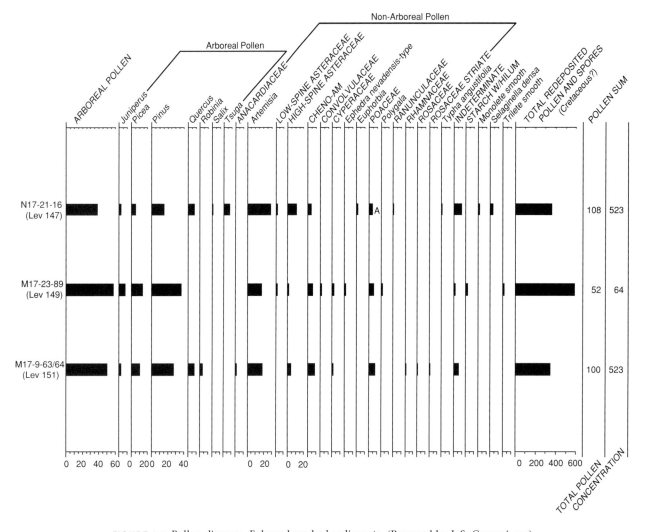

FIGURE 6.5 Pollen diagram, Folsom bonebed sediments. (Prepared by L S. Cummings.)

to avoid contamination. All instruments were washed between samples, and samples were protected from contact with modern charcoal. Sterile latex gloves were worn to prevent direct contact with the charcoal specimens.

Where possible, the light fractions were passed through a series of graduated screens (U.S. Standard Sieves with 2-, 1-, 0.5-, and 0.25-mm openings) to separate charcoal debris and to sort the samples initially. The contents of each screen were then examined. Charcoal pieces larger than 0.25 mm in diameter were broken to expose a fresh cross section and examined under a binocular microscope at magnifications up to 140×. Specimens were identified using manuals (Martin and Barkley 1973; Musil 1978; Schopmeyer 1974) and by comparison with modern and archaeological references. All of the elements necessary for genus- or species-level identification may not be present in a given cross section of wood charcoal, particularly fragments that are less than 500 μm in size. The identifiability is dependent on whether diagnostic cells are projecting into the visible plane under examination. If a series of cross sections is examined, however, the likelihood that diagnostic cells are present increases.

When dealing with very small fragments of charcoal, species-level identification was generally not possible. Pollen grains are also often difficult to identify to the species level. The inability to make taxonomic distinctions finer than the genus level constrains ecological interpretation, since different species within the same genus can have very different climatic tolerances. For example, Engelmann spruce (Picea engelmannii) is a subalpine species that today grows at elevations from 9,000 to 12,000 ft (3,000–3,650 m) and can tolerate mean annual temperatures from 20°F to 50°F (−7°C–10°C) and January temperatures as low as −4°F (−10°C); today, it grows no closer to Folsom than 100 km (Dick-Peddie 1993:51–52, 58; R. Thompson, Anderson, and Bartlein 1999a:76). In contrast, blue spruce (P. pungens) generally grows between 8,000 and 10,000 feet (~2450–3050 m), can tolerate mean annual temperatures that range from 25 to 72° F (−4°C–22°C) and drier settings, but requires warmer winters (Dick-Peddie 1993:51–52, 58; R. Thompson, Anderson, and Bartlein 1999a:79). Similar differences are also apparent in species of pine (e.g., bristlecone vs. ponderosa and piñon) that, along with spruce, were important genera in the vegetation, both past and present.

The interpretation of the oaks is complicated further by the fact that the dominant species present in the area today, *Quercus gambelii* (chapter 3), occurs both as an aggressive early successional species and as a typical form in climax lower montane forests (Dick-Peddie 1993:66, 125). Thus, even though it is not known that the genus of oak identified in the Holocene charcoal (see below) is indeed *Q. gambelii*, what its presence may be signaling in terms of local floristic history may not be clear.

RESULTS

Overall, pollen sums were low, ranging from 52 to 108 grains, with total concentration amounts from 64 to 523 grains (fig. 6.5). There is also a sharp decline in the amount of pollen present in the sample at the level of the bonebed (Level 149). As a result, and because there are only three pollen samples, any apparent trends in the changing percentages of each taxon by level may be not ecologically significant but, instead, the result of vagaries of preservation, sample size, and time averaging.

That noted, roughly half of the pollen assemblage (45%–60%) from the *f2* sediments at Folsom (table 6.3) is comprised of arboreal pollen, a figure characteristic of wooded settings (Fall 1997:1306). The genera that occur are the same as those characteristic of the modern community (e.g., juniper, spruce, pine, oak) and, by percentage, are dominated by pine (fig. 6.5). This taxon occurs in amounts comparable to that seen in Zone B-1 at Bellisle Lake, which is not surprising since pine is an overproducer of pollen and is readily subject to long-distance transport. It is also of interest to observe that the quantities of *Picea* pollen—over 10% of the total pollen—are elevated compared with those of pollen from more recent deposits in northeastern New Mexico (Scott-Cummings, unpublished).

The two dominant trees in the local environment today, *Quercus* (oak) and *Robinia* (locust), are present in the pollen samples from the bonebed, but only in very small amounts. *Quercus* pollen was identified in samples below and above the bonebed. *Robinia* pollen was observed only in the lowest sample. Very small amounts of two other arboreal taxa, *Salix* (willow) and *Tsuga* (hemlock), were recovered solely from the uppermost sample (N17-21-16). Although a species of willow occurs in the area today, hemlock is not part of the modern tree assemblage in New Mexico or, for that matter, across much of the American west (an anomaly discussed below).

Sagebrush, which is presently an important shrub in the local vegetation community, is represented by *Artemisia* pollen at frequencies of approximately 20% in the lower two samples but nearly doubles in the sample collected above the bonebed, possibly suggesting a change in local vegetation. A similar increase in high-spine Asteraceae pollen is observed and might represent an increase in shrubby plants and forbs

such as rabbitbrush and sunflower. Attesting to the relative patchiness of the environment throughout, Cheno-am pollen frequencies are relatively low (5%–10%) but present in all three samples (and slightly higher in the deeper sample), while Poaceae (grass) frequencies display a similar pattern.

Present in very small quantities are pollen from moisture-loving plants such as *Typha angustifolia* (cattail) and perhaps Ranunculaceae (buttercup family), as well as a variety of shrubs, forbs, and grass-like plants, including members of the Convolvulaceae, Cyperaceae (sedges), *Polygala*, *Euphorbia*, Rhamnaceae, and Rosaceae. All of these pollen types are present in small quantities (in all cases less than 5% of the assemblage), making their appearance in the sediments suggestive of local vegetation. Sedges might have grown with grasses, exploiting a similar habitat or might have grown under wetter conditions. Members of the Convolvulaceae and *Euphorbia* are considered to be weedy plants, growing in disturbed areas. *Polygala* is a large genus that includes both herbaceous and shrubby plants, usually occupies relatively dry habitats in the area today. Both Rhamnaceae and Rosaceae shrubs are noted in the area today. Indeterminate pollen, representing pollen grains that were too badly deteriorated to identify, makes up a relatively small portion of the pollen record.

In addition to pollen, a few spores were observed in the uppermost sample (N17-21-16), representing growth of a few ferns and moss as part of the local vegetation. Recovery of a few starch granules probably represents deterioration of seeds and/or roots in these sediments.

The results of the charcoal analysis are detailed in table 6.4, along with charcoal specimens previously identified at the site (Haynes et al. 1992). As noted in table 6.4, positive and species level identifications were not always possible. And, owing to the limited number of samples examined (17), quantitative trends within levels or time periods are again of limited reliability. However, there are a few observations to be made of these results.

The charcoal samples from below the bonebed, and from the bonebed itself, are uniformly conifers, notably pine and juniper, both of which were detected in the pollen from these levels. Spruce, also identified in the pollen, was not present as charcoal, though that may simply be a sampling phenomenon.

The charcoal analysis hints too at a change in the vegetation in the middle to late Holocene. Pine and other conifer species continue to be present in the samples from that time period, but by then so too is oak. Assuming that the absence of oak charcoal from the earlier deposits is not a sampling fluke, this indicates that a dominant species in the present plant community only became locally established long after the close of the Pleistocene. Its minor occurrence as pollen in the bonebed could reflect upslope transport from lower elevations (Fall 1992; Markgraf 1980). Although the appearance of oak charcoal might be a clue to when the modern vegetation community emerged, testing that hypothesis requires more data.

TABLE 6.4
Samples Submitted and Results of Charcoal Analysis

Sample	Mass (g)	Age (^{14}C Yr B.P. or Estimated)	Taxon	Comments
99-29	1.137	700 ± 40 (CAMS-74651)	Quercus sp.	Sample contained a large quantity of charcoal. Quercus was present as the dominant type. Charcoal-stained soil also was observed in the sample.
99-36	0.019 0.215	Ca. Late Holocene	Cf. Pinus Cf. Conifer	The Pinus charcoal was identified based on the presence of resin canals. The coniferous charcoal was very deteriorated, however, the sample is most likely Pinus. Estimated age based on stratigraphic position of sample between 99-29 and samples 99-24 and 99-44
99-24	0.001 0.048	Ca. Late to Middle Holocene	Quercus sp. Unidentified	Very small fragment of oak and an unidentified charcoal. Radiocarbon age of 840 ± 50 yr B.P. (CAMS-74650) rejected as too young.
99-44	0.015	Ca. Late to Middle Holocene	Pinus	Contained Pinus charcoal and charcoal-stained soil; the latter comprised the bulk of the sample. The charcoal stains contained no diagnostic cell structure. Radiocarbon age of 820 ± 40 yr B.P. (CAMS-74652) rejected as too young.
98-478	0.010 0.042	Ca. Middle Holocene	Quercus sp. Unidentified	From gravels at base of McJunkin arroyo and contained a variety of charcoal flecks including Quercus, and charcoal-stained soil. Sample came from same spot as charcoal that yielded radiocarbon age of 4,460 ± 50 yr B.P. (CAMS-57517).
97-49	0.004	4,640 ± 60 (CAMS-74644)	Pinus	Small fragments of charred Pinus recovered from a Giddings core along the paleochannel edge.
98-304	0.251	4,850 ± 50 (CAMS-57519) 4,910 ± 50 (CAMS-57518) average = 4,880 ± 35	Pinus	Feature 1 sample; charcoal fragments contained evidence of resin canals, a characteristic used in the identification of pine. The diameter of resin canals can be used to determine species-level identification; however, an insufficient number of resin canals were present in each specimen to make a determination.
99-58	0.027 0.008	9,820 ± 40 (CAMS-74647)	Pinus Conifer	On bedrock surface of valley wall. Both Pinus and conifer charcoal. The latter was badly deteriorated, but could also tentatively be identified as Pinus.
98-489	0.068	9,340 ± 50 (CAMS-57515)	Pinus	Sample was comprised of slightly vitrified pine charcoal and a quantity of charcoal-stained soil. The vitrified material has a glassy, shiny appearance due to fusion by heat.
99-45	0.002 0.004	9,440 ± 50 (CAMS-74653)	Pinus Unidentified	Included Pinus charcoal, as well as a number of unidentified fragments. The unidentified charcoal was badly deteriorated, thus preventing identification.
99-61	no data	10,420 ± 140 (CAMS-74648)	Unidentified	From reworked gravels in paleochannel. Although a large sample, the fragments were friable and lacked the diagnostic characteristics necessary for identification.
98-170	0.053	10,010 ± 50 (CAMS-74645)	Conifer	A family-level identification was deemed appropriate because the charred specimens were deteriorated, lacked evidence for intercellular spaces situated between the longitudinal tracheids, which are commonly associated with Juniperus, and lacked evidence of resin canals, which are common to the genus Pinus.

TABLE 6.4 (*Continued*)

Sample	Mass (g)	Age (¹⁴C Yr B.P. or Estimated)	Taxon	Comments
99-68	0.112	$10,510 \pm 50$ (CAMS-74649)	*Pinus*	From upper portion of tributary head-cut
99-11	0.004	Terminal Pleistocene, ca. 10,500	*Conifer*	A very small sample of charcoal in poor condition; the lack of diagnostic structure prevented identification to the genus level. The radiocarbon age ($14,800 \pm 2,500$ yr B.P.) rejected as too old based on context and stratigraphic position.
CS-70-7	No data	$10,760 \pm 140$ (AA-1709)	*Pinus*	North Bank, part of a sample of "single lumps or flecks widely dispersed throughout the [ƒ2] zone." From Haynes et al. (1992)
CS-70-7	No data	$10,850 \pm 190$ (AA-1711)	*Juniperus*	As above, from Haynes et al. (1992)
CS-70-7	No data	$10,890 \pm 150$ (AA-1710)	*Juniperus* or hardwood	As above, from Haynes et al. (1992)
CS-70-7	No data	$10,910 \pm 100$ (AA-1712)	*Juniperus* or hardwood	As above, from Haynes et al. (1992)
CS-70-7	No data	$11,060 \pm 100$ (AA-1708)	*Pinus*	As above, from Haynes et al. (1992)
98-319	0.025	$11,070 \pm 50$ (CAMS-57524)	Charcoal-stained soil	A majority of the charcoal was unidentifiable due to the absence of any diagnostic wood structure. Most likely, the unidentified materials represent charcoal dust that was sealed in sediments or charcoal that deteriorated in situ.
99-46	No data	Terminal Pleistocene?	Charcoal-stained soil	Sample contained only charcoal-stained soil. No diagnostic charcoal characteristics were visible, precluding any identification. Insufficient material present for radiocarbon dating.
99-60	No data	Terminal Pleistocene?	Charcoal-stained soil	Sample contained only charcoal-stained soil. The charcoal present exhibited no diagnostic cellular characteristics. Insufficient material present for radiocarbon dating.

NOTE: All specimens charred and in fragmentary condition. As samples are the same as those submitted for radiocarbon dating, additional information on provenience and age is given in Table 5.4. Species identifications by Kathryn Puseman, except where provided by Haynes et al. (1992).

DISCUSSION

The pollen and charcoal analysis largely yielded genera present in the area today, thus confirming results obtained in previous pollen work in the area (A. Anderson 1975; Scott 1972) and the record from Bellisle Lake. Yet, the inference that the Late Glacial vegetation did not differ significantly from that of the modern community arguably might simply result from taxonomic limitations; that is, some of the identified *genera* of pollen and charcoal might represent *species* not present in the area today. Is it possible, for example, that the species of *Picea* that occurs in the bonebed pollen is *P. engelmannii*, the high-elevation, cold-tolerant species of spruce, as opposed to *P. pungens*, which occurs in the area today?

Unfortunately, the question cannot be answered directly by looking at which genera are *present* in the assemblages. However, the answer may be hinted at by which genera are *absent*, using this negative evidence with due caution. Neither the pollen nor the macrofossil evidence shows the presence of subalpine, alpine, or tundra species. More specifically, arboreal genera that today characterize cooler-temperature, higher-elevation subalpine forests, such as species of *Abies* (white fir, subalpine fir) and the genus *Pseudotsuga* (Douglas fir) (Dick-Peddie 1993; R. Thompson, Anderson, and Bartlein 1999a), do not occur among the bonebed pollen and macrofossils. In turn, that the distribution and tolerance limits of most *Abies* species coincide with those of *Picea engelmannii* (Fall 1997; R. Thompson, Anderson, and Bartlein 1999a) suggests that the *Picea* species that occurs in the Folsom bonebed pollen is indeed *P. pungens* and not *P. engelmannii*. We might go even farther out on a limb to suggest that the species of pine recorded in the Folsom site bonebed are also not the high-elevation species—such as Bristlecone pine—but more likely the species characteristic of the region today: ponderosa and/or piñon pine.

To gain a sense of what that signifies in climate terms, table 6.5a presents temperature and precipitation ranges for the 11 conifer and hardwood tree species that occur in the Folsom area (data from table 3.6), as well as comparative data (table 6.5b) for five species that occur presently at higher elevations. Comparing these data statistically using *t*-tests (table 6.5c) shows that they differ significantly along two key dimensions: Higher-elevation taxa have a greater tolerance for cooler average annual temperatures and demand more moist settings. They also differ, though just outside a significance cutoff of $p = 0.05$, in terms of lowest limiting annual precipitation. In effect, the Folsom area is presently too warm and too dry to support these higher-elevation tree species. Given this, and their absence in the pollen record from the site, we infer that during the YDC similar climatic limits were in place.

The apparent absence of cold-tolerant trees at Folsom in Late Glacial times is of interest, and seemingly at odds with evidence elsewhere for cooler conditions during this period (Reasoner and Jodry 2000). Indeed, there is evidence that these cold-tolerant genera were able to survive at lower elevations throughout the Rocky Mountains and Sangre de Christos during the Late Glacial Maximum (LGM), and while their distributions extended upward during the post-LGM Bølling-Allerød warming, they again shifted downslope during the YDC (e.g., Fall 1997; Markgraf and Scott 1981; Reasoner and Jodry 2000). Whether these cold-tolerant taxa occurred at Folsom or on Johnson Mesa during the LGM when conditions were colder is not known; records from this period are lacking. What is apparent is that these taxa were absent from Folsom during the YDC, indicating that temperatures were by then already too warm for their survival. Elevation is likely a critical variable in explaining the differences in these areas: The sites displaying a strong Late Glacial vegetation response are at much higher elevations (>3,000 m) than Folsom (2,110 m) and, thus, more sensitive to the effects of decreased temperatures.

Turning from the absent genera to those that are present, the relatively high percentage of arboreal pollen indicates some woodland in the area, though the relative frequency of the spruce, pine, and juniper trees in that woodland is not known. *Pinus* pollen is more abundant than that of *Picea*, but given the difference between the two in pollen production, the latter might have been nearly as abundant as the pine. *Juniperus* pollen was less abundant than either of the others. Although all three of these species are to different degrees susceptible to long-distance transport, their occurrence as fragments of charcoal in the bonebed suggests a nearby presence. That trees were growing in the area is not surprising, given the Bellisle Lake evidence. Unfortunately, it cannot be ascertained whether these trees were growing in patchy, locally protected topographic settings, as forest in the uplands, or if there was a local woodland and just how

dense that woodland might have been—whether a closed forest or a more open parkland.

Quercus pollen was present in samples collected below and immediately above the bonebed, but only in small amounts, and was not in any of the charcoal samples from those older deposits. Given the vagaries of wind and precipitation owing to the local scarp effect of Johnson Mesa, and the potential for upslope pollen transport from lower elevations, it is not possible to conclude oak was present in Late Glacial times. Its first secure appearance is in the Middle Holocene. As *Q. gambelii* is limited by spring freezes (Fall 1997:1319), its occurrence may signal an increase in the growing season (a comparable signal is recorded by one of the snail species found above the bonebed, as noted below).

Artemisia pollen increases above the bonebed, suggesting the possibility that sagebrush became more abundant in the local vegetation after the Folsom occupation. The pollen record is also consistent with a reduction in both spruce and pine trees at this time, indicating a more open vegetation community. Grasses were present in all pollen samples examined, and also decrease slightly at this time. All of this may suggest an increase in scrub vegetation, which roughly coincides with the deposition of the *f3* shingle gravel, which on independent geomorphic grounds was thought to mark precipitation acting on a sparsely vegetated surface (chapter 5). That said, the charcoal data do not record a similar postbonebed decrease in arboreal taxa or an increase in open range plants (sagebrush, high-spine Asteracae). This difference, however, may simply be a consequence of taphonomic processes, notably the greater likelihood of preservation of more substantial woody plants.

Pollen of *Typha* (cattail), *Polygala* (milkwort), and Ranunculaceae (buttercup family), and possibly *Salix* (willow) as well (depending on which species of willow is present), hints at the presence of a nearby riparian community. However, because these taxa occur only in trace amounts, this inference requires further data and testing in order to be confirmed.

Possibly relevant to the question of the amount of moisture in this area at this time, however, is the presence of *Tsuga* pollen from above the bonebed. *Tsuga* is moisture-limited, only occurring in areas with a mean moisture index (MI) of 0.96 (R. Thompson, Anderson, and Bartlein 1999a). It thus demands far more humid conditions than any of the species that occur in the area today (MI = 0.618), and more than those taxa that occur in higher-elevation settings (MI = .756) (table 6.5). In fact, the closest occurrence of stands of *Tsuga* is in the Pacific Northwest; an eastern species of *Tsuga* comes no closer than the Mississippi valley (R. Thompson, Anderson, and Bartlein 1999a). If *Tsuga* was growing at Folsom in the Late Glacial, that would suggest a dramatically wetter, more humid climatic regime than previously supposed. But that possibility seems remote, as there is no other evidence to support precipitation at the levels required by *Tsuga*. Moreover, there is no evidence from any

TABLE 6.5
Temperature (°C) and Precipitation (mm) Ranges for 11 Conifer and Hardwood Tree Species That Occur in the Folsom
Area Compared to Four Species from Higher Elevations

Species	Temperature (°C)				Precipitation (mm)		
	January Lowest Limiting	Average Annual Low End	Average Annual High End	July Highest Limiting	Lowest Limiting Annual	Highest Annual	Moisture Index (AE/PE)
a. Folsom Area Today							
Abies concolor (white fir)	−10.7	−1.4	10.3	28	135	2,505	0.69
Cercocarpus cf. *breviflorus* (hairy mountain-mahogany)	−1.0	7.3	20.7	30.8	245	640	0.59
Juniperus monosperma (one-seeded juniper)	−9.2	1.6	22	30.8	200	830	0.51
Juniperus scopulorum (Rocky Mountain juniper)	17.9	−3.2	16.1	28.5	115	2,520	0.64
Picea pungens (blue spruce)	−12.2	−3.2	11.7	23.1	200	1,175	0.70
Pinus edulis (piñon pine)	−11.4	−0.6	17.8	30.3	135	830	0.46
Pinus ponderosa (ponderosa pine)	−14.7	−1.7	24.2	29.4	210	2,825	0.64
Populus augustfoilia (cottonwood)	−14.3	−3.9	18.1	28.3	145	1,175	0.61
Prunus virginiana (chokecherry)	−28.2	−5.5	18.7	27.6	155	4,370	0.85
Quercus gambelii (Gambel oak)	−12.0	−1.5	21.7	29.3	95	1,175	0.54
Robinia neomexicana (New Mexican locust)	−6.6	3.4	18.1	28.4	150	690	0.57
Average	−12.56	−0.79	18.13	28.59	162.27	1,703.18	0.62
SD	6.83	3.66	4.22	2.13	45.52	1,191.6	0.10
b. Higher Elevations							
Abies lasiocarpa (subalpine fir)	−30.0	−6.9	11.7	22.7	225	4,370	0.76
Picea engelmannii (Engleman spruce)	−18.7	−4.8	13.2	24.4	235	2,825	0.77
Pinus aristata (bristlecone pine)	−12.0	−3.2	3.8	14.2	395	820	0.79
Pinus flexilis (limber pine)	−17.7	−3.9	19.4	30.1	95	1,505	0.70
Pseudotsuga menziesii (Douglas fir)	−16.9	−3.9	24.8	28.9	235	4,370	0.76
Average	−19.06	−4.54	14.58	24.06	237.0	2778.0	0.76
SD	6.63	1.43	7.97	6.30	106.4	1,622.15	0.03
c. *t*-test of Data in a and b							
t-value	1.78	**2.181**	1.18	1.56	−2.02	−1.50	**−20.794**
Significance (2-tailed)	0.098	**0.047**	0.256	0.186	0.063	0.156	**0.014**

SOURCE: Data from Thompson, Anderson, and Bartlein (1999a, 1999b, 2000).

NOTE: Significant *t*-test results in bold.

other Late Glacial age pollen cores that *Tsuga* was present either in the Rocky Mountains to the west and northwest or on the High Plains of New Mexico, Texas, and Oklahoma to the east (Bryant and Holloway 1985; Fall 1997; Hall 1985; Reasoner and Jodry 1999; Vierling 1998). The trace amounts of *Tsuga* in the postbonebed pollen at Folsom are anomalous, and not indicative of the presence of this tree in the area. As to why it is present, a modern pollen rain study sheds some light on the matter. In a study of modern pollen, Hall (1990:57) recovered a single grain of *Tsuga*, "probably traveling 1300 km from the Pacific Northwest" in a high-elevation pollen trap located in north-central New Mexico. That pollen trap was located ~110 km west of Folsom. We suspect that the few grains of *Tsuga* in the post-bonebed sediments at Folsom resulted from just such a rare instance of very long-distance air transport.

Land Snails: Taxa, Distribution, and Habitats

With James L. Theler

Subfossil land snail shells are abundant in many Quaternary deposits on the Great Plains, where researchers have used them for more than a half-century as proxy indicators of past environmental conditions (e.g., Drake 1975; Frye, Leonard, and Glass 1978; Hibbard and D. Taylor 1960; Howard 1935; Neck 1986; D.D. Taylor 1960, 1965; D.D. Taylor and Hibbard 1955; Wells and Stewart 1987; Wyckoff et al. 1992). In this regard, the Folsom site was no exception, as terrestrial gastropods proved to be the most abundant animal remains recovered in the bonebed.

The majority of gastropod shells recovered from the deposits are identifiable to species on the basis of shell morphology. None are referable to an extinct taxa. Habitat and range distribution data from modern populations are therefore used as a source of proxy information to interpret subfossils of the same snail species in the reconstruction of Late Glacial environmental conditions. Researchers working with land snails generally agree that the factors exerting the greatest influences on land snail distribution and abundance are moisture and temperature (La Rocque 1966; D. Taylor 1960).

THE MODERN COMPARATIVE SAMPLE

Recent investigations at the Folsom site coincided with the Southern Plains Gastropod Survey (SPGS), initiated in 1995 to assess living land snail species and populations on the southern Great Plains (Theler, Wyckoff, and Carter 2004; Wyckoff, Theler, and Carter 1997). Land snails were collected at intervals along a 640-km transect from north-central Oklahoma to a western terminus near the Folsom site. Gastropods were collected from the base, slope, and top of Johnson Mesa. Both the north and the south escarpments of Johnson Mesa were sampled, offering contrasting habitat settings. The sampled southern exposure is open and dry, except for a spring seep located in a dense oak and locust woodland against the base of an escarpment near the top of

the mesa. This seep may represent a relict habitat. On the north face of Johnson Mesa, samples were collected from protected mesic niches in Bear Canyon and on the steep, wooded slopes above the canyon. The northern exposure maintains an array of montane-boreal snail taxa not found on the sampled southern face of the mesa. Altogether, these modern comparative collections produced a sampling population of nearly 8,000 individual snails across 21 species (Theler, Wyckoff, and Carter 2004).

In addition, two vegetation detritus samples for recovery of modern snails were collected at the Folsom site in the fall of 1997. The taxa recovered from these samples are shown in table 6.6, which for comparative purposes also includes taxa recovered from the archaeological deposits at the Folsom site.

As shown in table 6.6, four species of snails occur only in the archaeological deposits, and not in the modern samples: *Gastrocopta armifera, G. procera, Deroceras* cf. *D. aenigma,* and *D. laeve.* Correspondingly, there are three species that occur in the modern sample but that are not represented archaeologically: *G. pellucida, Pupilla blandi,* and *Zonitoides arboreus.* The remainders occur in both modern and archaeological samples, though not in all archaeological levels.

The data from the SPGS were supplemented by pertinent literature sources (e.g., Bequaert and Miller 1973; Hubricht 1985; Leonard 1959; Metcalf and Smartt 1997) and provide a basis for interpreting the environmental implications of snail species recovered from the Folsom deposits.

DERIVATION OF THE ARCHAEOLOGICAL SAMPLE

Early in the SMU/QUEST excavations, when land snails were spotted in the ground they were individually piece-plotted and separately bagged and collected, an exercise that proved to have remarkably few benefits, many costs (greatly increasing the tedium and decreasing the speed of excavation), and ultimately little payoff. As noted in Appendix A, snails that are readily visible and can be spotted in the course of excavation comprise only the very largest taxa and, thus, only a very small percentage of the gastropods present in the deposits. Many smaller-bodied taxa, as well as eggs and slug parts, pass undetected, with the result that the field sample of piece-plotted gastropods is unduly biased and of limited utility. It was not used here.

Instead, the snail data come from two separate samplings. The first sample is those gastropods recovered by water screening in the field. As described in chapter 4, all excavated sediment from the bonebed, collected in units 50 × 50 × 5 cm, was water-screened through nested 3.17- and 1.59-mm-mesh screens. After the material collected on those screens had field-dried, it was given a cursory sorting and picking in the field, with each class of items separately bagged, and then all material was returned to the laboratory at SMU, where the contents were again sorted and picked; gastropods from each screen were separately bagged.

All together, 182 of the 1.59-mm screens and 75 of the 3.17-mm mesh screens yielded snails, the bulk of those

TABLE 6.6
Occurrence of Modern Land Snails in the Vicinity of the Folsom Site and in Subfossil Form in the Archaeological Deposits at the Folsom Site

Taxon	Johnson Mesa		Folsom Site	
	North	South	Modern	Archaeological
1. *Cionella lubrica*	☐	☑	☐	☑
2. Gastrocopta armifera	☐	☐	☐	☑
3. *Gastrocopta holzingeri*	☐	☑	☑	☑
4. *Gastrocopta pellucida*	☑	☐	☐	☐
5. Gastrocopta procera	☐	☐	☐	☑
6. *Pupilla blandi*	☐	☐	☑	☐
7. *Pupilla muscorum*	☑	☑	☐	☑
8. *Pupoides inornatus*	☐	☐	☑	☑
9. *Vertigo gouldi*	☑	☐	☐	☑
10. *Vertigo modesta*	☑	☐	☐	☑
11. *Vallonia cyclophorella*	☑	☐	☑	☑
12. *Vallonia gracilicosta*	☑	☑	☑	☑
13. *Punctum minutissimum*	☑	☐	☑	☑
14. *Discus whitneyi*	☑	☐	☐	☑
15. *Oreohelix strigosa*	☑	☐	☐	☑
16. *Succineidae*	☑	☐	☑	☑
17. *Euconulus fulvus*	☑	☑	☑	☑
18. *Hawaiia minuscula*	☑	☑	☑	☑
19. *Nesovitrea electrina*	☑	☐	☐	☑
20. *Zonitoides arboreus*	☑	☑	☐	☐
21. *Vitrina pellucida*	☑	☑	☐	☑
22. Deroceras cf. D.aenigma	☐	☐	☐	☑
23. Deroceras laeve	☐	☐	☐	☑
No. of taxa	15	8	9	20

NOTE: Frequency of occurrence at Folsom site in parentheses. Taxa in bold are present archaeologically but were not documented in the modern samples. Table does not include tallies of juvenile genera, for which species could not be identified.

coming from the bonebed units N1033 E998, N1033 E999, and N1034 E999 within the M17 block (fig. 4.13). For this particular analysis, the gastropods recovered from the quads and levels in grid unit N1033 E998 (M17-19) were included—comprising a total of 33 levels/quads. Most of these snails were recovered from the 1.59-mm screen mesh; those snails recovered in the 3.17-mm mesh were included with the finer-mesh sample for the purposes of the analysis. This sample included 965 snail shells, comprising 11 taxa (data in table 6.7). We refer to this as the SMU sample.

The second sample is comprised of gastropods recovered by water-screening of previously unprocessed bulk sediment samples in the laboratory at the University of Wisconsin, La Crosse (UW-LC). The bulk sediment samples came from two excavation units, N1033 E998 and N1034 E998, both excavated in 1998. We refer to these as the UW-LC1 and UW-LC2 samples, respectively.[3] They were derived from systematic sampling of sediment from the southwest corner of each 5-cm level in each excavation unit, starting at Level 139 and extending into Level 152. The sediment from

N1034 E999 included separate 1.0- and 0.4-liter samples from each level. The data from these samples are given in tables 6.8 and 6.9.

At the UW-LC laboratory, these sediment samples were air-dried, measured for volume, and water-screened, with material larger than 0.425 mm retained in a No. 40 geologic sieve. All complete and potentially identifiable shell fragments were isolated by Theler from the water-screened residue under a low-power (6× to 10×) binocular microscope. Shells were sorted by taxon through comparison to standard reference guides and reference specimens housed at the UW-LC. Following identification, the shells were counted, cataloged by taxon, and stored in glass vials containing labels with pertinent taxonomic and provenance information. The taxonomy used in this report follows Turgeon et al. (1998) without reference to the prior synonyms. A discussion of taxonomy for many of the species considered here is given by Bequaert and Miller (1973) and Metcalf and Smartt (1997).

All together, the bulk samples processed at UW-LC had a total volume of 36.8 liters and produced 1,633 snails, with

TABLE 6.7
Frequency Distribution, by Level, of Snails from the SMU Sample

Taxon	139	140	142	143	144	145	146	147	148	149	150	151	Total
Cionella lubrica										3	4	2	9
Cionella cf. *C. Lubrica*									1		1	2	4
Gastrocopta armifera	1		2	10	5	2	9		1	1	3	6	40
Pupilla muscorum			4	11	12	21	13	5		1	1	5	73
Vallonia cyclophorella		1	4	4	3	3	8	3	1	3	1	6	37
Vallonia gracilicosta	5	6	43	54	46	50	55	30	6	28	70	144	537
Vallonia sp. juveniles	1		1			2	7	6		2	5		24
Discus cf. *D. whitneyi*						3	10	9	1	2	8	3	36
Oreohelix sp.	2	1	12	7	9	11	6	7		4	16	43	118
Succineidae		2	1	7	3	7	7	1			3	4	35
Euconolus fulvus									1			1	2
Euconolus sp. juveniles						1							1
Hawaiia minuscula		1	4	6	6	6	6	1			2	8	40
Nesovitrea electrina					2		1	1		1	2	2	9
No. snails	9	11	71	99	86	106	122	63	11	45	116	226	965
No. taxa	3	5	7	7	8	9	9	8	6	8	11	11	14
No. 50 × 50-cm quads	1	1	3	3	4	4	3	3	1	2	4	4	

TABLE 6.8
Frequency Distribution of Snails by Level in UW-LC1, N1033 E998

Taxon	139	140	141	142	143	144	145	146	147	148	149	150	151	Total
Cionella lubrica											1	2		3
Gastrocopta holzingeri					1	1		3			1		1	7
Gastrocopta procera			5											5
Pupoides inornatus			3											3
Pupilla muscorum	1	3	3	1	1	2	3		2		2			18
Vertigo gouldi								1	1	1	1		2	6
Vallonia gracilicosta	5	3	2		3	8	9	8	8	5	8	5	34	98
Vallonia sp. juveniles	6	3	8	4	11	16	20	15	15	14	11	23	78	224
Punctum minutissimum								1			1	8	21	31
Discus whitneyi								4				2		6
Discus sp. juveniles		1						2	17		1			21
Oreohelix strigosa			1	2	1							10		14
Oreohelix sp. juveniles		1					1	1					1	4
Succineidae	2		3			1	1	3	2		1			13
Euconulus sp. juveniles												2		2
Hawaiia minuscula		3	2			2	3	2	2		1	1		16
Nesovitrina electrina											2	1		3
Vitrina pellucida								3	3			4		10
Deroceras cf. *D. aenigma*							1				3		4	8
Deroceras laeve												1		1
Deroceras sp.													2	2
Subtotal	14	14	27	7	17	30	38	39	54	20	33	57	145	495
No. taxa	4	6	8	3	5	6	7	10	9	3	12	10	9	
Terrestrial juveniles		2	3			11	3	18	13	6	11	11	38	116
Total	14	16	30	7	17	41	41	57	67	26	44	68	183	611
Snail eggshells							2	15	10	4	14	3	102	150

NOTE: Each sample processed from 1 liter of sediment.

TABLE 6.9
Frequency Distribution of Snails by Level in UW-LC2 N1034 E999

Taxon	139	140	141	142	144	145	146	147	148	149	150	151	152	Total
Gastrocopta armifera		2	1		1		5							9
Gastrocopta holzingeri					1	2	15	3						21
Pupilla muscorum			1	2	6	4	31	8	1	1	2	3		59
Vertigo gouldi					2	1	1	1			1	4		10
Vertigo modesta											1			1
Vallonia cyclophorella					1						8	7	3	19
Vallonia gracilicosta	2		2	3	3	11	25	10	6	4	8	23	2	99
Vallonia sp. juveniles	1	1	2	3	2	18	56	11	13	10	29	49	20	215
Punctum minutissimum									1		3	11	4	19
Discus whitneyi									1	6		1	1	9
Discus sp. juveniles									1	32	1			34
Oreohelix strigosa							7	5						12
Oreohelix sp. juveniles		2	1	1	1	1					1			7
Succineidae		2				1	2	3				1		9
Euconulus sp. juveniles												1		1
Hawaiia minuscula					2		9	4	5		1	1	1	23
Nesovitriea electrina												2		2
Vitrina pellucida						1			6	8		5	1	21
Deroceras cf. *D. aenigma*												1	1	2
Deroceras laeve							1				1		4	6
Deroceras sp.					1									1
Subtotal	3	7	7	9	20	39	152	48	69	24	55	107	39	579
No. taxa	2	4	5	4	10	8	10	11	7	5	10	12	10	21
Terrestrial juveniles		1	3		2	6	27	11	12	2	5	35	2	189
Total	3	8	10	9	22	45	179	59	81	26	60	142	41	768
Snail eggshells						1	1	3	3	4	63	104	30	209

NOTE: Each sample processed from 1.4 liters of sediment.

an average density of 44 individuals/liter of sediment. In addition, 415 calcareous snail eggshells, with a density of 11 individuals/liter, were recovered. These samples contained a total of 20 gastropod taxa.

Although the SMU and UW-LC samples come from the same area and, in one case, from the same excavation unit (N1033 E998, there are significant differences between them as a result of the different procedures used in their derivation. The most obvious difference is in taxonomic richness. Analysis of the data in table 6.10, which compares the two in terms of sample size and richness, shows that the two are significantly different, as measured by the likelihood ratio chi-square statistic ($G^2 = 8.202$, df = 1, $p = 0.004$). As the Freeman-Tukey[4] deviates show, the SMU sample significantly underrepresents the number of taxa, while the UW-LC sample overrepresents the number of taxa, given their respective sample sizes.

These results are not surprising, given that the SMU sample, unlike the UW-LC sample, was picked without the use of a binocular microscope and without benefit of expertise in identifying tiny gastropod remains, let alone juvenile

TABLE 6.10

Comparison of Number of Individuals and Richness, by Archaeological Subsample, from Excavation Unit N1033 E998

	SMU	UW-LC
No. individuals	965	611
	(0.26)	(−0.31)
Fitness	**11**	**20**
	(−1.95)	**(2.01)**

NOTE: Freeman-Tukey deviates in parentheses. Values in bold significant at $p = 0.05$ level (±0.979).

snails, snail eggs, and slug parts. Thus, the SMU sample has fewer overall taxa, and these are dominantly large-bodied forms (e.g., *Gastrocopta armifera*, *Vallonia gracilicosta*). Rare and smaller genera, such as *Deroceras*, *Vertigo*, and *Punctum*, and the smaller species of *Gastrocopta* (e.g., *G. holzingeri*) are

TABLE 6.11

Contingency Table of Number of Individuals and Richness by Excavation Units N1033 E998 and N1034 E999

	N1033 E998 (UW-LC1)	N1034 E998 (UW-LC2)
No. individuals	611	768
Richness	20	20

NOTE: N1033 E998 = 1 liter of sediment. N1034 E999 = 1.4 liters of sediment. The differences are not significant.

TABLE 6.12

Contingency Table of Number of Individuals and Richness by Level in the Combined UW-LC1 and UW-LC2 Archaeological Subsamples

Level	Sample Size	Richness
139	10	3
140	13	5
141	23	8
142	8	3
143	6	**4 (1.49)**
144	30	8
145	37	8
146	117	**11 (–2.44)**
147	58	10
148	30	6
149	34	11
150	59	13
151	122	**12 (–2.36)**
152	19	**9 (1.76)**

NOTE: The difference is significant ($G^2 = 27.548$, df = 13, $p = 0.01$). Significant Freeman-Tukey deviates in parentheses and bold (significant at $p = 0.05$ level, ±1.335).

scarce or absent altogether. The SMU sample did not produce any species not otherwise found in the UW-LC sample.

Because the two UW-LC samples were derived from slightly different amounts of sediment (1 liter per level from N1033 E998, 1.4 liters per level from N1034 E999), an obvious question to ask is whether the larger samples of matrix yielded additional taxa. As shown in table 6.11, this does not appear to be the case: The richness values are not statistically different between the UW-LC1 and the UW-LC2 samples, as measured by the likelihood ratio chi-square statistic ($G^2 = .507$, df = 1, $p = 0.476$). Thus, the different volumes from which the respective samples were derived did not influence the number of taxa recovered. Because these samples are statistically comparable and come from adjoining units, they can be combined for analytical purposes. In combining them, juvenile forms are not included, save where their species designation is secure. Combining the two increases the sample size and thus provides more statistically secure results.

When these data are examined by excavation level there is a predictable tendency for levels with larger samples to have greater taxonomic richness. These data are shown in table 6.12 and figure 6.6. Regression of richness against the log of the number of individuals yields statistically significant r^2 values ($r^2 = 0.775$). However, there are clear exceptions to that general trend: Levels 143 and 152 have significantly greater numbers of taxa present than would be expected given their sample sizes, while the reverse is true of Levels 146 and 151, which otherwise have the largest sample sizes. What this suggests is that the relationship between sample size and richness is not entirely proportional and that beyond a certain sample size, ~50 individuals in these data, the species-richness curve flattens out, at which point one can assume that a further increase in sample size would only incrementally increase the number of taxa, perhaps by adding rare forms.

In fact, one might predict that there would be a statistical tendency for the rarest taxa to occur only in levels with larger sample sizes. That proves to be the case. The five rarest taxa in the combined UW-LC1 and UW-LC2 occur in levels with sample sizes that averaged 51.4 individuals, while the five most abundant taxa occur in levels that had sample sizes that averaged 40.4 individuals. Not surpris-

ingly, the rare taxa were found in fewer levels than the most abundant taxa, which tended to occur throughout the column. That is, the rare taxa were not concentrated at any one depth. This is not to say, however, that there was an absence of trends in the taxa by level, as discussed below.

What these data also suggest, from another angle, is that the sample is representative, though one must be wary of overinterpreting low species richness counts in levels with small sample sizes. As to why sample sizes are low in certain levels, the answer lies in the taphonomy of the Folsom deposits, not least in the changing depositional processes within those deposits. In an effort to assess the impact of these processes on the number and nature of the recovered snails, Arnold (1998, 1999) sieved sediment from each of the levels in N1033 E998 (UW-LC1) through four nested screens, dividing the matrix into size classes of >4 mm, 4 mm to 710 μm, 710 to 420 μm, and 420 to 250 μm. The sediment in each screen was then weighed to determine the relative percentage, by mass, of that textural class (Arnold 1998).

Figure 6.7 shows the number and richness of snails by level from UW-LC1 sample, alongside the proportion of the matrix in the corresponding level comprised of grains >4 mm in diameter, which are pebble-sized or larger, and the proportion of the matrix that was caught on the 710- 420-μm sieve, which represents the fraction of medium to coarse sands (data from Arnold 1999:table 2). As shown in figure 6.7, the upper levels (139–142) have far more coarse, pebble-sized and larger material, at least 10% by mass in these levels, while the lower levels (143 and below) have less of the coarse material and more of the finer-grained

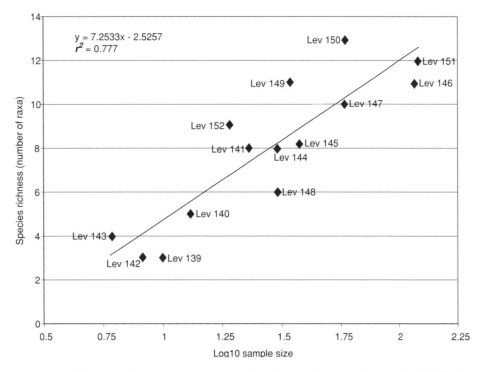

FIGURE 6.6 Plot of snail species richness against Log10 sample size in the combined UW-LC sample.

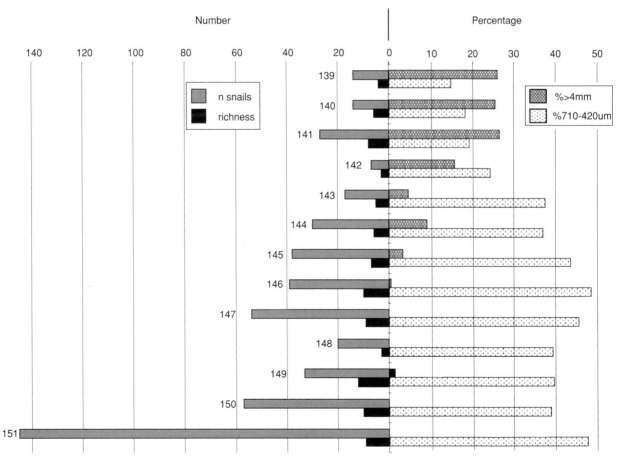

FIGURE 6.7 Left, plot of the number of snails and snail species richness by excavation level; right, the corresponding percentage of coarse (>4 mm) and fine (710-420 μm) sediment by excavation level.

sand-sized matrix (for a comparable discussion and diagram showing the position of bison bone relative to coarse matrix, see chapter 7:fig. 7.7).

The number and richness of snails recovered in the UW-LC1 sample mirrors that pattern: Snails are less frequent in the upper levels and increase in number and richness in the lower, finer-grained sediments. The distribution of *Vitrina pellucida* (*n* = 32) is illustrative here. This taxon was not recovered in the SMU sample, likely owing to its having a thin, fragile shell, which could not readily withstand field water-screening. *V. pellucida* comprises a large percentage (>20%) of the snail fauna in the lower portion of the bonebed, occurring at significant frequencies in excavation Levels 148 and 149. But it disappears in the increasingly coarser deposits above Level 145 (tables 6.8 and 6.9).

Still, the correspondence between clast size and snail size/frequency is not exact; note, for example, the lower number and richness of snails in the finer-grained sediments of Level 148. That said, it seems reasonable to conclude that snail recovery is partly a function of the coarseness of the depositional environment, for the obvious reason that fragile shells are less likely to survive when the grain size of the matrix contains clasts that are significantly larger than the shell. In such circumstances, even slight movement of the clasts will damage the shell. Any apparent frequency trends in those upper, coarser levels need to take this factor into account.

THE FOLSOM SNAIL FAUNA

There are some 20 taxa recovered from the *f2* sediments in the paleotributary at Folsom. Table 6.13 provides information on the modern distribution and habitats of these taxa, along with information on fossil distributions (information on the latter is not intended to be comprehensive). The archaeological specimens from the *f2* generally correspond to species inhabiting northern New Mexico today, save for those exceptions earlier noted, but the simple presence of a taxon in both the archaeological and modern snail faunas is only part of the story. Some of the taxa presently inhabiting the area are abundant, while others are rare and Folsom represents the margin of their range, as a result of limits imposed by moisture (e.g., *Oreohelix strigosa, Gastrocopta armifera*), elevation (e.g., *Vertigo gouldi, V. modesta, D. whitenyi*), vegetation (*Punctum minutissimum, Zonitoides arboreus*), and growing season (*G. procera*).

Archaeologically these taxa occur at varying frequencies, and they differ in their vertical distribution and abundance. Figures 6.8 to 6.10 illustrate these stratigraphic trends by showing what percentage by level is comprised of each particular snail species. These figures are from the combined data of UW-LC1 and UW-LC2, since these are the most taxonomically representative samples and provide the best information by level. Data from the SMU sample are not included because of the bias against smaller-bodied forms. Because a few taxa numerically dominate these deposits,

three separate figures are used to better highlight the relative contributions of less-abundant taxa. These figures are grouped by sample size. Only taxa that occur at frequencies of *n* = 15 or higher are plotted; the distribution of those that occur in smaller numbers are too susceptible to sample size vagaries to provide meaningful plots.

As these are plots of percentage values, they suffer from the usual statistical hazard of such measures—the problem of closed arrays. A percentage peak or valley of a particular taxon in a specific level might be more apparent than real, a function of the decline or increase of a more abundant taxa. To control for this effect, a contingency table of the frequencies of taxa by level was constructed, again using the combined UW-LC1 and UW-LC2 samples, and Freeman-Tukey deviates were calculated in order to identify whether and where frequencies of a taxa in a particular level were significantly higher or lower than would be expected (table 6.14).

Although partitioning the data by 5-cm excavation level is useful for analytical purposes, for discussion purposes reference is also made to the occurrence of snail taxa relative to the position of the bison bonebed. Those two scales can be relatively easily aligned: The average elevation of the bonebed in the M17 block, where virtually all of the samples discussed were obtained, was 97.57 m (±10 cm). That centers the bonebed in excavation Level 148. The modal occurrence of the bone, however, was in Level 149. The bison bone in this part of the site generally lay flat. Yet, bison bone is three-dimensional and, thus, was often not restricted to a single 5-cm level: The top of one bison cranium (M17-19-136) was first encountered in Level 146, and its base extended 2 cm into Level 150—a vertical span of some 25 cm (also chapter 7:fig. 7.6 and fig. 7.7). Overall, however, the great majority of the bone occurred between Level 146 and Level 149. Taken together, these data indicate that the base of the bonebed was approximately in Level 149, with larger elements projecting above and, to a lesser degree, below that level. Therefore, for purposes of discussion samples from Levels 139 to 145 are generally considered to be *above* the bonebed; samples from Levels 146 to 149, *within* the bonebed; and samples recovered in Levels 150 to 152, *below* the bonebed. The data are arrayed in this manner in table 6.15.

Gastropods were abundant in the levels below the bonebed (Levels 150–152), with the highest frequency (but not highest taxonomic richness) occurring in Level 151. These levels are, as noted earlier, comprised of fine-grained sediments in which snails ought to be well preserved, and taxonomic representation reasonably complete. These levels contain a suite of species with either exclusive occurrences or peaks in abundance, including *Cionella lubrica, Punctum minutissimum, Vallonia cyclophorella* (both of which occur in significantly high frequency peaks), and *Vertigo gouldi*, as well as the relatively rare taxa *Euconulus fulvus, Nesovitrea electrina*, and the slugs (*Deroceras laeve* and *Deroceras* cf. *D. aenigma*)—neither of which are present in the area today.

TABLE 6.13
Modern Habitat and Distribution of Snail Taxa Recovered Archaeologically at the Folsom Site

Taxon (Number of Archaeological Occurrences)	Modern Habitat and Distribution, and Fossil Distribution (Where Available)
Cionella lubrica (Muller, 1774), glossy pillar (*n* = 12)	Metcalf and Smartt (1997:25) report *C. lubrica* occurs at elevations >6,500 ft in forested montane habitats of New Mexico, where it prefers moist surfaces. Leonard (1959:191) found this species only in northeastearn Kansas associated with habitats with moist vegetation detritus. In the SPGS, this species occurred at elevations >7,000 ft at protected rock ledges and riparian woodlands on the Chase Ranch near Cimarron, New Mexico, and at a wooded spring seep at the base of the south face of Johnson Mesa near Folsom (Theler, Wyckoff, and Carter 2004). This may represent a disjunct population that has survived since the late Pleistocene. The species was not found in samples on the north face of Johnson Mesa.
Pupilla muscorum (Linnaeus, 1758), widespread column (*n* = 149)	A Holarctic species found from Alaska into the northern coterminous United States and at high elevations farther south (Metcalf and Smartt 1997:26). In northern New Mexico it can be found at elevations as high as 10,000 ft, and it is widespread today on the south and north faces of Johnson Mesa. *P. muscorum* tends to favor cool, dry climatic conditions. This species was widespread across lower elevations during the Pleistocene and is a common form in full glacial snail faunas on the Plains. In the Southern Plains region, this species is found as a disjunct population at Black Mesa in Cimarron County, Oklahoma (Metcalf 1984; Theler, Wyckoff, and Carter 2004). Some extant disjunct populations are comprised of smaller ecophenotypes (Metcalf and Smartt 1997:26).
Pupoides inornatus (Vanatta 1915), Rocky Mountain dagger (*n* = 3)	Metcalf and Smartt (1997:28, 135) report *P. inornatus* on rocky outcrops from three counties, Quay, San Miguel, and Union, in northeastern New Mexico, including the High Plains near Las Vegas, NM. It was not recovered during the SPGS (Theler, Wyckoff, and Carter 2004), and only one specimen was found in the modern leaf litter at Folsom. Other records include El Paso and Larimer counties, Colorado; Cheyenne County, Nebraska, and Cimarron County, Oklahoma (Bequaerrt and Miller 1973:178; Metcalf 1984:58, 60; D. Taylor 1960:74). There are two occurrences in Kansas, both believed to be specimens that arrived as riparian drift from Colorado (Leonard 1959:184). Its Pleistocene distribution was more extensive (Metcalf and Smartt 1997:27).
Gastrocopta armifera (Say, 1821), armed snaggletooth (*n* = 49)	Found across much of the eastern United States, westward to New Mexico. Metcalf and Smartt (1997:28) report *G. armifera* lives under brushy vegetation in riparian areas or on rocky hillsides and canyons bordering river valleys in New Mexico. It was not recovered from any modern samples at Folsom or on Johnson Mesa. It was found in a few protected upland locations across the southern Plains (Theler, Wyckoff, and Carter 2004). In Kansas, this species is widespread but declines in frequency from east to west (Leonard 1959:169–170, fig. 72), a pattern repeated in other Plains states (Hubricht 1985:71, map 44).
Gastrocopta holzingeri (Sterki, 1889), lambda snaggletooth (*n* = 28)	*G holzingeri* occurs across much of the eastern United States, from the Appalachians to the Plains. Its range extends westward from the Oklahoma Panhandle into northeastern New Mexico, primarily along the Dry Cimarron Valley (Metcalf and Smartt 1997:29). Metcalf and Smartt (1997:29) report *G. holzingeri* occurs in the litter under brushy or montane forest settings. It was found in protected upland settings at the eastern and western margins of the southern Plains (Theler, Wyckoff, and Carter 2004) and at a protected spring seep niche on the south face of Johnson Mesa. In Kansas, this species is found primarily in the northern portion of the state, in protected forested areas (Leonard 1959:175).

(continued)

TABLE 6.13 (*Continued*)

Taxon (Number of Archaeological Occurrences)	Modern Habitat and Distribution, and Fossil Distribution (Where Available)
Gastrocopta procera (Gould, 1840), wing snaggletooth (*n* = 5)	Metcalf and Smartt (1997:131) report *Gastrocopta procera* from both Colfax and Union Counties, but it did not occur at any of the SPGS sampling locales west of Black Mesa in Cimarron County, Oklahoma (Theler, Wyckoff, and Carter 2004). However, this snail can tolerate habitats from high montane forest, to lower, more arid juniper-shrub zones, to the southern High Plains to the east (especially Texas and Oklahoma), where it is one of the most common and widespread species (Metcalf and Smartt 1997:31). One researcher has made a good case for *G. procera* not living in regions with a growing season of <160 days (Baerreis 1980).
Vertigo gouldi (A. Binney, 1843), variable vertigo (*n* = 16)	A common snail of the Rocky Mountains, *V. gouldi* is widespread in the forested mountains of New Mexico, though in a limited range of elevation. The SPGS found *V. gouldi* abundant in moist deciduous litter in Bear Canyon and common at forested, protected locales on the north face of Johnson Mesa, but it was absent in samples from the southern exposure. It was not found at any of the upland habitats sampled in the southern Plains of Oklahoma (Theler, Wyckoff, and Carter 2004) and is represented only by fossils on the Great Plains to the east (Metcalf and Smartt 1997:33; see also Leonard 1959; Hubricht 1985:83, fig. 75).
Vertigo modesta (Say, 1824), cross vertigo (*n* = 1)	This high-altitude, high-latitude form presently ranges north into Alaska, Greenland, and Canada. In New Mexico, Metcalf and Smartt (1997:33) report that *V. modesta* lives in protected niches of deciduous litter (primarily aspen) in high forested mountains of New Mexico. The survey of Johnson Mesa found this species abundant on the mesic, forested slope of the north face and present in moist, deciduous litter at one station in Bear Canyon, also a northern exposure. It occurs only in fossil form in the eastern United States; it is absent as well from the Great Plains (Hubricht 1985:83, fig. 76; Metcalf and Smartt 1997:33).
Vallonia cyclophorella (V. Sterki 1892), silky vallonia (*n* = 56)	The most broadly distributed *Vallonia* in the western mountain states, *V. cyclophorella* ranges westward from the Rocky Mountains into Arizona and occurs widely in northeastern New Mexico, where it is the most common land snail. It is generally found in broken terrain—such as wooded, rocky, canyon slopes (Metcalf and Smartt 1997:34–35). At Johnson Mesa, *Vallonia cyclophorella* was found in small numbers at protected niches holding deciduous litter on the northern exposure. One individual was found in the mixed grass and deciduous litter adjacent to the Folsom site.
Vallonia gracilicosta (O. Reinhardt, 1883), multirib vallonia (*n* = 734)	Unlike *V. cyclophorella*, which occupies the Rocky Mountain region and westward, *V. gracilicosta* occurs more often in the eastern foothills of the mountains, but not much farther east than the 102nd meridian (Metcalf and Smartt 1997:34–35). Today, *V. gracilicosta* is probably the most common land snail in northeastern New Mexico, especially at higher elevations, as along the basltic flows and mesas of the Folsom region (Metcalf and Smartt 1997:35). It is widespread on the northern and southern exposures of Johnson Mesa, with highest densities in protected niches on the north face. This species has been recovered as a disjunct population at Black Mesa in Cimarron County, Oklahoma (Metcalf 1984; Theler, Wyckoff, and Carter 2004). Shells of this species were used in the stable isotope analysis.
Punctum minutissimum (I. Lea, 1841), small spot (*n* = 50)	Metcalf and Smartt (1997:39) report this species as widespread in the higher elevations of New Mexico, and six *P. minutissimum* were found in deciduous litter collected at Folsom. Nonetheless, only four individuals were found in samples from the north face of Johnson Mesa, and otherwise *P. minutissimum* was not recovered during the SPGS (Theler, Wyckoff, and Carter 2004), likely because it requires moist settings with good vegetation cover. In Kansas, it is restricted to woodlands in the northeastern portion of the state (Leonard 1959:136–137).

TABLE 6.13 (*Continued*)

Taxon (Number of Archaeological Occurrences)	Modern Habitat and Distribution, and Fossil Distribution (Where Available)
Discus whitneyi (Newcomb, 1864), forest disc (*n* = 51)	*D. whitneyi* is widely distributed across Alaska and much of Canada, the northern United States, and farther south along higher elevations (Metcalf and Smartt 1997:40). It was found living in mesic woodlands on the north face of Johnson Mesa. Hoff (1962:55, 60) found this species associated with the litter of aspen groves. Like *P. minutissimum*, *D. whitneyi* was not found during the SPGS, likely for similar reasons: its preferred habitat is damp and well protected. *D. whitneyi* is not reported living today on the Great Plains south of the Black Hills (Hubricht 1985:107, map 171). Metcalf and Smartt (1997:40–41) report that *Discus whitneyi* may hybridize with *Discus shimekii* (Pilsbry, 1890, 1948:604) in the Sangre de Cristo Mountains of New Mexico.
Oreohelix strigosa (gould, 1846), Rocky mountainsnail (*n* = 26)	Metcalf and Smartt (1997:41) report *O. strigosa* in moisture-retaining litter at higher elevations in the Sangre de Cristo Mountains, Colfax County, New Mexico. Pilsbry (1939:431) reported this species from adjacent Union County, from shells collected near Folsom by E. B. Howard in 1931 (Metcalf and Smartt 1997:41). A population was located in Bear Canyon and on the north face of Johnson Mesa during the SPGS; these are at the eastern margin of the species' present New Mexico range (Bequaert and Miller 1973:33–36, fig. 4). It is suggested that "inadequate moisture may be an important factor determining the lower limit of range in elevation" (Hoff 1962:53). During the Pleistocene, the species spread as far east as Illinois, where it occurs in Peoria Loess (Frest and Rhodes 1981).
Family Succineidae, sp. inderminate (*n* = 57)	The habitats of the genera and species in this family, which are "notoriously difficult" to identify (Metcalf and Smartt 1997:47), range from arid to stream-side settings (49).
Euconulus fulvus (O. Muller, 1774), brown hive (*n* = 2)	The SPGS found *Euconulus fulvus* on both the northern and the southern exposures of Johnson Mesa. This species is widespread today in the forested mountains of New Mexico, where it favors the detritus of deciduous trees and moist habitats (Hoff 1962:54). Metcalf and Smartt (1997:50) report this species for both Colfax and Union counties, New Mexico.
Nesovitrea electrina (A. Gould, 1841), amber glass (*n* = 14)	Metcalf and Smartt (1997:50) indicate *N. electrina* is a montane species found in vegetation detritus and leaf litter where soil is moist. Today *Nesovitrea electrina* is widespread on the northern exposure of Johnson Mesa. This species was not recovered during the SPGS (Theler, Wyckoff, and Carter 2004). In Kansas, Leonard (1959:112, fig. 44) noted this species in moist, wooded habitats and along stream margins in the northeastern part of the state.
Hawaiia minuscula (A. Binney, 1841), minute gem (*n* = 79)	*H. minuscula* occurs from Alaska and Canada south to Costa Rica; it is a common form in New Mexico, where it is widespread, ranging across the state from low to higher elevations (Metcalf and Smartt 1997:51). However, it is a rare species on Johnson Mesa, with the highest density at the protected spring seep on the south face. On the Southern Plains, *H. minuscula* is found in a wide range of habitats, including open grasslands, but it is most abundant in protected niches and at riparian woodlands under deciduous leaf litter (Theler, Wyckoff, and Carter 2004). In Kansas, this species is widespread (Leonard 1959:120, fig. 48).
Vitrina pellucida (Muller, 1774), western glass-snail (*n* = 31)	*V. pellucida* is presently found in the high mountains of New Mexico, usually in association with low vegetation (sedges and grasses) near standing water or in damp montane meadows (Metcalf and Smartt 1997:51). *V. pellucida* was found at the wooded spring seep on the south face of Johnson Mesa and was widespread on the north face. Theler recorded a disjunct population of *Vitrina pellucida* living at the marshy, well-vegetated spring head adjacent to the Hudson-Meng bison site in western Nebraska. Hubricht (1985:141, map 329) shows only two locales with this species in the Black Hills region of South Dakota.

(*continued*)

TABLE 6.13 (*Continued*)

Taxon (Number of Archaeological Occurrences)	Modern Habitat and Distribution, and Fossil Distribution (Where Available)
Deroceras laeve (O. Muller, 1774), meadow slug and *Deroceras aenigma* (Leonard, 1950) (*n* = 17)	*D. laeve* is a Holarctic species found across North America, from lower to higher elevations. In mountain settings, it occurs "mainly around springs and along streams and other bodies of water" (Metcalf and Smartt 1997:51). In modern sampling on the Southern Plains, *D. laeve* was found to be of low density and spotty in distribution, occurring most frequently in riparian woodlands. Sampling on Johnson Mesa did not yield this species. Metcalf and Smartt (1997:130) do not report *D. laeva* from Colfax or Union Counties, New Mexico. Leonard (1959:127) reports this species to be associated with woodland habitats and to require moisture. *Deroceras aenigma* is presumed to have required a similar habitat (Taylor 1960:80).

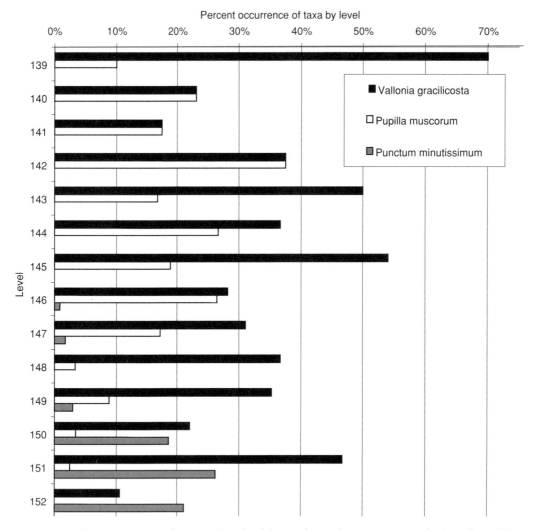

FIGURE 6.8 Percent occurrence by excavation level for snail taxa that occur at sample sizes of *n* > 50.

Percent occurrence of taxa by level

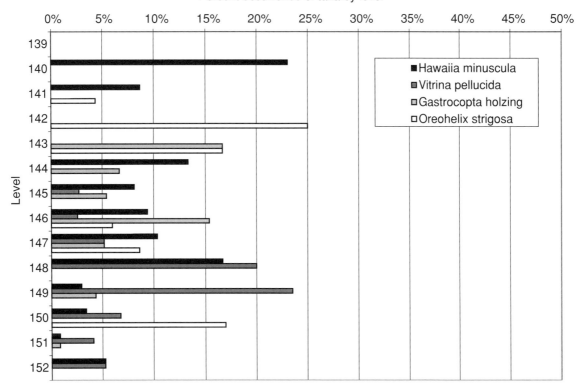

FIGURE 6.9 Percent occurrence by excavation level for snail taxa that occur at sample sizes of n = 25–50.

Percent occurrence of taxa by level

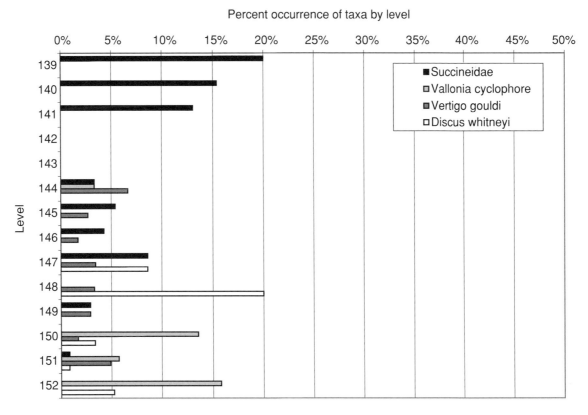

FIGURE 6.10 Percent occurrence by excavation level for snail taxa that occur at sample sizes of n = 15–25.

TABLE 6.14

Contingency Table of Species by Level from the Combined UW-LC1 and UW-LC2 Archaeological Subsamples

	139	140	141	142	143	144	145	146	147	148	149	150	151	152	Total
Vallonia gracilicosta	1.61	-0.64	-1.51	0.25	0.67	0.24	1.81	-1.23	-0.44	0.24	0.12	-1.77	**2.09**	**-2.09**	197
Pupilla muscorum	-0.12	0.89	0.56	1.42	0.35	1.67	0.88	**3.18**	0.77	-1.75	-0.68	**-2.61**	**-4.48**	**-2.37**	77
Punctum minutissimum	-1.13	-1.37	**-2.02**	-0.96	-0.77	**-2.41**	**-2.75**	**-4.09**	**-2.22**	**-2.41**	-1.19	**2.11**	**4.76**	1.46	50
Hawaiia minuscule	-0.94	1.59	0.44	-0.79	-0.63	1.19	0.39	1.01	0.97	1.64	-0.81	-1.01	**-3.47**	-0.08	39
Vitrina pellucida	-0.79	-0.96	-1.46	-0.66	-0.52	-1.75	-0.60	-1.43	0.03	**2.34**	**2.92**	0.50	-0.58	0.14	31
Gastrocopta holzingeri	-0.73	-0.89	-1.36	-0.61	0.94	0.51	0.26	**3.69**	0.20	-1.63	-0.37	**-2.56**	**-2.60**	-1.18	28
Oreohelix strigosa	1.55	1.41	1.59	-0.50	-0.39	0.03	0.55	0.30	1.52	-1.38	-0.09	**-2.19**	**-2.05**	-0.99	26
Succineidae	-0.68	-0.84	0.13	1.57	0.96	-1.55	-1.79	0.73	1.27	-1.55	-1.69	**3.04**	**-3.84**	-1.12	22
Vallonia cyclophorella	-0.53	-0.66	-1.02	-0.44	-0.34	0.17	-1.44	**-3.09**	**-1.96**	-1.24	-1.36	**2.84**	1.31	**1.85**	19
Vertigo gouldi	-0.46	-0.57	-0.90	-0.38	-0.30	1.05	0.14	-0.63	0.40	0.32	0.21	-0.36	1.25	-0.77	16
Discus whitneyi	-0.44	-0.54	-0.85	-0.36	-0.28	-1.04	-1.22	**-2.66**	**2.01**	**3.05**	-1.15	0.45	-1.32	0.68	15
Deroceras cf. D. aenigma	-0.31	-0.39	-0.62	-0.25	-0.19	-0.77	0.51	-2.04	-1.26	-0.77	**1.89**	-1.27	1.58	0.88	10
Gastrocopta armifera	-0.28	1.79	0.84	-0.23	-0.18	0.71	-0.83	1.78	-1.17	-0.71	-0.78	-1.18	**-1.96**	-0.49	9
Deroceras leave	-0.22	-0.28	-0.46	-0.18	-0.14	-0.58	-0.68	-0.19	-0.97	-0.58	-0.64	1.17	-1.65	**2.84**	7
Gastrocopta procera	-0.16	-0.21	**3.34**	-0.13	-0.10	-0.44	-0.52	-1.27	-0.75	-0.44	-0.48	-0.76	-1.30	-0.29	5
Nesovitrea electrina	-0.16	-0.21	-0.35	-0.13	-0.10	-0.44	-0.52	-1.27	-0.75	-0.44	1.66	0.66	-1.30	**1.85**	5
Cionella lubrica	-0.10	-0.13	-0.22	-0.08	-0.06	-0.28	-0.34	-0.87	-0.49	-0.28	1.10	1.65	-0.89	-0.18	3
Pupoides inornattus	-0.10	-0.13	**2.51**	-0.08	-0.06	-0.28	-0.34	-0.87	-0.49	-0.28	-0.31	-0.50	-0.89	-0.18	3
Euconulus sp. juveniles	-0.10	-0.13	-0.22	-0.08	-0.06	-0.28	-0.34	-0.87	-0.49	-0.28	-0.31	-0.50	1.84	-0.18	3
Vertigo modesta	-0.03	-0.04	-0.08	-0.03	-0.02	-0.10	-0.12	-0.35	-0.19	-0.10	-0.11	1.22	-0.36	-0.07	1
Total	10	15	24	9	6	31	38	110	53	30	34	60	122	19	566

NOTE: Taxa ordered by overall abundance. Levels in which a species occurs at a significant frequency, as measured by Freeman-Tukey deviates, are in bold (significant at $p = 0.05$ level, ±1.841).

TABLE 6.15
Significant Occurrences of Taxa by Levels, Combined UW-LC1 and UW-LC2 Samples

	Occur More Often Than Expected (Level)	Occur Less Often Than Expected (Level)
Above bonebed Levels 139–145 (*n* = 127)	*Gastrocopta procera* *Pupoides inornatus*	*Punctum minutissimum*
At bonebed Levels 146–149) (*n* = 239)	*Gastrocopta holzingeri* *Pupilla muscorum* *Discus whitneyi* (147–148) *Vitrina pellucida* *Deroceras* cf. *D. aenigma* (149)	*Vallonia cyclophorella* *Punctum minutissimum* *Discus whitneyi* (146) *Deroceras* cf. *D. aenigma* (146)
Below bonebed Levels 150–152 (*n* = 200)	*Vallonia cyclophorella* *Vallonia gracilicosta* (151) *Punctum minutissimum* Succineidae (150) *Nesovitrea electrina* *Deroceras laeve*	*Gastrocopta armifera* *Gastrocopta holzingeri* *Pupilla muscorum* *Vallonia gracilicosta* (152) *Oreohelix strigosa* Succineidae (151) *Hawaiia minuscula*

Common to nearly all of these forms is that they occur today at high elevations and in upland settings and, perhaps more significantly, in habitats that contain deciduous tree litter. Several of the species—such as *E. fulvus, N. electrina,* and *P. minutissimum* as well as the slugs—require or, at least, prefer moist settings (table 6.13).

Although one must be cautious in interpreting the scarcity of taxa, it is noteworthy that *Vallonia gracilicosta,* which is otherwise abundant throughout the deposit, is significantly underrepresented in the lowest levels, save for a spike in Level 151. Unlike the taxa common to these levels, *V. gracilicosta* prefers more open, less-vegetated habitats. *Pupilla muscorum,* which is also significantly underrepresented in Levels 150 to 152, prefers more montane (nondeciduous) forests (Metcalf and Smartt 1997:26).

The four levels of the bonebed (146–149) contain a large number of gastropods (*n* = 239), though the total is driven by the substantial sample (*n* = 117) from Level 146. Levels 148 and 149, coincident with the base of the bonebed, show a sharp decline in the number of recovered snails. Although the number of taxa similarly decline in those levels, they nonetheless remain proportional to sample size. Indeed, Level 149 has more species represented than one would expect given the sample size (see fig. 6.6). We see no obvious taphonomic explanation for the decline in the number of snails in these levels; after all, the fine-grained *f2* sediments in these same levels well preserve the bison bone. Perhaps, however, a highly localized event affecting the vegetation community reduced the snail population.

Leaving aside changes in the abundance and richness of the snail fauna, it is of interest to note that several of the species prevalent in the levels below the bonebed are significantly scarce above it, such as *P. minutissimum, V. cyclophorella,* and the *Deroceras* spp. The latter taxon has a peak at the base of the bonebed but quickly becomes rare up-section. In their place, the dominant gastropod taxa are *G. holzingeri* and *P. muscorum,* both of which were significantly underrepresented in the levels below the bonebed but have a significant frequency peak in Level 146, just at the top of the bonebed. *P. muscorum* is a montane-boreal species tolerant of cooler, drier, more open habitats (table 6.13). *G. holzingeri* can flourish under limited vegetation cover and appears to require less moisture than most montane-boreal species. This change in the snail fauna suggests the possibility that there was a shift from a better-watered, more heavily vegetated environment prior to the bison kill to a drier, more open setting characteristic of the area at the time the kill was made.

Another, possibly significant aspect of the snail fauna at the time of the Folsom kill is the change in frequency of *D. whitneyi,* which occurs at a significant frequency in Levels 147 and 148 but then becomes significantly rare in Level 146 and, afterward, disappears from the fauna. As noted (table 6.13), this is a high-elevation/high-latitude form (Metcalf and Smartt 1997:40).

Similarly, *Vitrina pellucida,* although never common in the snail fauna, peaks in Levels 149 and 148, but then it too slowly declines and then disappears from the fauna. However, in this instance, as discussed earlier, the decline may be attributable the vulnerability of this fragile shell in the increasingly coarser higher deposits.

Overall, the number of snails (*n* = 127) drops off considerably in the levels (139–145) above the bonebed and the taxonomic richness drops off as well. As observed earlier, this was the portion of the section marked by an upward increase in the coarseness of the matrix (fig. 6.7), driven by the massive downslope movement of shingle shale that covered portions of the site (chapter 5). The movement of this coarse sediment could have damaged or destroyed associated snail shells. The quality of the resulting shingle shale deposit as snail habitat is unclear.

There was a final spike in the number of shells high in the section: *G. procera* and *Pupoides inornatus* both occur only and at a significant frequency in Level 141, well above the bonebed. *G. procera* apparently requires at least 160 frost-free days (Baerreis 1980) and was not found in recent surveys near the Folsom site. It routinely occurs in fossil form during the Middle Holocene (e.g., Meltzer and Collins 1987:table 1; Neck 1986:table 24). The presence of *G. procera* in the upper levels at Folsom is taken to indicate an increase in warming, requiring as it does a growing season at least 30 days longer than that of the present, which is interpolated to be about 130 frost-free days (table 3.2). This "Plains-adapted" taxon (Metcalf 1997:84) is absent from the Folsom region today.

Likewise, *P. inornatus* favors more xeric habitats. Thus, the postbonebed levels are interpreted to have been deposited during a climatic period warmer than that of the bonebed and prebonebed levels, given the significantly low frequencies of *Punctum minutissimum* and the complete absence of the aridity-intolerant slugs.

Although the patterns are not as clear-cut as one might wish, the terrestrial snail species in the Folsom site sediments show apparent shifts in local habitat conditions, and the emphasis here is on "local," for the snails represent microscale ecological variation. In most cases, these shifts were probably responses to changing regional climate. The prebonebed levels contained a group of cool-adapted, montane-deciduous snail species that indicate a moist, well-vegetated habitat. During the period in which the bonebed formed the environment appears to have become a drier, more open setting. That warming and drying trend continues in the postbonebed assemblage, culminating in the occurrence of *G. procera*.

The Folsom gastropod sample includes many taxa that occur today at higher elevations. Dillon and Metcalf (1997) ran a series of vertical collecting transects on several mountains in New Mexico. The closest of their transects, on Lake Peak, is located in the Sangre de Christo Mountains west/southwest of Folsom, and ranged in elevation from 1,950 to 3,658 m (6,400 to >12,000 ft). Nearly a dozen of the species recorded on the Lake Peak Transect were ones recovered at Folsom, and examination of their elevation ranges is instructive (table 6.16), for it shows that almost all of these species are today routinely found at elevations higher than that of the Folsom site. Note particularly the modal elevation of these taxa compared to the elevation of Folsom, at ~2,109 m (~6,919 ft). This reinforces the general point made earlier that the

TABLE 6.16
Modern Elevation Ranges (Feet Above Sea Level) of the Snails Recovered at the Folsom Site

Taxon	Maximum	Minimum	Mode
Hawaiia minuscula		8,000	
Oreohelix strigosa	10,000	8,000	9,600
Vertigo gouldi	8,400	8,000	8,000
Vertigo modesta	11,600	8,000	8,400
Euconulus sp. (cf. *fulvus*)	10,000	7,600	8,000
Nesovitrea electrina	10,000	7,600	9,200
Punctum minutissimum	10,000	7,600	8,400
Vitrina pellucida	12,000	7,600	9,600
Cionella lubrica	8,000	6,800	7,600
Discus whitneyi	8,400	6,800	8,000
Vallonia cyclophorella	11,200	6,400	8,000

SOURCE: Based on the Lake Peak transect in the Sangre de Christo Mountains, ordered by *minimum* elevation. Data from Dillon and Metcalf (1997:table 1).

NOTE: Elevation of the Folsom site is 6,900 ft

occurrence of these species at Folsom was a consequence of regional downward shifts in temperature during Late Glacial times.

The assemblage from the Folsom site can also be compared with snails recovered at other well-dated terminal Pleistocene localities, including several pene-contemporaneous archaeological sites such as Clovis (New Mexico), Lubbock Lake (Texas), and Plainview (Texas) (Drake 1975; Howard 1935; Hester 1972; Neck 1986; Pierce 1987) (fig. 1.1). The Folsom snail fauna in many respects is similar; almost all of the terrestrial snail species living on the Southern High Plains also occurred in the Folsom area (Metcalf 1997). Yet, it differs in one very significant respect: At Folsom, no aquatic snails were found in any level or context. They are absent in spite of the fact that the bonebed was situated within one drainage (the paleotributary) and alongside another (the paleovalley). This is not to say that moisture-loving forms are lacking at Folsom; the slugs in the excavation levels below the bonebed prove otherwise. Rather, it shows that there were neither permanently flowing streams, nor standing water, nor aquatic vegetation in the paleotributary before, during, or after the bison kill occurred. We do not have a sample from the paleovalley so cannot determine whether there was surface water in it.

This absence of freshwater gastropods is in sharp contrast to, for example, the dozen or so Folsom-age localities along Blackwater Draw and at Clovis examined by Drake (1975), wherein aquatic forms (e.g., members of the genus *Gyraulus* and *Stagnicola*) comprised the vast majority of the recovered snails (Drake 1975; also, Hester 1972; Metcalf 1997:table 1). To be sure, terrestrial forms present at Folsom—such as *G. armifera, G. holzingeri, G. procera, E. fulvus,*

Hawaiia minuscula, V. cyclophorella, V. gracilocosta, and *N. electrina*—were also found at these sites, but they were in the minority. Neck (1986:64–65) found a similar mix of aquatic and terrestrial forms in terminal Pleistocene-aged sites along Blackwater Draw, Running Water Draw, Seminole Draw, and Mustang Draw, all on the Southern High Plains of eastern New Mexico and western Texas, though no quantitative data were provided on the relative proportions of each.

What is similar between Folsom and these High Plains sites is that *Zonitoides arboreus,* a "reliable indicator of woodlands" (Metcalf 1997:84), is absent from the archaeological deposits at Folsom, and extremely scarce on the High Plains (Neck [1986:table 25] reports it in the terminal Pleistocene deposits at the Plainview site). Yet, as noted, *Z. arboreus* is present today in the Folsom region. All this suggests that the environment at the Folsom site at the time of the kill was cooler and drier, and did not have the density of the woodland that is present today. This result is consistent with the pollen and charcoal records.

Land Snails: Carbon and Oxygen Stable Isotopes

With Meena Balakrishnan

The presence of particular land snail species in the Folsom site bonebed provides a rough gauge of the climatic and environmental conditions in Late Glacial times. The composition of snail shells can provide further information on those conditions, through the analysis of stable carbon ($\delta^{13}C$) and oxygen ($\delta^{18}O$) isotopes recorded in shell aragonite. The geochemical controls on shell aragonite have been examined through studies of the effects of climate, environment and feeding strategies on modern land snail physiology and shell production (e.g., Balakrishnan and Yapp 2004; Goodfriend 1992; Goodfriend and Ellis 2002; Goodfriend and Margaritz 1987; Lécolle 1985; Margaritz, Heller, and Volokita 1981; Yapp 1979).

These and other studies have shown that land snails are moisture and temperature limited and tend to be active after rains and when ambient temperatures are between 10°C and 27°C (50°F and 81°F). Activity and growth—including, of course, the precipitation of the shell aragonite—effectively cease above and below those temperatures (Balakrishnan 2002:27–28, 67). The snails' dominant oxygen input comes via ingested water, and thus $\delta^{18}O$ values in land snail shell appear to represent the combined effects of four variables: temperature, relative humidity, $\delta^{18}O$ of ambient water vapor, and $\delta^{18}O$ of snail-ingested water (Balakrishnan and Yapp 2004:2021; also Goodfriend, Margaritz, and Gat 1989). The factors are not equally weighted, however; a 1°C change in temperature yields only a ~0.1 per mil (‰) change in $\delta^{18}O$, while a fractional (0.01) decrease in relative humidity can produce an ~0.4‰ change in $\delta^{18}O$ (Balakrishnan 2002:88). Thus, $\delta^{18}O$ values in land snails, particularly in a cold-winter region like Folsom, are primarily a record of summer climate (Balakrishnan 2002:67) and

will be more sensitive to shifts in relative humidity than changes in temperature. Assuming that seasonal air mass movements were comparable in Late Glacial times, they will also primarily reflect the isotopic composition of meteoric water from the Gulf of Mexico (chapter 3; Nativ and Riggio 1990a, 1990b).

In turn, $\delta^{13}C$ values are a function of the carbon that comes from oxidation of ingested organic matter; that is, with the photosynthetic character of the vegetation at the locality, whether C_3, C_4, or mixed (Balakrishnan 2002:13–14, 37, 67; also Goodfriend and Magaritz 1987). There is some variation in $\delta^{13}C$ values across species as a result of physiological differences and perhaps selective feeding strategies (Balakrishnan and Yapp 2004:2021; Goodfriend and Ellis 2002:2000; Stott 2002).

Carbon and oxygen stable isotopic composition of land snails recovered from Folsom can potentially provide a more precise measure of local climate and vegetation than the simple presence of a taxon. Accordingly, samples of shell aragonite from the site were analyzed by Balakrishnan (2002); she also examined modern snail shells collected by Theler and colleagues during the SPGS (Theler, Wyckoff, and Carter 2004; Wyckoff, Theler, and Carter 1997). As temperature, relative humidity and precipitation, and relative proportion of C_3 and C_4 plants varied over the course of the SPGS transect, this makes it possible to assess the relative contribution of these variables, as well as others (e.g., altitude), to stable isotopic composition and, thus, to better understand the differences in isotopic composition between Late Glacial and modern snails in this and other geographic localities (Balakrishnan 2002; Balakrishnan et al. 2005b). These results, and others stemming from this study, are presented by Balakrishnan and Yapp (2004) and Balakrishnan et al. (2005a, 2005b). Only the analytical results directly relevant to the Folsom site are discussed here, and with a different emphasis than in those publications.

The snails from the Folsom bonebed reported here came from the M17 and M15 excavation blocks (fig. 4.13), specifically grid units N1033 E998, N1034 E998, and N1034 E999, which are adjoining squares in the northern portion of the M17 block, in the densest part of the bonebed excavated from 1998 to 1999. The snails were extracted from standard sediment samples taken from each excavation level within those units. The extractions were done by Arnold (1998, 1999), Balakrishnan (2002), and Theler, as described above. Isotopic analyses were performed only on fossil shells that showed no evidence of diagenetic change in the aragonite. The details of the laboratory procedures for the isotopic analysis are given by Balakrishnan (2002:122–123) and Balakrishnan et al. (2005a). All isotope values are reported relative to the PDB standard,[5] are expressed as parts per mil (‰), and have an analytical uncertainty of ±0.1‰.

Only fossil shells of *Vallonia gracilicosta* were analyzed as part of this study. This species is the most common land

TABLE 6.17
$\delta^{13}C$ Values (‰) for Land Snails from N1033 E998, N1034 E998, and N1034 E999, Levels 138 to 152

Level	$\delta^{13}C$ value (‰)						Average	SD	CV
138	−7.3						−7.3		
139	−6.9	−6.6	−6.1				−6.5	0.40	6.19
140	−6.7	−3.9					−5.3	1.98	37.36
141	−5.6	−4.8					−5.2	0.57	10.88
142	−6.9	−6.6					−6.8	0.21	3.14
143	−6.4						−6.4		
144	−7.4	−6.7	−5.7	−5.0	−4.6		−5.9	1.16	19.81
145	−7.1	−7.0	−6.9	−6.7	−4.5		−6.4	1.09	17.00
146	−6.6	−5.9	−5.6	−4.6	−4.1		−5.4	1.01	18.78
147	−6.9	−6.8	−6.4	−6.2	−5.5		−6.4	0.56	8.80
148	−8.0	−7.4	−6.7	−6.6	−5.4	−4.7	−6.5	1.23	19.01
149	−7.9	−6.5	−5.7	−5.3	−5.0		−6.1	1.16	19.12
150	−7.5	−7.5	−7.3	−6.6			−7.2	0.43	5.91
151	−7.6	−7.0	−7.0	−6.4	−6.3	−5.0	−6.6	0.89	13.65
152	−7.1						−7.1		

SOURCE: Data from Balakrishnan (2002:152–153).

NOTE: CV, coefficient of variation.

snail recovered from the deposits and occurs throughout the excavation levels (table 6.13, fig. 6.8). Focusing on one species helps avoid potential complications attendant with species-specific variability (Balakrishnan 2002:68, 122)[6] and makes it possible to assess whether and to what degree isotopic composition varies over the time represented by the excavated levels. Because of variation in $\delta^{18}O$ values between adult and juvenile *Vallonia*, the latter were not used.[7] Comparative modern $\delta^{13}C$ and $\delta^{18}O$ data on *V. gracilicosta* are available from the SPGS (Balakrishnan 2002; Balakrishnan et al. 2005b) and are used here for reference purposes.

CARBON ISOTOPES

The vegetation of the Folsom area today has a mix of C_3 and C_4 vegetation, with more of the former at higher elevations (chapter 3). The modern land snails collected during the SPGS reflect this: $\delta^{13}C$ values of *V. gracilicosta* from the nearby Owensby sampling locality, on the eastern edge of Johnson Mesa, just 75 m (250 ft) higher than the Folsom site (fig. 3.1), range from −11.7‰ to −8.4‰ (Balakrishnan 2002:107). These values fall squarely within the range of observed $\delta^{13}C$ values of snails in C_3 biota (Goodfriend and Magaritz 1987). Including all snail species from Owensby and from two, more distant SPGS localities at comparable elevations, the C.S. Ranch and Chase localities, yields a similar range of $\delta^{13}C$ values, from −13.7‰ to −8.2‰ (Balakrishnan 2002:127).

In fact, across the SPGS transect there is a strong linear relationship between elevation and $\delta^{13}C$ values in snail shells ($r^2 = 0.6367$ [data from Balakrishnan 2002:104–107]), such that at higher altitudes $\delta^{13}C$ values become more negative. This relationship, of course, is primarily a function of variation in the relative amount of C_3-to-C_4 vegetation, which generally increased with elevation along the SPGS transect (Balakrishnan 2002:82-83; Balakrishnan et al. 2005b).

The carbon isotopic composition of the snails from the Folsom bonebed ranged from a low of −8.0‰ to a high of −4.1‰ (table 6.17). Mean values, and variation about the mean, were consistent across levels, independent of sample size (recall that *V. gracilicosta* was scarce in the lowest levels). These $\delta^{13}C$ values indicate a significant amount of C_4 vegetation in snail diet in Late Glacial times and thus, presumably, in the local vegetation at the site, and are consistently and significantly more positive than $\delta^{13}C$ values seen in the modern snails of this region (fig. 6.11).

In fact, the range of $\delta^{13}C$ values in Late Glacial *V. gracilicosta* from Folsom are most comparable to the $\delta^{13}C$ values in a modern sample of *V. gracilicosta* obtained at an elevation of 1,312 m (4,300 ft) on Black Mesa, Oklahoma, some ~800 m (2,615 ft) *below* the Folsom site and ~100 km (60 miles) east as well (fig. 6.11). It appears that the vegetation at Folsom in Late Glacial times was not significantly different from the C_4-dominated grassland present today on the semiarid Southern High Plains.

Moreover, there are no significant excursions in the carbon isotopic composition over the time interval represented by the excavated deposits (Balakrishnan 2002:128; Balakrishnan et al. 2005a). *V. gracilicosta* had access to and was feeding on similar vegetation throughout that interval.

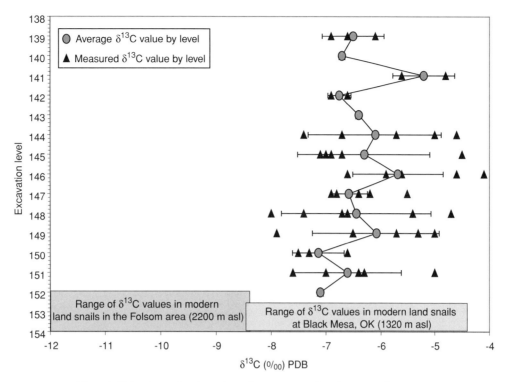

FIGURE 6.11 Plot of $\delta^{13}C$ values of archaeological specimens of *Vallonia gracilicosta*, by excavation level. The $\delta^{13}C$ range for modern *V. gracilicosta* samples from the Folsom area at 2200 m asl is shown in the upper left. The $\delta^{13}C$ range for modern *V. gracilicosta* samples from the Black Mesa area, Oklahoma, at 1320 m asl, is shown in the lower right.

Further indication of the lack of change in the forage is evident in the measures of central tendency in this sample: The mean for all the archaeological snails is −6.28‰ and the coefficient of variation (CV) is −16.67. These measures are isotopically more positive and less variable than those in the modern transect sample (mean = −8.4‰, CV = 22.39), though the difference in the CV values just misses being statistically significant at the 0.05 level ($z = 1.500$, $P = 0.067$, using the significance test of Zar [1999: 141–145]).

This evidence for a comparatively higher percentage of C_4 vegetation at the end of the Pleistocene is supported by $\delta^{13}C$ values ($n = 7$) in soil carbonate nodules examined from several units and levels across the bonebed (fig. 6.12) (Balakrishnan 2002:131; Balakrishnan et al. 2005a), indicating common sources of carbon. However, two other soil nodules yielded significantly more negative values (−15.3‰ and −16.7‰), which may reflect a contribution of ^{13}C-depleted CO_2 in the local soil budget, perhaps as a result of near-surface oxidation of small amounts of biogenic methane from the decay of underlying bison carcasses (as speculated by Balakrishnan et al. 2005a).

Studies of soil carbonates and other proxy indicators from nearby regions similarly support an increase in relative C_4 productivity between 11,000 and 10,000 ^{14}C yr B.P. (Connin, Betancourt, and Quade 1998; Holliday 2000a; Humphrey and Ferring 1994; Nordt et al. 2002). Together, these data suggest that landscapes were open, C_4 grasses

being relatively shade intolerant, and overall conditions were relatively dry, but there was still sufficient summer rainfall to support the growth of warm-season, C_4 grasses (Nordt et al. 2002:185–186).

OXYGEN ISOTOPES

The $\delta^{18}O$ values for the snails from the deposits at Folsom are provided in table 6.18 and illustrated in figure 6.13. There are several observations to make of these data. First, and looking across all excavation levels, there is a wide range of $\delta^{18}O$ values, from a low of −6.9‰ (Level 142) to a high of 2.7‰ (in Levels 149–151), a span of 9.6‰. To put this figure in context, consider that the $\delta^{18}O$ values for the modern *V. gracilicosta* collected on nearby Johnson Mesa range from −3.1‰ to −1.0‰, a span of just 2.1‰ (Balakrishnan 2002). A nearly identical isotopic range is characteristic of snails at much lower elevations; if one includes all $\delta^{18}O$ values for *V. gracilicosta* in the SPGS, the overall range is −3.8‰ to −0.7‰ (mean = −2.20‰, CV = −47.70).

In this respect the $\delta^{18}O$ values present an almost mirror image of the $\delta^{13}C$ values: Where the modern $\delta^{13}C$ values shifted, becoming more negative across space and up elevation, modern $\delta^{18}O$ values are relatively constant across space and up elevation. In contrast, the archaeological $\delta^{13}C$ values stayed relatively constant and within a fairly narrow range over the total amount of time represented by

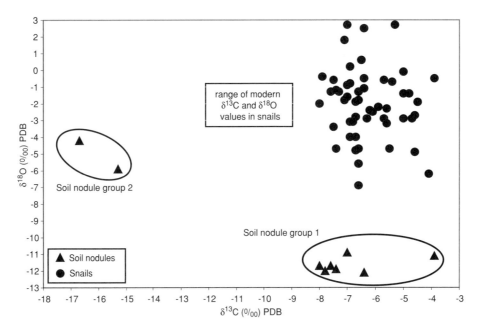

FIGURE 6.12 Plot of $\delta^{13}C$ by $\delta^{18}O$ values for archaeological specimens of *Vallonia gracilicosta* (*circles*) and soil carbonate nodules (*triangles*). The intersecting ranges of $\delta^{13}C$ and $\delta^{18}O$ values for modern *V. gracilicosta* samples is shown in the rectangle.

the excavation levels at Folsom, while the archaeological $\delta^{18}O$ values are highly variable, not only within specific slices of time, represented by the excavation levels, but also over the total amount of time represented in the archaeological deposits.

In the time span represented by the archaeological deposits—many centuries, if not a millennium or more, all during the highly variable climate of the Late Glacial—there would be considerable variation in the climate signature reflected in $\delta^{18}O$ values. However, even within specific excavation levels, which presumably represent much shorter intervals of time, the $\delta^{18}O$ values still range across as much as 5.6‰ and, also, yield proportionately high coefficients of variation. It is notable, in this regard, that the lowest variation in $\delta^{18}O$ values occurs in Level 139, which dates to the Early Holocene.

Among the factors that might be driving that variability are microhabitat differences, such as variation in ground cover, drainage, etc., and fine-grained temporal differences. The latter are difficult to tease out of the archaeological record, simply because that record is itself time-averaged, and a single excavation level may not represent an instant in time but, instead, can incorporate snails deposited over weeks, months, or years. Not all the snails in the same level are necessarily the same age in real time, despite being contemporaneous in archaeological time. Thus, there is a reasonable possibility that some of the variation in the $\delta^{18}O$ values within specific levels reflect different annual signatures. Resolving the source of variability will require precisely dated snail samples, which may put the hypothesis safely out of reach of resolution, given the inherent coarse-

ness of the radiocarbon chronology, for which the best that might be hoped for is a measure of archaeological contemporaneity. Even so, obtaining additional isotopic data would at least help refine the pattern and clarify the variability evident in the present analyses.

In any case, because of the high variability in $\delta^{18}O$ values within the excavation levels, average values are instead used, so as to reduce the microenvironmental and fine-grained temporal "noise" (Balakrishnan 2002:68). As shown in figure 6.13, there are discernible trends in these average values. The highest (most positive) values occur in the lowest levels (152–150) and are higher than modern values. The $\delta^{18}O$ values tend to become more negative up through the bonebed levels and into those immediately above the bonebed, ultimately ranging from 1.8‰ in Level 152 to -3.30‰ in Level 144. Above that level, the $\delta^{18}O$ values "zigzag" upward within more positive $\delta^{18}O$ values, between -3.9 and 0.5‰, and generally within the modern range (-3.8 to -0.7‰ [Balakrishnan 2002:126, 144]). As Balakrishnan et al. (2005a) observe, the hinge point in that overall trend (Level 144) corresponds roughly to the level where the depositional environment changes over from fine-grained *f2* sediments to the coarser material of the *f3* stratum (see also fig. 6.7).

The overall decrease in $\delta^{18}O$ values through the column was 5.8‰. Such a decrease could conceivably have been driven by changes in temperature, relative humidity, or the source of meteoric water. To be sure, during the YDC temperature shifts were significant, but triggering such a wide shift in $\delta^{18}O$ values would require an increase in temperature of the order of ~22°C, which would put

TABLE 6.18

δ¹⁸O Values (‰) for Land Snails from N1033 E998, N1034 E998, and N1034 E999, Levels 138 to 152

Level	δ¹⁸C value (‰)						Average	SD	CV
138	−0.8						−0.80		
139	−4.7	−3.1	−2.5				−3.43	1.14	33.12
140	−4.0	−0.5					−2.25	2.47	109.99
141	−1.4	−3.2					−2.30	1.27	55.34
142	−6.9	0.2					−3.35	5.02	149.86
143	−1.1						−1.10		
144	−4.9	−4.7	−2.9	−1.9	−0.6		−3.00	1.84	61.19
145	−4.8	−1.9	−1.8	−1.6	−0.8		−2.18	1.53	70.05
146	−6.2	−2.7	−2.3	−2.2	−1.3		−2.94	1.89	64.38
147	−4.7	−4.0	−3.1	−2.4	−0.5		−2.94	1.62	55.09
148	−5.6	−2.9	−2.8	−2.0	−1.2	−0.7	−2.53	1.73	68.45
149	−2.9	−1.4	−0.4	0.6	−2.7		−0.28	2.11	753.14
150	−3.4	−1.8	−1.3	−0.6			−1.78	1.19	67.04
151	−2.9	−1.3	−0.9	−0.1	−2.5	−2.7	−0.00	2.21	
152	−1.8						−1.80		

SOURCE: Data from Balakrishnan (2002:152–153).

NOTE: CV, coefficient of variation. CV values in several cases are anomalously high, which is in part a statistical result of values that range from negative to positive, and have averages that approach 0 (e.g., Levels 149 and 151). If all the values were normalized to the same sign, the CV values would decrease significantly.

ambient temperatures well beyond the tolerance range of the snails (Balakrishnan 2002:135; Balakrishnan et al. 2005a).

Alternatively, that shift could have been caused by a decline in the δ¹⁸O values of precipitation, which might have resulted from one of several causes: (1) a lowered summer temperature; (2) an increase in the amount of summer precipitation; (3) a change in the δ¹⁸O of the water vapor; or, finally, (4) a change in relative humidity (Balakrishnan et al. 2005a). As argued elsewhere, the first two possibilities seem unlikely. Lowering summer temperature to the degree necessary to cause the observed decrease in δ¹⁸O would also lower the temperature beyond the thermal limits of snails, and there is no evidence in any of the other paleoclimatic indicators of a precipitation increase substantial enough to trigger a 5.8‰ swing in δ¹⁸O values (Balakrishnan et al. 2005a).

A decline in the δ¹⁸O values of the water vapor transported into the Folsom area could have been caused by several mechanisms, including a change of the source region, changes in condensation history during transport, and changes in the δ¹⁸O of the source water. Of these, there is little evidence that the source region varied from what it is today—the Gulf of Mexico (Yu and Wright 2001)—or that there were radical changes in condensation history during transport from the Gulf into the Folsom region.

It is the case that there was a change in the isotopic composition of the Gulf of Mexico over the Late Glacial, owing to the inflow of δ¹⁸O depleted waters from the melting North American Ice Sheets (NAIS) (Aharon 2003).[8] Deglaciation was marked by a series of meltwater pulses into the Gulf (Fisher 2003; Marshall and Clarke 1999; Teller, Leverington, and Mann 2002), manifest in negative δ¹⁸O excursions in planktonic foraminifera, such as the −5.1‰ shift during MWF-4 (12,250 to 11,200 ¹⁴C yr B.P.; Aharon 2003, also Kennett, Elmstrom, and Penrose 1985; Spero and Williams 1990). Changes of this sort could produce corresponding decreases in the δ¹⁸O of summer rains at Folsom (Balakrishnan et al. 2005a). However, the Folsom bonebed falls squarely within the Younger Dryas Chronozone (11,000-10,000 ¹⁴C yr B.P.), when glacial meltwater was diverted down the St. Lawrence Seaway into the North Atlantic (Broecker and Kennett 1989; Fisher 2003; Saucier 1994; Teller, Leverington, and Mann 2002). During this period δ¹⁸O values in the Gulf of Mexico display a 3‰ *positive* swing (Aharon 2003; Kennett, Elmstrom, and Penrose 1985; Spero and Williams 1990).

It is nonetheless possible that the negative shift in the early part of the snail oxygen isotope record at Folsom, which predates the bison kill, is correlated with the latter part of the MWF-4 isotopic event in the Gulf of Mexico. If that is so, perhaps the more positive δ¹⁸O values partway up the bonebed correlate with the input of isotopically heavier water in Younger Dryas times. Testing that possibility is presently difficult, owing to the vagaries of radiocarbon dating during the YDC, as well as the lack of finer radiocarbon resolution of the bonebed and the core sequences from the Gulf of Mexico.

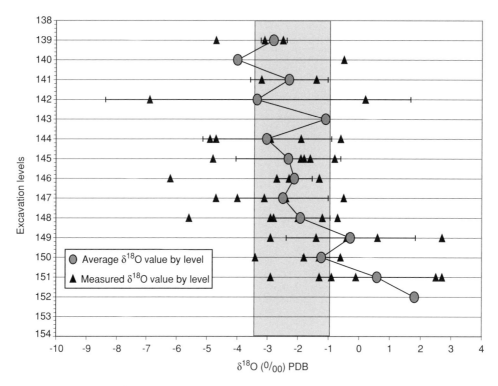

FIGURE 6.13 Plot of $\delta^{18}O$ values of archaeological specimens of *Vallonia gracilicosta,* by excavation level. The $\delta^{18}O$ range for modern *V. gracilicosta* samples from the Folsom area is shown in gray.

As a variation on the theme of changes in the $\delta^{18}O$ of the source, water vapor from the Pacific Ocean, which comes into the area primarily in the winter and spring (chapter 3), has $\delta^{18}O$ values that are considerably more negative than those derived from the summer moisture that originates in the Gulf of Mexico (Nativ and Riggio 1990a, 1990b). One might hypothesize that snails that lived and died in spring and received their moisture from winter snow melt would have a different isotopic signature than those active later in the summer. Alternatively, if there were episodic increases in the contribution of Pacific moisture during the YDC, as Amundson et al. (1996) posit for the variability of $\delta^{18}O$ values in Wind River Basin soil carbonates, they could also contribute to the downward shift in $\delta^{18}O$ and, perhaps, to the wide range of $\delta^{18}O$ values (the "noise") observed in each level.

Although the *Tsuga* pollen testifies to the possibility of storm tracks from the Pacific bringing oxygen-depleted moisture into the area, the snail data fail to support any significant contribution from this source. If there was an increased contribution of Pacific moisture (however derived), one would expect more negative $\delta^{18}O$ values for the snails. In fact, those values ought to be more comparable to those of the soil nodules (fig. 6.12), which reflect an annual precipitation signature that has values in the range of $-10‰$ to $-12‰$. The summer rain at present has a $\delta^{18}O$ value of $-6.7‰$ (Balakrishnan et al. 2005b). However, as is evident in fig. 6.12, the $\delta^{18}O$ values for the snails are more

positive than the values in the soil nodules, at least those in nodule cluster 1. The soil nodule cluster 2 values are unusual, also evident in their carbon signature, and reflect summer oxygen values.

Finally, and as has been shown (Balakrishnan and Yapp 2004), variation in oxygen values can be driven by changes in relative humidity. Were there an increase in relative humidity of ~0.01, it would cause a decrease of about $0.4‰$ in snail $\delta^{18}O$ values (Balakrishnan et al. 2005a). As Balakrishnan et al. (2005a) note, to account for the $5.8‰$ swing in $\delta^{18}O$ values, there would have to have been an increase in relative humidity of the order of ~0.13. Although such a shift is certainly possible, it was probably not the sole cause of that isotopic excursion (Balakrishnan et al. 2005a).

More likely, the $5.8‰$ decrease in $\delta^{18}O$ values was driven by some combination of increased relative humidity, lower summer temperature, and lower $\delta^{18}O$ values from the Gulf of Mexico moisture source (Balakrishnan et al. 2005a). Teasing apart the relative contributions of each will be no easy task, though one can constrain values for certain variables.

Returning to the snail taxa in these levels may shed some light on the matter. Recall that in the lowest, prebonebed levels, the fauna was dominated by a group of cool-adapted, montane-deciduous snail species indicative of a moist, well-vegetated habitat. Slugs were present in these levels and absent from the overlying levels of bonebed, suggesting a decrease in relative humidity up the section. In effect, the snail habitat data point to a very different interpretation than the isotope data. What accounts for the discrepancy?

Several possibilities come to mind: It may be that *V. gracilicosta* and, say, the slugs were at the site at different points in time, despite being deposited in the same archaeological horizon (the "time-averaged" phenomenon). Depending on the life span and seasonal occurrence of *Vallonia* and *Deroceras,* each may be accurately recording the climatic conditions as they existed when these individuals were alive. Or it may be that climatic conditions were in sufficient flux during the period of time represented, perhaps several centuries or more from Level 152 up through the bonebed, that the isotopic signature is providing a representative look at longer-term Younger Dryas climatic fluctuations. Finally, there is the less likely possibility that *Vallonia,* which is widespread throughout the section but less frequent in the lowest levels, came in only during brief favorable moments.

Bison Bone: Carbon and Nitrogen Stable Isotopes

Examining the isotopic signature of vertebrate remains has become a routine source of data and evidence in Quaternary studies, including on the Plains. Commonly, these studies have focused on carbon isotopes, with the aim of reconstructing diet, but also to inform on climatic and environmental conditions (e.g., Chisholm et al. 1986; Connin, Betancourt, and Quade 1998; Gadbury et al. 2000; Heaton 1999; Hedges, Stevens, and Richards 2004; Jahren, Todd, and Amundson 1998; Larson et al. 2001; Leyden and Oetelaar 2001; Lovvorn, Frison, and Tieszen 2001; McKinnon 1986; Meltzer and Collins 1987; Stafford 1984; Stevens and Hedges 2004; Tieszen 1991, 1994; Tieszen, Reinhard, and Forshoe 1997a). Less common, but nonetheless important, are studies that examine patterns in nitrogen, oxygen, and hydrogen isotopes in bone and teeth (e.g., Connin, Betancourt, and Quade 1998; Fricke and O'Neil 1996; Gadbury et al. 2000; Hedges, Stevens, and Richards 2004; Jahren, Todd, and Amundson 1998; Sponheimer and Lee-Thorp 1999; Stevens and Hedges 2004). The work described here focuses on carbon and, to a lesser degree, nitrogen isotopes in the bison remains from Folsom, as well as from the Archuleta bison (chapter 4). To put these results in a larger context, comparative data from other localities in different settings, Paleoindian and otherwise, are included as well.

The analytical foundation of bone isotope studies is discussed in detail in many places and need not be repeated here (see Ambrose 1990, 1991; Chisholm et al. 1986; Connin, Betancourt, and Quade 1998; Ehleringer, Cerling, and Helliker 1997; Heaton 1999; Hedges, Stevens, and Richards 2004; Larson et al. 2001; Schoeller 1999; Tieszen 1991, 1994; Tieszen, Reinhard, and Forshoe 1997a; van Klinken 1999). However, a brief review and summary of some essential aspects of carbon and nitrogen isotope studies are in order, beginning with the stable carbon isotopes, ^{13}C and ^{12}C.

The $^{13}C/^{12}C$ ratios ($\delta^{13}C$) recorded in bison bone, horn, and teeth are driven primarily by diet and reflect foraging selection and floristic composition, notably the relative contribution of C_3 to C_4 plants (Chisholm et al. 1986; Connin, Betancourt, and Quade 1998; Jahren, Todd, and Amundson 1998; Larson et al. 2001; Tieszen 1994). As noted in chapter 3, bison consume mostly grasses, along with some sedges. The former include both warm-season C_4 and cool-season C_3 types. The relative contribution of each to bison diet varies along several dimensions, including herd structure, competition, fire history, forage quality, season, and patch composition and diversity (e.g., Chisholm et al. 1986:199; Knapp et al. 1999:42–43; Larson et al. 2001:51–52; Tieszen 1994:266, 276; Vinton et al. 1993), but on a large scale the key variable is latitude.

Yet, though $\delta^{13}C$ values directly record diet, assuming that a cow-calf herd of bison will be relative generalists in their foraging—the "bison as lawn mowers" model of Larson et al. (2001:51; also, chapter 3)—they can indirectly provide a measure of climate and environment (Larson et al. 2001; Stevens and Hedges 2004). This reflects the fact that the underlying distribution of C_3 and C_4 plants is primarily controlled by temperature and, thus, varies by latitude (chapter 3; see also Connin, Betancourt, and Quade 1998; Ehleringer 1978; Larson et al. 2001; Lauenroth, Burke, and Gutman 1999:fig. 7; Ode, Tieszen, and Lerman 1980; Stowe and Teeri 1978; Teeri and Stowe 1976:fig. 1; Tieszen 1994; Waller and Lewis 1979). There is, however, a longitudinal component as well, since the mean annual precipitation and proportion of precipitation falling in summer also affect the distribution of C_3 and C_4 plants (Lauenroth, Burke, and Gutman 1999:fig. 6; Paruelo and Laurenroth 1996; Tieszen 1994:268–270). Along a given longitude, C_3 grasses decrease with latitude, while C_4 grasses increase, and C_4 grasses decrease east to west (Lauenroth, Burke, and Gutman 1999:fig. 7; Paruelo and Laurenroth 1996:figs. 2, 6; also Mole et al. 1994). Additional factors effecting grassland composition include available light or a "canopy effect" (Cerling and Harris 1999; Heaton 1999; Ode, Tieszen, and Lerman 1980; Stevens and Hedges 2004:980; Tieszen 1994:264), nutrients, edaphic factors, and changes in atmospheric CO_2 (Collatz, Berry, and Clark 1998; Cowling and Sykes 1999; Ehleringer, Cerling, and Hellinker 1997; Huang et al. 2001; Heaton 1999; Stevens and Hedges 2004:980; Tieszen 1994:263–264).

The last factor is relevant in this case, for as Ehleringer and others (1997) have shown, lower atmospheric CO_2 favors the growth of C_4 over C_3 plants, even at cooler temperatures (also Collatz, Berry, and Clark 1998; Cowling and Sykes 1999; Huang et al. 2001; cf. Nordt et al. 2002). Atmospheric CO_2 values are currently ~350 ppmV (parts per million by volume), and the C_3/C_4 "crossover" is at a daytime growing-season temperature of ~21°C to 25°C, depending on the type of C_4 plant (Ehleringer, Cerling, and Hellinker 1997:292). During the YDC, atmospheric CO_2 values were in the range of ~245 to 265 ppmV (Monnin et al. 2001; Raynaud et al. 2000; also McElwain, Mayle, and Beerling 2002), and the C_3/C_4 crossover temperature was on the order of 16°C to 18°C (Ehleringer, Cerling, and Hellinker 1997:fig. 3; also Collatz, Berry, and Clark 1998).[9]

Although the δ^{13}C values of a particular C_3 or C_4 species can vary seasonally and annually over several parts per mil as a result of these several factors (Heaton 1999:table 4; Mole et al. 1994; Tieszen 1994:268–272), the two groups form a bimodal distribution of δ^{13}C values, with a mean of −26.5 ‰ for C_3 plants and −12.5 ‰ for C_4 plants (Smith and Epstein 1971; see also the discussion by Chisholm et al. 1986; Ehleringer, Cerling, and Hellinker 1997; Larson et al. 2001; Steuter et al. 1995:table 2; Tieszen 1994:263–364).

There is, in turn, an ∼5 ‰ ^{13}C isotopic enrichment that occurs from plant to bone collagen (Krueger and Sullivan 1984; Tieszen 1994),[10] and thus the expected δ^{13}C values in a vertebrate diet comprised solely of C_3 plants will average −21.5 ‰, while an animal consuming just C_4 plants will have δ^{13}C values that average −7.5 ‰ (Chisholm et al. 1986; Jahren, Todd, and Amundson 1998; Tieszen 1994). In order to ascertain the relative amount of C_4 plants in the diet represented by a particular δ^{13}C value, simple interpolation can be used, following the procedure of Chisholm et al. (1986) as well as others (e.g., Leyden and Oetelaar 2001; Lovvorn, Frison, and Tieszen 2001). In equation form,

$$\text{Percentage } C_4 \text{ plants} = 1 - (\delta^{13}C_{obs} - \delta^{13}C_{4mean})/$$
$$(\delta^{13}C_{3mean} - \delta^{13}C_{4mean})$$

or, from the equation describing the obtained regression line,

$$\text{Percentage } C_4 \text{ plants} = (\delta^{13}C_{obs} + 21.5)/0.14.$$

The resulting estimates are assumed to have an error factor of 5% (Chisholm et al. 1986:197).

δ^{13}C values can show spatial, temporal, or species variability within plants (Heaton 1999:646.647; Mole et al. 1994; Tieszen 1994) and can vary by skeletal part within animals (Gadbury et al. 2000; Tieszen 1994:275), and so it may not always be possible to confidently ascribe small scale differences in isotopic values to diet, as opposed to other sources of variation. That said, unlike fossil plants and, to a lesser degree, snails, both of which provide "snapshots" of the local vegetation from a very brief span of time (Heaton 1999), the isotopic values in bone collagen, both δ^{13}C and δ^{15}N, have a longer turnover time and represent a multiyear average of the animal's foraging (Chisholm et al. 1986; Heaton 1999; Hedges, Stevens, and Richards 2004:960; Leyden and Oetelaar 2001; Matheus 1997; Stevens and Hedges 2004; Steuter et al. 1995:759; Tieszen 1994:275). Thus, even though the nature of the forage consumed by bison may change seasonally or annually (Chisholm et al. 1986; Gadbury et al. 2000; Larson et al. 2001; Peden et al. 1974; Plumb and Dodd 1993; Steuter et al. 1995; Tieszen 1994; Vinton et al. 1993), which might be detectable in biogenic hydroxyapatite in teeth and other parts of the skeleton with more finite and/or seasonally specific growth periods

(Gadbury et al. 2000; Larson et al. 2001; Tieszen 1994:275; see discussion by Cerling and Harris 1999), such will not be evident in samples derived from bone collagen.

The matter is complicated still further in bison, since this animal is large, mobile, and can forage over a wide and floristically diverse area in the course of a season, year, or lifetime (Bamforth 1988:83–84; Chishom et al. 1986; Hanson 1984:102–103). Hence, its δ^{13}C signature may not accurately reflect the character of the vegetation in the region where it died, raising a problem of equifinality. δ^{13}C values in bison recovered archaeologically that are significantly different from those in a modern sample from the region could result from either (1) significant environmental change, assuming that the archaeological bison had not foraged extralocally and are recording the local vegetation (e.g., Morlan 1994:761), or (2) long-distance or seasonal movements by bison between habitats with more or less C_3 and C_4 grasses, in which case δ^{13}C values may reflect not environmental change, but only feeding strategies (see Chisholm et al. 1986; Jahren, Todd, and Amundson 1998:474; Leyden and Oetelaar 2001).

Nitrogen stable isotope composition, ^{15}N/^{14}N (δ^{15}N), is a function of diet and environment, with N values mediated by the herbivore's metabolic processes (Hedges, Stevens, and Richards 2004:962). The environmental variables of greatest importance are aridity and the amount of rainfall (Heaton et al. 1986; Schwarcz, Dupras, and Fairgrieve 1999:629)—as precipitation decreases, δ^{15}N values increase (Grocke, Bocherens, and Mariotti 1997; Schwarcz, Dupras, and Fairgrieve 1999:fig. 1; Stevens and Hedges 2004:983; Tieszen 1994:276). This is true of both plants and animals, and occurs in a sufficient number of arid and semiarid environments to lead to the suggestion that the relationship is "universal" (Schwarcz, Dupras, and Fairgrieve 1999:635)[11] and the publication of regression equations and bivariate plots by which δ^{15}N values can be converted to annual precipitation in millimeters (e.g., Schwarcz, Dupras, and Fairgrieve 1999:635; also Grocke, Bocherens, and Mariotti 1997).

The mechanics of the relationship between rainfall and δ^{15}N values in plants is seemingly straightforward, a result of rising δ^{15}N of soil nitrogen, which occurs with the evaporative loss of ammonia from the soil that is depleted with δ^{15}N with respect to nitrate, ammonium ions, and organic nitrogen (Schwarcz, Dupras, and Fairgrieve 1999:630; Stevens and Hedges 2004:982). Less clear is why rainfall and δ^{15}N values in animals conform to that same relationship (Grocke, Bocherens, and Mariotti 1997; Schwarcz, Dupras, and Fairgrieve 1999:629). It has been argued that such values could reflect food shortages and tissue cycling, rather than a direct effect of water shortage and stress (Ambrose 1991; Grocke, Bocherens, and Mariotti 1997; Tieszen 1994:276). However, although bison often lose substantial weight over the course of the winter, there is no evidence that such seasonal food shortages are associated with seasonal ^{15}N enrichment. Data from Wind Cave bison show that the annual range of $\delta^{15}N$ values is a mere 0.6‰ (Tieszen 1994:table 6). Moreover, $\delta^{15}N$ values and precipitation are correlated, even when drinking water is abundant (Schwarcz, Dupras, and Fairgrieve 1999:634).

DATA AND METHODS

Eight samples of bone and teeth from the Folsom site were analyzed for stable isotopes. The data for this study were obtained from samples from both the North and the South Bank excavation areas. All but one of these (specimen N22-3-1, an antilocaprid femur) were from bison; of the seven bison specimens, one (L23-8-9, a bison carpal) was not a part of the Paleoindian kill (chapter 5).

Extraction and analysis of the collagen from the Folsom site bison bone were done at the Stafford Research Laboratories and at the Alaska Quaternary Center, as previously described (chapter 5). Following extraction, stable carbon and nitrogen isotope values were determined using a mass spectrometer. All carbon values are reported relative to the PDB standard; nitrogen values are reported relative to the AIR standard. Analytical precision is ±0.1‰.

Unfortunately, carbon:nitrogen ratios, percentage yield, and other indicators of degradation or contamination (see van Klinken 1999) are available for only those samples analyzed by Matheus at the Alaska Quaternary Center, which includes one of the Folsom specimens and the Archuleta bison. The percentage yield of the former was 13%, and the C:N ratio was 2.8:1; the yield of the Archuleta specimen was 9%, and the C:N ratio 2.9:1 (pure collagen is about 3.2:1 [Matheus 1997]). In both cases the extracted material looked collagenous, and these results suggest that both had sufficient collagen for analysis (Paul Matheus, personal communication, 2003; see discussion by Ambrose 1990; Stafford, Brendel, and Duhamel 1988; van Klinken 1999). Although comparable data are lacking for the other Folsom samples, it is reasonable to assume that these too had sufficient collagen for analysis.

RESULTS

The Folsom site bison $\delta^{13}C$ and $\delta^{15}N$ values are provided in table 6.19 and plotted in figure 6.14. For comparative purposes, the plot includes results from isotopic analyses of the Archuleta bison, later Holocene bison recovered from Bone Cliff, a recently discovered locality upstream of the Archuleta bison (Mann 2004), and Historic-period bison, around 100 to 300 ^{14}C yr B.P., from the site of Mustang Springs on the Southern High Plains of West Texas (Byerly and Meltzer 2005). These comparative localities are used in the absence of isotopic data from modern bison in the area.[12]

Several observations can be made of these data, leaving aside for the moment the two anomalously low $\delta^{13}C$ values (−17.17‰ and 17.9‰ [CAMS-74659 and CAMS-96034, respectively]).

1. There is a wide range—some 3.4‰—in the $\delta^{13}C$ values of the Folsom Late Glacial age bison, which is wider than commonly seen in single herds (e.g., Tieszen 1994:275) or even relative to mean values over time at a single site (e.g., McKinnon 1986).
2. These results are well within the isotopic range expected in animals with a diet comprised of between 60% and 80% C_4 plants. This suggests a greater amount of C_4 grasses in this locale than occurs there at present (52%).
3. The $\delta^{13}C$ values for the Folsom Late Glacial bison (−13.1‰ to −9.7‰) overlap almost precisely with the values obtained from the Bone Cliff bison, which range in age from 800 to 2,400 ^{14}C yr B.P. This indicates that Late Glacial forage and foraging patterns were not significantly different than in this same area in later Holocene times.
4. Finally, and perhaps somewhat surprising, even given the snail isotopic evidence, the $\delta^{13}C$ values in the Folsom Late Glacial bison are virtually identical, save for a slight negative offset, to that obtained from the Historic-period bison remains from the Mustang Springs site. Yet, Mustang Springs is at 823 m in elevation (2,700 ft), nearly 1,310 m (4,300 ft) lower than Folsom, experiences significantly higher average annual temperature, and has much hotter summers, which combine to produce higher evaporation rates and thus lower effective precipitation. Today and historically the Mustang Springs area is short-grass plains, dominated by C_4 grasses.

The two outlying $\delta^{13}C$-depleted values, in contrast, are in the range of a bison with a diet of mostly C_3 plants. There are several potential explanations of these anomalous isotope values: (1) these samples could have come from different episodes, not otherwise detectable within the precision of radiocarbon dating, perhaps the result of multiple kills/deaths at the site over a relatively brief period of time (a possibility first raised by Oliver Hay in 1929; chapter 7); (2) they could be from animals that at the moment of their death were part of the same herd but that, prior to that time, had been a part of different cohorts and had fed in very different patches; or (3) the bones from which they were derived could have had a different diagenetic history, the $\delta^{13}C$ depletion a consequence of taphonomic processes.

Given the evidence at hand, the first possibility seems unlikely. As discussed in chapter 5, there is no archaeological or stratigraphic evidence to indicate that this is anything more than a single kill episode, and thus a single herd (also Brown to Hay, January 12, 1929, OPH/SIA). The second possibility seems more plausible, since it is known that animals with different foraging histories can be present alongside one another. Previous studies have documented examples of bison cohorts that have comparably wide-ranging $\delta^{13}C$ values (e.g., Larson et al. 2001:47–49; cf. Tieszen 1994).

TABLE 6.19

Carbon and Nitrogen Isotope Values for Folsom Site Bison and Antilocaprid Specimen, Ordered Chronologically

Site/Provenience	Lab. No.	Taxon	Element	Age (^{14}C Yr B.P.)	$\delta^{13}C$	% C_4 Diet	$\delta^{15}N$
Bone Cliff	CAMS-105769	Bison	radius	815 ± 35	−10.65	77.50	5.4
Bone Cliff	CAMS-105770	Bison	mandible	830 ± 40	−9.7	84.29	5.4
Bone Cliff	CAMS-105771	Bison	unidentified	830 ± 40	−12.98	60.86	5.00
Folsom, North Bank (L23-8-9)	CAMS-74654	Bison	carpal	9,220 ± 50	−9.39	86.50	9.41
Archuleta Creek	CAMS-96033	Bison	axis	10,190 ± 30	−10.80	76.50	9.28
Folsom, North Bank (CS70-11)	SMU-179	Bison	"fragments"	10,260 ± 110	−12.5	64.29	—
Folsom, South Bank (M17-19-106)	CAMS-74656	Bison	humerus	10,450 ± 50	−13.10	60.00	9.13
Folsom, South Bank (M17-24-310)	CAMS-74658	Bison	tibia	10,450 ± 50	−11.13	74.07	9.24
Folsom, North Bank (R-20-14(5))	CAMS-96034	Bison	thoracic vertebra	10,475 ± 30	−17.90	25.71	6.6
Folsom, South Bank (M17-24-141)	CAMS-74657	Bison	molar	10,500 ± 40	−9.97	82.36	9.32
Folsom, North Bank (L23-8-10)	CAMS-74655	Bison	humerus	10,520 ± 50	−11.54	71.14	10.12
Folsom, North Bank (Q21-8-52)	CAMS-74659	Bison	tibia	10,520 ± 50	−17.17	30.93	6.83
Folsom, North Bank (N22-3-1)	CAMS-74934	Antilocaprid	femur	9,270 ± 50	−19.06	17.43	7.3

SOURCE: Data from Stafford Research Laboratories (T. Stafford), Alaska Quaternary Center (Paul Matheus). Data on SMU sample from Hassan (1975:table 16).

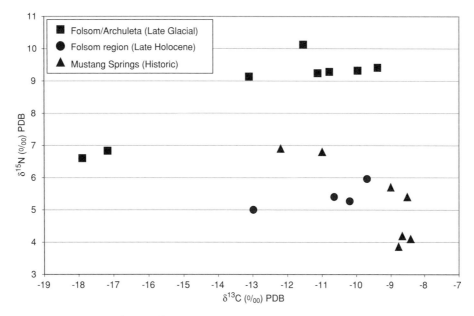

FIGURE 6.14 Plot of $\delta^{13}C$ by $\delta^{15}N$ for Late Glacial age bison from Folsom site and the Archuleta locality (*squares*); Late Holocene bison from the Folsom area (*circles*); and Historic bison from the Mustang Springs site, west Texas (*triangles*).

Whether the Folsom herd was comprised of individual animals from different cohorts with different foraging histories cannot be precluded, given that the life histories of these individual animals are not known, but then it cannot be demonstrated either, given the data at hand.

The third possibility, that there were diagenetic differences between these samples, garners some support in the observation that the two $\delta^{13}C$ outliers were both recovered from the North Bank, in geological contexts indicative of redeposition. Having been buried in the paleovalley, the bones were subject to a very different hydrological history than those found in the paleotributary. That said, a third specimen from the paleovalley had a $\delta^{13}C$ value consistent with those from the paleotributary (−11.54‰ [CAMS-74655]). However, this particular specimen was found some 27 m upstream and, perhaps more important, was at a higher elevation and in a different portion of the paleovalley than the other two. Were there a "canopy effect" in the immediate area of the present channel, along which all three of these North bank specimens were recovered, the upstream specimen should have been isotopically depleted as well.

Jahren, Todd, and Amundson (1998:472) found a similar disparity in $\delta^{13}C$ values across the Hudson-Meng bison bonebed, which they attributed to whether the bone was recovered from lower and wetter and more rapidly buried areas or from portions of the site that were higher, better drained, and exposed longer prior to burial. However, they suggested that the most reliable $\delta^{13}C$ values would be found in the lower and wetter settings, and this would seem not to be the case at Folsom, where perhaps other aspects of their complex taphonomic history are coming into play. Ultimately, resolving the cause of these outliers cannot be done with the data at hand, and will require further data and analysis

The $\delta^{15}N$ values for the Folsom Late Glacial specimens are relatively high: Most are in the range of 9‰ to 10‰, save for the same two outlier specimens, which returned much lower $\delta^{15}N$ values (<7‰) (table 6.19). Again leaving these aside, it is noteworthy that the Folsom Late Glacial bison yielded substantially higher $\delta^{15}N$ values than the late Holocene Bone Cliff specimens and the bison from Mustang Springs (fig. 6.14). The high values from the Folsom Late Glacial bison could result from one or a combination of factors, most especially a period of water or dietary stress, leading to an enrichment of $\delta^{15}N$ in tissue (Leyden and Oetelaar 2001:18; Schwarcz, Dupras, and Fairgrieve 1999; Stevens and Hedges 2004:982–983). If that is indeed the case, it corroborates other data, notably the $\delta^{18}O$ record from the snails, which suggest a relatively dry Late Glacial environment. $\delta^{15}N$ values of this magnitude are comparable to those reported in areas that receive ~300 mm of precipitation per year (Schwarcz, Dupras, and Fairgrieve 1999:fig. 1). However, though the temptation to use $\delta^{15}N$ values to gauge aridity and (indirectly) biotic productivity is strong, this measure must be used with some caution, as it has been shown that diagenetic processes can strongly influence those values, at least in plants (DeNiro and Hastorf 1985).

Little need be said of the stable isotope results of the antilocaprid remains, included here for the sake of completeness, save that the $\delta^{13}C$ value obtained is significantly lower than all values obtained for bison and is in keeping with this species' highly selective C_3 feeding (Larson et al. 2001:51; Schwartz and Ellis 1981:350).

DISCUSSION

The Folsom Late Glacial bison seemingly had a diet comprised primarily of C_4 plants, which implies a very different grass flora than is here at present. But before accepting this evidence, it is necessary to consider the potential problem of equifinality raised above: namely, Does that isotopic signature reflect the Folsom environment at the time of the kill with the bison herd feeding on locally available C_4 grasses, or had these animals spent most of their time foraging in C_4 grasslands elsewhere, say, at lower elevations, where C_4 grasses were more common than in the Folsom area? The $\delta^{13}C$ values cannot resolve this question, since even if this was a nonresident herd, this does not preclude the possibility the vegetation in the vicinity of the Folsom site contained a significant C_4 component. Corresponding data on strontium isotopes might (e.g., Hoppe et al. 1999), but such are not available here.

However, the isotopic evidence from snails can help resolve the matter, at least in part. Snails, of course, are not as wide ranging or as mobile as bison. Nor do the $\delta^{13}C$ values in their shells record vegetation consumed over many years of foraging. Thus, snail remains testify directly to the character of the local forage, at relatively brief periods of time. And, of course, the $\delta^{13}C$ values in the Folsom snail isotopes testify to a C_4 dominant vegetation. Clearly, then, the isotopic signature in the bison could have been derived locally, but of course that still does not preclude the possibility that these bison had spent much of their time foraging elsewhere in C_4 grasslands (a matter that might be explored using trace element analyses).

Paruelo and Laurenroth's (1996) response function was earlier used to derive a ballpark estimate of the relative amount of C_4 grasses present in the Folsom area today. The resulting value, ~52%, is lower than the estimated 60% to 80% of C_4 plants evident in the diet of the Late Glacial Folsom bison. Assuming that the carbon isotopic signature in the Folsom bison is "locally" acquired (as the snail evidence would suggest), and that these bison foraged in this environment in a relatively nonselective manner, this provides additional support for the claim the Folsom in Late Glacial times supported a different flora than at present, one with a greater proportion of C_4 grasses. This line of evidence, in turn, affirms the earlier conclusion that summer rains were sufficient to sustain C_4 grass productivity.

Looking across a broader geographic region, the mean $\delta^{13}C$ value for the Folsom bison (-13.47‰ which includes the two outliers; the mean without them is -11.43‰) is significantly more positive than the mean $\delta^{13}C$ values for bison from other kill sites of comparable age, including the various Paleoindian components at the Agate Basin site, Carter/Kerr-McGee, Hudson-Meng, and Jones-Miller (data from Jahren, Todd, and Amundson 1998; Lovvorn 2001: table 1; Tieszen, Reinhard, and Forshoe 1997b). The bison in these sites had a diet that was primarily based on C_3 vegetation (60%–90%). This is not surprising, given that these sites are at least 600 km farther north from Folsom, at lati-

tudes >43°N (compared to Folsom at ~37°N), where C_4 grasses are more limited (Teeri and Stowe 1976:fig. 1).

More comparable to Folsom are the data from Blackwater Locality No. 1, Lubbock Lake, and the Cooper sites on the Southern Great Plains in eastern New Mexico, western Texas, and Oklahoma, respectively (data from Bement 1999b:164; Connin, Betancourt, and Quade 1998:table 1; Stafford 1984:table 2). These three sites are at comparable latitudes, though lower elevations, than Folsom, and each has Folsom-age bison bone $\delta^{13}C$ values that overlap with those from Folsom—and, for that matter, Mustang Springs.

The individual and average data from most of these sites are plotted in figure 6.15, which shows the results just noted, as well as the strong relationship between $\delta^{13}C$ values and latitude. That such a relationship obtains is wholly expected since, as noted in chapter 3, modern bison diets closely mirror the latitudinal shift in the relative percentage of C_3 and C_4 grasses on the landscape (compare fig. 6.15 with Tieszen 1994:fig. 13; also Gadbury et al. 2000; Steuter et al. 1995:764).

With regard to the $\delta^{15}N$ evidence, the possibility that the Folsom $\delta^{15}N$ values signal a dry period is not necessarily precluded by the very low $\delta^{15}N$ values from the Mustang Springs site, which is at present in a much drier setting than Folsom. The low values in the Mustang Springs samples may reflect other processes, such as grass fires, that may have altered nitrogen cycling (P. Matheus, personal communication, 2003).

The apparent depletion in $\delta^{15}N$ by 2‰ to 3‰ between the Folsom Paleoindian samples and the Late Holocene specimens from Bone Cliff is the mirror image of a shift seen in European large mammals (horse, cattle, and deer) over the course of the YDC (Hedges, Stevens, and Richards 2004:963; Stevens and Hedges 2004:979). There, $\delta^{15}N$ values were lower during Late Glacial times and became increasingly more positive into the Holocene (Hedges, Stevens, and Richards 2004:fig. 1; Stevens and Hedges 2004). The difference might be attributable to the effects of cooler temperatures in those higher latitudes, increased water availability from meltwater, and nitrogen cycling in postglacial soils, although the precise causes are uncertain (Hedges, Stevens, and Richards 2004:963–964).

Summary: The Late Glacial Environment of Folsom

It is now possible to assemble a consistent picture of the climate and environment of the Folsom site at the time of the Paleoindian occupation and address the questions raised first in chapter 3 and, again, at the outset of this chapter. Several of those pertained to temperature and precipitation: In general, multiple lines of evidence show that conditions were cooler and drier during Paleoindian times than today. The snail fauna includes nearly a dozen species that presently occur at higher elevations, and snail $\delta^{18}O$ values appear to be driven in part by lower summer temperatures.

But neither the Bellisle pollen nor the pollen and plant macrofossils from the bonebed give evidence that it was

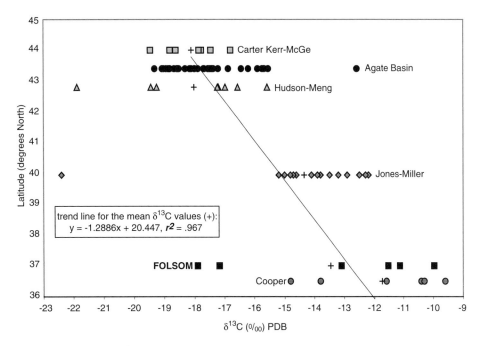

FIGURE 6.15 Plot of $\delta^{13}C$ for bison specimens from select Plains Paleoindian sites, arrayed north (top) to bottom (bottom). Mean values for each site shown as +.

cold enough long enough to trigger the downslope movement of tundra, alpine, or even subalpine species, such as *Abies*, or species like *Pinus aristata*, granting the limits to the taxonomic resolution of pollen. That there is a discrepancy between the vegetation and the snail records, the latter showing the presence of higher elevation forms, the former not, is hardly surprising. The snail fauna would have been much more sensitive and responded much more rapidly to temperature changes than would the tree species. Such changes do, however, play out at higher-elevation settings farther north (e.g., Fall 1997)

Precipitation was also lower than at present, as is evident in the aeolian character of the *f2* sediment, the absence of any aquatic snails, the $\delta^{18}O$ values in the snails, and the high $\delta^{15}N$ values in bison bone. Though conditions were drier than at present, there is no compelling evidence of long-term drought (brief, episodic drought, of the sort envisioned by Holliday [2000] cannot be ruled out). There was sufficient precipitation to support tree growth, and sufficient summer precipitation to support the growth of warm-season C_4 grasses. There does not appear to have been an increase in the contribution of winter precipitation, judging by the lack of $\delta^{18}O$ depleted values in snail isotopes.

There were nonetheless scattered wet places on the landscape, as indicated by the presence of cattail and *Ephedra* in the bonebed pollen and the aquatics in the Bellisle pollen record. The anomalous occurrence of *Tsuga* (hemlock) pollen immediately above the bonebed is not the result of Pacific Northwest precipitation levels, for which there is no evidence, but instead likely results from long-distance trans-port from an unusual Pacific storm track pushing into the area, which might also have been a factor (albeit a minor one) in contributing to the $\delta^{18}O$-depleted values in snails in the higher levels of the bonebed.

The nature and structure of the vegetation community can be reconstructed from the pollen, macrofossil, and snail and bison isotope data, with the caveat that there are limitations to their use: notably, the lack of quantitative and statistically representative pollen/macrofossil data and taxonomic identifications no finer than the genus level. Those notwithstanding, it is reasonable to conclude from the pollen evidence that the tree genera present on the landscape in Late Glacial times were the same as those in the area today. This answers in the negative the question of whether the area would have provided sufficient MAST trees or other forage suitable for humans, since the arboreal component today is inadequate to the task (chapter 3).

There were fewer trees on the landscape than presently, as is apparent in the snail and bison isotopic evidence of a substantial C_4 component in the vegetation. There is no evidence in the $\delta^{13}C$ values of a carbon-depleted canopy effect, save perhaps for the two bison bone isotope outliers on the North Bank, but more data are needed to resolve the cause of these outlying values. The pollen data also indicate the presence of open parkland, covered with grasses and shrubs. The relative density of woodland is not known from the pollen/macrofossil record, but the snail fauna sheds light on this issue. The woodland snail *Zonitoides arboreus* is absent from the deposits at Folsom, though it is present today in the region. From this, it is reasonable to conclude that the surrounding landscape was more open than closed, especially in

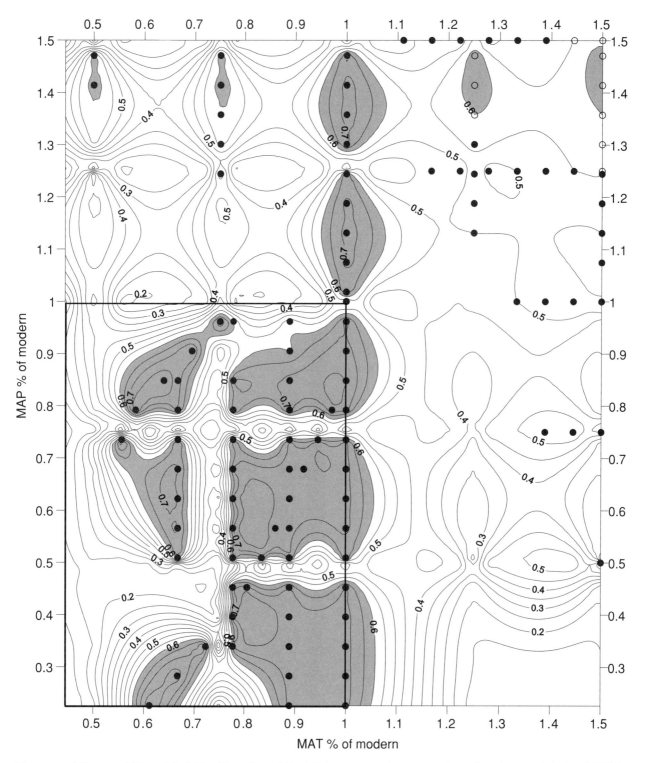

FIGURE 6.16 A "topographic" model plotting Mean Annual Precipitation expressed as a percentage of modern precipitation (MAP%) against Mean Annual Temperature expressed as percentage of modern temperature (MAT%), with Seasonality of Precipitation (SEA) forming the contour lines. Areas on that surface that can be expected to produce > 60% 64 grass cover are shown in gray.

the later (upper) portion of the deposit—this based on pollen/macrofossils and snail isotope evidence.

The Bellisle and bonebed pollen hint, and the snail and bison bone $\delta^{13}C$ isotope evidence confirm, that there was a substantial grass component to the vegetation, with C_4 grasses occurring in greater proportion than at present. Using $\delta^{13}C$ to C_4 cover interpolations, and assuming Folsom bison diets provide a representative sample of the vegetation on the landscape (chapter 3), the C_4 species comprised 60% to 80% of the grass cover (chapter 3; using formulas from

Paruelo and Lauenroth 1996). The vegetation of the Folsom area during the Late Glacial thus more closely resembled the modern biota of the much warmer and drier High Plains than it does the vegetation that presently surrounds the site. This extends the observation made by Connin, Betancourt, and Quade (1998:187) of a high incidence of C_4 plants across much of the Late Glacial American Southwest.

The underlying cause of that C_4 dominance may be attributable in part to lowered atmospheric CO_2 during Late Glacial times, which would have favored the growth of C_4 over C_3 plants, even at temperatures cooler than present (Ehleringer, Cerling, and Hellinker 1997; also Collatz, Berry, and Clark 1998; Cowling and Sykes 1999; Huang et al. 2001; cf. Nordt et al. 2002).[13]

Lowered atmospheric CO_2 is not likely the sole factor in play, of course. As Paruelo and Lauenroth (1996) demonstrate, the amount of C_4 grass cover is, within limits and under present atmospheric CO_2 values, a predictable function of mean annual temperature (MAT), mean annual precipitation (MAP) and seasonality of precipitation (SEA) (Paruelo and Lauenroth 1996:fig. 7). It is possible using their regression equation to model how various changes in each of these three factors would affect C_4 cover. The results of such a model are illustrated in figure 6.16, which is a climatic "contour map" created by varying MAT, MAP, and SEA, above and below the values that characterize the Folsom area today (MAT = 9°C, MAP = 442 mm/year, and SEA = 48%).[14] Once the contour plot was created, points on that "surface" where various combinations of MAT, MAP and SEA yielded C_4 cover between 60% and 80% were plotted as solid dots.

Doing so reveals that were MAT and/or MAP free to vary, high levels of C_4 are possible at MAT and MAP values higher than at present, though with SEA values much lower than at present. Of course, none of the data assembled in this chapter suggests that either MAT or MAP were higher than present during the YDC. Indeed, they point to the opposite conclusion. The "climate space" of greater interest is therefore the box in the lower left of fig. 6.16, which bounds the area wherein both MAT% and MAP% are equal to or lower than present values. Within that space, virtually all of the solid dots fall at or above SEA contour levels of 60%. In effect, in order to produce 60% to 80% C_4 cover under Late Glacial temperature and precipitation conditions, the relative amount of summer precipitation would have had to be at least 12% higher than at present. But this model does not incorporate the effects of atmospheric CO_2 values, so the degree to which the higher C_4 cover in Late Glacial times is a function of a seasonal shift in the amount of precipitation, lowered atmospheric CO_2 values, some combination thereof, or of still other variables, remains to be resolved.

What is known is that there was a local abundance of C_4 grasses, which enabled bison to inhabit the Folsom area during the Late Glacial; they need not have been exotic to the region or just passing through. But could bison have overwintered there? If the Folsom area received snowfall in amounts comparable to today, and if cooler temperatures reduced melting, ensuring that snow stayed on the ground longer into the spring, then the early-season C_3 grasses on which bison depend may not have been abundant or readily available (L. Todd, personal communication). Under such circumstances, bison would likely have abandoned the area for lower, less snow-covered settings. If they did, their movement may not be detectable in their bone isotopes, since they could have foraged on C_4 grasses in those settings. If, however, the area received less snowfall than at present, they could have overwintered. But were that the case, there ought to be a greater contribution of C_3 grasses to their diet, and commensurately lower $\delta^{13}C$ values, than observed. Thus, the current evidence suggests that bison did not overwinter in the area.

There is a consistent signal across all lines of data of a significant change in climate a few centuries after the bison kill at Folsom. That change is marked by an erosional hiatus in the bonebed, the Bellisle Lake sediment core, and other localities across the Folsom area (Mann 2004). It begins soon after 10,200 ^{14}C yr B.P. and, at Folsom, is followed several centuries later by deposition of the *f3* unit, a massive sheet wash that, along with the snail oxygen isotope record, indicates greater moisture than before. This may signal the intensification of the summer monsoon. At roughly the same time, however, there is also a trend evident in the Bellisle pollen and in snail species (e.g., *G. procera* and *P. inornatus*) of increased summer temperature, which likely raised evaporation rates and, in certain settings such as Bellisle Lake, may have offset gains in precipitation.

The larger climatic trend, then, is a shift from cool and dry conditions marking the YDC in this region toward warmer and wetter conditions in Early Holocene times. This is a pattern broadly consistent with records elsewhere in the southern Rocky Mountains (Betancourt 1990; Elias 1996; Fall 1997; Markgraf and Scott 1981; Mayer et al. 2005; Vierling 1998).

Notes

1. To test for the presence of pollen in the sediments of these lakes, samples were initially obtained by bucket auger from one of the dry lake basins on the eastern end of Johnson Mesa—this on the reasoning that if pollen was preserved in a dry lake, the chances were good it would also be preserved in one of the more permanent lakes, which could then be cored using standard palynological field techniques. This preliminary effort did indeed yield a few badly damaged pollen grains and a warrant for further effort.

2. During one especially monumental tug-of-war between those of us on the platform and the stubborn lake bottom clays, which had a deep and tenacious hold on the coring device, the coring platform suddenly cracked down the seam between its two halves. The platform abruptly became V-shaped but, fortunately, did not break apart. All hands on deck scrambled to the outer edges—sailors call it hiking—the platform was forced back to horizontal, and coring resumed.

3. A third set of samples was processed by Theler at UW-LC from at, above, and below an exposed bison sacrum (1997 FS 47) recovered from the 1997 test excavation unit, N1033 E1004. However, because the vertical control on this set of samples was not as precise and did not match up with the excavation levels used in 1998 and 1999, this set is not included in the subsequent discussion. In general, the snails from this set matched those recovered in 1998, though it also included a single specimen of an additional species, *V. parvula*, that does not otherwise occur in any of the archaeological or modern samples from the site.

4. Freeman-Tukey deviates provide a means of identifying the specific cells in a contingency table that show a significant difference between observed and expected values. The values center on 0 (no significant difference) and, by their sign, identify which cells have observed tallies that are larger [+] or smaller [−] than would be expected (Freeman and Tukey 1950). Significance levels are calculated by table.

5. $\delta^{13}C = [(R_{sample}/R_{standard})-1] \times 1,000‰$, where
$R = {}^{13}C/{}^{12}C$ or ${}^{18}O/{}^{16}O$.

6. $\delta^{13}C$ values for *Vallonia* and *Gastrocopta* from the same locality were correlated, for example, but the $\delta^{18}O$ values were not (Balakrishnan 2002:84).

7. Adult $\delta^{18}O$ values reflect climatic signatures from a longer period of time than in juvenile shells and, thus, are arguably better climatic indicators since they would average out the effects of any unusual weather (Balakrishnan 2002:85). $\delta^{13}C$ values do not vary significantly between adults and juveniles, suggesting that feeding strategies remain constant over the course of the snail's lifetime (Balakrishnan 2002:85).

8. The $\delta^{18}O$ of Laurentide Ice Sheet meltwater is not known, but indirect evidence puts it from −25‰ to as isotopically light as –38‰ SMOW (Aharon 2003).

9. The current $\delta^{13}C$ of CO_2 is −8.0‰. $\delta^{13}C$ values during the YDC were ~6.77, based on ice core data (Leuenberger, Siegenthaler, and Langway 1992:table 1; see also Cerling and Harris 1999:348; Monnin et al. 2001; Spero and Lea 2002; Tieszen 1994:263; Van De Water, Leavitt, and Betancourt 1994).

10. Enrichment in bone apatite is ~12‰ of ingested forage (Larson et al. 2001:36.37; see also Cerling and Harris 1999; Connin, Betancourt, and Quade 1998).

11. Cormie and Schwarcz (1995) report otherwise for North America, but their study focused on white-tailed deer, in habitats with >10% C_4 grasses, and given that deer are not necessarily denizens of the open, semiarid High Plains, the results could represent animals that fed in wooded riverine settings. The results might look different if applied to, say, modern bison populations. The relationship between precipitation and $\delta^{15}N$ appears to break down in wetter environments, where there is no differential tendency for nitrogen conservation as a function of rainfall.

12. Although there is a small herd on a ranch at Trinchera Pass, they are grain-fed through the winter, so any isotopic analysis would not be informative of natural forage patterns.

13. A paper by Koch, Diffenbaugh, and Hoppe (2004), which appeared after this chapter was written, reports isotopic evidence indicating similarly high percentages of C_4 plants in the south-central United States in the Late Glacial. In an interesting analysis, they explore—in a more sophisticated analysis than was possible in the context of this chapter—the relative effects of climate and atmospheric CO_2 in producing those high values. It was gratifying to see that despite differences in approach we arrived at similar results and conclusions.

14. To create the plot, a grid of calculations was made for temperatures that potentially ranged from 4.5°C to 18°C (0.5 to 2 times modern), with precipitation ranging from 100 to 800 mm (~0.23 to 1.8 of modern) and seasonality ranging from 0% to 100%. To put all variables on a comparable scale, the plot does not use absolute numbers for MAT and MAP but, instead, normalizes these as percentage of modern: thus values of 1.0 along the age *X* and *Y* axes mark present conditions at Folsom, values of 0.5 mark temperature and precipitation values half of what they are at present, etc. SEA is already expressed as the percentage of precipitation that falls during June, July, and August. For display purposes, the axes only extend to 1.5% of modern. These values were chosen to encompass the full range of potential values that might conceivably have occurred during the YDC, and to err on the side of caution the end members are likely much drier and much wetter than any climatic conditions that were experienced in the Folsom region in Late Glacial or Holocene times.

SEVEN

The Faunal Assemblage and Bison Bonebed Taphonomy

DAVID J. MELTZER, WITH LAWRENCE C. TODD,
WITH A CONTRIBUTION BY
ALISA J. WINKLER

It was the bison bones at Folsom that caught McJunkin's eye, and until the midsummer of 1926, that was all the Colorado Museum was interested in as well (chapter 2). That changed, of course, when the first projectile point came out of the ground, and this became an archaeological site. Even so, the site continued to be referred to as the Folsom "bone quarry," testimony to the sheer numbers of bison bones, which constitute by bulk the largest class of material from the site. Much of the early analytical attention was on just what kind of bison these were; questions about how the hunters exploited these animals seemed self-evident—this was a kill site, after all. And yet, within any such site, much more can be learned about hunters and their prey than merely the names of each.

Our analysis of the faunal remains recovered from the various excavations at the Folsom site therefore proceeded along several lines of inquiry. We sought to understand (1) the patterns of Paleoindian use of the bison, by examining butchering and processing patterns, carcass dispersion, element selectivity, and patterns of bone transport; (2) the characteristics of the animals killed here, by examining their numbers, the age and sex composition of the herd, and their diet and health; and (3) the postdepositional taphonomic processes, by examining the spatial distribution and density of remains, the diagenetic and geomorphic processes that may have affected the preservation and positioning of exposed and then buried carcasses and individual bone elements, and modification by carnivores or scavengers.

Although many of the questions of interest to us were not asked, or answered, during the original research, the large sample of bison remains collected during those earlier investigations is useful for our purposes and can be readily combined with our data to address more contemporary analytical questions. We do so below.

Before proceeding, it should also be noted that the vast majority of the faunal remains recovered from the site, both in the original work and in our more recent investigations, are of bison. However, other mammalian species, primarily small mammals, were also recovered. There was some question in the 1920s as to whether the Folsom group hunted deer, as a few elements of *Odocoileus hemionus* were apparently recovered (Roberts 1940:59); our more recent investigations also yielded a single fragmentary element of a cervid. We have found no compelling evidence that any animals other than bison were exploited by Paleoindian groups at Folsom. For this reason, the focus here is on bison. Nonetheless, for the sake of completeness the non-bison remains are briefly discussed at the end of this chapter.

Early Views of the Folsom "Bone Quarry"

The 1920s Folsom excavations were conducted by paleontologists, for whom the site was less a source of information about Late Glacial hunter-gatherers or bison paleoecology and more a place to acquire fossil material for museum display (chapter 2).[1] It was telling that they referred to Folsom as a "bone quarry," for that was their purpose in being there—to mine the site for fossils and, once their presence became known, for artifacts as well. Because the bison remains appeared to be a species new to science, much of their analytical attention was devoted to the question of taxonomy. Knowing the species and its place in bison evolutionary history would help determine its geological age and, thus, the age of the associated artifacts (Figgins 1927a:16; Figgins to Hay, December 21, 1926, DIR/DMNS). As Kirk Bryan put it, "To get results with these ancient men we must learn more about the Pleistocene vertebrates" (Bryan to Wetmore, August 16, 1928, NMJ/SIA).

Although far less attention was paid to other matters, the original investigators did make useful observations on the site taphonomy (though they did not use that term), as well as on bone preservation. And they occasionally commented on how and on how many bison may have been killed, butchered, and processed here. They also made passing observations on additional taxa found. Little of this information was detailed in publications, but it can be gleaned from unpublished correspondence.

Questions of Bison Taxonomy

From the outset, Figgins (1927a:16) had little doubt that the Folsom bison were "of a race quite new to science." His colleagues agreed. Beyond that, however, there was less agreement about whether the Folsom bison were a new species or possibly a new genus, and whether there was more than one species or perhaps even more than one genus present at the site. The way Cook saw it there was evidence "quite clearly, of two, and possibly three species of Bison among that material; unless there is a most remarkable range of variation both in type and size in that species, which seems unlikely" (Cook to Schwachheim, July 10, 1928, HJC/AGFO; also Figgins to Hay, December 10, 1926, DIR/DMNS). Cook's was an extreme view, but he was not alone in having it.

That two bison species, let alone two genera, might be present at Folsom reflected in part uncertainty about how to gauge the significance of the anatomical and morphological variability among the specimens, but also something of the taxonomic practices of the time. The period from 1928 to 1933, as McDonald later observed, saw the "most concerted surge of taxonomic splitting" than any period before or since. During this brief time, two new genera, eight new species, and six new subspecies were named (McDonald 1981:40). That this was the period when the bison from Folsom and Lone Wolf Creek were first described is no coincidence, for the eruption of taxonomic names is in large part due to Cook, Figgins, and Hay. Their publications alone (e.g., Figgins 1933b; Hay and Cook 1928, 1930) account for both of the new genera and nearly half of the new species and subspecies that McDonald tallied.

Although Cook was the Colorado Museum's "Honorary Curator" of Paleontology, the analysis of the Folsom and Lone Wolf Creek bison would not be left entirely to him (Figgins to Cook, November 30, 1926, HJC/AGFO; Figgins to Hay, November 30, 1926, December 21, 1926, DIR/DMNS). This was at Figgins' direction, for he believed that because artifacts were found with those bison, there was "a double importance" to correctly assessing their specific identity and antiquity (Figgins to Hay, December 21, 1926, OPH/SIA). Doing the work jointly with Oliver Hay would avoid "all possible question regarding the geological age to which the *Bison* belongs, as well as to remove all doubt as to its specific determination; Dr. Hay being the best informed upon this group of animals"(Figgins 1927a:16).[2] Or so Figgins said publicly. Privately, he told Cook that having Hay's participation would give their claims sufficient paleontological credibility that the only criticism that could possibly remain would be "whether or not someone is a crook and that the arrowheads were "planted," and Figgins was confident he could effectively rebut that insinuation (Figgins to Cook, December 28, 1926, HJC/AGFO).

Hay and Cook's (1928) initial report on the Folsom bison was put together in late December of 1927 and published just a month later. The speed with which this paper moved from inception to publication was in part an effort to stave off Barnum Brown's naming the new bison species. Hay and Cook's first paper was thus little more than a list of three new species of bison and one new species of mammoth from Folsom, Lone Wolf Creek, and Frederick. It was, as Hay put it, "just enough to hold priority" (Hay to Figgins, December 16, 1927, DIR/CMNH; see also Brown to Hay, December 12, 1927, VP/AMNH).

The Folsom specimen was named *Bison taylori,* in honor of Frank Taylor, the late president of the Board of Trustees of the Colorado Museum of Natural History. This animal was, in their view, distinct from their recently excavated Lone Wolf Creek bison, which Hay and Cook believed was another new species, *B. figginsi.* In a subsequent, more detailed description of the Folsom and Lone Wolf Creek material completed in mid-1929 (Cook to Figgins, May 9, 1930, HJC/AGFO), Lone Wolf Creek was elevated to a new genus, *Simobison figginsi* (Hay and Cook 1930:23–24).[3] Folsom remained within the genus *Bison*. Table 7.1 details the changing taxonomic designations of the bison from these two sites.

As far as Hay and Cook were concerned, there was little question of the "distinctness" of the specimens from those two sites. They based their identifications primarily on cranial attributes, notably the shape of the skull, which was narrower in the Lone Wolf Creek specimen, and on the geometry of the horn cores, which were shorter, broader, and flatter in Lone Wolf Creek (Hay and Cook 1930:27). Little taxonomic weight was placed on postcranial material, simply because it was oftentimes difficult to link cranial and postcranial material from a single individual.

The Folsom bison was considered more closely related to modern *B. bison* than the specimen from Lone Wolf Creek. The most significant contrast between the Folsom and modern bison was almost entirely a matter of size: "In all cases, the bones of the Folsom skeleton are larger than the corresponding ones of *Bison bison*—on average, 21% larger," they estimated (Hay and Cook 1930:27).

The anatomical contrast Hay and Cook perceived between the Folsom and the Lone Wolf Creek bison was likely accentuated by several factors: the perceived gap in the presumed ages of those two sites (chapter 2); the absence or inaccessibility of a large sample of comparative material, making it difficult for Hay and Cook to appreciate the morphological variability within a population and, thus, leading them to place undue weight on anatomical differences in their small sample; their reliance on crania and especially horn cores, which vary substantially within bison taxa; the attendant difficulty of linking cranial and postcranial materials; and, perhaps, Hay's not being able

TABLE 7.1
Changing Views of the Taxonomic Identity of Bison from Folsom and Lone Wolf Creek

Site	Hay & Cook (1928)	Hay & Cook (1930)	Figgins (1933)	Barbour & Schultz (1936)	Skinner & Kaisen (1947)	McDonald (1981)
Folsom	Bison taylori (CMNH 1236)	B. taylori (CMNH 1236)	Stelabison occidentalis taylori (CMNH 1236)	B. antiquus taylori	B. antiquus figginsi	B. antiquus occidentalis
		Bison sp. indet. (CMNH 1237)	Bison oliverhayi (CMNH 1240)			
Lone Wolf Creek	Bison figginsi (CMNH 574)	Simobison figginsi (CMNH 574)	—	—	B. antiquus figginsi	B. antiquus occidentalis

NOTE: Type specimen number in parentheses, if from these sites.

to spend much time with the actual specimens, since the fossils were in Denver and he was in Washington. Hay had to depend on photographs and measurements of the specimens, though he was able to direct which measurements were to be taken and the camera angles for the photographs (see, e.g., Hay to Figgins, April 25, October 24, 1927, DIR/CMNH).

Hay and Cook not only saw significant differences between the Folsom and the Lone Wolf Creek specimens, but also perceived species-level variation within the Folsom faunal assemblage—particularly among a group of specimens that appeared to be smaller than *B. taylori*. Hay thought the smaller form was similar to *B. occidentalis,* like Williston's (1902) Kansas specimen.[4] Cook was uncertain about that assignment but was sure the differences were "hardly attributable to individual variation" (Cook to Hay, July 7, 1928, March 12, 1929, OPH/SIA; Hay to Brown, June 8, 1928, VP/AMNH; Hay to Cook, March 11, 1927, DIR/CMNH). Brown, in sharp contrast, considered the smaller form merely a female of *B. taylori* (Brown to Hay, June 9, 1928, August 27, 1928, VP/AMNH; see also Brown to Figgins, December 17, 1928, DIR/DMNS; Brown to Frick, August 8, 1928, VP/AMNH; Figgins to Brown, December 13, 1928, DIR/DMNS, and January 15, 1929, VP/AMNH; Figgins to Hay, April 1, 1929, OPH/SIA).

Throughout the winter of 1928 to 1929, Brown and Hay sparred over the question of whether there was one or two species of bison at Folsom. Brown insisted the differences in the skeletal materials were proportionate to the differences between male and female *B. bison.* Hay, in turn, tried to convince Brown that the outward projection of the rim of the eye orbits, present on the smaller specimen, was a feature found only in bulls, and not cows, and thus

the smaller animal at Folsom could not be a *B. taylori* cow. Each urged the other to look at more specimens: Brown suggested that Hay needed to pay more attention to sexual dimorphism. Hay replied that Brown should examine the skulls of modern bison cows, to see that they lacked that orbit projection. Neither convinced the other (Brown to Hay, December 17, 1928, January 10, 12, March 22, April 16, 1929; Hay to Brown, December 17, 1928, January 9, 11, March 11, 21, 1929; all VP/AMNH). At one point, an exasperated Brown wrote: "I do not think it possible under the circumstances that there were two species of bison killed together at this place, it is so unmistakably a slaughter of a herd at the same time that there can be no confusion in regard to the occurrence"(Brown to Hay, January 12, 1929, OPH/SIA). A meeting in February 1929 in Washington did not reconcile their differences, although Brown thought Hay might ultimately come around to his way of thinking (Brown to Figgins, February 1, 1929, DIR/DMNS). He was mistaken. Hay dismissed Brown's arguments with a wave, saying that if somebody reanalyzes the material later and "shows that I am wrong, the credit will be his" (Hay to Brown, March 21, 1929, VP/AMNH; also Hay to Cook, March 30, 1929, OPH/SIA). Privately, however, he wrote to Cook:

> So far as I can see he [Brown] bases his ideas on the supposition that they both belonged to one herd and were killed on the same day. To identify species thus is to ignore structure and to identify on matters which can not be proved, and if proved would not justify the conclusion. There may have been, certainly were, two species in that region. One herd might have been slaughtered on one day, the other next day, the next week, or month, or century, without any geological evidence

being left of the time that passed, or there may have been individuals of two species mingled together in one herd, so far as I know. (Hay to Cook, March 1, 1929, HJC/AGFO; also Hay to Cook, March 30, 1929, OPH/SIA)

Brown tried to convince Hay to postpone publication until all the bison remains recovered during the 1928 season had been worked up, for that collection included female bison (Brown to Hay, January 12, 1929, VP/AMNH). Hay demurred for lack of time (Hay to Brown, March 21, 1929, VP/AMNH), and then privately inquired of Figgins whether Brown had exerted any pressure to delay the appearance of his report, which was slotted to appear in the Colorado Museum's *Proceedings* series (Figgins to Hay, April 1, 1929, OPH/SIA). Hay went ahead with publication, but he did offer Brown a compromise: The smaller "type" specimen (CMNH 1237) was only identified as *Bison* sp. indet., and not as *B. occidentalis* (Hay and Cook 1930:30), apparently because Hay could not fully convince Cook on the point (according to Figgins 1933b:20). Cook had done what Brown had recommended, and examined skeletons of bison bulls and cows, and while he assured Hay he agreed with him, it was "always possible that individuals may find somewhat divergent opinions on the same evidence" (Cook to Hay, March 12, 1929, OPH/SIA).

For his part, Figgins leaned toward Brown's position but assured Hay that he had only limited knowledge of prehistoric bison, and thought it possible that two species could co-occur at the kill (Figgins to Hay, April 1, 1929, OPH/SIA). His professed limited knowledge notwithstanding, a few years later Figgins (1933b:17) published his own taxonomic assessment of the bison in the Colorado Museum collections, one "of the largest . . . ever brought together." Like Hay, he saw two different forms among the Folsom bison, and not only identified them as completely new species, but thought one was a new genus. His basis for doing so were pillars that he spotted on the upper third molars of some of the specimens. Believing that this attribute was not merely an abnormality but instead a taxonomically distinct trait "markedly at variance" and otherwise absent from other "races of *Bison*," he used it to define the new genus, *Stelabison occidentalis* (Figgins 1933b:18).[5]

An examination of the type *B. taylori* skull from Folsom revealed that it had that pillar on the upper third molar and, thus, was a member of *S. occidentalis*. Since the pillar occurred in that specimen in a slightly different form and size, Figgins (1933b:19–20) referred it to a new subspecies, *S. o. taylori*. Figgins did not blame Oliver Hay, who by then had died, for missing such an obviously important anatomical feature, since Hay had only Cook's measurements and photographs to draw up the descriptions of the *B. taylori* type:[6]

This doubtless accounts for [Hay's] omission of reference to so prominent a character as an outer pillar in molar 3, and which appears to have escaped the attention of the junior author. [Cook] (Figgins 1933b:20)

As to the other bison specimen from Folsom that Hay and Cook had left unspecified, Figgins sided with Hay. It was distinct from the other Folsom bison in lacking pillars on the third molars and differed in cranial proportions as well. He therefore assigned it to a new species, *B. oliverhayi* (Figgins 1933b:21). For reasons unclear, at the end of that same paper he identified the specimen as *B. b. oliverhayi* (Figgins 1933b:33). It may have just been a typographical error.

Further study of the type material was not undertaken for another decade, but during that time Barbour and Schultz (1936) used the Folsom material to compare the taxonomic identity of the bison remains from their excavations at the Scottsbluff site (Nebraska). Examining these remains, as well as the type specimens of *B. antiquus,* suggested to them "that *B. taylori* and *B. antiquus* are two very closely related species. Perhaps it would be best to retain the name *taylori* only as a variety since *antiquus* has priority" (Barbour and Schultz 1936:435). They almost went as far as to synonymize *B. antiquus* and *B. occidentalis,* believing that the two species were very similar, but resisted the impulse so as to distinguish between northern (*occidentalis*) and southern (*antiquus*) forms (Barbour and Schultz 1936:435).

A fuller examination of the great majority of bison specimens, archaeological, paleontological, and historic, which included a reanalysis of the Folsom material as well as a thorough revamping of bison taxonomy, was undertaken in the 1940s by Morris Skinner and Ove Kaisen (1947:132), the latter the son of Folsom's 1928 field foreman, Peter Kaisen. Their now-classic work stemmed the tide of taxonomic splitting (McDonald 1981:41). They quickly disposed of the external pillar Figgins found to be so taxonomically significant as "an erratic tendency toward a 'mutation,'" and synonymized *Stelabison* with *Bison* (Skinner and Kaisen 1947:160). Seeing "no difference of specific value" between *B. taylori* and *B. figginsi,* they joined them together under *B. antiquus*. But because those two forms possessed "slightly longer and more posteriorly direct horn cores," Skinner and Kaisen felt that they warranted their own subspecies. After a footnote of worry over whether *B. taylori* or *B. figginsi* would be the more appropriate name for the subspecies, they chose the latter, assigning all the Folsom and Lone Wolf Creek specimens to *B. antiquus figginsi* (Skinner and Kaisen 1947:181–182).[7] As to Figgins' *B. oliverhayi*, their study of the holotype in comparison with the holotype of *B. taylori* showed that the differences between the two "were only those to be expected in individual variants of one population." Since *B. oliverhayi* and *B. taylori* were synonymous, then *B. oliverhayi* could be considered *B. antiquus figginsi* (Skinner and Kaisen 1947:182).

Ultimately, then, by the late 1940s, the discussion reached the conclusion, voiced by Brown two decades earlier, that the individuals within this archaeological population were all members of the same species. There's been little taxonomic dispute since then, at least with regard to the Folsom bison. McDonald's (1981) comprehensive reanalysis of the North American bison comes to the same conclusion. He too synonymized all of the prior genera/species created by Cook, Figgins, and Hay, because they were "based either on minor and taxonomically insignificant individual variations or distorted reconstructions of badly damaged specimens" (McDonald 1981:94).

McDonald differed slightly from Skinner and Kaisen, however, in placing the Folsom and Lone Wolf Creek bison in a different subspecies: *B. antiquus occidentalis* (McDonald 1981:85, table 37). In his view, this was a single, widespread, highly variable taxon, whose distribution ranged from Puebla, Mexico, north to the Yukon Valley of Alaska, between 10,990 and 5,440 years B.P. (McDonald 1981:85–95). Radiocarbon dates on the Folsom site bison bone (chapter 5) fall comfortably within in the early part of this range, and we accept McDonald's taxonomic assessment.

The Structure of the Bonebed

The structure of the bonebed as it appeared in the 1920s is not altogether clear. Save for the 1928 plan map prepared by the American Museum (fig. 4.2) and Kaisen's accompanying tally (table 7.2), and some sketch maps drawn by Frank Figgins and Schwachheim in earlier years, there are few spatial data on the distribution and patterning of skeletal elements. Indeed, Brown (1928) himself grumbled that it was "impossible to say exactly what the position of the skeletons were during the Colorado Museum excavation, as no quarry chart was kept, but presumably the bones were very close together, some of them in skeletal relationship [articulated]." In fairness to the Colorado Museum crew, the "quarry chart" (fig. 4.2) prepared by Brown's crew is also limited in information it provides (chapter 4).

Nonetheless, there are clues in the archives to the distribution and disposition of the bison remains uncovered during the original excavations. For one, it was observed that bison remains were more heavily concentrated in the area of the Colorado Museum excavations than in the more extensive area of the American Museum's excavations. If, indeed, the distribution of the skeletal remains coincides with where the animals fell, this suggests that carcasses were densely packed in this relatively small corner of the site. As it happens, this would be the upstream junction of the paleotributary and paleovalley, and may bear on the question of how and where the bison were trapped and killed (chapter 5).

As to how the remains occurred, the *B. taylori* type specimen (CMNH 1236) was originally identified by Hay and Cook (1928) as a "complete skeleton," but Hay and Cook (1930:27) subsequently corrected themselves:

Examination of field records and data shows that there is no certain association of the type skull and jaws, No. 1236, with the balance of the skeleton. While they were found in close proximity, so also were parts of several other bison skeletons, and so mixed with these as to make the definite association of skeletal parts impossible. A few strings of vertebrae, and some limb and foot bones were so found assembled [articulated] as to make certain their individual association.

This was a point echoed by Figgins. The bison at the site were for the most part "largely disarticulated" (Figgins to Hay, February 18, 1927, DIR/DMNS), though certain elements, most commonly ribs and vertebrae, occurred in articulated position (F. Figgins, field notes, DMNS).

That pattern appears to have been true of the remains uncovered during the 1928 season as well. Although Kaisen reported to Brown that "we are finding nearly complete skeletons. Connected some. . . . Have found four skeletons more or less complete but skulls are gone [on] some of them and two had lower jaws" (Kaisen to Cook, July 8, 1928, HJC/AGFO), his field notes suggest otherwise. And the summary tally Kaisen prepared that accompanies the plan map of the recovered bison (table 7.2) indicates a pattern of incomplete skeletons and mixed elements.

Kaisen also observed that there was variability across the site in the distribution of skeletal elements. When he shifted excavations from the east to the west side of the site in mid-July, 1928, he found

. . . a lot of bones all along on the west side, the side you [Brown] found the skull. With that skull was found the jaws very close to the skull & to east another skull with 3 lower jaws close to skull. . . . You know that small skull Ernie was working on? Close to that was found a large part of a small bison. . . . All along on the west side to where this small skull was on the south is [a] tangle of bones and very numerous. A very large and a small individual are found badly mixed up. . . . How much more than two animals there is I don't know yet, but I saw two other jaws. The limb bones and feet are numerous. It must have been a centipede. (Kaisen to Brown August 19, 1928, VP/AMNH; punctuation added)

Kaisen does not specifically contrast this with the bones on the east side of the site, but certainly the impression one gains from the "quarry chart" drawn in 1928 is that the skeletons found on the east side were more complete (table 7.2; items 4 and 7). Their quarry chart, of course, notes that the "skeletons" on the west side of the site were "more or less mixed" (fig. 4.2). We too observed piles of mandibles in this area of the site, associated with nonarticulated crania, as discussed below.

On the North Bank, Kaisen recorded that the bone layer was discontinuous and that, where bone occurred, it was "piled in a heap." One such heap yielded "some ribs limb bones foot bones and a jaw." He suspected that these elements may have simply washed out from the opposite bank. That inference was based in part on the observation that the floor of the "quarry" formed "a gradual slope from north to south" (Kaisen to Brown, August 29, 1928, VP/AMNH). Kaisen was unaware that the North Bank was a part of the paleovalley or that bison could have drifted in from upstream in that drainage.

Evidence for the slope of the surface on which the bones were found in 1928 can be derived from the data

TABLE 7.2

Kaisen's Record of the Composition, Depth, and (Occasionally) Location of the Bison Remains from the 1928 Excavations at the Folsom Site

Item No.	Description (Verbatim from Kaisen Notes, with Notes in Brackets)	Depth	Location
1	Lower jaws and cerv[ical] vert[ebra]	5 ft	
2	Part of foot—fair		
3	Part of foot—fair		
4	Most of a skeleton—look[s] to be most of the vert[ebral] column, a lot of ribs and some limb bone, foot bones, back of skull, one horn, 1 lower jaw. Bones are the best found in the quarry. The long spines are broken off in the middle.	9 ft	
5	Large part of vert[ebral] column, about 20 vert[ebra]. Part of feet and limb bones. Bones are not so good but will clean up fairly well.	5.5 ft	
6	Rib - may possibly belong to No. 4		
7	Lower jaws, upper teeth, small part of back of skull. Vert[ebral] column up to mid sacrum, some broken ribs on right side, the head of the ribs seems in place.Some limb bones, both scapula. Bones not good but can be worked up. Bone looks better on under side. (*from note in margin*: 1 metapodial in No. 5 belongs to No. 7. 1 radius with string around belongs to No. 7.)	5.5 ft	
8	Part of foot in 2 [partial?] blocks	6 ft	
9	Tibia and foot bones	7 ft 8 in.	
10	Humerus and metapodials	8 ft 6 in.	
11	Part of skeletons. Vert[ebra], sacrum, humerus, ulna & radius, and foot bones. Bones not so good.	6 ft 5 in.	
12	Two skeletons mixed up together [and?] skull, some foot bones close to west bank of Quarry may belong to No. 12. Bones are marked No. 12?	9 ft 3 in	19'N 25'W
13	Pelvis, Vert[ebra], scapula found close to No 11. Could perhaps some of it belong to No. 11 but doubtful.	6 ft 5 in.	
14	Lower jaw, ulna & radius, [perhaps?] other bones. Arrow taken up with the bones. Arrow left is [sic] found.	9 ft 3 in.	
15	Arrow C and parts of skeleton. Arrow left is [sic] found. Bones and arrow taken up together in one block. Part of skeleton was as bad as to be useless, and that part was not collected. Skeleton laying from 4 ft north to arrow.	5 ft 3 in.	10'5"N
16	Skull of young bison and a part of skeleton. A series of vert[ebra], perhaps 20 or so. The[y] are bad but might be used to restore a young Bison and some limb bones.	5 ft 2 in.	9'8"S 14'W
17	Neck and part of dorsals. May possibly be used for No. 16.	5 ft 2 in.	
18	2 skulls and jaws found near No 12. A lot of scattered bones but look like there are a large part of skeleton. Quite a no [number] of ribs. It is large. 3, 4, 5, 6, 7 probable are associated. One skull was badly broken up but may work very well. A part of No 23 is mixed up with it.	5 ft 10 in.	
19	Upper teeth associated		
20	A small bison mixed in with No 18. A lot of associated material.	5 ft 10 in.	
21	A mixed up lot. Some are associated limb bones, foot bones, vert[ebra], jaws & ribs	5 ft 5 in.	
22	Lower jaws	5 ft	
23	= 12? Ribs, humerus, femur (Vert[ebra] 12), jaws, North side of creek	11 ft 9 in.	
24	Jaws large	9 ft 8 in.	63'N 20'W

SOURCE: From *Record of Folsom Bison Quarry*, by P. Kaisen, Department of Vertebrate Paleontology, American Museum of Natural History.

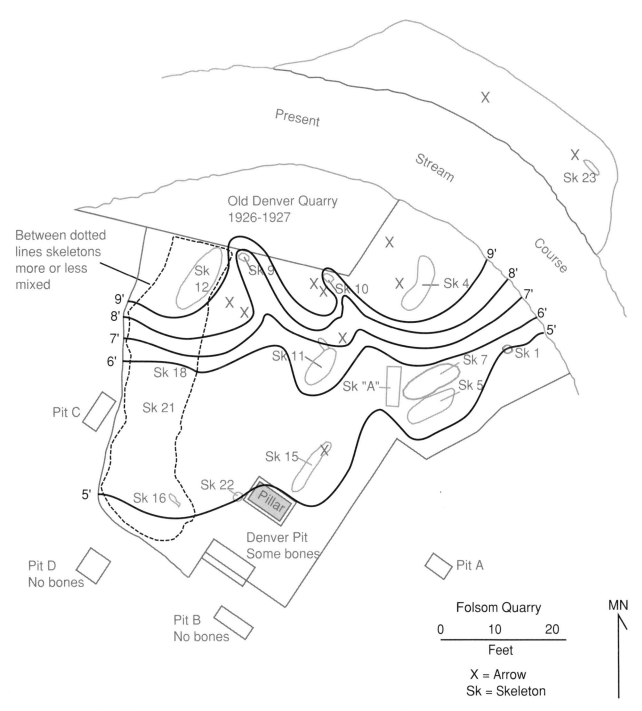

Present

Stream

Course

X

X
Sk 23

Old Denver Quarry
1926-1927

Between dotted
lines skeletons
more or less
mixed

Sk
12

Sk 9

X

X X

X Sk 10

X

Sk 4

9'

8'

7'

6'

5'

9'
8'
7'
6'

X X

X

Sk 18

Sk 11

Sk 7

Sk 1

Sk "A"

Sk 5

Pit C

Sk 21

5'

Sk 16

Sk 22

Sk 15

Pillar

Pit D
No bones

Denver Pit
Some bones

Pit A

Pit B
No bones

Folsom Quarry

MN

0   10   20

Feet

X = Arrow
Sk = Skeleton

FIGURE 7.1 AMNH 1928 excavation plan map showing contours in feet below the modern surface of the floor on which the bison bones were resting. The data on depths below surface are provided in Table 7.2 and Table 8.2.

provided by Kaisen on the depth of the skeletal remains and projectile points found on the site. The skeletal data are provided in table 7.2 and used in figure 7.1 to create contours of the surface of the bonebed, superimposed on the 1928 AMNH plan map.

In general, skeletal elements recovered on the North Bank of the arroyo were deeper than those on the South Bank. The bonebed was within 1.5 m of the surface at the upper end of the paleotributary, and nearly 3.65 m below the sur-

face some 25 m away, at the point where the paleotributary joins the paleovalley—a vertical difference of just over 2 m. Our recent excavations support this: On average, the elevation of bone on the North Bank was 2.5 m below the level of the bonebed on the South Bank some 15 m away (chapter 5). Elevations recorded on skeletal remains recovered in 1928 suggest that the surface on which they rested was relatively level over a long reach, but, as shown in figure 7.1, began to drop off toward the junction with the paleovalley.

Taphonomy and Bone Preservation

There were also apparent differences in the condition of the bone across the site. One of the reasons Kaisen shifted over to the west side was that the faunal remains on the east side were running out and, in any case, were proving "too bad to collect" (Kaisen to Brown July 8, 1928; Kaisen to Cook, July 8, 1928, HJC/AGFO). Yet, poorly preserved bone was not restricted to the east side of the site. According to Cook (1928b:38), the bones uncovered during the 1926 to 1927 Colorado Museum excavations on the west side "were extremely fragile, and great care and skill were required to collect, prepare and restore them for mounting." Similarly, Kaisen reported that

> the bones are very bad on the top. Just like powder. But the undersides are good. It looks like they might have been exposed on the upper side for some time before they was covered over. . . . That is all along the west side. . . . On the south the bones are very bad and broken. (Kaisen to Brown August 19, 1928, VP/AMNH, punctuation added)

Brown was interested in this "weathered bone," for he realized it might bear on the question of whether all the animals were killed at the same instant. If there were significant differences in weathering, that might "clear up the matter of whether there were not skeletons lying around the waterhole before the large kill" (Brown to Kaisen, August 27, 1928, VP/AMNH). He would ultimately conclude that all the bones were from a single episode (Brown 1928b).

When Barbour and Schultz (1936:435) examined the Folsom bone as part of their analysis of the Scottsbluff bison, they too were struck by its condition: "It is unfortunate that the material from the Folsom quarry was crushed and so poorly preserved. Much Plaster of Paris was used in the restoration of both the Colorado Museum and American Museum collections of bison skulls from the Folsom Bison Quarry. Because of this, it is almost impossible to get accurate measurements of the skulls." It appears, however, that they focused on the crania, which may well have been in worse condition and likely to have received more attention, as well as plaster and plastic wood, from the excavators and preparators.

In fact, the Folsom bison bone from the 1920s fieldwork is in excellent condition, even granting the excavation, preservation, and conservation techniques in use at the time. In a minority of cases, alterations made in the field or by museum preparators obscure the bone surface. For example, excess shellac was occasionally applied on-site; in the museum broken or missing portions of bone elements were sometimes repaired or replaced with plaster or plastic wood, which might then be sanded. Fortunately, the latter practice mostly occurred on elements destined for the display skeletons and was rare on the research specimens.

Clark Wissler thought, based on his examination of the site in August of 1928, that the bone surfaces were not weathered and, therefore, "could not have been uncovered long" before they were buried (Wissler, Field Diary, August 2–10, 1928, ANTH/AMNH). We concur.

As to other factors that may have affected the condition and preservation of the faunal remains, Cook (1927a:245) believed he saw evidence for postmortem attrition and damage in the apparent evidence of trampling, marked by a scapula with "a plain footmark stamped out of it and driven down into the matrix below, where another bison had stepped on it while it lay in the mud." We saw the very same feature on a scapula uncovered during our excavations but are inclined to attribute that damage to the susceptibility of this very thin platy bone to natural breakage from any of a variety of sources, including rodent tunneling. It seems highly unlikely such damage could have taken place when the animal was alive and its carcass yet to be cleaned of hide, flesh, and muscle. It should only occur if the bone lay exposed on the surface and was trampled during later use of the site.

However, there is no evidence we have seen—or that was evident during the original investigations—of more than a single kill occurring on this spot. Brown was quite explicit on the point:

> The bones all show the same degree of fossilization, were lying as carcasses would lie, close together in a kill, and were evidently all killed at the same time. Only in one or two cases were bones lying on top of each other, although they were evidently lying on an irregular surface. (Brown 1928b; also Brown to Hay, January 10, 1029, OPH/SIA)

Others concurred (e.g., Wissler, field diary, August 2–10, 1928, ANTH/AMNH).

Numerical Matters I

Little attention was paid to quantitative measures of the assemblage in the 1920s, and in fact there exists no contemporary inventory of the number skeletal elements recovered.[8] Contemporary analytical attention to quantitative matters, such as it was, focused instead on how many individual animals were represented at the "bone quarry."

After his first look at the skeletal material in the fall of 1927, Brown (1928a:826) supposed that the Colorado Museum had recovered remains of eight animals, seven bulls, and one cow. Half a year later, he revised that estimate upward to at least 16 animals, based on the presence of 32 calcanea in the Colorado Museum collections that, added to the 14 whole and partial skeletons his crews excavated in the summer of 1928, put the total number of animals killed at Folsom at 30 (Brown 1928b; Brown to Kaisen, September 4, 1928, VP/AMNH). A few months later (December, 1928), he put the number still higher, at "40 to 50 skeletons" (Brown 1929; also, Brown to Hay, January 10, 1929, OPH/SIA). It is unclear on what evidence he did so. But like the stock market, Brown's estimates fell after 1929, to just 30 to 40 specimens (Brown 1936). That estimate was echoed by Cook (1931b:102), who put the number of skeletons at "more than 30 individuals" (in their reports on the Folsom fauna, Hay and Cook [1928, 1930] give no estimates of the number of animals).

Bison Killing, Butchering, and Processing

There was a difference of opinion among the original investigators as to whether the kill had taken place at Folsom or elsewhere. Cook (1928b:40) raised the possibility that an animal "might be shot in countless places by an arrow, and still escape his human enemy to carry the points away in his wounds." Modern buffalo "like to wallow and roll in mudholes" and, more important, "wounded, feverish animals would naturally seek cool waters to lie down in when suffering from such wounds" (Cook 1928b:40). Believing, as he did, that "swampy, marshy conditions" characterized the Folsom site at the time (Cook 1927a:244), Cook (1928b:40) surmised that it may have been the last refuge of the dying bison: "No doubt some of the animals whose skeletons we find here did just this [fled to this cool, wet area with spears lodged in their bodies] and, lying down, never got up again." Thus, Folsom was not the location of the kill, but the final resting place of the dying prey.

Brown and Frank Roberts thought otherwise: "It seems more than likely that a number of people surrounded the herd of buffalo, shooting them down at close range. . . . Possibly brush fires were used on one side to hold the herd; a hundred yards below the quarry there are many exposures of charcoal, not campfire sites, that are probably contemporaneous with the kill" (Brown 1928b; see also Roberts 1936:14–15). More possibilities and details of what Brown and others were thinking are evident in the field notes of Wissler for the week in August of 1928 when he was on the site. As he was there with Brown, as well as Kirk Bryan and Earl Morris, his thoughts may reflect their joint discussions (Wissler, field diary, August 2–10, 1928, ANTH/AMNH).

Bison hunting technique:

1. Simultaneous kill. Proof—position.
2. Spot not a natural trap. Proof—no other bones.
3. Only one kill at the spot. Proof—see 1.
4. Probable method: Some means of holding the bison. Pound (ice), bog, snow, fire. Two men, or one, could have done it, except in case of pound. Points suggest two men. Possibility of ambush, at water hole, over a period of a day or two.
5. A bison hunting culture represented? [Proof] Uniqueness of points.
6. Problem—to find that culture.
7. Other evidences of man—the fire in the back of the stream.

Once the animals were killed, Brown and others thought the carcasses were processed for their meat as well as their hides. The latter inference was based on the observation that "very few tail bones were present, indicating that the tails were taken off with the hides" (Brown 1928b; also Bryan 1937; Wissler, field diary, August 2–10, 1928; cf. Wheat 1972:102).

What Roberts thought at the time of his 1927 and 1928 visits was not known, but his later excavations at Lindenmeier gave him the opportunity to consider the nature of the kill at Folsom. Like Cook, he believed that the kill took place around a water hole or marshy spot and that, after the kill, "as much of the flesh as could be carried away had been removed from the carcasses, [and] they were left to sink in the mire" (Roberts 1936:14).

On the Utility of the Collections from the Original Investigations

The bison remains acquired during the original Colorado and American Museum excavations are inadequate for some analytical tasks, such as fine-scale spatial analyses. They are nonetheless of considerable value for many other analyses, for several reasons. For one, these are large assemblages: From 1926 to 1928, more than 3,600 skeletal elements were recovered from Folsom, of which 3,115 elements are identifiable. In addition, they appear to be relatively complete. As best we can determine, all or nearly all of the bone from these excavations found its way into one of these two institutions, and since then the collections have remained mostly intact. There were some exchanges of material, but it was not a wholesale practice and took place mostly early on between the Colorado and the American museums, as each institution sought to piece together entire skeletons for display.[9] There is no record of skeletal elements being loaned or given to other institutions, and even if the latter did occur, as was not unusual in those days, the numbers must have been relatively small and should not unduly bias the sample.

That said, because of the exchanges between Denver and New York, it might be impossible to determine exactly how many elements came from the 1926 to 1927 excavations as opposed to the 1928 excavations. Still, the tallies should be reasonably accurate, and the combined total from the two institutions ought to be very close to the number of specimens recovered during the original three years of excavations.

Although the Folsom faunal collections at these institutions are well curated, there is occasionally ambiguity as to whether a specific element came from Folsom. For example, at the Denver Museum some of the Lone Wolf Creek specimens are in drawers with Folsom bone, but these are problems relatively easy to correct, in part given the very distinctive look and feel of the Folsom site bone.

A New Look at an Old Bison Bonebed

Assessing the Sample

The broader tactics we used in the field for the collection of the bison bone are detailed in chapter 4 and appendix A. As noted, skeletal elements, segments, and portions were identified and recorded using a coding system slightly modified from one developed for use on Paleoindian bison bonebeds on the Plains (Todd 1987a:table 6.2; also Hill 2001; Kreutzer 1996). Because those codes are used in many of the tables that follow, they are listed here in table 7.3.

We also collected data in the field to address a variety of taphonomic questions. These included observations of the following:

- the surface condition of the bone that might be attributable to exposure, animal activity, geomorphic processes, and, of course, human butchering and processing;
- the long-axis orientation and inclination of each bone and bone fragment, to determine whether the bones are positioned in ways that suggest postdepositional orientation or transport by water;
- the horizontal and vertical distribution of shingle shale across the excavation blocks in the tributary, to ascertain what effect episodes of sheet wash may have had on the underlying bonebed in the M17 area and whether the movement of the shingle across the bone deposit was responsible for removing bison remains from the M15 area; and
- data on the sediments in the various excavation areas, to determine whether the presence or absence of bone might be explained by differences in preservation potential as mediated by sediment chemistry.

The bison remains we recovered were subject to the same examination and analysis as conducted on the bone curated at the various museums (as detailed in chapter 4).

In the portion of the bonebed uncovered in our excavations, the bone was distributed in a roughly oval pattern (fig. 7.2A) and contained several hundred elements, including three virtually complete crania, a series of articulated vertebrae, and a miscellany of other skeletal parts. Additional faunal remains were recovered in a few other excavation units on the South Bank and in several test units on the North Bank. In addition, approximately 200 bones, primarily small (<5 cm) fragments, were mapped and collected over the several years during and following our excavations as they eroded out onto the surface of the site. Virtually all of these are small, fragmentary, and, thus, often unidentifiable. All together, 687 skeletal elements were recovered in our excavations and surface mapping, of which 380 were identifiable as bison specimens and thus analytically useful here. Most of the unidentifiable fragments are presumed to be bison. By itself, this is a relatively small sample, and it came from what may have been the margin of the bonebed. As a result, it is not necessarily representative or sufficient for certain analytical procedures.

Therefore, the investigations reported on in the remainder of this chapter also include, where the data are available and appropriate, the results of our reanalysis of the faunal collections from the Colorado and American museum investigations, as well as the cranium fragment collected by E. Baker in 1936,[10] a radius collected by C. V. Haynes in 1970, and the complete cranium recovered in 1972 by Trinidad State Junior College. We believe combining these separate samples is justified. They were all found in a common stratigraphic context; were derived from what appears to be the same, single event; and show no evidence of intrasite stratigraphic differences suggestive of multiple kills (chapter 5 and below).

That said, however, the AMNH map and Kaisen's field notes implied that the 1928 excavations yielded more or less complete skeletons. Yet, our excavations did not. That apparent difference raises the question of whether our respective excavations took place in previously undetected or spatially distinct activity areas within the bonebed, or whether that difference might be more apparent than real. With regard to the latter, in the 1920s the analytical focus was on skeletons, rather than skeletal parts and their taphonomic history (chapter 4), and as a result, a cluster of bones was commonly referred to or mapped as "a skeleton," rather than as individual elements.

Since no maps of sufficient detail exist from the original investigations to enable us to assess the precise position of the skeletal elements, we turned to our inventories of the bone recovered from 1926 to 1928 and from 1997 to 1999 (table 7.4) to examine whether statistical differences exist in the frequency distribution of specimens, as a proxy measure of the similarity of the respective assemblages.

When the number of identified specimen (NISP) values from the CMNH, the AMNH, and our own excavations are plotted (fig. 7.3) and statistically compared, it is immediately clear that the several assemblages are, in fact, quite similar. The Spearman rank-order correlations among the several assemblages range from $r_s = 0.653$ (CMNH × QUEST), to $r_s = 0.709$ (AMNH × QUEST), to $r_s = 0.873$ (AMNH × CMNH) ($t = 5.91$, 6.89, and 12.27, respectively, all correlations being statistically significant at $p = 0.001$; $n = 49$).

The primary difference evident among the several samples is their size: The AMNH remains are from a much more extensive excavation area and comprise a faunal sample 1.65 times larger in terms of NISP (although not in terms of minimum number of individuals; MNI) than the CMNH sample, and roughly five times larger than the sample recovered in our excavations. Despite the apparently more destructive nature of the 1920s excavations, there was no appreciable difference in the relative recovery rate of complete elements. However, there is a significant difference in terms of which elements tend to occur more often than would be expected in each of the collections (table 7.4). The initial conclusion to be drawn from this observation is that, despite these having been referred to as "skeletons" on the 1928 plan map, the differences in element frequencies, including the significant underrepresentation of some elements and overrepresentation of others, indicate that the 1920s excavations did not yield many whole skeletons.

Looking further into table 7.4, we can see interesting patterns in the cell-by-cell differences in the occurrence of specific elements. For example, phalanges and metapodials are

TABLE 7.3
Descriptive Codes Used in the Folsom Faunal Analyses

(a) Element Codes Used in the Folsom Analyses

Cranium/Mandible

CRN—cranium	MR—mandible	HY—hyoid	TFR—unidentified tooth fragment	HS—horn sheath

Axial Skeleton

AT—atlas	AX—axis	CE3-7—specific cervical	CE—unidentified cervical	TH1-14—specific thoracic
TH—unidentified thoracic	LM1-5—specific lumbar	LM—unidentified lumbar	SAC—complete sacrum	CA—caudal
VT—unidentified vertebra	CS—costal cartilage	SN—unidentified sternal element	MN—manubrium	ZY—xyphoid process
RB1-14—specific rib	RB—unidentified rib			

Appendicular Skeleton, Forelimb

SC—scapula	HM—humerus	RD—radius	UL—ulna	RDU—radius-ulna
MC—metacarpal	MCF—5th metacarpal	CP—unidentified carpal	CPU—ulnar carpal	CPI—intermediate carpal
CPR—radial carpal	CPS—fused 2nd and 3rd carpal	CPF—4th carpal	CPA—accessory carpal	

Appendicular Skeleton, Hindlimb

IM—os coxae	PV—complete pelvis	FM—femur	PT—patella	TA—tibia
LTM—lateral malleolus	MT—metatarsal	MTS—2nd metatarsal	CL—calcaneus	AS—astragalus
TRC—fused central and 4th tarsal	TRS—fused 2nd and 3rd tarsal	TRF—1st tarsal	TR—unidentified tarsal	

Appendicular Skeleton, Other

PHF—1st phalanx	PHS—2nd phalanx	PHT—3rd phalanx	PH—unidentified phalanx	SEP—proximal sesamoid
SED—distal sesamoid	SE—unidentified sesamoid	MP—unidentified metapodial		

Fragments

US—totally unidentified fragment	LB—unidentified long bone	FB— unidentified flat bone	CB— unidentified cancellous bone

---

(b) Portion Codes Used in the Folsom Analyses

Cranium

PAR—parietal	FN—frontal	ZGO—zygomatic	LC—lacrimal	INV—incisive
NSL—nasal	OCC—occipital	JUG—jugular process	TMP—temporal	PAL—palatine
MX—maxilla	HC—horn core	PET—petrous	SR—skull roof (FN & HC)	BRC—braincase (FN & OCC)
SKO—other skull fragment				

Mandible

HRM—horizontal ramus	DRM—dentary ramus	RAM—ascending ramus	DAM—DRM & RAM	SYM—symphysis
TW—tooth row	BRD—ventral border	COR—coronoid	DYS—dyastema	

(continued)

TABLE 7.3 (*Continued*)

Teeth (Except for IC, Use for CRN and MR)

P2-P4—specific premolar	PML—unidentified premolar	M1–M3—specific molar	ML—unidentified molar	DP2–DP4—specific deciduous premolar
DML—unidentified deciduous premolar	IC—incisor	EN—tooth enamel	RT—root fragment	

Os Coxae (IM)

IL—ilium	IS—ischium	PB—pubis	AC—acetabulum	ACS—AC & IS
ACL—AC & IL	ACP—AC & PB	ICL—cranial IL	ILD—caudal IL	ISC—cranial IS
ISD—caudal IS	PBS—pubic symphysis			

Long Bones and General Codes

PR—proximal	PRS—proximal end & <½ of the shaft	PSH—proximal end & >½ of the shaft	DS—distal end	DSS—distal end & <½ of the shaft
DSH—distal end & >½ of the shaft	EP—unidentified epiphysis	PRE—proximal epiphysis	DSE—distal epiphysis	DF—diaphysis
DPR—proximal diaphysis	DDS—distal diaphysis	DFP—diaphysis & fused PRE	DFD—diaphysis & fused DSE	SH—long bone shaft
FK—bone flake, <½ the circumference of SH	HE—head	CDL—condyle	IFK—bone impact flake	US—unspecified

Scapula (SC)

GN—glenoid	GNB—glenoid & blade	GS—glenoid & spine	BL—blade fragment	SP—spine fragment

Ulna (UL)

ANC—trochlear notch	OLC—olecranon portion	PRE—olecranon tuber

Vertebrae

CN—centrum	CNN—CN & neural arch	CNS—CN & dorsal spine	CNW—CN and wings (AT only)	CNT—CN minus TSP
CAN—CN and AP	NAS—neural arch & SP	DRE—dorsal spinous EP	SP—dorsal spinous process	TSP—transverse spinous process
AP—articular process	AEP—anterior epiphysis	PEP—posterior epiphysis		

Hyoid (HY)

ANG—angle	BDY—body/blade

(c) Segment Codes Used in the Folsom Analyses

CO—complete	PR—proximal	DS—distal	LT—lateral	ME—medial
CR—cranial (anterior)	CD—caudal (posterior)	DR—dorsal	VN—ventral	L—left
R—right	EG—edge	PL—posterolateral	PM—posteromedial	AM—anteromedial
AL—anterolateral	FO—fore	HD—hind	IN—interior	EX—exterior
END—end	CDL—condyle	FR—unidentified fragment	NAP—unspecified segment	EN—tooth enamel
OCL—occlusal surface (of teeth)	HB—split rib blade	ACT—acetabular (IM)	TM—teres major	DT—deltoid tuberosity (HM)
SF—supracondaloid fossa (FM)	MO—minor trochanter (FM)			

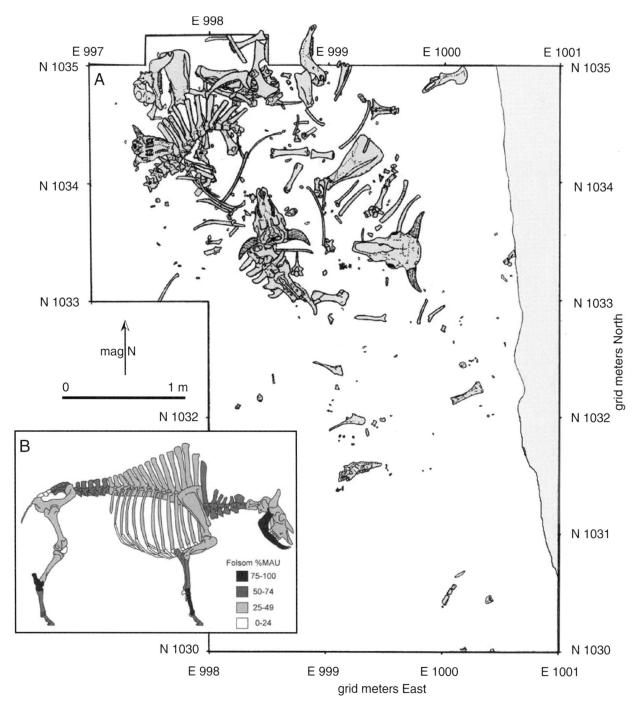

FIGURE 7.2 A. Portion of the bonebed exposed in the excavations in the M17 and adjoining excavation blocks, 1998–1999. The area shown in gray was excavated in 1928. B. Folsom bison skeletal element frequencies (from all excavations) presented as %MAU.

consistently overrepresented in the CMNH collection and significantly underrepresented in the SMU/QUEST collection (note too the significant underrepresentation of astragalii), as measured by their Freeman-Tukey deviates. Two possible explanations come to mind, and these need not be mutually exclusive. First, the pattern may reflect preservation and collection; notably, that higher density and more easily identifiable elements tend to appear more readily in collections made under field techniques of the 1920s. There is a significant overrepresentation of tooth fragments, ribs, and thoracic vertebrae in our inventory, compared to the numbers from the 1920s excavations. These are items that easily fragment and would not have been of particular interest in the 1920s, given the goals and nature of their field techniques.

In apparent exception to that pattern, however, some otherwise fragile and possibly easily overlooked elements,

TABLE 7.4
NISP from Excavations and Collections Made in 1926–1927 (CMNH), 1928 (AMNH),
and 1997–2002 (QUEST), Ordered in Decreasing Total Frequency

Element	Code	CMNH	AMNH	SMU/QUEST	Total
Rib	RB	**154 (−2.27)**	261(−1.33)	**109 (5.78)**	524
Thoracic vertebra	TH	89 (−1.48)	152 (−0.61)	55 (3.50)	296
Proximal sesamoid	SEP	87 (−0.34)	148 (0.74)	23 (−0.96)	258
1st phalanx	PHF	**85 (1.84)**	108 (0.17)	**4 (−5.08)**	197
2nd phalanx	PHS	**90 (2.71)**	95 (−0.69)	**4 (−4.90)**	189
Cranium	CRN	**28 (−4.29)**	**114 (2.84)**	17 (−0.02)	159
3rd phalanx	PHT	**69 (2.13)**	74 (−0.77)	**7 (−2.67)**	150
Mandible	MR	53 (0.14)	78 (−0.25)	18 (0.48)	149
Lumbar vertebra	LM	48 (0.60)	61 (0.85)	17 (0.89)	126
Cervical vertebra	CE	**26 (−3.04)**	**86 (2.07)**	14 (0.14)	126
Distal sesamoid	SED	31 (−0.06)	54 (0.78)	**5 (−1.66)**	90
Os coxae	IM	**38 (2.59)**	**25 (−2.07)**	5 (−0.85)	68
Calcaneus	CL	25 (0.57)	33 (−0.23)	6 (−0.28)	64
Astragalus	AS	29 (1.56)	31 (−0.30)	**1 (−2.84)**	61
Metatarsal	MT	23 (0.54)	31 (−0.11)	5 (−0.48)	59
Tibia	TA	26 (1.45)	25 (−0.85)	4 (−0.76)	55
Fused 2nd & 3rd carpal	CPS	20 (0.29)	30 (0.20)	4 (−0.72)	54
Radial carpal	CPR	22 (0.97)	26 (−0.25)	3 (−1.09)	51
Metacarpal	MC	21 (0.76)	28 (0.13)	**2 (−1.67)**	51
Radius	RD	16 (−0.14)	25 (−0.14)	7 (0.79)	48
4th carpal	CPF	16 (−0.06)	30 (0.92)	**1 (−2.22)**	47
Ulnar carpal	CPU	14 (−0.48)	29 (0.84)	3 (−0.86)	46
Intermediate carpal	CPI	17 (0.27)	27 (0.47)	2 (−1.44)	46
Fused central & 4th tarsal	TRC	20 (1.05)	24 (−0.01)	**1 (−2.13)**	45
Femur	FM	13 (−0.48)	28 (0.98)	2 (−1.30)	43
Accessory carpal	CPA	17 (0.90)	20 (−0.18)	2 (–1.10)	39
Fused 2nd & 3rd tarsal	TRS	13 (−0.02)	22 (0.37)	3 (−0.46)	38
Scapula	SC	**6 (–2.18)**	27 (1.49)	4 (0.10)	37
Ulna	UL	14 (0.34)	20 (0.06)	3 (−0.41)	37
Vertebra	VT	18 (1.43)	**10 (−2.40)**	**8 (1.74)**	36
Caudal vertebra	CA	**7 (−1.70)**	25 (1.22)	4 (0.15)	36
Humerus	HM	**6 (−1.78)**	**25 (1.59)**	2 (−0.78)	33
Lateral malleolus	LTM	9 (−0.61)	21 (0.90)	2 (−0.72)	32
Axis	AX	7 (−0.65)	16 (0.56)	3 (0.22)	26
Tooth fragments	TFR	**0 (−4.89)**	**6 (−2.18)**	**18 (5.22)**	24
Sacrum	SAC	5 (−1.08)	14 (0.49)	4 (0.92)	23
5th metacarpal	MCF	6 (−0.03)	11 (0.46)	1 (−0.56)	18
Metapodial	MP	**13 (2.22)**	**5 (−1.63)**	**0 (−1.97)**	18
2nd metatarsal	MTS	4 (−0.46)	8 (0.05)	3 (0.99)	15
Atlas	AT	3 (−0.81)	10 (0.89)	1 (−0.25)	14
Patella	PT	5 (0.14)	8 (0.24)	1 (−0.25)	14
Radius/ulna	RDU	4 (−0.31)	10 (0.89)	**0 (−1.67)**	14
Hyoid	HY	**8 (1.61)**	3 (−1.46)	1 (−0.08)	12
1st tarsal	TRF	5 (0.63)	4 (−0.74)	2 (0.74)	11
Dew claws	DC	**7 (1.98)**	**1 (−1.86)**	0 (−1.12)	8
Sternal element	SN	**5 (1.85)**	**0 (−2.44)**	0 (−0.78)	5
Costal cartilage	CS	2 (1.20)	0 (−1.31)	0 (−0.37)	2
Manubrium	MN	1 (0.86)	0 (−0.78)	0 (−0.20)	1
Xyphoid process	ZY	1 (0.86)	0 (–0.78)	0 (−0.20)	1
Subtotal of identified specimens		1,226	1,889	381	3,496
Unknown		386	181	305	872
Total, all specimens		1,612	2,070	686	4,368

NOTE: Freeman-Tukey deviates in parentheses. Bold cells are significant at $p = 0.05$ level (± 1.583).

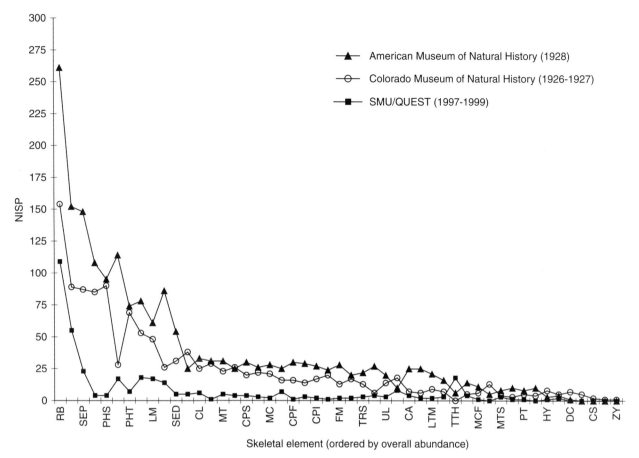

FIGURE 7.3 Histogram showing the comparative skeletal element frequencies (NISP) from the Colorado Museum, American Museum, and the SMU/QUEST excavations.

such as hyoids, costal cartilage, dew claws, and sternal elements, occur at a much higher than expected frequency in both the Colorado and the American museum collections. Yet, this too may reflect the excavation practice of the 1920s, especially among paleontologists, of jacketing large blocks of sediment—note the high representation of crania in the AMNH collection—for subsequent excavation under controlled laboratory conditions.

This suggests a second possible explanation, notably that such differences in element frequency may also represent meaningful spatial patterns in prehistoric activity. Thus, and as can be seen in many sites, there are differences in skeletal completeness and element representation across a bonebed (Johnson 1987; Kreutzer 1996; Todd 1987a). Again, and unfortunately, the absence of detailed provenience information precludes us from demonstrating this, however, the significant differences in element occurrences may be signaling that fact. Sternebrae, dew claws, and hyoids are overrepresented in the 1920s excavations, despite their fragility and small size. Yet, our excavations, which involved more careful and more intensive recovery methods, produced only a single hyoid, no dew claws, and no sternal elements. Also significantly underrepresented in our excavations are the extremely dense and easily identifiable elements, such

as phalanges and astragalii. A final observation on the data in table 7.4: In all collections mandibles occur at about the frequency one would expect by chance. Clearly, this element was not differentially distributed.

Although there are potential intrasite differences in spatial patterns, as well as recovery, collection, and curation, no site is completely homogeneous spatially, and those differences that occur are not so significant as to warrant separate analytical treatment. Since we seek to understand the site as a whole, and lack spatial and provenience data to treat intrasite variability, it seems statistically appropriate to combine the faunal assemblages from the various excavations to provide a fuller and more representative picture of the assemblage, with the explicit caveat that even this combined sample may represent only a portion of the total bonebed that once existed.

Exploring the Taphonomic History of the Folsom Bison Bonebed

Skeletal preservation in the area of our excavation is excellent. Cortical surfaces of the bone show only minimal deterioration, and the assemblage contains elements not often preserved in Paleoindian contexts, including a complete

TABLE 7.5
Weathering Stages of Folsom Bison Bone

Wx Stage	Description (Todd et al. 1987:123)	CMNH	AMNH	SMU/ QUEST	Total	Percentage	Cumulative Percentage
0	Unweathered, fatty	0	0	0	0	0	0
1	Unweathered, dry	999	142	159	1,300	75.93	75.93
2	Limited surface weathering, some longitudinal cracking	97	146	53	296	17.29	93.22
3	Lght surface flaking, deeper cracking	42	53	8	103	6.02	99.24
4	Patches of fibrous bone with moderate flaking and cracking	2	5	4	11	0.64	99.88
5	Deep cracking and extensive flaking	0	2	0	2	0.12	100.00
6	Bone falling apart	0	0	0	0	0.00	100.00
	Total	1,140	348	224	1,712		

calf cranium, still possessing horn cores and incisives (fig. 4.16). Among the only other Paleoindian localities with comparable bone preservation are the Folsom component at the Agate Basin site (Hill 2001:58), and the Cooper site (Bement 1999b). There is little a priori reason to expect that the Folsom bison remains spent a great deal of time exposed on the surface or otherwise experienced significant postdepositional or postburial modification.

Nonetheless, before assuming that the character of a faunal assemblage is solely the result of human activity, it is necessary to explore the possible effects of natural processes on this assemblage (Todd 1987a:110–112). We therefore turn to a consideration of bonebed taphonomy, to explore what effects, if any, natural processes such as surface exposure, carnivore action, and fluvial or other transport processes may have had on the carcasses and bones while they were exposed on the ground surface, and of what sedimentary processes (if any) might have affected the bones once they were buried. It is important to stress, of course, that our understanding of the site's taphonomic history is necessarily derived from the rather limited window our excavations provided, supplemented by information that could be gleaned from the analysis of the collections from the 1920s.

HOW LONG DID THE BISON BONE LAY ON THE SURFACE PRIOR TO INITIAL BURIAL?

The subaerial weathering indicated on the bone in the Colorado and American museum collections, as well as that recovered in our more recent investigations, was tallied using the weathering stages of Behrensmeyer (1978) as modified by Todd (1987a:123). Of the 1,712 identifiable

elements for which data are available, the great majority (1,300 elements, or 75.9%) fall into weathering stage 1, while 99.2% of the bones were in weathering Stages 1 through 3 (table 7.5).

Of the bones that show surface deterioration at a greater than expected frequency, that is, Stage ≥ 2 ($n = 412$), weathering is dominantly unilateral; that is, one side was exposed, while the other was buried, and thus one side would be at Stage ≥ 2, while the other would be at Stage 1 (a pattern that appears to occur generally [Behrensmeyer 1978:154]). Only two elements, a thoracic vertebra and a metatarsal found in 1928, were weathered on all surfaces, suggesting that they either were not completely buried or were buried and then reexposed at a later time (e.g., Frison and Todd 1986:39; Hill and Hofman 1997:74).

Additional, but somewhat ambiguous information on the rapidity of burial comes from root etching on the bone (Lyman 1994:375–376). Ambiguity arises from uncertainties about the mechanical and chemical aspects of root etching, the kinds of vegetation and ground cover that produce it, and even whether the bones must be buried, and how deeply, before root etching occurs (Lyman 1994:375–377; also Hill 2001:26). Hill and Hofman (1997:75) observe that in faunal assemblages buried deeply (e.g., Waugh and Twelve Mile Creek), root etching is minimal, while there is more intensive root etching in those assemblages found closer to the surface, such as at Lipscomb. A similar pattern may occur at Cooper, where the bones in the upper part of the deposit show greater etching (Bement 1999b:table 31). Of course, depth of burial is not the same as rate of burial (cf. Hill and Hofman 1997:75). Nonetheless it seems reasonable to argue that if bone is in a vegetated environment likely to produce etching, then such ought to appear if the

TABLE 7.6
Root Etching on Folsom Bison Bone

Area of Bone Surface Showing Root Etching	n	Percentage	Cumulative Percentage
0%	240	16.43	16.43
1%–10%	1,130	77.34	93.77
11%–20%	39	2.67	96.44
21%–30%	23	1.57	98.02
31%–40%	10	0.68	98.70
41%–50%	5	0.34	99.04
51%–60%	9	0.62	99.66
61%–70%	2	0.14	99.79
71%–80%	3	0.21	100.00
>81%	0	0	
Total	1,461		

TABLE 7.7
Root Etching by Weathering Stage on Folsom Bison Bone

Amount of Root Etching	Wx Stage 1	Wx Stage 2	Wx Stage 3	Wx Stage 4	Wx Stage 5	No Data	Total
0%	186	42	8	4	0	0	240
1%–10%	897	162	57	4	2	8	1,122
11%–20%	17	17	3	0	0	2	37
21%–30%	4	13	6	0	0	0	23
31%–40%	5	3	1	1	0	0	10
41%–50%	2	2	0	0	0	0	5
51%–60%	3	4	1	0	0	0	9
61%–70%	4	1	0	0	0	0	2
71%–80%	1	3	0	0	0	0	3
>81%	0	0	0	0	0	0	0
Total	1,117	247	76	9	2	10	1,451

bone was within the root zone for any appreciable period of time. Obviously, this is contingent on the depth of the root zone, which can vary considerably. The data for root etching on the Folsom bone are reported in table 7.6; the classes of etching represent the area of the bone surface covered by etching (root etching stages after Todd 1993:fig. 75; also Hill 2001:268)

As is apparent from table 7.6, the Folsom faunal assemblage shows relatively little evidence for root etching. On nearly 95% of the bone, no more than 10% of the area shows root etching. It is of some interest to cross-tabulate root etching versus bone weathering. Presumably, if the amount of root etching is a function of the rapidity of burial, it ought to correlate with the weathering stage, on the assumption that bone that lay close to or exposed on the surface for an extended period of time, and thus potentially subject to longer surface weathering, would also have been

in relatively close contact with the surface vegetation. These data are listed in table 7.7.

Contingency table analysis of the data in table 7.7, excluding the "No Data" column and the ">81%" row under "Amount of Root Etching," indicates that root etching and weathering stage are not independent, as measured by the likelihood ratio chi-square statistic ($G^2 = 95.033$, df = 32, $p = 0.000$). Those results are likely influenced by the fact that the vast majority of the bones for which we have data both are lightly weathered (Stage 1) and have little root etching ($\leq 10\%$).

In order to probe the relationship further, we collapsed these data into two columns, one of weathering Stage 1, the other with all bones showing weathering Stage ≥ 2. Doing so offset the large number of bones in weathering Stage 1 and reduced the number of cells with no bones. These data are listed in table 7.8. Examining these data using Freeman-Tukey

TABLE 7.8
Root Etching by Weathering Stage 1 Versus
Weathering Stages ≥2

Amount of Root Etching	Wx Stage 1	Wx Stage ≥2	Total
0%	186 (0.11)	54 (−0.13)	240
1%–10%	897 (1.13)	**225 (−2.12)**	1,122
11%–20%	**17 (−2.35)**	**20 (3.13)**	37
21%–30%	**4 (−4.24)**	**19 (4.12)**	23
31%–40%	5 (−0.95)	**5 (1.49)**	10
41%–50%	3 (−0.32)	2 (0.78)	5
51%–60%	4 (−1.12)	**5 (1.64)**	9
61%–70%	1 (−0.26)	1 (0.73)	2
71%–80%	**0 (−2.20)**	**3 (1.79)**	3
Total	1,117	334	1,451

NOTE: Freeman-Tukey deviates in parentheses. Values in bold are significant at $p = 0.05$ level (±1.31).

deviates shows that root etching tends to occur at significantly higher than expected frequencies in weathering Stages ≥2. This suggests that there is a relationship between the duration for which a bone was exposed on or close to the surface and the amount of root etching.

One might further anticipate that weathering and root etching would be greatest on the skyward (exposed or upper) surface of the bone (Behrensmeyer 1978:153–154), on the assumption that root etching would occur on the surface facing the roots of the plant, just as weathering should occur on the skyward surface. A co-occurrence of heavier weathering and root etching proves to be the case in other bison bonebeds, such as Cattle Guard (Jodry 1999a:74–77) and Cooper (Bement 1999b:134–135). Yet, we found the opposite pattern at Folsom. In virtually all the cases where there was at least one face where weathering was Stage ≥2, root etching routinely occurred on the less-weathered, downward side of the bone.

That this difference exists is not wholly unexpected. The relative amount of weathering and root etching will be dependent on several factors, including the duration of subaerial exposure, the rapidity and depth of burial (e.g., a slow building of thin sediment layers versus a rapid burial by a thick blanket of sediments), the size of the element and the amount exposed on the surface—as opposed to the amount of depth buried, and the precipitation and vegetation regimes. Thus, in the case of the Folsom bonebed, we might hypothesize that the initial surface exposure of these carcasses was long enough to allow slight weathering on the upward surface and initiate root etching on the undersides.

Elements can also trap soil moisture beneath them and effectively prevent evaporation in that highly localized area, which would promote or preferentially attract roots (L.R. Binford, personal communication, 2002). We see evidence of this in the intact scapula excavated from 1998 to 1999,

which showed root etching primarily on the undersurface of the blade. However, the minimal degree of root etching, and in turn the limited amount of weathering, suggests that the carcasses were exposed on the surface for a relatively brief period prior to burial, perhaps no more than a few years, or were otherwise protected in a heavily vegetated or shaded area until burial (Behrensmeyer 1978; Todd 1987a). The former seems more likely, given the absence of ancillary evidence for significant vegetative ground cover; this does not conflict with the evidence for root etching, as that may have developed at any time after the bones were buried. Once sedimentation was initiated it was rapid and deep. More than 20 cm of sediment buried the bonebed, as indicated by the limited weathering and relatively little root etching on one of the intact crania (cf. Jodry 1999a:76)

There is slight variation in weathering among the bones, which raises the possibility that this variation corresponds to the spatial differences in the condition of the bone across the site (Kreutzer 1996; Todd 1987a:112; Todd, Witter, and Frison 1987:71). This variation was, as noted earlier, remarked on by Kaisen in 1928, who saw the bone on the eastern side of the site as being in "poorer condition" (Kaisen to Brown, July 8, 1928; Kaisen to Cook, July 8, 1928; VP/AMNH). We think it reasonable to interpret that comment as referring to bone showing more surface erosion and weathering. Testing this matter is of some interest, as it raises several alternative possibilities about the taphonomic history of the bonebed: notably, that certain areas were exposed for longer periods, or perhaps reexposed and further weathered at a later time; or that the absence of the shingle "armor" in different parts of the site influenced preservation; or that there were differences across the site in moisture regimes, snowdrift locations, vegetation, or shade.

Yet, our attempts to test these alternative possibilities directly were unsuccessful; our excavations on the eastern side of the site did not yield any faunal remains. And, unfortunately, we were not able to identify within the AMNH collections specific elements that came from that side of the site. However, one element (fig. 7.4) that we believe comes from the east side clearly demonstrates the extreme differences in weathering between the upward and the downward surfaces.

In addition, because only the 1928 excavations extended to that side of the site, we tried using a proxy measure to assess whether there are differences in bone weathering across the site, by comparing the surface weathering between the collections from 1926 to 1927, restricted to the western side of the site, and those from 1928 (table 7.9).

Analysis of the data in table 7.9 indicates that there is a significant difference in weathering stages between 1926–1927 and 1928 ($G^2 = 295.61$, df = 4, $p = 0.000$). The collection from 1926 to 1927 is significantly overrepresented by Stage 1 bone and underrepresented in all other stages; the reverse is true for the 1928 assemblage. Yet, it is best not to overinterpret this result, for two reasons. First, the 1928 assemblage contains elements from across the site,

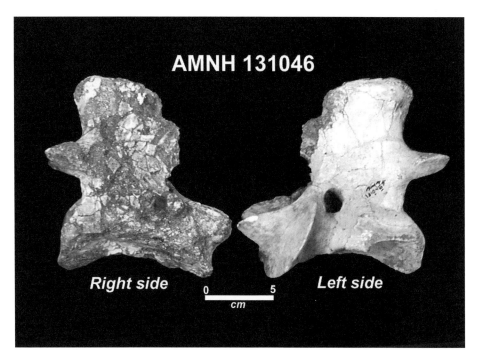

FIGURE 7.4 Bison axis from the 1928 American Museum excavations (specimen AMNH 131046), showing contrasting weathering on opposite sides. (Photo by L. C. Todd.)

TABLE 7.9
Weathering Stages of Folsom Bison Bone,
Partitioned by Collection

Wx Stage	CMNH 1926–7	AMNH 1928
1	**999 (4.09)**	**142 (−8.81)**
2	**97 (−7.56)**	**146 (9.10)**
3	**42 (−4.05)**	**53 (5.15)**
4	**2 (−1.59)**	**5 (1.94)**
5	**0 (−1.67)**	**2 (1.45)**
Total	1140	348

NOTE: Freeman-Tukey deviates in parentheses. Values in bold are significant at $p = 0.05$ level (± 1.24). All but two cells are significant at the 0.01 level (± 1.63).

TABLE 7.10
Contingency Table of Bone Weathering
Stage by Area of Site

	Wx Stage 1	Wx Stage 2	Wx Stage 3	Wx Stage 4	Total
South Bank	147	47	8	4	206
North Bank	12	6	0	0	18
Total	159	53	8	4	224

NOTE: The difference is not significant ($G^2 = 2.757$, df = 3, $p = 0.431$).

so the significant disproportion might not result just from the poorer condition of bone on the eastern side. Second, and more important, the disparity might reflect a sampling vagary: Although we have weathering data on more than 95% of the elements from the 1926 to 1927 assemblage, we unfortunately have weathering data on less than 20% of the 1928 assemblage. Were the data on the 1928 collection more complete, the sample sizes would even out, and perhaps so too would the apparent statistical disparity in surface weathering. The bottom line is that we cannot confirm or explain the differences in the condition of the bone across the site observed by Kaisen in 1928.

For the same reason, among the 1920s collections we were also not able to discriminate which elements were recovered from the North Bank, and so we cannot use these data to speak to the question of whether there are differences in surface weathering between the paleotributary and the paleovalley. However, we do have data from our recent investigations relevant to this point. Elements we recovered from the North Bank do not show appreciably greater surface weathering than those we found on the South Bank, suggesting that they too were not exposed on the surface for very long prior to their initial burial (table 7.10). Obviously, there is other evidence that these bones were subsequently reexposed, but surface weathering indicates that the later exposure was not of long duration either.

CARNIVORE MODIFICATION AND RODENT GNAWING

Additional evidence that the bone lay exposed on the surface for only a brief period comes from the relative scarcity of evidence for modification of the Folsom bison bone by

TABLE 7.11A
Frequency of Elements Showing Carnivore and/or Rodent Modification, Ordered by Total NISP

Element	Modification Absent (NISP)	Modification Present (NISP)	Total (NISP)	% NISP Modified	Casper Site % NISP Modified
HM	**22 (–1.44)**	**9 (3.59)**	31	29.0	37.0
RDU	11 (–0.60)	**3 (1.85)**	14	21.4	22.1
TA	36 (–0.77)	**7 (2.51)**	43	16.3	3.3
PT	11 (–0.34)	2 (1.31)	13	15.4	10.0
IM	38 (–0.44)	**5 (1.72)**	43	11.6	30.6
FM	37 (–0.30)	4 (1.33)	41	9.8	36.0
MT	50 (–0.18)	4 (0.95)	54	7.4	0.9
RB	375 (–0.14)	21 (0.73)	396	5.3	9.0
TRS	34 (0.14)	1 (–0.30)	35	2.9	0.0
UL	33 (0.14)	1 (–0.26)	34	2.9	15.8
MR	128 (0.27)	3 (–1.25)	131	2.5	0.2
PHT	140 (0.32)	**3 (–1.46)**	143	2.1	0.0
MC	48 (0.21)	1 (–0.73)	49	2.0	0.0
AS	59 (0.26)	1 (–1.04)	60	1.7	0.0
TH	236 (0.53)	**3 (–2.93)**	239	1.3	3.9
PHF	192 (0.58)	**1 (–3.59)**	193	0.5	0.0
Total	1,450	69	1,519		

TABLE 7.11B
Comparative Site Data on Carnivore Gnawing

Site	Type	NISP Carnivore Modification/Total NISP (%)	Reference
Casper	Kill	132/705 (18.7%)	Todd et al. (1997:table 3)
Folsom	Kill	69/1,450 (4.5%)	This book
Jones-Miller	Kill	462/1,941 (23.8%)	Todd (1987:table 4)
Agate Basin, Folsom level	Processing	11/109 (1.1%)	Hill (1994:50)
Agate Basin, Hell Gap level	Processing	32/648 (4.9 %)	Hill (2001:150, 156)

NOTE: Freeman-Tukey deviates in parentheses. Values in bold are significant at $p = 0.05$ level (± 1.34).

scavengers, which in this case might have included medium to large carnivores (perhaps canids or ursids) or rodents. These animals can modify bone in different ways, at different times, produce different signatures, and have very different effects on an assemblage. Not all of that modification occurs while the bone is fresh: Fossorial rodents can impact an assemblage long after the elements are buried (e.g., Bement 1999b:65–67).

Observations of carnivore and rodent actions are combined here, because there is too little evidence of gnawing by any class of animal to warrant breaking them out separately. Only 69 of 1,519 elements (~4.5%) examined show tooth marks or other evidence of gnawing or chewing, and many of those from the 1920s collections are ambiguous and not securely referable to carnivore or rodent activity (table 7.11A). The relative scarcity of such evidence is in marked contrast to the situation at other Paleoindian kills, such as Casper (Todd et al. 1997), Cattle Guard (Jodry 1987; Jodry and Stanford 1992), and Jones-Miller (Todd 1987b) (table 7.11A and B). In this regard, Folsom bears a resemblance to several late fall and winter processing and habitation localities, such as the Folsom and Hell Gap levels at the Agate Basin site and the Horner II bone assemblage (Frison, Todd, and Bradley 1987:367; Hill 1994, 2001; Todd 1987a:151).

If we focus attention on just those elements of the bison skeleton that were modified by carnivores or scavengers (table 7.11A), and compare their frequency distribution to the overall frequency distribution of those elements, the difference is statistically significant ($G^2 = 69.84$, df = 15, $p = 0.000$). In effect, whether or not an element was modified by the action of carnivores or scavengers was not a function

TABLE 7.12
Frequencies of Elements Showing Modification by Carnivores and/or Rodents
by Weathering Stage

	Carnivore or Rodent Gnawing Present	Carnivore or Rodent Gnawing Absent	Total
Weathering Stage 1	44	1,256	1,300
Weathering Stage ≥2	19	393	412
Total	63	1649	1,712

NOTE: The difference is not significant. $G^2 = 1.265$, df = 1, $p = 0.261$.

of how abundant that element was on the site. If one also includes the other 32 elements that show no evidence of modification at all and repeats the analysis, the difference remains statistically significant, and even strengthens the conclusion that sample size does not drive the occurrence or frequency of gnawing.

Gnawing occurs across 16 bone elements, primarily those comprising the appendicular portion of the skeleton. To explore whether there is nonrandom patterning to which of these 16 elements are being modified by carnivores and scavengers, we examined Freeman-Tukey deviates for the data shown in table 7.11A. The results reveal that several elements are being modified by carnivores and rodents significantly more often than would be expected by chance. These elements are the innominate, the long bones of the forelimb (humerus and radius/ulna), and the hind limb (the femur, which missed being significant by 0.01, and the tibia). This is in keeping with patterns of carnivore consumption and gnawing documented elsewhere (Binford 1981:60–77; Haynes 1980; Kreutzer 1996:126–127; Lyman 1994:205–215; Todd 1987a:197).

Not surprisingly, given the density and greasiness of long bone epiphyses relative to long bone shafts (Binford 1981:51; Blumenschine and Marean 1993:282–283; Marean and Frey 1997:707–709; Marean and Kim 1998), gnawing was most commonly observed on the projecting and articular ends of those bones. Gnawing of the Folsom bison humerii, for example, is concentrated on the proximal end, where tooth marks, grooves, and furrows are present on several specimens; we address below the impact of carnivore gnawing on our various element counts (cf. Blumenschine and Marean 1993:294; Marean and Frey 1997; Marean and Kim 1998; Marean and Spencer 1991). In at least one instance, the tooth marks appear to be in the size class of a coyote.

Surprisingly, ribs (RB) and thoracic vertebrae (TH), although the two most abundant elements in table 7.11 (NISP = 396 and 239, respectively), do not show an unusually high incidence of gnawing. In fact, thoracic vertebrae are significantly "undergnawed," given their sample size. Clearly, both of these elements would have provided considerable meat; unlike, for example, the phalanges (PHF and

PHT) which are also significantly undergnawed given their sample size but have far less available meat.

Binford (1981:65) observes that carnivores commonly gnaw the spines and processes of vertebrae, so one hypothesis to explain this pattern is that the spines of the thoracic vertebrae at Folsom, where the rich hump meat would have been, had already been removed by human hunters so was unavailable to scavengers. Although more than 88% ($n = 165$) of the thoracic vertebrae at Folsom have portions of their spines present, more than half of these (85/165) lack their extreme dorsal edge, making it difficult to discern the incidence of gnawing on these elements. The same pattern is evident on the ribs as well.

We also examined whether individual bone elements or portions thereof that were exposed for longer periods (e.g., Stage ≥2 weathering) were more subject to carnivore or rodent modification than bone elements buried more rapidly (Stage 1 weathering). These data are reported in table 7.12. The overall relationship between weathering stage and gnawing proves not to be significant ($G^2 = 1.265$, df = 1, $p = 0.261$). This indicates that bones exposed for longer periods of time were not, in fact, more likely to have been modified by carnivores or rodents. Unfortunately, and again because we lack provenience data on the bone from the original investigations, we cannot determine whether there was a difference in the incidence of carnivore modification across the bonebed, as, for example, Todd (1987a:197) observed at Horner II.

While the gnawing present in the Folsom assemblage indicates that carnivores and rodents had access to the bones, this was obviously not an assemblage accumulated by carnivores. Moreover, it does not appear as though non-human scavengers were at the site in appreciable numbers or were able to gain access to all of the bison carcasses. Yet, unlike the situation at, for example, Olsen-Chubbuck (Wheat 1972) or the Cooper site (Bement 1999b), where the tight packing of the carcasses likely restricted access by carnivores (Todd 1987a:197), at Folsom the carcasses were widely dispersed and could have been easily accessed, if they were exposed on the surface for a long period of time. This might suggest that carnivores were not on this particular spot on the landscape in large numbers. Perhaps this, in turn, suggests that prey abundance in the Folsom area may have been low or unpredictable, and hence there was not a large resident population of carnivores or scavengers.

FLUVIAL PROCESSES: BONE ORIENTATION AND TRANSPORT

The effects of flowing water on bone exposed on a surface or washed out by fluvial action will vary depending on the competence of the current; the structural density, size, mass, shape, and initial orientation of the bone; the degree of articulation of the elements; the surface gradient; the presence of obstructions and local channel conditions, etc. (Frison and Todd 1986:54–80; Lyman 1994:171–187). Other factors being equal, in cases of very low-energy flow across a surface bone elements can remain essentially undisturbed. As the competence of the flow increases, bone can be aligned by the flow, often in predictable ways. As Voorhies (1969:66) observed, long bones generally become oriented parallel to the current with their heavier ends upstream, while smaller or lighter bones, or long bones with ends of approximately equal weight, are rolled broadside down the bed, often at right angles to the flow.

Small bone and bone fragments are especially susceptible to movement, even under conditions of relatively gentle slope wash, which would not otherwise move or reorient larger bone. Thus, water-floated assemblages will also show size sorting. Large bones, such as crania and innominates, might obstruct the "flow" of smaller elements, which would then collect around those larger pieces, forming a lag deposit (Frison and Todd 1986:53–58, 65; Kreutzer 1988:225, 1996:109; Lyman 1994:179–180). In very high-energy alluvial settings where bone may only stop when encountering an obstruction, or along the edges of rivers or streams where currents may roil and eddy in the shallower depths and/or around obstacles along the bank, those stream-oriented, or stream-orthogonal, patterned alignments may in turn disappear, as may certain elements (Frison and Todd 1986:65–67; Lyman 1994:172–173).

In effect, patterning in bone orientation potentially reveals whether and how elements in a deposit have been moved by current, but a lack of orientation is not necessarily indicative of the absence of water transport. Put more succinctly, an item can be reoriented without having been transported or transported without being oriented. To assess whether the absence of orientation indicates an absence of transport, one must examine additional evidence: notably, the inclination of the bone, evidence for rolling and abrasion, the associated sediments, etc. (Lyman 1994). In the absence of any evidence for patterning in orientation or ancillary evidence of bone transport, one can presume that the remains were buried with a minimal amount of disturbance.

That would appear to be the case for bone found in the paleotributary. This was not an active drainage during Folsom times. There is some hint, however, that the aeolian deposits that filled the paleotributary in the Late Glacial may

have been redeposited by a gentle flow of water washing into this topographic low (chapter 5). If, indeed, there was water flowing across the surface of the paleotributary, it might have shifted the position of bone as it lay exposed on that surface. Yet, given the absence of evidence for abrasion, the lack of evidence for size sorting of the bones, and the patterned weathering and root etching, if there was movement by fluvial action, it was minimal (e.g., Frison and Todd 1986:40).

Although the effect of water in the paleotributary may have been subtle, in the paleovalley there is stratigraphic evidence to expect substantial movement of bone by fluvial action (chapter 5). Indeed, it appears that some elements were eroded out of primary context and redeposited downstream, based on the observation of breakage, loss of projecting articular ends, etc.

Rose diagrams enable us to assess whether fluvial action has influenced the position of the bone, by revealing whether there is preferential orientation to the elements (Kreutzer 1988; Lyman 1994:178). Data on the orientation of the bone elements were tallied in 10° increments between 180° and 360°; that is, bidirectionality was collapsed onto a single scale to remove mirror imaging (Lyman 1994:178). These data are solely from the SMU/QUEST excavations; the data and maps from the 1920s do not allow post hoc determination of bone orientation.

In order to explore the possibility that the size of the element might determine its susceptibility to movement (Kreutzer 1988:225–226), the bones were partitioned into several size classes. These classes were derived partly from an examination of modality in the data, but also somewhat arbitrarily; that is, we assumed that a bone or bone fragment less than 50 mm in length was small, at least by bison bone standards, and its susceptibility to transport or movement would be much greater than that of larger elements. The largest element in our sample was a cranium, with a maximum length of 713 mm. As it happens, 227 of the 423 elements (54%) for which data are available are ≤50 mm in maximum length. Obviously, the distribution of size classes is positively skewed. For exploratory and analytical purposes, two classes of large bones were established: one with a maximum length of >300 mm; the other, >100 mm (which includes elements >300 mm). Two classes of smaller bones were also created: one with a maximum length of <100 mm; the other with a maximum length of <50 mm. In addition, bones from the South Bank and North Bank were separately plotted and examined, to see whether patterning in orientation differed between the paleotributary and the paleovalley, the a priori hypothesis being that bone from the paleovalley would show alignment based on fluvial processes. These data are reported in table 7.13.

None of the resulting diagrams reveal any obvious alignments or show a significant number of bones loading heavily on a narrow directional axis; only one diagram is shown here—that of "South Bank, all bone" (fig. 7.5). To ensure that more subtle alignments are not lurking in the data, a series of likelihood ratio chi-square statistics (G^2) was calculated for each of the groups in table 7.13. Expected values were

TABLE 7.13
Frequencies of Specimens Grouped by Orientation (10°) Classes, South and North Banks

| Orientation (deg) | All Bone | Large Bone | | Small Bone | | North Bank |
		>300 MM	>100 MM	<100 MM	<50 MM	All Bone
180–189	19	2	5	13	11	2
190–199	29	3	9	19	17	0
200–209	22	2	9	12	8	1
210–219	22	3	7	15	12	1
220–229	24	3	10	14	10	0
230–239	15	3	4	11	7	1
240–249	19	0	3	15	12	0
250–259	16	1	5	11	10	0
260–269	19	3	6	13	7	1
270–279	34	2	12	21	16	0
280–289	18	2	3	15	12	3
290–299	9	1	4	7	3	1
300–309	21	4	7	12	9	1
310–319	24	1	5	18	14	3
320–329	19	4	8	12	8	3
330–339	14	2	5	9	7	6
340–349	15	4	4	11	7	2
350–359	22	3	4	16	13	1
Total	361	43	110	244	183	26

derived by dividing the sample size by the number of possible classes (18). Thus, in the case of "South Bank, All Bone," the expected value was 361/18 = 20.056.

In no instance was a significant G^2 value returned. That is, all orientations were evenly distributed around the compass and not aligned. Two of the groups, "South Bank, all bone" and "South Bank, bone <50 mm," had a single cell with a significant Freeman-Tukey deviate. In both cases, it was the same cell: the interval from 290° to 299°, which in both instances occurred significantly less than in all other intervals. A single significant Freeman-Tukey deviate in two separate cases, however, does not a preferred orientation make. All that said, visual examination does show one subtle trend, and that is for a slight loading on a west-to-east axis. Although obviously not significant in a statistical sense, this may prove of some relevance in the manner of the movement of the shingle shale, as we discuss below.

It was not expected that there would be a preferred orientation to the bone in the paleotributary. However, it is surprising that the North Bank bone also fails to show a nonrandom orientation, given the erosional and depositional history of this part of the site (chapter 5). Two possible explanations come to mind: First, the sample size may simply be too small to be representative. Shipman (1981:71; also Lyman 1994:178) suggests that at least 72 bones are

necessary to adequately gauge patterns of orientation, and our sample falls below that minimum. Alternatively, perhaps here on the edge of the paleovalley a complex pattern of the flow against the valley wall did not align the bone with the downstream current (chapter 5). Resolving which of these is correct will require additional data.

The degree of inclination of the bones (table 7.14) further supports the supposition that there has been little postdepositional disturbance in the paleotributary but more in the paleovalley. For the 361 elements from the paleotributary for which data are available, the mean inclination is 15.21°, although the mode is just 6° and fully 75% of the elements had inclinations ≤20°. In contrast, bones from the paleovalley had an average inclination of 24.7° with a virtually identical mode (24°); only 37% had inclinations of 20° or less (fig. 5.10). The difference in inclination between the two areas is significant, as measured by the likelihood ratio chi-square statistic ($G^2 = 20.897$, df = 9, $p = 0.013$). Freeman-Tukey deviates indicate that on the North Bank there are significantly fewer bones in the 1° to 10° inclination class, and significantly more in the 21° to 30° class, than would be expected by chance.

The low long-axis inclinations of the bone in the paleotributary further suggest, along with the absence of scratch marks on the bone surface, that animal trampling, such as movement, dispersal, and breakage, was minimal (Fiorillo

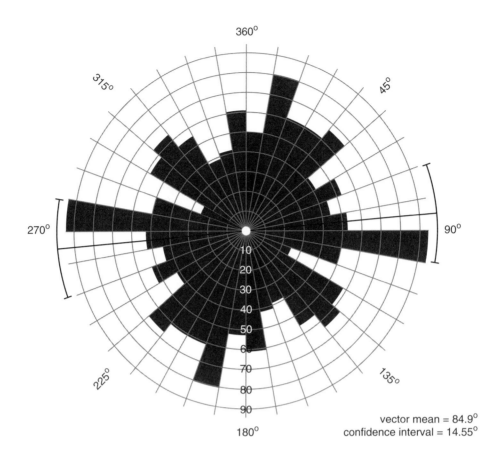

360°

315°

45°

270°

90°

10
20
30
40
50
60
70
80
90

225°

135°

vector mean = 84.9°
confidence interval = 14.55°

180°

FIGURE 7.5 Rose diagram of the orientation of bone elements in the M17 and adjoining blocks on the South Bank.

1989). Given the demonstrable evidence of high-energy hydraulic reworking of sediments in the paleovalley, we are not inclined to attribute the higher angles of the bone in that setting to trampling.

POSTBURIAL COLLUVIAL PROCESSES IN THE PALEOTRIBUTARY (SOUTH BANK)

Once the bones in the paleotributary were covered with *f2* sediments, they were not completely protected from reexposure and/or further destruction. As briefly described in chapter 5, the *f3* shingle shale washed across the top of the *f2* and, in places, cut down very close to and perhaps into the underlying bonebed (fig. 7.6). In order to assess whether this process removed bone from certain parts of the site, we now provide a more detailed examination of the horizontal and vertical dispersal of the *f3* clastics in the M17 and M15 excavation blocks within the paleotributary.

The data for this analysis come from the shingle shale recovered in the water-screening, which, as a reminder, was done separately for each 50 × 50 × 5-cm quad, for all excavated quads in both the M17 and the M15 blocks. The terms coarse and fine, as used below, refer to the shingle shale from the 3.175- and 1.587-mm screens, respectively. The shingle shale caught on the individual screens was separately bagged

and weighed, and its volume measured by use of graduated cylinders. There is a very high correlation ($r^2 = .99$) between the mass and the volume of the shingle shale. Both measures are used, however, as they inform on different aspects and can be put to different analytical purposes. Mass can provide a measure of the competence of the slopewash, while volume can show the relative contribution of shingle shale to the overall matrix.

In the M17 Block

Overall, there are data available from 897 of the excavated quads in M17 and its contiguous blocks, which, at 4 quads per level, equal ~224 levels, distributed over 17 units. We do not have complete records on all 897 quads; coarse fraction data are available from 814 quads, while fine fraction data are available from 820 quads. They are not all the same quads. The maps and statistics that follow are based on the combination of the coarse and fine fractions that come from the 738 quads for which both kinds of data are available.

The first observation to be made on these data is that shingle shale comprises a very small component of the overall matrix: In the M17 block, the coarse fraction on average accounts for less than 5% of the matrix by volume, while the fine fraction usually comprises less than 2% of the

TABLE 7.14
Contingent of Specimens Grouped by Inclination (10°)
Classes, South and North Banks

Inclination in 10° increments	South Bank	North Bank
0	5 (0.24)	0 (−0.50)
1–10	165 (0.53)	**4 (−2.33)**
11–20	105 (0.21)	5 (−0.65)
21–30	39 (−0.89)	**9 (2.56)**
31–40	26 (−0.18)	3 (0.86)
41–50	10 (−0.30)	2 (1.15)
51–60	4 (0.24)	0 (−0.41)
61–70	5 (0.24)	0 (−0.50)
71–80	2 (−0.35)	1 (1.09)
81–90	0 (0.00)	0 (0.00)
Total	361	24

NOTE: Freeman-Tukey deviates in parentheses. Values in bold are significant at $p = 0.05$ level (±1.31).

matrix in any given quad. The contribution of both coarse and fine fractions thus averages less than 7% by volume. There are, of course, quads where shingle shale comprises a greater proportion of the matrix, but not surprisingly those tend to be in the higher excavation levels, and well above the bonebed, as is clear in figure 7.7, which shows the position of the bonebed and the density of shingle shale by level. Note that on the right side the paired bars at each level represent the average percentage shingle by level (upper bar), and the frequency of quads at those levels with more than 33% shingle (lower bar). Thus, the bars are not on the same scale.

The density of shingle shale (coarse + fine fraction) in the M17 block diminishes with depth (fig. 7.7). Shingle shale peaks at over 20% of the matrix in Level 141 but then rapidly declines to just over 2% of the matrix in Level 148. As is apparent from figure 7.7, the contiguous bonebed, as opposed to the occasional isolated element, was encountered in many areas of the block by the top of Level 144 and disappears by the base of Level 151. As previously discussed (chapter 6), the mean elevation of the bonebed in this area of the site is 97.57 m (SD = 10.4 cm), which centers it in excavation Level 148. However, the mean elevation in this instance is somewhat misleading, given the slight elevation dropoff in the bonebed evident in figure 7.7 between grid meters 1033.5 and 1034 north. That dropoff notwithstanding, most of the skeletal material recovered was at nearly the same elevation, and generally covered a vertical span of approximately 20 cm. All of this, incidentally, suggests that this was a relatively level portion of the site, in contrast to the steeper slopes evident in the contours of the northern part of the bonebed recovered in the 1920s excavations (fig. 7.1).

In only 38 of these 897 quads, the amount of shingle exceeds one-third (33%) of the volume, and those quads too

are less frequent with depth (fig. 7.7). There are no quads in which shingle exceeds two-thirds (66%) of the volume of the matrix, although a few come close: Four quads have >60% shingle; in those quads, the amount of shingle shale by mass averages 9,000 g (19.9 lb).

In general, although the coarse and fine fractions are correlated ($r_p = 0.82$), the total amount of shingle shale, by mass and by volume, is driven more by the coarse fraction. On average, 60% of the shingle by mass is the coarse fraction, and the correlation of coarse mass:total mass is $r_p = 0.99$; the correlation of fine mass:total mass is $r_p = 0.89$. There is, in virtually all cases, a much greater contribution of coarse to fine shingle shale: The average coarse fraction by quad is ~750 g (maximum, 7,776 g), and the average fine fraction is ~260 g (maximum, 2,781 g).

What all this indicates, in mechanical and taphonomic terms, is that the slopewash was of sufficient competence to move relatively large masses of shingle shale. This was not a gentle surface flow, nor does it appear that there were significant obstacles blocking the downslope movement of the shale fragments.

In an effort to better understand the process of the shingle shale buildup over the bonebed, the coarse and fine fractions by *volume* for each quad were combined. The combined figures were then used to construct contour maps of the density of shingle shale (as a percentage of the matrix) by volume by level (fig. 7.8). The contour interval for each map is 5%, and the values range from 0% to just over 60% density.

This sequence of maps, arrayed from Level 140 to Level 151 (upper left to lower right), illustrates several points. First, they graphically confirm the decline in the density of shingle shale by depth. However, they also show that the shale was not evenly spread across the block, but occurred at higher densities in some parts. Further, despite the overall decline of the shale with depth, there were relatively higher-density pockets of shingle shale in some of the lower levels—specifically, in the southern half of the block. What this implies is that the surface across which the shingle shale was distributed was irregular. Although it is conceivable that the dispersal of the shale shaped that surface, this would imply greater erosion of the upper surface of the *f2* than was observed. We conclude instead that the surface was already irregular when the slopewash of the shale took place. Since this was a surface of bison carcasses barely covered by aeolian sediment, this is not unexpected.

Furthermore, and as mentioned, the underlying distribution of bison bone coincides with the overlying distribution of shingle shale (figs. 7.6 and 7.7). We suggest that the bones served as an obstacle to sheet wash that, as is apparent from the rose diagram and, of course, from the relative location of the valley wall, was washing across the paleotributary in an approximate west-to-east flow. All of this caused the shingle to pile up in this area on and around the bones and, especially, we suspect, around the larger elements and articulated units, thus both burying and armoring the bonebed. The bonebed provided the foundation of its own preservation.

FIGURE 7.6 Close-up image showing the relative stratigraphic position of the bonebed, the top of which is at the level of the tip of the horn core on this bison cranium, and the overlying *f3* shingle shale. (Photo by D. J. Meltzer.)

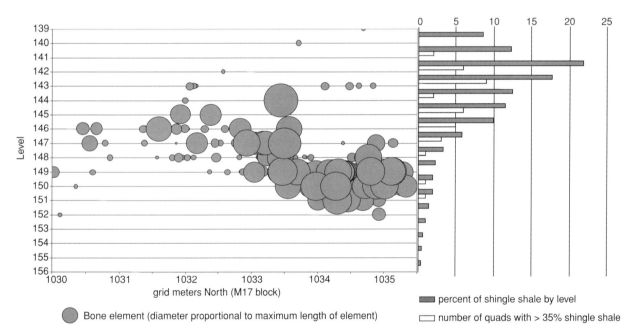

Bone element (diameter proportional to maximum length of element)

percent of shingle shale by level

number of quads with > 35% shingle shale

FIGURE 7.7 Backplot showing the elevation of bison bone elements in the M17 and adjoining blocks, and corresponding density of coarse and fine shingle shale, by excavation level. The diameter of the circles is proportional to the size of the bone element.

The position of the shingle shale relative to the level of the bonebed is perhaps more easily seen in figs. 7.9 and 7.10, which illustrate the density contours of the shingle shale in profile, in a series of west-to-east cross sections at 0.5-m intervals from the south to the north end of the M17 block (fig. 7.9), and the corresponding perpendicular series of south-to-north cross sections, also at 0.5-m intervals, but this time from the western to the eastern sides of the M17 block (fig. 7.10). These cross sections are oriented to reflect the general direction of slope wash from west to east off the valley wall and from south to north down the axis of the paleotributary, respectively. Thus, the presumed paleotopographic slope in both sets of figures declines from left to right. The two figures are set to the same scale, there is a threefold (3×) vertical exaggeration, and the contour interval is the same as in figure 7.8.

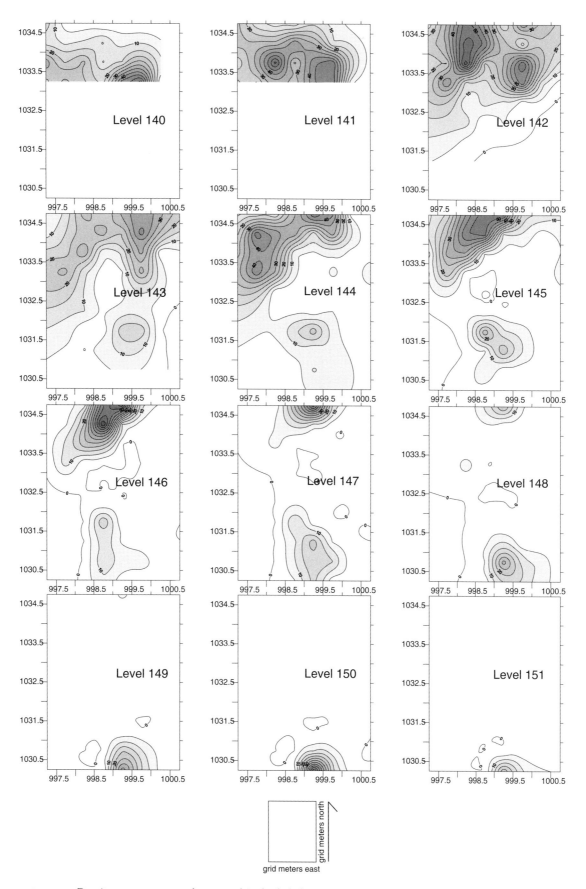

FIGURE 7.8 Density contour maps of percent shingle shale by excavation level in the M17 block, from Level 140 (upper left) to Level 151 (lower right). Darker areas have greater concentrations of shingle shale.

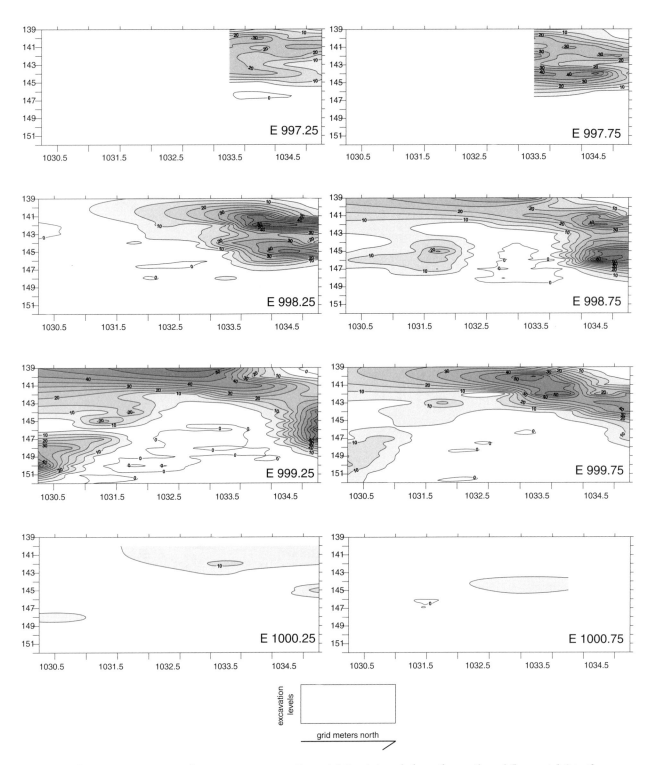

FIGURE 7.9 Density contour map of west to east cross-sections at 0.5 m intervals from the south end (lower right) to the north end (upper left) of the M17 block, showing the percent shingle shale by excavation level. Darker areas have greater concentrations of shingle shale.

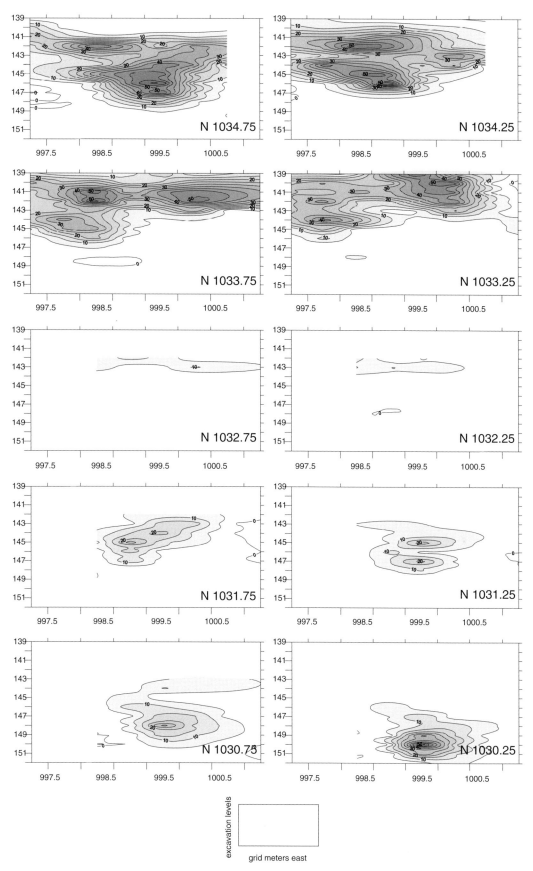

FIGURE 7.10 Density contour map of south to north cross-sections at 0.5 m intervals from the western side (upper left) to the eastern side (lower right) of the M17 block, showing the percent shingle shale by excavation level. Darker areas have greater concentrations of shingle shale.

It is clear from these figures that the lower boundary of the shingle shale is well above the bonebed layer, save for the southern half of the block. There is also evidence in figures 7.9 and 7.10 of more than one pulse of slopewash carrying shingle shale out across the tributary. Visible in both figures is an undulating layer of lower-density shingle shale, sandwiched between zones of higher-density shingle shale. That band is most visible on the more western lines of the south-to-north cross section and the more northern lines of the west-to-east cross section. It appears, from the shape of the lower portion of the shingle shale deposit, that the first pulse filled in topographic low spots over the bonebed. That was followed by a brief pause in the process, during which time more sediment than shingle shale was deposited. In turn, another pulse of shingle shale dispersed across the surface, which, because those topographic lows were now filled, was more evenly and widely spread.

The lower reach of the shingle shale in the southern half of the block corresponds with a relative scarcity of bone in that same area (fig. 7.2A). This raises the question of whether bison carcasses were present in that area, but were lost as a result of erosion due to the shingle shale, or whether no carcasses and few bone elements were deposited there originally. The data from this area cannot well resolve that question. However, the data from the M15 block 5 m to the south may bear on the issue.

In the M15 Block

Fewer excavation units were placed in the M15 block. Therefore, we have data on shingle shale from only 158 quads, for the majority of which ($n = 98$) we have data on just the coarse fraction. There are data on corresponding fine and coarse fractions from just 57 quads.

As in the M17 block to the north, the shingle shale here comprises a very small component of the matrix. In fact, the deposition of the shingle shale was even lighter here. On average, the coarse fraction comprises less than 1.5% of the matrix by volume; the fine fraction, less than 0.6%. There is only a single quad in the M15 block where the percentage of shingle by volume is greater than 9%, and that is the northwest quad of N1024 E997, Level 139, where the percentage by volume is 21.2%. In keeping with the pattern seen in the M17 block, that concentration occurs high in the profile and well above what would have been the elevation of the bonebed, were it to have extended into the M15 block at the same elevation at which it occurs in the M17 block. Because the contribution of shingle shale to the M15 block matrix is so light, maps at the same contour interval as used in figures 7.9 and 7.10 are essentially blank, as there are only 11 quads in which the percentage of shingle data by volume is greater than the first visible contour threshold (5%). Hence, we do not include here the corresponding density contour maps for this area.

Also in keeping with the M17 block, an identical percentage (60%) of the gravel in the M15 block is comprised of the coarse fraction. What this suggests is that both areas received slopewash of the same relative composition, but far less of it reached the M15 block, likely because that area is farther from the valley wall. These data indicate that if there once were bison carcasses in that area of the site, they were not physically removed by slopewash of *f3* shingle shale.

SEDIMENT CHEMISTRY AND BONE PRESERVATION

But was bone once present in the southern part of the M17 block and the M15 block, and then removed by chemical weathering, rather than mechanical disturbance? To test that possibility, Fourier-transform infrared spectroscopic (FTIR) analysis was undertaken by Todd Surovell on sediment samples obtained from excavation blocks and an auger transect extending from the M15 to the M17 block. The primary goal was to assess whether the presence/absence of bone in these different parts of the site could be a result of differences in sedimentary conditions influencing preservation, by using the relative concentration of calcite as a proxy for diagenetic history. If calcite is present in sediments, other things being equal, bone ought to be as well, and vice versa. The study incidentally provides useful information on the sediment mineralogy (for analytical details and results, see appendix D).

The FTIR analysis shows that site sediments and the local Smoky Hill Shale bedrock are indistinguishable, at least with respect to the mineral species present. All samples are comprised of varying amounts of calcite, clay, and quartz, with only occasional clasts of basalt—the volcanic component is inconsequential. Calcite concentrations, measured as the ratio of calcite:silicate values, however, vary across the site, depending on the relative acidity of the depositional environment. Calcite is rare or absent in samples from the McJunkin *(m2)*, indicating relatively acidic conditions during the deposition of this unit. The calcite stringers that occur in the McJunkin are likely the result of secondary deposition. Below the McJunkin into the *f2*, calcite concentrations increase, with the greatest relative amounts between the elevations of 97.5 and 98 (appendix D). This correlates well with the stratigraphic position of the bonebed, which is on the lower side of that calcite peak, at a mean elevation of 97.57. Likewise, Goldberg and Arpin (1999) found high concentrations of $CaCO_3$ in the several samples they examined from the *f2*.

Within the *f2*, however, there is horizontal variation in the concentration of calcite. It occurs at the highest concentrations on the northern and southern extremes of the sampled areas, that is, within the M15 and M17 blocks. In the central portion of the transect, where calcite concentrations are relatively low, there must have been greater leaching of carbonates.

What this suggests, in general, is that mineralogical conditions within the *f2* in both the M15 and the M17 blocks were comparable. Therefore, the virtual absence of bone from the M15 block (two small radius fragments) and the southern part of the M17 block is not simply a consequence of poor preservation. Rather, it seems reasonable to infer that bison bone was never present in this part of the site.

THE AREA AND EXTENT OF THE BISON KILL

These data help delimit the edge of the bison kill and the bonebed in this part of the paleotributary. Since bison bone is essentially absent in the M15 block and relatively light in the southern portion of the M17 block, and in neither area is there evidence of chemical or physical agencies that would have removed the bone, we infer that the southern edge of the bonebed likely occurred between our M17 and our M15 blocks; the intervening M16 block has not been excavated.

Of course, this is only one corner of the bonebed. Nonetheless, combined with data available from the 1928 excavations, we can provide a ballpark estimate of the original size of the kill area. If this area extended from, say, the southern edge of the M17 block to the mouth of the paleo-tributary, that is a south-to-north distance of ~30 m. Given the west-to-east distance across the paleotributary and as shown in the 1928 plan map (~20 m), the carcasses were scattered across an area of ~600 m^2.

The overall extent of the kill, however, must be greater than that since bison remains were also found in the paleo-valley, almost 30 m north of the M17 block (fig. 4.17). Needless to say, one cannot assume that bison carcasses were distributed continuously across that distance. Nonetheless, these dimensions suggest that the kill possibly extended over a total area of ~800 m^2. At best, however, that is a very rough approximation and makes some critical assumptions (not least about the accuracy of the 1928 plan map) that may not be supportable. Even so, a kill area of this size is well within the range seen ethnographically. Indeed, O'Connell, Hawkes, and Blurton-Jones (1992) show that 800 m^2 may be on the small side, even for a kill of more than 30 large mammals. Were there additional excavations across the entire area, the total number of bison and the estimated kill area might be larger.

Regardless, at this scale and given the current estimate of the animals killed here (MNI = 32; see below), the carcasses were widely dispersed. Using the more conservative estimate of the area of the bonebed (600 m^2), and assuming that a dead bison covered ~3 m^2, this would yield a density of 1 carcass every 6.25 m^2. Hofman (1999a:128) estimates the density as 1 carcass every 10 m^2. The difference between our estimates is attributable to our using carcass area rather than MNI and using a different measure for the overall extent of the site. By either measure, there was ample room to move around while butchering their carcasses, which readily explains the lack of large articulated units, as discussed below.

CRUSHING

Despite the thick blanket of sediment and gravel that ultimately built up over the bonebed, only eight bones from the entire 1926 to 1928 excavations show evidence of crushing. Six of those eight were thoracic vertebrae, which had

TABLE 7.15

Measurement Data on Folsom Humerii
in the AMNH Collections

Specimen No.	HM7 (mm)	HM11 (mm)	Sex
131623	91.5	103.7	Cow
130480	91.7	100.6	Cow
130693	91.7	106.4	Cow
131578	93.9	104.9	Cow
130694	94.4	106.2	Cow
130319	97.1	109.7	Cow
130314	97.3	109.9	Cow
130294	100.3	109.9	Cow
130293	100.9	108.2	Cow
Average	95.42	106.61	
130972	103.7	115	Bull
130447	104.3	118	Bull
33801	105.1	119	Bull
33801	105.6	121	Bull
130618	105.8	121	Bull
130275	107.4	118.1	Bull
131282	110.3	120	Bull
131239	113.3	121	Bull
Average	106.93	119.14	

crushed spines (AMNH 131589, 131592, 131593, 131594, 131595, 131610). The other two were femur shafts (CMNH 1245, 7581). The weight of sediment over the bonebed, therefore, seems not to have had much impact on the character of the bone.

However, in the paleovalley, the bison cranium recovered during the 1972 excavations did show significant compression crushing. The left orbit and maxilla, though still a part of the cranium, had been distorted and displaced laterally several centimeters. This is in marked contrast to the complete and undistorted crania recovered from the paleotributary on the South Bank. The North Bank overburden is substantially thicker; at least 3 m of McJunkin clays buried the cranium found in 1972 by Trinidad State Junior College.

The Folsom Bison Herd: Age, Sex, and Season of Death

Bison display strong sexual dimorphism in body size and skeletal robustness. As a consequence, metric data on certain postcranial elements make it possible to distinguish bison cows from bulls (Todd 1987a). We obtained metric data on several such elements but focus here on humerii, of which 22 measurable specimens are available in the Folsom assemblage. Of those, 17 were from skeletally mature animals, on which sex determinations are most securely based (Todd 1987a:162–163). These data are provided in table 7.15 and indicate that of those 17 specimens, 9 were from cows and the remainder were from bulls. This tally may include paired elements and should not be considered MNI values.

TABLE 7.16
Age Group Data on the Folsom Mandibles

Age	NISP	MNI
Group 1, calves	4	3
Group 2, yearlings	8	4
Group 3	6	3
Groups 4–6, mature	29	15
Total	47	25

The relative proportion of cows to bulls is affirmed by the cranial data. Although many of the crania are fragmentary, there are two reconstructed bull crania, one each from the Colorado and American museum excavations, and four cow crania, one each from the American Museum and Trinidad State excavations and two from the SMU/QUEST excavations (fig. 7.2A). As with the humerii, nearly two-thirds of the cranial remains come from sexually mature cows.

In order to determine the demographic structure of the herd, we examined mandibular molar eruption and wear patterns on all available specimens. These data are shown in table 7.16 (also Todd, Rapson, and Hofman 1996:170). Of the 25 animals represented, 40% are in Group 3 and under and, thus, at or below the earliest stage of breeding. Very young animals are well represented, with calves and yearlings having a combined MNI of 7 (Todd, Rapson, and Hofman 1996:170). The remaining 60% (MNI = 15) are in Group 4 and above, which probably includes both young bulls—those less than seven years old—and all cows (Berger and Cunningham 1994:162). This profile is typical for a cow–calf herd (Berger and Cunningham 1994).

We estimate that the ~32 bison in this cow-calf herd were killed in the fall, based on patterns of dental eruption and wear (Todd, Rapson, and Hofman 1996:169–170). These indicate ages of 0.4–0.5 years for the Group 1 animals and 1.4–1.5 years for the Group 2 animals. At that time of the year, fat reserves in cows are greater than at other times (Jodry 1999a:47; Speth 1983; Todd 1991:232–233). This was, as best we can determine, a relatively healthy herd. We observed only a very few dental or postcranial pathologies in all of the collections of Folsom bison bone.

Furthermore, based on the relatively uniform weathering of the bones examined in the extant collections, the within-cohort uniformity of eruption and wear patterns of mandibular molars, and the stratigraphic and depositional context of the faunal remains in the limited area of the bonebed we have been able to examine, we believe, as Brown did in 1928, that all the remains come from a catastrophic death assemblage, which we obviously interpret as the killing of a single herd. This is typical of Folsom sites (Stanford 1999:301).

The mean metric data for the humerii of the Folsom cows and bulls from table 7.15 are plotted in figure 7.11 alongside comparable data from a range of sites that include *B. antiquus* as well as *B. bison* (comparative data from Hofman and Todd

2001:table 1); these data illustrate that the Folsom bison fit comfortably within the size range of other *B. antiquus* specimens. However, we note that since this is a single-event kill of a cow-calf herd, it may not include the largest and most dominant bulls—which would have been a part of smaller bull herds—so the data may underestimate the maximum size of the bulls in this population. Were such included here, they might approach the size of the Agate Basin Folsom-age bulls, which, Hill (2001:104) suggests, were being procured at different locations at different times from the cows.

We observed earlier that the modern taxonomic designation of the animals from the Folsom site is *B. antiquus* and that the age of these specimens fit well within the radiocarbon and distribution range noted by McDonald (1981). The postcranial metric evidence provided here affirms that taxonomic assignment.

Numerical Matters II

A total of 3,397 identifiable *B. antiquus* bone and tooth elements (NISP) have been recovered from all the excavations in the Folsom bonebed (table 7.4). These are the basis for the subsequent derived measures of MNE, MAU, and MAU%, the data for which are reported in table 7.17 (also fig. 7.2B).

These tallies do not include detailed shaft fragment counts, as advocated by Marean (e.g., Marean and Fry 1997; Marean et al. 2001). Although we readily accept his point that ignoring shaft fragments runs the risk of undercounting elements, and certainly applies in some situations, we do not think it a critical issue here, for three reasons. First, we have documented that the influence of carnivores on this assemblage is slight, so there is little reason to think we are missing a significant number of long bone epiphyses. Second, we have excellent preservation of the bone, which makes it possible to say with confidence that carnivore modification is slight, and therefore we are not likely missing and/or undercounting any elements. Finally, we worry that because of collection and curation practices, many such fragments would have been missed or discarded, and hence calculations based on the remaining sample would be potentially misleading.

The data in table 7.17 form the basis for our estimates of the MNI represented in the faunal assemblage. We calculate that at least 32 bison comprise this assemblage, based on counts of astragali (AS; a count of fused second and third carpals [CPS] yielded an MNI estimate of 31).

As is evident from table 7.17 (fig. 7.2B), the relative proportion of elements varies. Certain elements occur at relatively high percentages, including the mandibles, carpals, tarsals, metapodials, and phalanges. With the exception of the mandibles, all of these elements are of higher density and lower utility (Emerson 1993; Kreutzer 1996; Wheat 1972:102–103). The elements that occur at lower percentages include the upper limbs, ribs, hyoids, first tarsals, second metatarsals, dew claws, and, as Brown observed in 1928, tails or caudal vertebrae. There are at least two possible explanations, which may not be mutually exclusive, to

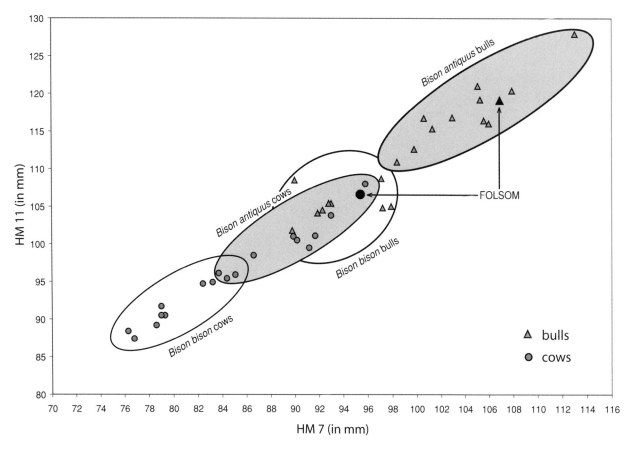

FIGURE 7.11 Plot of measurements of distal depth (HM11) against distal articular breadth (HM7) of the humeri of the mature Folsom site bison bulls (*triangles*) and cows (*circles*). Comparative data from archaeological *Bison antiquus* (range of bulls and cows shaded) and modern *Bison bison* (range of bulls and cows not shaded). (Comparative data from Hofman and Todd 2001; measurements following Todd 1987a.)

explain these patterns of skeletal element representation, and we address them in turn.

First, these patterns could be the result of density-mediated preservation. Lyman (1985, 1994) and others have long cautioned that patterns of skeletal element representation may reflect differential preservation of high-density elements (see also Lam, Chen, and Pearson 1999; Rogers 2000). To test this issue, NISP and (separately) complete element counts were plotted against bison bone density values using density data from Kreutzer (1992) and Lyman (1994:tables 7.3, 7.6). Statistical analysis shows no correlation between the two; the Spearman rank-order correlation between MNE and average bone density is $r_s = -0.011$ ($t = -0.05$, not significant; $n = 25$). We therefore conclude that element frequency varies independently of bone density, a conclusion not unexpected given the excellent preservation of fragile bone elements.

Second, these patterns could result from butchering and processing activities on-site and transport off-site. There are fewer long bones, such as femora, humerii, radii, and tibia, and the assemblage appears to be dominated by low–meat yielding skeletal parts traditionally considered low-utility elements, such as lower legs and mandibles (e.g., Wheat 1972:102–103; also Emerson 1993). These are parts generally discarded in the course of butchering. Subsequent statistical

analysis of the assemblage indicated that there was no correlation ($r_s = -0.014$, $t = -0.07$, not significant) between ranking of element utility, using values derived from Emerson (1993), and recovered elements (MNE). However, MNE counts are driven by skeletal abundance, and in this instance the correlation is dampened by a few highly fragmented outliers of high-utility elements, notably ribs and thoracics.

To account for these numerical vagaries, we plot (fig. 7.12), MAU% against element utility. Although in this instance the correlation is again statistically insignificant, the plot reveals relevant anatomical patterning that does reflect prehistoric patterns of utility-related transport. Specifically, we observe, as shown in the box on the left side of figure 7.12, that elements of low utility (<25) tend to be common, while higher-utility elements (>25) in general occur at a MAU% of less than ~60% and, in this instance, no less than ~30%.

In table 7.18 we parse these data in a different manner, by dividing the bone elements into four ranked utility classes (utility values of 1%–24%, 25%–49%, 50%–74%, and 75%–100%), then examining the observed frequencies in each of those classes. The expected frequencies are modeled on the assumption that a completely preserved kill of 32 bison would be expected to yield 32 crania, 64 mandibles, 160 cervical vertebrae, 896 ribs, etc. MNE counts (table 7.17)

TABLE 7.17
Summary Inventory and Element Counts of Folsom Site Bison Bone

Element	Code	NISP	MNE	MAU	MAU%
Axial skeleton					
Cranium	CRN	159	11	11	36.1%
Mandible	MR	149	57	28.5	93.4%
Hyoid	HY	12	7	3.5	11.5%
Atlas	AT	14	14	14	45.9%
Axis	AX	26	24	24	78.7%
Cervical vertebra	CE	126	88	17.6	57.7%
Thoracic vertebra	TH	296	187	13.36	43.8%
Lumbar vertebrae	LM	126	93	18.6	61.0%
Sacrum	SAC	23	15	15	49.2%
Caudal vertebra	CA	36	33	2.2	7.2%
Rib	RB	524			
Proximal rib	RB PR		280	10	32.8%
Costal cartilage	CS	2	1		
Forelimb					
Scapula	SC	37	26	13	42.6%
Humerus	HM	33			
Complete	HM CO		20	10	32.8%
Proximal	HM PR		23	11.5	37.7%
Distal	HM DS		30	15	49.2%
Radius	RD	48			
Complete	RD CO		32	16	52.5%
Proximal	RD PR		39	19.5	63.9%
Distal	RD DS		39	19.5	63.9%
Ulna	UL	37	32	16	52.5%
Accessory carpal	CPA	39	39	19.5	63.9%
4th carpal	CPF	47	47	23.5	77.0%
Intermediate carpal	CPI	46	46	23	75.4%
Radial carpal	CPR	51	51	25.5	83.6%
Fused 2nd & 3rd carpal	CPS	54	54	27	88.5%
Ulnar carpal	CPU	46	46	23	75.4%
Metacarpal	MC	51			
Complete metacarpal	MC CO		41	20.5	67.2%
Proximal metacarpal	MC PR		43	21.5	70.5%
Distal metacarpal	MC DS		40	20	65.6%
5th metacarpal	MCF	18	18	9	29.5%
Hind limb					
Os coxae	IM	68	29	14.5	47.5%
Femur	FM	43			
Complete	FM CO		20	10	32.8%
Proximal	FM PR		28	14	45.9%
Distal	FM DS		20	10	32.8%
Patella	PT	14	14	7	23.0%
Tibia	TA	55			
Complete	TA CO		32	16	52.5%
Proximal	TA PR		24	12	39.3%
Distal	TA DS		29	14.5	47.5%
Lateral malleolous	LTM	32	32	16	52.5%
Talus	AS	61	61	30.5	100.0%
Calcaneus	CL	64	49	24.5	80.3%
Fused central & 4th tarsal	TRC	45	45	22.5	73.8%

TABLE 7.17 (*Continued*)

Element	Code	NISP	MNE	MAU	MAU%
1st tarsal	TRF	11	11	5.5	18.0%
Fused 2nd & 3rd tarsal	TRS	38	38	19	62.3%
Metatarsal	MT	59			
Complete metatarsal	MT CO		48	24	78.7%
Proximal metatarsal	MT PR		49	24.5	80.3%
Distal metatarsal	MT DS		46	23	75.4%
2nd metatarsal	MTS	15	15	7.5	24.6%
Feet					
1st phalange	PHF	197	178	22.25	73.0%
2nd phalange	PHS	189	182	22.75	74.6%
3rd phalange	PHT	150	147	18.375	60.2%
Proximal sesamoid	SEP	258	258	16.125	52.9%
Distal sesamoid	SED	90	90	11.25	36.9%
Dew claw	DC	8	8	1	3.3%
Subtotal bison NISP		3397			

NOTE: RDU (NISP = 14) are factored into the MNE, MAU, and MAU% counts for RD and UL. Other identifiable bison elements (e.g., MN, MP, SN, ZY) are not included in these measures. MNE, minimum number of elements, MAU, minimum number of animal units, MAU%, percent MAU.

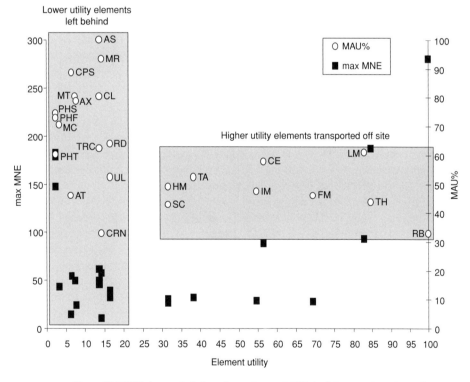

FIGURE 7.12 Plot of MAU% (*open circles*) and maximum MNE (*solid squares*) against element utility values. Vertical shaded rectangle identifies lower-utility elements left on site; horizontal shaded rectangle identifies lower-utility elements taken off site. (Element utility values from Emerson 1993.)

TABLE 7.18
Expected and Observed MNE Values, Partitioned by
Element Utility Classes

Element Utility	Expected MNE	Observed MNE	Elements Included
1%–24%	**1,376** **(−4.15)**	**943 (5.40)**	AS, AT, AX, CL, CPS, CRN, MC, MR, MT, PHF, PHS, PHT, RD, TRC
25%–49%	192 (1.00)	**78 (−1.42)**	HM, SC, TA
50%–74%	288 (0.42)	137 (−0.55)	CE, FM, IM
75%–100%	**1504 (3.66)**	**560 (−5.51)**	LM, TH, RB

NOTE: The difference is significant ($G^2 = 93.455$, df = 3, $p = 0.000$). Freeman-Tukey deviates in parentheses. Values in bold are significant at $p = 0.05$ level (±1.20). Expected and observed MNE values derived by summing values for all elements in that utility class.

can serve as observed values in this instance by assuming, for reasons given, that it is appropriate to treat this as a single fauna that cannot be subdivided or aggregated in any other fashion. Contingency table analysis (table 7.18) reveals that there is a disparity in element representation by utility: Low-utility elements are significantly overrepresented in this assemblage, while high-utility elements are significantly underrepresented (as measured by Freeman-Tukey deviates).

Curiously, we only have one-third the number of crania that we might expect from a kill of this size. Crania, of course, are not readily transported elements, nor are they of especially high utility, unless of course the brain is being exploited (Wheat 1972:102). Yet, none of the cranial fragments or (obviously) any of the six complete crania recovered from the site are broken in such a way as to indicate that brains were being utilized (e.g., Frison 1991:fig. 5.4; Walker 1980:fig. 72). Even when crania are broken to gain access to the brain, they are almost always left behind (Ewers 1955:170). Why, then, are crania underrepresented at Folsom? Several hypotheses come to mind. First, crania, as large bulky items of low utility, can interfere with butchering activities and, thus, may have been selectively removed from the immediate kill area—but not too far, given their mass. Evidence in support of this hypothesis is in the relatively higher concentration of skulls in the margins of the bonebed where the 1990s excavations took place, and not in the central portion of the bonebed where the 1920s excavations occurred. This would further explain the disparity in the number of crania between the 1920s excavations (MAU%, 32.3) and the 1990s excavations (MAU%, 85.7) and the fact that of the six complete crania recovered, three were from the significantly smaller 1990s excavation area.

Second, despite their mass, crania are relatively fragile elements, the tops of which would have been exposed for longer periods than other elements (Todd and Rapson 1999).

Since there was rapid, but not instantaneous and universal burial, crania would have been exposed longer to weathering and deterioration and to variable preservation across the site. Finally, excavation techniques used in the 1920s may have fragmented crania badly enough to have affected relative MNE counts. Such is evident in a comparison of the ratio of NISP:MNE for crania from the AMNH (~23:1) and SMU/QUEST (2.6:1) excavations. For these reasons we suspect that the scarcity of crania may be more apparent than real, and their numbers may not accurately represent their original occurrence and frequency. Although we do not exclude cranial elements from our subsequent analysis, we do not give them undue analytical weight.

Segmentation and Use of the Carcasses by Folsom Hunters

On the presumption that there was selective use of the carcasses, we turn our attention to the manner in and intensity with which the carcasses were segmented. None of the elements recovered from the site were broken for marrow. There is virtually no evidence of bone impact fractures. Nor do any elements show signs of on-site processing for bone grease. In effect, the nutritional value of each carcass was not completely exhausted, in keeping with the general pattern seen at other Paleoindian kills (e.g., Bement 1999b; Hill and Hofman 1997; Todd et al. 1997; Wheat 1972; cf. Hill 2001:54) and in contrast with, for example, late Prehistoric kills (e.g., Bartram 1993:121; Frison 1982b; Todd 1991; Todd et al. 1997).

The butchering was relatively thorough: Bison remains uncovered during our excavations were from mostly disarticulated skeletons, save for a few articulated and/or conjoined skeletal elements (figs. 7.2A and 7.13). We cannot say for certain whether the observed patterns of skeletal disarticulation and scattering resulted from dismemberment during butchering or through diagenesis, because it is dangerous to make direct linkages between disarticulation and intentional dismemberment (Todd 1987b). However, the articulation patterns coupled with evidence of selective removal of high-utility elements from the Folsom bison carcasses provide strong circumstantial evidence that disarticulation was primarily the result of intentional carcass segmentation.

Yet, despite the apparent thoroughness with which the bison carcasses were taken apart, the recovered bones show remarkably few cutmarks on their surfaces. Indeed, of the sample of ~1,500 bones examined for this purpose, we observed only 7 elements with cutmarks (less than 1%), and of those, only 4 were definite. Obviously, given the excellent condition of the bone surfaces, butchering marks would be visible, if present. However, as recent studies have shown, cutmarks are not a necessary consequence of butchering (e.g., Egeland 2003), nor are they common in Folsom-age bison kills (e.g., Bement 1999b; Hill 2001:83; Hill and Hofman 1997; Todd, Hofman, and Schultz 1992).

Not surprisingly, those few elements with cutmarks are those that would be expected to show signs of butchering:

FIGURE 7.13 String of articulated thoracic and lumbar vertebrae in M17 block. This is the only significant articulated series of elements recovered In the SMU/QUEST excavations. Note the nearby scapulae. (Photo by D. J. Meltzer.)

mandibles (n = 5), ribs (n = 2), and an ulna (n = 1). While of very limited evaluative strength, the location of at least one group of cutmarks on the olecranon process of an ulna (fig. 7.14) suggests that a forelimb joint targeted for dismemberment was the humerus-radius/ulna articulation. Support for this interpretation is provided by skeletal element frequencies (fig. 7.12). Upper forelimb elements (scapula and humerus) are represented at a lower frequency than the radius-ulna, carpals, metacarpals, and phalanges. The scapula-humerus may have been an easily removed and handily transported carcass segment. Indeed; it is necessary to remove those portions of the carcass in order to fully access ribs and thoracic vertebrae. Along these same lines, other bones in the "high-utility elements transported off-site" category (fig. 7.12) also make sense in terms of a series of transportable anatomical packages: upper hind limb (femur and tibia), vertebral segments, and rib slabs.

Those packages can be seen in the relative frequency of elements left on site and can be used to infer butchering patterns. Folsom-age sites are often characterized by what has been termed a "gourmet butchering pattern" (Todd 1987b, 1991; also Binford 1978b; Wissler [1910:41–42] called it "light butchering"), in which there are a large number of low-utility elements and a correspondingly low number of high-utility elements. Although the pattern at the Folsom site proper suggests gourmet butchering, it does not conform to that "idealized" form; few sites do (see discussion by Hill 2001:72–74). Instead, we see not just attention to the highest-utility elements and only those elements (e.g., those with utility indexes higher than, say, 70) but, instead, a more eclectic use of high-utility elements, ranging from 25 to 100 (e.g., the right-hand box in fig. 7.12). In effect, once above a

utility value of ~25, there seems to have been little element-specific selectivity or gourmet usage.

Ribs may be the exception to this pattern. These are the highest-utility elements within the bison carcass and, correspondingly, have the lowest representation within the assemblage (fig. 7.12). Aside from the ribs, the remainder of the elements with utility values >25 do not appear to have been removed in a significantly greater or lesser proportion than the mean value for this group (average MAU%, 48.1). Importantly, none of these elements, ribs included, were entirely removed from the site.

All of this leads to the conclusion that we are seeing a pattern of selective processing and removal from the immediate kill area of 40% to 70% of the more desirable elements (utility >25) within the carcasses. Accordingly, the majority of what was left behind were low-utility skeletal parts, having been stripped of about half the high-yield elements. This is a pattern characteristic of initial field butchering and processing in a kill area (e.g., Cannon 2003; Hill 2001; Jodry 1999a; Kreutzer 1996; Wheat 1972).

We infer that this processing took place essentially where the animals were dropped, on the assumption that heavy elements like crania, of which at least half a dozen complete specimens were found, would not be moved any significant distance (see also Roberts 1936:15–16; Wheat 1972). In the absence of bonebed maps from the 1920s, we cannot say whether there were more subtle differences in the character of this initial butchering across the kill area. That said, the 1928 plan map does indicate that the west side of the bonebed yielded "more or less mixed" skeletons and, apparently, a higher density of bison remains (chapter 4). Given the overall structure of the faunal assemblage, and in

FIGURE 7.14 Cutmarks on the olecranon process of a bison ulna from the Colorado Museum excavations. (Photo by D. J. Meltzer.)

the absence of additional spatial information, this might be interpreted as an area in which animals were simply more closely spaced, rather than being an area in which different kinds of activities were taking place.

The observed patterns of butchering leave unanswered many details of the bison processing and indirectly raise questions about other aspects of the site and occupation—such as whether there exists or once existed other areas to the site, a matter puzzled over almost from the outset of investigations (e.g., Wissler 1928, Roberts 1936:14–15). Ethnographic and ethnoarchaeological evidence suggests the rough guideline that it takes a minimum of 1 to 2 hr of processing time per animal (L. R. Binford, personal communication, 1999; Ewers 1955:160; Wheat 1972:109–110, 116; Wissler 1910:41). This figure varies, of course, depending on the size of the animals, the spatial scatter of the carcasses, the extent of the butchering, the size of the labor force working on the task, the skill of the individual doing the processing, the available tools, whether carcasses were processed serially or in parallel, and even the temperature—among other factors (Cannon 2003; Frison 1991:141, 299–302).

The butchering of the bison at Folsom was fairly thorough, and ample time was spent ensuring the removal of heavy, lower-utility elements and the preparation (and mass reduction) of higher-utility elements for transport (Cannon 2003; Meltcalf and Barlow 1992). Taken as no more than a rough estimate, one can argue that the field processing of ~32 Folsom bison must have taken at least two to three days. Presumably, the hunters would have camped nearby

during that time to protect the kill from scavengers (Frison 1991:301; Roberts 1936).

What activities might have taken place in an associated camp area depends in turn on what followed the initial processing of the carcasses in the kill area. The relative scarcity of high-utility elements in the bonebed suggests that those parts were removed, but to where is unknown. There are at least two possibilities, each of which has different implications for the kinds of activities that might have taken place and the amount of time that might have been spent on-site.

- The high-utility elements were not processed on-site but transported off-site in meat/bone packages (e.g., rib racks or other segments) for subsequent processing. This implies that only the minimum amount of time and activity necessary to initial butchering and processing was spent on-site.
- The meat stripping/drying of the high-utility elements took place on-site, in an as yet undiscovered area. This implies more activities and a longer period of residence, but how much more and how much longer would depend on whether the group stayed only for the time it took for meat stripping and drying or whether they chose to make an extended stay in the area, and perhaps also on whether this was a logistical task group or a residential group.

Which strategy was taken depended on several factors, many the same as those determining the time that it takes to

TABLE 7.19
Non-bison Remains from the Folsom Site

Taxon	Elements Reported (CMNH No.)	NISP	Reference
Sorex sp.	Cranium, L and R mandible, axis, scapula, L and R humerus, L ulna. L and R innominate, sacrum, R femur	11	This book
Canis cf. *Occidentalis* (syn. *C. lupus*) (gray wolf)	None specified		Brown (1928b)
Citellus (syn. *Spermophilus*) sp. indet. (ground squirrel)	Humerus (CMNH 1249)	1	Hay and Cook (1930:36)
Cynomys ludovicianus (blacktail prairie dog)	L ramus, including several molars and premolars (CMNH 1248)	4	Hay and Cook (1930:35)
Thomomys fulvus (syn. *T. bottae*) (Botta's pocket gopher)	Part of a skull, 1 incisor, 1 premolar, 2 molars (CMNH 1247)	4	Hay and Cook (1930:36)
Microtus sp.	Cranium, L and (2) R scapula, (2) L and (2) R humeri, (2) R and L ulnae, 3 radii, 33 vertebrae, 2 manubria, 12 ribs, L and R innominate, sacrum, L and R femur, (2) L and R tibia, 2 metapodials, and various fragments	113	This book
Lepus californicus (blacktail jackrabbit)	Femur, tibia, and humerus fragment (CMNH 1142)	3	Hay and Cook (1930:36)
Odocoileus hemionus (blacktail deer)	R femur shaft, R cannon bone shaft (CMNH 1244 & 1245)	2	Hay and Cook (1930:30)
Cf. *Antilocaprid*	Femur	1	This book

butcher the animals, but factored in as well are transport costs, which include the number and size of carcasses, the number of available carriers, and the distance to the next camp(s) (Bartram 1993:121, Cannon 2003; Emerson 1993:139–140, 150; Hill 2001:53–55, 77–78; Meltcalf and Barlow 1992; see also Roberts 1936:15; Wheat 1972:101ff; Wissler 1910:41–42). Large animal kills like this one permit transport decisions based more on body part utility than do small animal kills, since whole bison cannot be transported—unlike, say, whole rabbits. It is possible, even advantageous, to be more selective with regard to what is carried off-site when the carcasses are large and/or abundant (Cannon 2003; Emerson 1993:139). If temperatures are relatively warm or at least dry, groups could more easily butcher the animals and dry the meat, which would significantly reduce its weight and make it easier to carry (Bartram 1993:121, 131–132).

Unfortunately, resolving whether the Folsom hunters packed their spoils out soon after they completed the initial field butchering, or whether they lingered for a time, will not be possible until associated camp or habitation area can be located, or its presence altogether precluded.

The Non-bison Remains from Folsom

With Alisa J. Winkler
Other animals besides bison were found during the original investigation. Howarth and Figgins reported fragments of

"an unidentified form and a deer midway between the size of a black-tail and an elk" (Figgins to Brown, July 14, 1926, VP/AMNH; also Roberts 1940:59), as well as a variety of bones from other taxa. These were sent to Oliver Hay, who identified them as *Odocoileus hemionous*, *Cynomys ludovicianus* (blacktail prairie dog), *Thomomys fulvus* (Botta's pocket gopher, *T. bottae*), *Citellus* (syn. *Spermophilus*) species indeterminate (ground squirrel), and *Lepus californicus* (blacktail jackrabbit) (Hay to Brown, undated, but ca. September 15, 1928, VP/AMNH; the fauna is reported by Hay and Cook 1928, 1930:8).

A slightly different list of taxa is provided by Brown (1928b), who also reports *C. ludovicianus*, *T. fulvus*, *L. californicus*, and *O. hemionus* as coming from the site. He does not include *Citellus* but adds *Canis* cf. *occidentalis* (table 7.19). Further, Brown (1928b) reports that only "single jaws" of *C. ludovicianus*, *T. fulvus*, and *L. californicus* were found, which does not cleanly match Hay and Cook's (1930) description of the recovered elements.

Unfortunately, it is not clear from these reports whether both sources are referring to the same specimens or whether they are each reporting on faunal remains independently recovered in their respective excavations (in 1926–1927 and in 1928), especially since specimens moved back and forth between institutions and Brown's paper (1928b) was not a detailed report on the fauna. And, of course, they each report the presence of a taxon not reported by the other.

The extant collections do not help resolve the matter. At the Denver Museum, the Folsom faunal collection ostensibly includes remains of 16 other species in addition to bison (Drawer 7 of Quarter Unit 70). Yet, some of those elements are clearly are out of place in this site and collection (e.g., *Equus, Marmota*). Based on this and other evidence we suspect that faunal material from different sites was mixed together in one of the "Folsom" drawers. Fortunately, it was not difficult to identify the Folsom site bison bone. A query of the American Museum's Collections database yielded no mammals other than bison from the Folsom site, though a dozen other species from a "Folsom Cave."

Regardless, it is evident that very few non-bison remains were recovered from the site (Brown 1928b). Even if these represent two independently derived samples, they are likely a sample of no larger than a dozen identifiable specimens.

For the most part, there seemed to the original investigators little reason to assume that most of these remains were associated with the Paleoindian activities on site (Cook to Hay, February 11, 1930, OPH/SIA). Regarding the rodent remains from the site, Hay and Cook admitted the possibility that "they may have been intrusively included, at a somewhat later date than when the bison lived here," even though no evidence was found to indicate this (Hay and Cook 1930:35).

Hay and Cook (1930) reported deer from the site, as had others (Brown 1928b; Figgins to Brown, July 14, 1926, VP/AMNH; Howarth to Cook, May 18, 1928, HCP/AFNM; also Figgins 1927b:232), which Cook believed might "show some signs of association" (Cook to Hay, October 12, 1928, OPH/SIA). Once again, all may be reporting on the very same element—that is, merely echoing each other—or they may have independently recovered different elements from the same individual animal or different elements from different individuals.

The correspondence suggests that deer bones were recovered not during the 1928 season (Brown to Hay, September 10, 1928, VP/AMNH) but in earlier seasons. Howarth, in fact, reported deer from the site even before excavations had begun, and Figgins learned from his crew in 1926 that they had found "a much smaller species of animal—a considerable part of the skeleton—and Frank [Figgins] sent in a toe bone of it, which proves to be either a deer or an antelope. It is large for the latter but narrow for the former" (Figgins to Cook, August 3, 1926, HJC/AGFO; also Figgins to Brown, July 14, 1926, VP/AMNH).

Where those remains were found is not specified. According to Howarth, the deer remains he recovered in the spring of 1926 came from the site "on the northeast corner of the pit on the east side of the creek under that large bunch of overhanging brush" (Howarth to Cook May 18, 1928; Cook to Howarth, May 23, 1928, HCP/AFNM). Howarth was off in his orientation by ~60°—the small sketch accompanying his letter places the bones on the North Bank, in the northwest corner of the site, an error due to the fact that he had the arroyo draining north-south and not northwest-southeast.

Unfortunately, the purported deer specimens cannot be relocated. The one *Odocoileus* specimen in the Denver Museum collections labeled as coming from Folsom (catalog no. 2099) is identified as an *Odocoileus* femur, but the element is neither a femur (it is a tibia) nor, obviously, from an *Odocoileus*.

Evidence bearing on the question of whether deer were exploited at Folsom emerged during the 1999 season on the North Bank of the site. A 53.5-mm portion of a femur (specimen N22-3-1) from a medium-sized mammal was recovered from the back wall of unit N1055 E1003 (fig. 4.18). Granting the lack of provenience of elements recovered in the 1920s, this appears to be the same general area of the site where Howarth reported finding the deer bone (it is beneath that same brush overhang).

The specimen is a left proximal shaft fragment (caudal segment), including the partial distal end of the minor trochanter, and displays green bone fractures. Based on the robusticity of the minor trochanter remnant, the specimen is from a fully mature individual. The individual would have been comparable in size to a fully mature female mule deer or a fully mature male pronghorn. The specimen is clearly not an elk and seems too small to fit Figgins' description of "a deer midway between the size of a black-tail and an elk" (Figgins to Brown, July 14, 1926, VP/AMNH). However, it is not certain on what basis Figgins scaled his description; that is, whether it was based on careful comparative measurements or was just a rough estimate from memory.

In the shape of the distal minor trochanter and of the shaft from lateral ridge to the base of trochanter, the specimen we recovered more closely resembles a pronghorn than a deer. Therefore, it may be not a cervid but, instead, an antilocaprid. Although it is impossible to prove the point, it may be that the femur shaft found in 1999 came from the same individual whose partial remains were found in the 1920s. It has not escaped our notice that a right femur shaft was recovered during the Colorado Museum excavations (table 7.18), while the one we recovered was a left. Were this a matched set, then this recent discovery would resolve the question of whether deer (or antelope) were exploited at Folsom, for specimen N22-3-1 was not found in the Paleoindian-age *f2* sediments, but instead came from the overlying *f3* stratigraphic unit. And as discussed in chapter 5, the radiocarbon date on this specimen is 9,270 ± 50 ^{14}C yr B.P., making it substantially younger than the Paleoindian presence on-site.

Other non-bison fauna recovered during the 1999 investigations include a variety of small mammals (NISP = 124) recovered from the water screening operations. The specimens came from a single excavation unit, N1031 E998, from levels below the bison bonebed. They were identified to the genus level where appropriate. Cranial remains probably could be identified to the species level with adequate comparative collections not available to us. Two genera are present: the microtine rodent *Microtus* sp., the vole or meadow mouse, and *Sorex* sp., the long-tailed shrew.

Five species of *Microtus* are known from the extant fauna of New Mexico (Bailey 1931; Findley et al. 1975). Of these, the Folsom *Microtus* is unlikely to belong to *M. pennsylvanicus,* based on the morphology of the M2 (the Folsom M2 lacks an additional posterior triangle), or to *M. ochrogaster,* based on the Folsom M1 having five and not three closed triangles. The Folsom *Microtus* should be compared closely with *M. montanus, M. mexicanus,* and *M. longicaudus.* It is of interest that none of these five extant species of *Microtus* are currently known from the study area (Bailey 1931; Findley et al. 1975). Only *M. pennsylvanicus* and *M. longicaudus* are reported from Colfax County, to the west/southwest.

Findley et al. (1975) describe five species of *Sorex* in the extant fauna from New Mexico. None of them are reported from the Folsom area, and only *S. vagrans* and *S. palustris* are known from western Colfax County. The Folsom shrew is assigned to *Sorex* based on the presence of alveoli for five unicuspids in the maxilla. Unfortunately, the specifically diagnostic unicuspid teeth are missing.

The specimens described here may represent the remains of two individuals of *Microtus* and one of *Sorex.* Preservation is excellent—many delicate elements are present, and breakage is relatively minor. Almost all elements, excepting the tiny phalanges, are represented. Pitting or etching of the bone was not observed. There is no evidence of cultural modification of the bones—for example, burning or butchering. The few taxa and individuals recovered, and their excellent and relatively complete preservation, suggest that these Folsom small mammals may have died in and been excavated from their burrows. Species of *Microtus* and *Sorex* have been reported to utilize shallow burrows (Bailey 1931; Nowak, 1999).

Summary: The Folsom Bison Bonebed

Approximately 32 *B. antiquus,* members of a cow–calf herd, died in the fall at the Folsom site approximately 10,500 years ago. This is not a natural death assemblage, as evidenced by (1) the lack of pathologies, which suggest a relatively healthy herd; (2) the age structure, which indicates catastrophic rather than attritional death; (3) the landscape position, which is not next to a water hole or likely spot of natural death; and (4) the lack of evidence of carnivore predation (i.e., Todd 1987a:195). Instead, we see clear evidence of human activity in the creation of this assemblage, evidenced by the (1) projectile points embedded between the ribs, (2) the point:animal ratios (28:32), (3) the evidence of skeletal element removal (fig. 7.2B), and (4) the presence of cutmarks (fig. 7.14), albeit rare.

The evidence of carcass use, which reflects initial butchering and processing, took place primarily where the animals were dropped. The carcasses were dismembered with reasonable thoroughness. High-utility elements, such as ribs, vertebrae, and upper limbs, were mostly removed, though to where is uncertain. Discarded on-site were the low-utility elements (lower limbs, mandibles, and crania) and those unutilized high-utility elements that, for want of transport capability, need, or time, were left behind. There is no evidence of on-site marrow or brain extraction, although because some high-marrow content bones were removed, such may have taken place elsewhere (cf. Hill 2001:230 ff.).

Overall, we interpret this pattern as fitting the general model of gourmet butchering seen in other Folsom-age sites, but, as expected, the fit is imperfect, since there are variations on that theme, as recently demonstrated in taphonomically oriented analyses of bonebeds of this age (e.g., Jodry 1999b). Specifically, there is not an invariant focus on just the highest-utility elements (e.g., ribs) to the exclusion of all lower-utility elements, though as we noted, ribs—which have the highest utility—were removed at the highest frequency. Instead we see a more general use of higher-utility carcass "patches" or "configurations" (*sensu* Metcalf and Barlow 1992; Rogers 2000) such as limb groups like the humerus/scapula and femur/tibia.

In some instances, those patches may have been exploited not because of their own utility or food value (e.g., tibia), but because they were still attached to other, higher-utility elements (e.g., femur). Binford (1978b) refers to these as "riders." Rider transport may occur more frequently at kills made by hunter-gatherers en route elsewhere, who had little residence time on-site (e.g., Hill 2001:53–54).

In the case of other carcass patches, we observed the removal of elements that are not riders, such as the humerus/scapula segments. In this instance, their removal from the carcass is quite understandable in light of the fact that these are "structured" resources (Metcalf and Barlow 1992; Cannon 2003). That is, front leg elements must be detached in order to reach higher-utility elements such as ribs. Therefore, their handling costs are "amortized" over a broader range of utility values and their removal is more cost-beneficial than were these elements in these patches selected on the basis of their own utility (see the discussion in Cannon 2003). Once detached, these bone segments could have been removed for the cost of their transport, and that may have tipped the balance in the decision of whether such otherwise medium-utility elements were taken and used.

Having argued all this, however, we note that detailed empirical grounding for the relative costs of structured versus unstructured resource use has yet to be developed (Cannon 2003:8). But the data we have assembled here suggest that such patterns in resource use are occurring, and that future research along these lines would be quite profitable.

The bone lay exposed on the surface of the paleotributary for only a brief period of time after the animals died, during which there was only minimal scavenging by carnivores and no trampling by other animals. The overall lack of patterning to the orientation, the low angles of inclination, and the several articulated units clearly suggest that the bone was not moved or transported by natural processes any significant distance. In fact, we infer that the location of the bone piles in this area of the site roughly approximate the position of the animals at the time of their death. Once in place, the larger elements and articulated units served as microtopographic barriers influencing the density and distribution of the sheet wash of shingle shale that subsequently blanketed the bonebed. The sheet wash was initially gentle, as it did not move or cause major realignments of these elements but, instead, simply moved some of the smaller, disarticulated, and more mobile elements up against and around obstacles like crania. Figure 7.2A illustrates a calcaneus butted up against a scapula and multiple small bone fragments that came to rest upslope, but not downslope, of the large cranium.

Apropos of the search for the associated camp or processing area, we note that in this portion of the site, which contains bone elements in primary context as well as sheet wash coming off the valley wall, we found no evidence of any tools or retouch flakes or any other sign of occupational debris. What this suggests is that there was no camp or processing area to the west or upslope of this portion of the bonebed: that is, within the catchment area of that sheetwash.

In sharp contrast, skeletal elements found in the paleovalley, though also not exposed on the surface for very long or ravaged by scavengers, have inclination values that evince postdepositional transport. In many cases this led to breakage, loss of projecting articular ends (fig. 4.17), and the like. That the skeletal elements from the paleovalley do not show appreciably greater surface weathering suggests that they too were rapidly buried, even after having been reexposed and moved down the valley (fig. 5.10).

Notes

1. A Folsom bison was on mounted display at the Denver Museum from the late 1920s almost to the end of the twentieth century. It has now been relegated, headless, to a storage corridor in the Museum (fig. 4.20).

2. Figgins had originally left the matter of collaboration with Hay entirely up to Cook's "personal preferences and decision" (Figgins to Cook, November 30, 1926, December 6, 1926, HJC/AGFO). But just a month later he changed his mind and directed Cook to work with Hay (Figgins to Cook, December 28, 1926, HJC/AGFO). Cook tells a slightly different version of the story (Cook to Gidley, November 26, 1930).

3. Cook explained his taxonomic philosophy in Cook to Hay, February 20, 1928, OPH/SIA.

4. Cook, in fact, had initially suggested that the Folsom specimens were "closely related" to *Bison occidentalis* (Cook 1927a:244), but that included all the specimens from the site.

5. The type specimen of *Bison occidentalis* was not available to Figgins, save through published descriptions. The type specimen, moreover, lacked dentition (Figgins 1933:18). However, given that crania with that distinctive external pillar were otherwise similar to the type specimen, Figgins believed that the *B. occidentalis* was part of the same group. Hence, he subsumed *B. occidentalis* within his newly created genus, *Stelabison occidentalis*.

6. Figgins also did not fail to point out that Cook's measurements of the type skull were inaccurate, and he therefore appended his own (Figgins 1933:20). Cook took an equally dim view of Figgins' efforts at taxonomy but did not think they were "worth that much time and printer's ink" to correct (Cook to Schultz, November 27, 1931, HJC/AGFO). By then, of course, Cook's connection with the Colorado Museum had been severed, in what Cook interpreted as a vendetta by Figgins (see Chapter 2), so there was no love lost between the two one-time colleagues, and this might have exacerbated the sniping.

7. The competition between *B. taylori* and *B. figginsi* was nip-and-tuck: *B. figginsi* had "paragraph priority," but the holotype was crushed and reconstructed, and thus not a good candidate for a type specimen (Skinner and Kaisen 1947:181; see also McDonald 1981:94). On the other hand, *B. taylori* had been "popularly accepted, primarily because the holotype was found in association with artifacts of the well-known Folsom man." Still, in the end, taxonomic priority and *B. figginsi* won out (Skinner and Kaisen 1947:182).

8. The first tallies were only made seven decades later by Todd and Hofman, who in 1990 recorded the collection at the American Museum of Natural History (Todd 1991; Todd, Rapson, and Hofman 1996). They were only able to examine and tally the mandibles in the Denver Museum collection. In 1999 and 2000, Meltzer and Todd finally and fully inventoried the 1926–1928 collections at both the American and the Denver museums.

9. We occasionally found notes in specimen drawers that read, for example, "sesamoids with Denver bison skeleton" (this from Case 6-093, Drawer 1, at the AMNH).

10. We were not able to examine the cranium fragment collected by E. Baker in 1936, which is curated at the Maxwell Museum, University of New Mexico. However, Bruce Huckell kindly provided photographs and a description of the specimen, which consists of a portion of the right orbit, frontal, and horn core (Huckell, personal communication, 2003).

EIGHT

Artifacts, Technological Organization, and Mobility

DAVID J. MELTZER

The great majority of artifacts recovered from the Folsom site are projectile points. Found scattered across a bonebed, a few deeply embedded in bison skeletons, there is little doubt what these points were used for: They were weapons for killing very large game.

To be sure, these hunters must have had additional artifacts in their toolkit. Other sites of this age are replete with a variety of formal and non-formal tools, such as end and side scrapers, gravers, burins, ultrathin and other bifaces, cores, and preforms (e.g., Jodry 1999a; Root, William, and Emerson 2000; William 2000). But then these other sites are localities in which a wider range of activities took place. At Folsom, as the faunal remains attest, we have evidence only of where the kill and initial butchering took place. Were there associated camp and/or processing areas where more intensive butchering, meat and hide preparation, tool production, or other activities occurred, there would almost certainly be a wider variety of tools represented. This is not to say the Folsom kill area was completely devoid of other artifacts: Four flake tools were recovered from the site over the years, but unfortunately none in situ, and two of those have long since disappeared without study. Practically speaking, the analytical attention in this chapter must be on the projectile points, though not to the complete exclusion of those other tools.

Yet, while granting the limits to what can be learned of a group's adaptive technology when viewed through the narrow window of their weaponry (chapter 1; Bamforth 2002), projectile points can nonetheless provide an important vantage on Folsom mobility strategies and technological organization. Paleoindian projectile points in general, and Folsom points in particular, were routinely made of high-quality cryptocrystalline stone, they were carefully curated through multiple uses at many places across the landscape, and the broken and/or exhausted specimens were ultimately discarded far from where the stone used in manufacture was obtained.

That tendency on their part provides us with the analytical opportunity to explore a variety of issues and questions (chapter 1), such as (1) where and how these groups procured their stone, as a means of probing the scale of their mobility and something of their settlement strategies; (2) how they organized their technology, and maintained their weaponry over multiple uses(s) in time and space, to understand how they grappled with logistical disparity between where stone was procured and where it was put to use; (3) the patterns of breakage and discard in their points, to possibly glimpse how their weaponry was assembled (hafted), propelled (trust or thrown), and ultimately discarded (lost or abandoned); and (4) how or whether these points vary, as a means of perhaps detecting standardization in manufacture, form, and stylistic attributes that might lurk within this functional class.

Such issues were not of pressing concern to the original investigators at the Folsom site. Rather, and typically for the period, much of their attention was on identifying the basic features of the points, how they were mounted and used, and what they might represent in terms of a prehistoric "culture." As more such points were found, and found to vary, a long discussion also evolved over how best to define the type. These matters are explored in the first part of this chapter and in appendix E.

Folsom Projectile Points—First Impressions

I have never seen such a point before or the peculiarly marked material from which it is made. (J.D. Figgins to Hay, November 8, 1926, OPH/SIA)

To a generation of archaeologists raised on the belief that Pleistocene-aged artifacts ought to look primitive (Paleolithic), the finely made Folsom fluted points first discovered (fig. 2.10) were a decided surprise (Howard 1935: 123; Roberts 1941:105). Points of this sort had not been seen prior to 1926, or if seen—as some eastern fluted points had been—were not accorded any particular significance or assigned any great antiquity (e.g., Beauchamp 1897; Holmes 1897b).

Nowadays, the key attribute by which Folsom points are instantly recognized is, of course, their fluting. However, the diagnostic quality of fluting was not immediately appreciated when these points were first observed in 1926 and 1927, perhaps in large part because the initial sample of points was so small. Indeed, the fact these points were fluted seems not to have been even noticed. In their seminal papers on the site, Figgins' and Cook's observations centered on the similarity in form between the projectile points of Folsom and those of Lone Wolf Creek, the latter, of course, not being fluted at all (the Lone Wolf Creek specimens were later assigned to the Plainview type [Wormington 1948:10]). The only differences Figgins observed between the two were in the shape of their blades, those from Folsom having "decidedly more tapering at the point," and in their overall workmanship, those from Folsom being "quite superior" (Figgins 1927a:232, 234, 1928:82; also, Cook 1927a:244). The absence of fluting from the Lone Wolf Creek points went unnoticed by Figgins or, if noticed, was not accorded any significance.[1] Even Carl Schwachheim, who possessed a large artifact collection from the region, on the basis of which Figgins could state that Folsom points were indeed unusual, made no comment on the fluting. The day the first point appeared he merely observed he'd found a "part of a broken spear or large arrow head" (appendix B, July 14, 1926).

It was only after the 1927 season, and with five projectile points in hand, that Figgins realized that the "hollowed sides of the artifacts" were "not accidental, but the result of remarkable skill in chipping" (Figgins to Hay, September 7, 1927, DIR/DMNS, OPH/SIA; also Cook to Loomis, December 20, 1927, HJC/AGFO; Figgins to Nelson, September 27, 1927, DIR/DMNS). Indeed, by then he considered that "broad spall" to be the "chief character of that culture" (Figgins to Kidder, October 17, 1927, DIR/DMNS). Roberts, upon seeing the points in September 1927 (also Brown 1928b), similarly reported that they

> . . . are quite different from the forms found on the surface in that part of the country. They measure about 2 inches in length and about one in width and are very finely chipped. Down the center, on each side, from point to tang, is a groove formed by the knocking out of a single flake on each side. This resulted in a point similar in its nature to modern bayonets. (Roberts to Fewkes, September 13, 1927, FHHR/NAA)

Important Points about Folsoms

Roberts invoked the bayonet metaphor mostly, it appears, to more easily describe, to someone who had never seen fluting before, the look of the point (Nelson, grappling with the same problem and lack of a term, described a flute scar as a "flake bed down the midrib region" [Nelson to Figgins,

December 20, 1927, DIR/DMNS]). But Roberts' choice of metaphor bespoke the question of whether fluting had that function, or what its purpose might be, a matter that prompted considerable discussion (which continues to this day [e.g., Ahler and Geib 2000, 2002; Amick 1999a; Collins 1999b; Ingbar and Hofman 1999; Osborn 1999; Wilmsen and Roberts 1978]).

Coloring contemporary views of the function of fluting was the realization that it was a costly technology. Detaching a flute, Cook observed, was no easy task, and one that was surely accompanied by a high failure rate: "It is not easy to understand how this could be done without [the flute] breaking out crosswise and ruining the point" (Cook 1928b:40; also Figgins 1934:3). Under the circumstances, Cook reasoned, Folsom knappers must have had a "higher degree of skill in the art of stone working than that possessed by most races" (Cook 1928b:40) and/or that fluting was the work of specialists (Cook 1931b:103; for recent arguments along these lines, see Ahler and Geib 2000; Bamforth 1988, 1991; Bamforth and Bleed 1997; Bradley 1982).

Moreover, removing flutes "so thinned the points that they became extremely fragile," which perhaps helped explain "the rarity of perfect [complete] specimens" (Roberts 1935:17), a scarcity noticed by virtually all contemporary commentators (e.g., Brown 1928b:4, Howard 1935:122; Renaud 1931:13; Roberts 1935:17; also Collins 1999b:24). Without question, these points broke frequently. Brown was struck by the scarcity of complete specimens from Folsom. He supposed that was understandable in light of human behavior: "If the habit of the people was that of modern Indians, [after the kill] they subsequently recovered the unbroken arrows, for we have found only one perfect arrow out of the total of 16 excavated" (Brown 1928b:4). Roberts (1935:17), for his part, did not suppose that complete points had been removed but, instead, that the large number of broken points simply reflected the consequences of fluting, which weakened the points and made them more vulnerable to breakage when used (also Renaud 1931: 12–13). For E.B. Renaud (1931:15), who saw fluting as the final stage of a historical process that began with slight basal thinning of otherwise unfluted "Yuma" points, fluting was clear evidence that technological evolution was the harbinger of its own demise.

Yet, despite the fact that fluting appeared to be costly, possibly required production by specialists, and so weakened the point that it increased the likelihood of catastrophic failure in use, it nonetheless appeared to be ubiquitous among Folsom points, if in part only by definition. It therefore seemed to contemporary observers that there must have been a benefit or purpose to offset those costs and that this was not merely a decorative or stylistic feature. Cook (1928b:40) supposed that the "double grooving" of the points served two purposes and, like Roberts, invoked a military analogy, but as something more than a descriptive aid: "It offered a good seat on which to secure the shaft of the split stick on which it served as a point or head; and, second, such a point permitted a wound to bleed more freely. Bayonets similarly shaped were used in the World War." Cook did not notice the contradiction that if fluting had served the first purpose, then this effectively prevented it from having served the second purpose. If the flutes were buried beneath the bindings and elements of the haft, they could not have enhanced bleeding.

Assuming that fluting was functional, or conveyed a technological advantage, Renaud supposed that it was intended to facilitate hafting and/or to reduce the weight of the stone, so as to allow it to travel farther with more velocity (Renaud 1931:11–12; also Roberts 1935:17–18). Others suggested that fluting helped to "improve penetrating qualities, to permit the point to break off in the animal, to allow the head to slip out of the fore-shaft, and to promote bleeding" (Roberts 1935:18; also Brown 1928b:825).

The discussion of fluting and its potential importance in hafting in turn prompted the further observation that in many Folsom points the "base and edges for about one-third the length of the blade were smoothed" (Roberts 1935:20). Grinding was evidently related to hafting, but whether it was an incidental by-product of rubbing against a wooden or bone handle or a deliberate dulling of the edges in anticipation of hafting was at first unclear (Roberts 1935:20–21). However, most soon came to view the process as deliberate, intended to prevent "cutting the lashings with which [the points] were attached to the end of the shafts" (Roberts 1940:63; also Figgins 1935:6; Renaud 1934:3).

It also quickly became apparent that edge and basal grinding were not diagnostic of Folsom points alone, for when other Paleoindian lanceolates were discovered over the next decade, they too were ground (Figgins 1934:4; Roberts 1935:21, 1940:62–63). But then so were other "totally unrelated points" (Roberts 1940:62). Grinding was an unreliable criterion for establishing historical affinity (Figgins 1935:6).

It seemed self-evident that Folsoms "were weapon points and . . . especially well adapted for that purpose" (Renaud 1934:4; also Howard 1935:114). As additional Folsom points were found in non-kill localities, it came to be appreciated not all were necessarily used just for this purpose:

> It must not be forgotten that in primitive industries many artifacts may have secondary uses. The sharp edge of a spear point, for instance, may be utilized for cutting or scraping. It may also be that some of the points were blades or hafted knives such as are seen in other cultures. (Renaud 1934:4)

Given that these were primarily hafted weapons, and fluting may have enhanced the hafting, what might that reveal about the nature of the armature to which they were hafted? Figgins (1927a) initially identified the points as arrowheads, although Schwachheim believed they were instead the "link between the spear or lance & the notched

[arrow] points of a later day" (appendix B, August 29, 1927). He tried unsuccessfully to convince Figgins on the matter:

> Reasons: if they were arrows the shaft [stem] would be broken off and the arrow would be whole or broken, while the six or seven inch lance head would be broken off even with the end of the shaft thus [Schwachheim provides an illustration] and the shaft and the rest of the lance head would be lost and no doubt never found. While if they were arrows and broken the workmanship on the barbs would still be shown. The break wouldn't come in the broad part of the head. (Schwachheim to Figgins, December 23, 1926, DIR/DMNS)

Figgins considered the matter but concluded that the points were "far too slender and delicate for the heavy shaft that would be required for even a small deer." Spear points he was familiar with were always "heavier and cruder" (Figgins to Schwachheim, January 12, 1927, DIR/DMNS). He decided to stick with "arrowhead."

Roberts and Kidder, after examining the in situ point in September, 1927, came away convinced otherwise (Roberts to Fewkes, September 13, 1927, FHHR/NAA). As Kidder later explained to Figgins, the points were simply too large to be arrowheads and, instead, were lance points or "tip darts to be thrown with the atlatl" (Kidder to Figgins, November 19, 1927, December 15, 1927, DIR/DMNS; also Brown 1928b; Howard 1935:106; Roberts 1935:21; Wormington 1939:9). Figgins averred he had described them as arrowheads because of what he perceived as differences of opinion among the archaeologists on the topic and his own "personal lack of experience" (Figgins to Kidder, January 6, 1928, DIR/DMNS).

Nonetheless, on one matter Figgins (1928:82) was certain: These points were of "unique workmanship and embody a perfected design to attain the least possible resistance to entrance and the greatest accuracy" in hunting. Brown concurred: "Unlike the triangular form of modern arrowpoints, which wedge from the moment they enter the skin until they penetrate beyond the barbs, this type of point makes the greatest incision at the initial impact, thus, with the same impetus, penetrating farther into a resisting body than the triangular point" (Brown 1928b:825; also Crabtree 1966).

Folsom Manufacture and Technology

The Folsom site points themselves shed little light on the question of how fluting was accomplished (Cook 1928b:40), though Figgins early on suspected that the removal of the flutes "was practically the last process in fashioning these points" (Figgins to Kidder, October 17, 1927, DIR/DMNS). It was not until there was a larger sample of points from additional sites, which included early-stage manufacturing failures from Lindenmeier, that the process of fluted point manufacture began to come into focus. That production process was first outlined by E.B. Renaud (1934), then elaborated by Roberts (1935, 1936) based on his work at Lindenmeier. Though the details differ, our understanding of the production process today is not significantly different (e.g., Ahler and Geib 2000; Akerman and Fagan 1986; Boldurian and Hubinsky 1994; Bradley 1991, 1993; Crabtree 1966; Flenniken 1978; Frison and Bradley 1980; Gryba 1988, 1989; Nami 1999; Sollberger 1977, 1985, 1989; Sollberger and Patterson 1980; Tunnell 1977; Tunnell and Johnson 2000; Winfrey 1990).

Although lacking evidence of production, Folsom provided the first glimmer of the kind and source of stone used in point manufacture. As recognized early on, a variety of high-quality stone such as "flints, chalcedony, agate, jasper," all of which belonged to the "general class of silicates," was used to make the assemblage (Cook 1928b:40). Where the particular stone used by the Folsom hunters had been acquired was the subject of only cursory discussion. In November of 1930, Charles Gould sent Barnum Brown a package of Alibates agatized dolomite (Brown to Figgins, November 20, 1930, DIR/DMNS). Gould, then the Oklahoma State Geologist (recall his role in the Frederick site controversy; chapter 2), had worked in the Texas Panhandle several decades earlier and been the first to map and describe the Alibates Formation (Gould 1906; also Banks 1990:91). After comparing the stone from Alibates with the American Museum's Folsom points, Brown wrote:

> Six of ours are definitely jasper that could have been derived from material similar to that you have sent; one is probably jasper, and the eighth is petrified wood. Five the Colorado Museum collected are jasper of the same sort; one is quartzite; another probably petrified wood, the other two I cannot speak about authoritatively. (Brown to Gould, November 17, 1930, copy in DIR/DMNS; see also Brown to Figgins, December 19, 1930, DIR/DMNS)

Brown was puzzled as to why Gould thought the stone was agatized dolomite, rather than a jasper. Gould replied that it was because it formed a constituent part of a ledge of dolomite that outcropped over parts of the Panhandle of Texas (fig. 3.2), and so when he first saw it he "gave it the name 'agatized dolomite' and the name has simply stuck. It probably is a jasper as you suggest" (Brown to Gould November 17, 1930, DIR/DMNS; Gould to Brown, November 21, 1930, DIR/DMNS; see LeTourneau 2000:434).

Gould copied Brown's letter identifying the American Museum's points as Alibates, then sent it along with a stone sample to Figgins, that he might compare it with the Folsom site points in his possession (Figgins to Gould, November 24, 1930, DIR/DMNS). Figgins too saw that "four of the artifacts from the Folsom quarry are unquestionably the same as the Texas 'dolomite'" (Figgins to Brown, December 5, 1930, VP/AMNH).

Gould identified the best source of Alibates as being "on the bluffs overlooking the south Canadian River" and described its geological position and range of variation,

"from the pure gray dolomite through the impure cherty material to the agatized (or should we say jasperoid) material that I sent you, Mr. Brown, and Dr. Nussbaum" (Gould to Figgins, December 1, 1930, DIR/DMNS). Although Brown, Figgins, and Gould must have been aware of the great distance between Folsom and Alibates (~265 km), its implications for Folsom mobility were not discussed, not surprisingly given the times.

Nonetheless, Figgins thought it would be useful to go to Alibates to look for springs and, from them, nearby Folsom campsites, reasoning that "water and the dolomite are the all important factors." After finding the area where the greatest evidence of quarrying was apparent, one should "begin digging" (Figgins to Gould, December 4, 1930; DIR/DMNS).

Defining a Point Type/Defining a Culture

In their initial papers on the site, neither Figgins (1927a) nor Cook (1927a) offered a formal type description of a Folsom point. It was only half a dozen years later that Figgins (1934:3) provided one:

> The best Folsom types have wide spalls removed from the sides, beginning at the base and sometimes extending quite to the tip, producing a hollowed or "fluted" effect. The bases are concave, often to a depth of a quarter or an inch or more and thus forming ear-like backward projections. In practically all cases Folsom artifacts are widest forward of a point midway of their length, and their width sometimes equals, or exceeds, half of their length.

Importantly, Figgins was not prescribing a definition. He was instead identifying characteristics of the "best" Folsom points, by which, presumably, he meant the most instantly recognizable and least equivocal. By 1934 Figgins well understood that there was variability in the morphology and technology of Folsom points, as a consequence of differences in lithic raw material, reworking of the point, and "the personal equation" in knapping skill (Figgins 1934, 1935:2–3). That variability, in Figgins' view, would be most apparent in the concavity of the base, the edge retouch, and fluting. If a point was thin, it might not be fluted at all or might be fluted on only one side; this was, he observed, the case with the very first point found at the Folsom site (Figgins 1927a:fig. 3, 1934:3). Thus, the absence of fluting did not necessarily justify the exclusion of a particular specimen from assignment to the Folsom category, as long as it had the other "typical" attributes (Figgins 1935:2–3).

Figgins' thoughts in 1934 as to what constituted a Folsom point were in large part a reaction to typological arguments being made by others, especially E. B. Renaud, and appeared in the midst of a protracted debate over Paleoindian point typology (e.g., Cook 1931b; Cotter 1939; Figgins 1934, 1935; Howard 1935, 1943; Renaud 1931, 1932, 1934; Roberts 1935,

1936, 1937, 1939, 1940; Wormington [1948] provides a contemporary summary; see also LeTourneau [1998a]). That debate began several years after the original Folsom discovery, as hundreds and then thousands of similarly large, lanceolate projectile points were reported. At first, the newly discovered specimens were lumped in with Folsom points, mostly because they were alike in size and shape. But it was soon obvious that many of them, particularly the unfluted "Yuma" points found in great numbers on the Dust Bowl–scoured High Plains of eastern Colorado (e.g., Cook 1931b; Renaud 1931, 1932, 1934), were an awkward fit with Folsom.

Then, too, even larger and less finely made fluted points appeared across eastern North America (Shetrone 1936) and from the newly discovered sites of Clovis, New Mexico, and Dent, Colorado, where they were associated with mammoth remains (Figgins 1933a; Howard 1935). Though these points were also fluted, they were so obviously different from those at Folsom that to call them that would render the label meaningless. So they were given other names, no less awkward, such as "Generalized Folsoms" and "Folsom-like" (e.g., Cotter 1937; Figgins 1934, 1935; Howard 1935: 105–123; Renaud 1931, 1932, 1934; Roberts 1935:8–9, 1937:161, 1939, 1940:56; Shetrone 1936).

By the mid-1930s, terms like Folsom and Yuma had become "catch-all" categories that contained a wide variety of dissimilar forms that, as Roberts (1936:21) complained, had little chronological and cultural meaning. The resolution of the so-called "Folsom-Yuma Problem" (Howard 1935:110; Roberts 1936:21, 1940:61) would consume archaeological attention for more than a decade.

The details of that debate do not directly bear on the Folsom site so need not concern us in this chapter. Yet, the issue warrants discussion in a book on Folsom, for it illustrates the role of the Folsom site's assemblage in that debate; it helps explain why, when a formal type definition finally emerged, none of the points from the site was identified as the type specimen (Wormington 1939; but see Howard 1943:229; Roberts 1935:5, 1939:533); and it highlights, in sharp historical relief, why such types are at their root ill-defined concatenations of stylistic, functional, and technological attributes (Meltzer 2001). The latter, in a backhanded way, helps explain why the vagaries of point typology are of less concern to the present generation of Paleoindian archaeologists (Amick 1999a; Ingbar and Hofman 1999; Sellet 2001). Recognizing the importance of this topic, but also the unnecessary detour it would cause here, I refer readers to appendix E, where the Folsom-Yuma Problem is more appropriately discussed.

The Folsom material was so very different from any archaeological assemblage then known that it was immediately hailed as a separate "culture" or "race" by the paleontologists who excavated here (Brown 1928a:825; 1928b; Cook 1928b:39; Figgins 1927a:231, 1934:2). Their reasons for designating it so were not well articulated but included Folsom's high degree of skill in stoneworking (Cook 1928b:40,

1931:102; Figgins 1928); its geographic distribution on the western Plains (Renaud 1931:16); the subsistence based on hunting of extinct bison and, before Clovis points were differentiated, possibly other extinct species such as mammoth (Brown 1928a:825; Figgins 1928:19, 1934:5; Renaud 1931:16); the relatively recent Asian origin (Brown 1928a: 824; Figgins 1934:5); and, of course, its great antiquity (Cook 1928b:40, 1931:102; Figgins 1928:19, 1934:5). But did these characteristics make Folsom a culture, or even prove useful in defining one?

For a paleontologist a single skeleton or specimen could serve to define a new species. They apparently reasoned that using a single site or small set of sites to define a culture was little different. Yet, a skeleton once had a skin that enclosed it; archaeological sites did not have corresponding "cultural" skins. Archaeologists of the 1930s were themselves not without sin in using the term "culture," as their colleagues in social anthropology, most especially Julian Steward, were quick to point out (e.g., Steward and Setzler 1938). But even archaeologists of the time could see that the manner in which Brown, Cook, and Figgins were using the term *culture* was inappropriate (Roberts 1935:9). Roberts (1935:9) complained that there was too "much loose talk and writing about the 'Folsom Race,' the 'Folsom Culture,' and 'Folsom Man,' when actually all that was known was the characteristic point."

Roberts was not exaggerating either. In speaking of Folsom points, Brown (1928b:825) observed, "These artifacts are probably darts, antedating the bow and arrow, and if they are recognized as distinctive, may be designated as the 'Folsom culture.'" From a strictly anthropological point of view, Roberts (1935:9) lectured his paleontological colleagues, "It is still incorrect to speak of a 'Folsom Culture' because the remains so designated probably should be considered only as one aspect of a basic, widespread early hunting pattern which may have extended across . . . the continent." For Roberts, Folsom was not a culture but instead a "definite complex of associated implements" (Roberts 1935:9, also Howard 1935: 106; Roberts 1940:55–56). Figgins (1935:5), who had earlier advocated a Folsom culture, would come to agree.

Julian Steward may have wished for more discussion from Roberts of just what the Folsom culture (lowercase "c") may have been and what it represented, but given the evidence in hand in the 1930s, Roberts was acting quite reasonably, and certainly within accepted archaeological practice.

A Tally of Fluted Points Recovered, 1926–1928

It is difficult to ascertain how many projectile points or other artifacts were found at Folsom during the original investigations from 1926 to 1928. Contemporary sources variously record 14, 16, 17, and 18 projectile points (Schwachheim's diary, Brown [1928a:128] and Figgins [1929:9], Cook [1931b:102], and Brown [1936:813], respectively). The differences in the totals may in part have to do with the fact that not all of the points were found in the

TABLE 8.1

Data on Location and Depth (in Feet ['] and Inches ["]) of Projectile Points Found during the 1928 Field Season

Specimen No.	AMNH Point [a]	Depth (Below Surface)	Horizontal Location Relative to the Pillar
7	A	9'	[No data provided]
8	B	9'	[No data provided]
9	C	5'3"	10'5" North
10	D		[Found in the backdirt]
11	E	7'6"	25' North/4'10" West
12	F	9'3"	28'6" North/9'9" West
13	G	8'	20'10" North/9' West [original]
			20'10" North/17' West [overwritten]
14	H	8'	28'6" North/19'9" West [original]
			20'6" North/20'9" West [overwritten]
15	I	9'3"	19'10" North/27'5" West [coordinates reversed? If so, should be 27'5" North/19'10" [or 9'10"?] West)
16	J	11'9"	53' North /5'4" West
17	K	11'9"	

[a]AMNH designations linked to specimen and collection numbers in Table 8.2.

SOURCE: As recorded by P. Kaisen (from Kaisen, "Record of Folsom Bison Quarry,"1928, VP/AMNH).

field; some were recovered while working up the plaster-jacketed bison material in the laboratory:

> As Mr. Figgins has no doubt told you he ran into another Folsom point in position beside the skull of one of the Folsom bison which he was helping Reinheimer clean up in the laboratory a few days ago. It is but a tip [Sn 6] but at least that much more evidence. (Cook to Brown, November 4, 1929, VP/AMNH; see also Figgins to Brown October 23, 1929, VP/AMNH)

This particular point tip was photographed with the skull, but for reasons unclear, a midsection from another point (Sn 15) was added alongside, giving the false impression the two fragments refit. The photograph was then published with the erroneous caption, "First fragmentary Folsom point found associated with bones of extinct bison" (Wormington 1957:fig. 3).

Unfortunately, the number of points and point fragments cannot be checked against the actual specimens, as several points have since disappeared and are known only from Schwachheim's diary or published photographs (appendix B, figs. B.1 and B.2; also Howard 1935:plate XXXIII; Wormington 1957:fig. 7). Those losses are partly attributable to the sometimes lax curation procedures of nearly a century ago, which was worrisome even at the time. Cook grumbled to Figgins about the "unusual courtesy [that] has been shown to Mr. Barnum Brown in permitting him to have practically all the arrows found in this Folsom deposit, and carry them about the country with him while on lecture and other trips" (Cook to Figgins, December 17, 1928,

HJC/AGFO). Recall from chapter 2 that Brown (1928a:824), ever the showman, opened his talks on human antiquity in America by triumphantly raising the points for his audience to see. Several of those well-traveled points ended up in Brown's desk drawer at the American Museum, where they remained until his death in 1950. Only then were they transferred to the American Museum's anthropology collections (G. G. Simpson, December 15, 1950, Anthropology Accession Records, AMNH).

No provenience data are available for the projectile points recovered during the 1926 and 1927 seasons, save that they came from the relatively small area excavated by the Colorado Museum. Better, but still incomplete provenience information was recorded by P. Kaisen on the points recovered during the 1928 field season (Kaisen, "Record of Folsom Bison Quarry," unpublished field notes, 1928, VP/AMNH). Those data are listed in table 8.1 and, for historical consistency, are also referred to in this section by their AMNH letter. The horizontal and vertical distances, where recorded by Kaisen, are presumably measurements relative to the datum pillar (chapter 4). The points listed in table 8.1 are shown on the 1928 plan map of the site prepared by the American Museum (fig. 8.1). Unfortunately, that map does not label the points by the letters (A–K) with which they were designated in the field, or their specimen number, but simply by X's on the map. There are only 10 of those, since one of the points, AMNH D, was found on the backdirt. When a point was found can help convey where it was found, whether in the South or the North Bank, in the excavations as opposed to in the backdirt, during the original excavations or after, and so on.

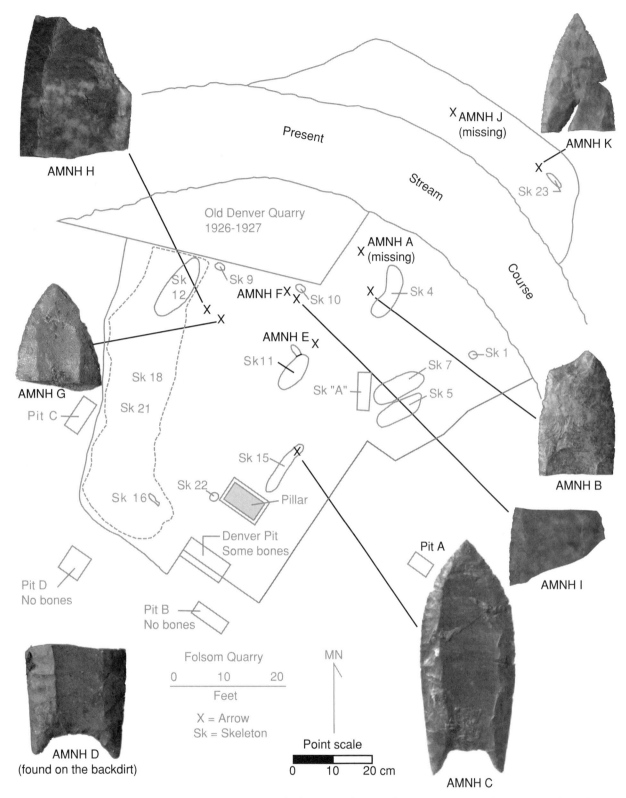

AMNH H

AMNH J
(missing)

AMNH K

Present

Stream

Course

Old Denver Quarry
1926-1927

AMNH A
(missing)

Sk 9

Sk 12

X X

AMNH F X X Sk 10

Sk 4

X

Sk 1

AMNH G

AMNH E X

Sk 11

Sk 7

Sk "A"

Sk 5

Sk 18

AMNH B

Pit C

Sk 21

Sk 15

Sk 22

Sk 16

Pillar

AMNH I

Denver Pit
Some bones

Pit A

Pit D
No bones

Pit B
No bones

MN

Folsom Quarry

0   10   20
Feet

X = Arrow
Sk = Skeleton

Point scale

0   10   20 cm

AMNH D
(found on the backdirt)

AMNH C

FIGURE 8.1 AMNH 1928 excavation plan map, showing the location of projectile points recovered that season. (Photos by D. J. Meltzer.)

Based on these sources, most of the points (AMNH A, B, C, E, J, and K) can be readily assigned to an *X*, and the remaining ones can be designated by process of elimination—granting certain errors in the 1928 map and the recorded proveniences. Their positions are shown in figure 8.1.

Taking these positions into account, along with what is known of the contours of the bonebed (fig. 7.1), it is clear that the majority of the points were found in the deeper portion of the western half of the site, virtually all at depths of more than ~2.1 m (7 ft) below the 1928 datum surface. That does not mean, however, that they were in deeper portions of the bonebed, as we do not know how thick the bonebed was in this area (evidence from our nearby excavations puts it at about ~25 cm thick) (chapter 7). The historic accounts suggest that there was a higher density of skeletal material in this area. Nonetheless, it is not possible to use these data to argue that the points were deep within a pile of carcasses and were therefore beyond the reach of the Folsom hunters, as one can argue for other Folsom and later kills in which carcasses were deeply stacked (e.g., Hofman 1999a; Wheat 1972).

Other Classes of Artifacts from Folsom

If only imprecise information is available on the number and provenience of projectile points recovered from the site, even less is known about other classes of artifacts that may have been found during the initial investigations or in the decades afterward. The original reports do not mention any tools other than projectile points. None may have been found, either because such were not present in the kill area (Roberts 1939:534) or because the work was conducted by paleontologists not using screens, who might overlook nondiagnostic artifacts like small flake tools. Or perhaps flake tools were found but were simply not of particular interest and thus were not reported.

The earliest and (essentially) only published record of other tools having come from the Folsom site was by Roberts, who, in the late 1930s, made passing reference to a "portion of a nondescript flake knife, and one example of a generalized type of scraper" (Roberts 1939:534, 1940:59). The current whereabouts of these two tools is unknown.

Roberts does not appear to be referring to the quartzite knife found by Ele Baker in November 1936, which is today in the possession of the Baker family. That specimen, according to photographs taken during that visit (fig. 8.2), was recovered from the North Bank, at the top of what appears to be the buff-colored sediments of unit *f2*. Judging by the known height of individuals in the photographs (T. Baker, personal communication, 2004) and the presence of the same fence posts seen in the 1928 photographs, the specimen came from a depth of ~3.9 m below the 1936 surface, at the downstream end of the 1928 profile wall, perhaps close to where Projectile Point K and skeleton 23 appear in figure 8.1. The artifact was apparently found next to fossil bone, perhaps the partial bison cranium recovered on that same visit (chapter 7).

Although the two flake tools described by Roberts appear to be lost, two specimens meeting that description appear in an undated but vintage 35-mm slide labeled "Folsom" provided to me by my colleague Fred Wendorf. They are shown alongside the quartzite knife found by Baker (fig. 8.3). Unfortunately, only a limited amount of information can be gleaned from the slide, since the image is not detailed and the colors are off, based on the known color of the Baker knife (of which, more below).

Was a Cache of Folsom Points Found Nearby?

As part of a National Park Service Theme Study, Erik Reed visited Folsom in the late 1930s and reported that a cache of "ten or a dozen Folsom points" had been found "on the rimrock [Johnson Mesa] west or north of the [Folsom] bison-quarry." The cache was said to have been discovered "by a Mr. Jim Macey of Clayton, New Mexico, who has the specimens in his barber-shop in that town" (Reed 1940:4–5). That a Folsom cache might have been found in the vicinity of the site is intriguing, given the remarkably sparse local archaeological record, the possibility that the cache might have been associated with the kill or at least with transit through this area, as well as the fact that no other Folsom cache is known to exist here or elsewhere on the Plains (Collins 1999a, 1999b; Meltzer 2002; Wyckoff 1999; cf. Osborn 1999:199).

However, it is not at all certain that the cache was genuine. Richard Louden, a longtime avocational archaeologist residing in the region, recalled having seen a cache of Folsom points in the late 1930s, about the time Reed reported this one (Louden, personal communication, 1998). The cache Louden saw was then in the possession of William Ross, President of Trinidad State Junior College (Trinidad, Colorado). Louden was deeply suspicious of its authenticity, since all of the points seemed to him "too perfect" and all were of the same material: Alibates agatized dolomite. He surmised that the cache had been manufactured by Marvin McCormick, a well-known flintknapper who lived in nearby Pritchett, Colorado, specialized in producing Paleoindian points, used Alibates almost exclusively, and whose primary livelihood during the 1930s Depression was making and selling stone tools (John Whittaker, personal communication, 1998; also Whittaker and Stafford 1999). In those same years E.B. Howard remarked that it was quite curious that many "excellent arrowheads and other artifacts" were being made for the tourist trade, but he had "yet to meet anyone who can produce a Folsom point" (Howard 1935:110). He had obviously not encountered McCormick or his work.

Unfortunately, it is not possible to say whether the cache Louden saw and the one reported by Reed (1940) were one and the same. The current whereabouts of this cache (caches?) is not known. But Louden has heard of no other cache of Folsom points from this area. Given the reasonable doubts that might be raised about its authenticity, the disappearance

FIGURE 8.2 A. Ele Baker pointing to the spot on the North Bank where he found the quartzite knife, 1936. B. Wider view taken on that same visit, showing the position of the find spot relative to the surface. (Photos courtesy of the Maxwell Museum.)

may not be that profound a loss, save for those interested in McCormick points and technology.

Investigating Folsom Assemblage Variability

As stated at the outset of this chapter (also chapter 1), there were multiple questions guiding the recent analysis of the artifact assemblage from Folsom. It is not a large assemblage or a very diverse one, being comprised of just over two dozen projectile points and a handful of flake tools. Gathering the data for the analyses was something of a logistical challenge (though not an insurmountable one), since the artifacts are scattered in public and private collections in more than half a dozen locations across the country. This being the Folsom site, several of the projectile points are on permanent museum display, and though casts are available, they are otherwise inaccessible. An equal challenge is ascertaining just how many artifacts have come from the site, a task made difficult by the unknown number of points that may have been collected by site visitors over the years and by the apparent disappearance of a few of the known specimens.

I therefore begin this section with an accounting of the projectile points from the site, before moving on to an examination of lithic raw material patterning, projectile point morphometrics, and patterns of hafting, maintenance, breakage, loss, and discard in this class of artifacts. I then turn attention briefly to the few other tools from the site, before closing with what this assemblage reveals of Folsom mobility strategies and technological organization.

Assembling an Analytical Sample

Overall, the Folsom site assemblage is comprised of at least 28 projectile points and/or point fragments, some of which have disappeared and are known only from photographs or entries in Schwachheim's diary. A full listing is provided in

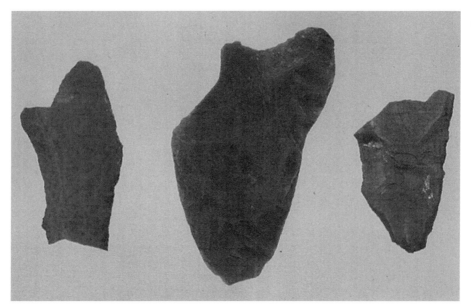

FIGURE 8.3 Three of the four flake tools known from the Folsom site. The center specimen is the Baker knife; the whereabouts of the other two specimens are unknown. The Baker specimen is 9.5 cm long. (Photo courtesy of F. Wendorf.)

table 8.2, which is based on my own tracking of points in public and private collections, checked against a similar inventory provided by Adrienne Anderson (personal communication, 2001). The list is arranged and numbered by order of discovery, as best as that could be determined. It is so organized for several reasons, not least that in several cases the date of discovery is all that is known of points that have since disappeared. The points are referred to throughout this chapter using the specimen numbers (hereafter Sn) in table 8.2 rather than by their field or museum catalog numbers, since not all of the points have corresponding catalog numbers, and this provides a common and less cumbersome numbering system.

Recall that less than a decade after the original excavations were complete, Renaud (1934:2) put the total number of projectile points from Folsom at 19, a total that apparently included specimens recovered from the site's backdirt in the years immediately following the excavations (the same total is given by Roberts [1935:17] and Wormington [1939:6]). Such discoveries were made in 1931 by Brown (Sn 18), in 1933 by Howarth (Sn 19, Sn 20), and in 1934 by E.B. Howard (Sn 21) (see Howarth to Cook, February 26, 1932, and Howarth to Cook, July 25, 1933, HJC/AGFO; Howard 1943; Roberts 1935:17). In later decades, there were other visitors who collected artifacts at the site, most notably Homer Farr, longtime Superintendent of nearby Capulin Volcano National Monument. Farr found at least three fluted points on the site, two of which are currently on display at the Capulin Monument Visitors Center (Sn 22, Sn 23). The third is in a private collection (Sn 24), given by Farr to the father of the present owner. Several additional points found at Folsom by site visitors have come to the attention

of the archaeological community (Sn 25, Sn 27, Sn 28); others, surely, have not.

Of the 28 points from the site, 6 have disappeared (Sn4, Sn 7, Sn 16, Sn 19, Sn 20, Sn 26). Fortunately, casts and photographs are available for two of the missing specimens (Sn4, Sn 16). Of those 24 points and point fragments, I was able to examine 17 original specimens and casts of the remainder (fig. 8.4A and B). That includes the casts of the two missing specimens (noted above), as well as five specimens currently on museum display or otherwise unavailable (Sn 3, Sn 9, Sn 11, Sn 12, Sn 28). Measurements taken from casts should be accurate to within a millimeter or two and should not significantly bias the metric data. However, it was oftentimes impossible to determine certain critical attributes from the casts, such as the occurrence and extent of edge and basal grinding.

Of the 24 points and point fragments known, all appear to represent separate projectile points and do not conjoin, save Sn 13 and Sn 21, the quartzite tip and base. The tip was found in July 1928 and originally designated AMNH G. The base was found in 1934 by E.B. Howard "near the bottom of the dump on the left-hand side facing the datum point" (Howard to Brown, January 7, 1937, VP/AMNH). Soon thereafter Howard borrowed the quartzite point tip to see if the two refit. He initially had trouble joining them, but ultimately concluded:

There is no question now that they are part of the same specimen. The reason the two casts we tried to fit together at the Museum that day would not go together, is that the break was at an angle. When the middle line of the tip was lined up with the middle line of the base, they fitted all right. There is some difference in the weathering which indicates that they were separated for

TABLE 8.2

Complete Inventory of Projectile Points from the Folsom Site, 1926 to Present

Specimen No. (Fig. No.)	*Collection*	*Descriptive Notes (Published Photographs)*
1 (2.10, 8.4A)	DMNS 1391/3	Found July 14, 1926. Lacks base, but includes a portion of haft area, as indicated by presence of small amount of edge grinding (Figgins 1927:fig. 3, left; Howard 1935:plate 33, top row, second from right; Wormington 1957:figs. 6, 7, top row, second from left). *Schwachheim diary entry:* "Found part of a broken spear or large arrow head near the base of the fifth spine taken out. It is about 2 inches long & is of a dark amber colored agate & of a very fine workmanship. It is broken off nearly square & we may find the rest of it. I sure hope we do." (See appendix B, fig. B.1)
2 (2.10, 8.4A)	DMNS 1261/1A	Found October, 1926. Lacks base; edge grinding absent. Blade portion was "refit" to small midsection wedge in CMNH laboratory (Figgins 1927:fig. 3 right, fig. 4; Howard 1935:plate 33, top row, third from right; Wormington 1957:fig. 5).
3 (1.4, 8.4A)	DMNS 1262/1A	Found August 29, 1927. Reworked, and complete except for missing corner. This is specimen examined in situ in September, 1927 (Howard 1935:plate 33, top row, first on left). *Schwachheim diary entry:* "I found an arrow point this morning it is of a clear colored agate or jasper. It is not exposed the full length but it is hollow on the sides and looks something like this [sketch]. The point was near a rib in the matrix. One barb is broken off." (See appendix B, fig. B.2)
4 (8.4A)	Original specimen is missing; cast at AMNH	Found August 29, 1927. Base only, with missing corner (Howard 1935:plate 33, bottom row, second from right; Wormington 1957:fig. 7, bottom row, third from right). *Schwachheim diary entry:* "Since noon Mr. [Floyd] Blair found another not in place but in the loose dirt; is much the same shape 1 inch wide at break and ¾ at base. Shaped like this but more of it [sketch]. Made of dark red flint."
5 (8.4A)	DMNS 1391/2	Found 1927. Lacks base, but includes a portion of haft area, as indicated by presence of small amount of edge grinding (Howard 1935:plate 33, top row, third from left; Wormington 1957:fig. 7, bottom row, second from left).
6 (8.4A)	DMNS 1263/1A	Found 1927. Point tip, with only small part of flute visible on one face (Wormington 1957:fig. 3 incorrectly shows this tip refit to midsection Denver 1391/1).
7	None; specimen is missing	Found June 25, 1928. **AMNH A**. Whereabouts of original unknown, nor does a cast exists. Specimen known from a sketch in Schwachheim's *diary,* which shows a point midsection, longitudinally split. Also photograph on file, AMNH 411138 and 411139. *Schwachheim diary entry:* "Glenn [Streeter] found a piece of broken arrowpoint something like this" [sketch].
8 (8.4A)	AMNH 20.2.5871	Found June 27, 1928. **AMNH B**. Lacks base and tip shows signs of reworking and impact fractured (Wormington 1957:fig. 7, bottom row, first on left). *Kaisen field notes:* "found near No 4 skeleton" *Schwachheim diary entry:* "Mr. Brown found a broken point today. It is shaped like all the others. Found 9 feet 1½ inches below the surface" [sketch].
9 (8.4A)	AMNH 20.2.5865	Found July 13, 1928. **AMNH C**. Complete, but with excavator breaks (Howard 1935:plate 33, top row, first on right; Wormington 1957: fig. 7, top row, first on left).

(*continued*)

TABLE 8.2 (*Continued*)

Specimen No. (Fig. No.)	Collection	Descriptive Notes (Published Photographs)
		Schwachheim diary entries: "Ernie [Kaisen] struck another arrowpoint of jasper. It is identical in shape and chipping with the rest. He broke the point off but the rest is in the bank in place. Not yet exposed" [sketch] [July 13, 1928]. "Ernie's point is about 6 inches from the bones & is 2 3/8 inches in length. It is complete only where he broke it with the pick" [July 19, 1928].
10 (8.4A)	AMNH 20.2.5867	Found July 16, 1928. **AMNH D**. Base only, with lateral snap occurring just beyond haft area. Found in backdirt in 1928 (Howard 1935:plate 33, bottom row, third from right; Wormington 1957:fig. 7, bottom row, second from right). *Kaisen field notes:* "Found on dump." *Schwachheim diary entry:* "Mr. [Pete] Kaisen found a broken point on the dump. It seems to be a little wider than the others but the same shape as all the others. It was broken off about 1 1/4 inches from the base & has lime formed on it" [sketch].
11	AMNH 20.2.5866	Found July 17, 1928. **AMNH E**. Nearly complete, but with impact fracture and burinated tip, and slight damage to base (Howard 1935:plate 33, bottom row, third from left; Wormington 1957:fig. 7, bottom row, third from left). *Schwachheim diary entry:* "I found another point today. It looks like agate and is about 2 inches long with the point broken off. It looks something like this. [sketch] In shape it is just like the others & the workmanship is very fine."
12	AMNH 20.2.5868	Found July 23, 1928. **AMNH F**. Reworked, but otherwise complete. Impact fractured tip (Howard 1935:plate 33, bottom row, first on left; Wormington 1957:fig. 7, top row, second from right). *Kaisen field notes:* "A heavy rain washed around. Where it had worked [in back of] bones, and on uncovering the paper that had covered the bones arrow was showing. Over 1 inch sticking out. Point had break, struck with a digger [?]. Schwachheim was the one that [?] the paper of[f?]." *Schwachheim diary entry:* "We opened up everything this morning & while doing so I found another point which the rain had washed out. It has been struck by a pick or tool & the point broken off. It is of nearly clear agate & the same shape as the others. Only about ½ is exposed [sketch]. I hope the base is there."
13 (8.4A)	UM 34-30-1 (tip)	Found July 27, 1928. **AMNH G**. Point tip, found by AMNH and refit to base found by Howard in 1934 (Howard 1935:plate 33, bottom row, first on right; Wormington 1957:fig. 7, bottom row, first on right). But see text. *Schwachheim diary entry:* "I found the point of a broken point something like this [sketch] & this evening . . ." [continued onto **AMNH H**].
14 (8.4B)	AMNH 20.2.5872	Found July 27, 1928. **AMNH H**. Lacks base and tip (which is impact fractured), but includes a portion of haft area, as indicated by presence of edge grinding. Excavator breaks (Howard 1935:plate 33, top row, second on left; Wormington 1957:fig. 7, top row, first on right, but shown upside down). *Schwachheim diary entry:* ". . . Mr. Kaisen found the widest one of them all. It is of jasper & the very finest of work. He struck it & broke it into 5 pieces. It is shaped about like this [sketch] & it is sure fine stone."

TABLE 8.2 (*Continued*)

Specimen No. (Fig. No.)	Collection	Descriptive Notes (Published Photographs)
15 (8.4B)	DMNS 1391/1	Found July 30, 1928. **AMNH I**. Midsection only (Wormington 1957:fig. 3 incorrectly shows this midsection refit to point tip Denver 1263/1A). *Schwachheim diary entry:* "Mr. Kaisen found another fragment of an arrow point [sketch]. It is of white agate & a very fine stone."
16	None; specimen is missing	Found August 28, 1928. **AMNH J**. From North Bank. Tip broken, and apparently reworked, but otherwise complete; whereabouts of original unknown (Howard 1935:plate 33, bottom row, second from left; Wormington 1957:fig. 7, top row, third from right). *Kaisen field notes:* "North of creek." *Schwachheim diary entry:* "Ernie [Kaisen] found the base of another arrow today. This one is No 10 for this season and the lowest one found, 11' 9". It is the only one found so far on the north side & is of an agate or jasper, red with a few clear stripes" [sketch].
17 (8.4B)	AMNH 20.2.5869	Found August 29, 1928. **AMNH K**. From North Bank. Point tip and blade; point broke above the haft area; fluted on one face only. Excavator breaks (Wormington 1957:fig. 7, top row, third from left). *Kaisen field notes:* "Found in a drape and near bones of No. 23." *Schwachheim diary entry:* "Ernie [Kaisen] found another point this evening. It is the same material as his last one & nearly fits on it. This specimen is No. 11 [sketch]. He struck it with a pick & broke it in 3 pieces."
18 (8.4B)	AMNH 20.2.5870	Found in 1931 by B. Brown on the backdirt pile (Howarth to Cook February 26, 1932, HCP/AFNM). Lacks base and tip, but includes a portion of haft area, as indicated by presence of small amount of edge grinding. Possibly reworked. Remnant flute visible on one face only.
19	None; specimen is missing	Found in 1933 by F. Howarth. Specimen reportedly "a good arrowhead . . . not the typical Folsom point but exactly the same type as the one found by Brown a year or two ago" (Howarth to Cook, July 25, 1933, HJC/AGFO).
20	None; specimen is missing	Found in 1933 by F. Howarth (same visit on which he found Specimen 19). This pieced described as a "piece of chipped flint which was evidently a broken point" (Howarth to Cook, July 25, 1933, HJC/AGFO).
21 (8.4B)	UM 34-30-1 (base)	Found in 1934 by E. B. Howard (Howard to Brown, December 19, 1941, VP/AMNH). Point base, refit by Howard to tip found by AMNH in 1928 (Howard 1935:plate 33, bottom row, first on right). But see text.
22 (8.4B)	CAVO-115	Found by Homer Farr. Lacks base, but includes a portion of haft area, as indicated by presence of grinding. In the 1970s, Farr reported that this point was "found about 20 years later [than CAVO-116] by Robert Hoke and me partly embedded in the shoulder blade of a prehistoric animal" (Farr to Peters, March 17, 1972, CAVO). There are discrepancies in this account, and Anderson (personal communication) records that Farr told Steen, during their 1950s visit to the site that yielded CAVO-116, that he and Hoke had found CAVO-115 the year before.
23 (8.4B)	CAVO-116	Found by Homer Farr, longtime Superintendent of Capulin Monument. Complete, but reworked. In the 1970s, Farr claimed to have discovered this point "in situ in 1926 in the skull of a prehistoric Taylor bison" (Farr to Peters, March 17, 1972, CAVO). There are discrepancies in that account, however. There is also a record this specimen was found in the company of C. Steen in the 1950s (Anderson, personal communication).

(*continued*)

TABLE 8.2 (*Continued*)

Specimen No. (Fig. No.)	Collection	Descriptive Notes (Published Photographs)
24 (8.4B)	B. Burchard collection	Found by Homer Farr, date unknown. Given to father of present owner sometime in the 1950s. Specimen is complete, but reworked.
25 (8.4B)	A. Brown collection	Found in 1969 in the "toss dirt" at the site by the present owner (A. Brown, personal communication).
26	None; specimen is missing	Found in 1970 by an individual on a commission visiting the site when it was being considered for State Monument status. She apparently refused to give it to a public institution (A. Anderson, personal communication).
27 (8.4B)	D. Brown collection	Found in the 1970s at the site by the father of the present owner (D. Brown, personal communication). Lacks base, but includes a portion of haft area, as indicated by presence of edge grinding.
28 (8.4B)	DMNS A2006.1	Found on the backdirt at the site in 1994. Point base, with evidence of impact damaged tip; missing corner (Dixon and Marlar 1997: figs. 1 and 2).

NOTE: Institutional collections identified as follows: AMNH, American Museum of Natural History; CAVO, Capulin Volcano National Monument; DMNS, Denver Museum of Nature and Science; UM, University Museum, University of Pennsylvania.

a long time, and were not broken at the time of excavation. There is other evidence: both pieces are the only pieces from the site made of quartzite; the fluting or groove is only on one side, and a flaw along one of the lateral ridges, marking an irregular line along one edge of the fluting, checks on both halves. (Howard to Brown, December 19, 1941, VP/AMNH; see also Howard 1943:228)

Because the points apparently cross-mended, the tip was donated to the University of Pennsylvania Museum (Howard 1943:228). Some time later the pieces were joined with plastic wood.

When this composite specimen was recently examined, the two joined parts were clearly not a seamless whole. The refitting was badly done and the parts were affixed as though they were congruent segments—which they are not, as Howard observed. With the permission of the University Museum's Curator of Collections, the two pieces were separated and the plastic wood was removed. Doing so revealed that the widest part of the tip is, in fact, wider than the widest part of the base by ~2 mm. Moreover, the base does not "flare" out, as it ought to in order to match up with the tip, were the two to be joined directly, though there is some slight edge damage where the base broke, which might account for this. There is a similar difference (~2 mm) in the width of the flute scars of the two segments, which seems greater than the variation one normally sees in scar morphology (Ahler, personal communication, 2005). Indeed, there appears to be traces of a flute scar on *both* faces of the tip, though indistinctly so on one face. Further, the tip broke in a hinge fracture, while the base broke flat, possibly in a radial fracture, although no point of impact point is vis-

ible. For that matter, although both are of quartzite, the two differ in grain size (the base appears finer-grained); it seems unlikely that a piece this small would differ that much in texture, but it is possible. Were they part of the same, highly varied piece of stone, there ought to be more variation within each fragment, but there is not. Both are quite uniform in texture.

If these are two pieces of the same point, there must be a midsection fragment that fit between them, and across which the texture of the stone changed significantly, the flute scar angled and narrowed toward the tip, while the segment itself expanded. Alternatively, the fragments are from two separate points. Although Occam's razor would suggest that the only two projectile points of white quartzite from the site, one a projectile point tip and the other a base, must be part of the same piece, that might not apply here. There are simply too many gaps between the fragments, literally and figuratively. Until a midsection filling those gaps is found, I will count them as separate specimens.

The four missing points (Sn 7, Sn 19, Sn 20, Sn 26) can, with caution, be included in the overall tally of projectile points recovered from the site. Based on Schwachheim's diary sketches, comments by Howarth (Howarth to Cook, July 25, 1933, HJC/AGFO), and Anderson's records, it appears that two of those missing specimens (Sn 19, Sn 26) were complete or nearly so; the other two were fragments. Unfortunately, one can only speculate whether the two fragments (Sn 7, Sn 20) would refit onto any of the extant broken points. However, I assume that the two complete or nearly complete specimens represent separate points and include them in the tally. Therefore, I estimate that a *minimum* of 26 separate projectile points has been recovered from the Folsom bison kill.[2]

Metric and nonmetric data of varying quality and completeness are available for 24 of the 26 points from the site, from examination of either the originals or casts. The dimensions measured on each point are listed in table 8.3 and shown schematically in figure 8.5; the resulting values for the Folsom site points are listed in table 8.4.

In addition to these measures, observations were also made of the lithic raw material, evidence of reworking or resharpening, the presence of impact fractures, evidence of basal grinding (recorded as presence/absence), and breakage class (using codes discussed below). These observations are listed in table 8.5.

In order to embed the Folsom site sample into a larger analytical context, I compiled comparative metric and nonmetric data on fluted points from a series of other Folsom localities (fig. 1.1, table 8.6). These sites are Blackwater Locality No. 1 (including the Mitchell Locality [Boldurian 1981, 1990; Broilo 1971; Hester 1972]), Cooper[3] (Bement 1999b), Elida (Hester 1962; Warnica 1961), Hot Tubb (Meltzer, unpublished data), Lake Theo (Buchanan 2002; Harrison and Killeen 1978), the Lindenmeier Coffin Collection (Gantt 2002), Lipscomb (Hofman and Todd 1990; Schultz 1943), Lubbock Lake (Johnson 1987), Shifting Sands (Hofman, Amick, and Rose 1990; R. Rose, personal communication, 2002), and Waugh (Hill and Hofman 1997; Hofman 1995). These sources provide data on ~470 finished projectile points. Although there are many sites with published measurements on unfinished Folsom points and preforms (e.g., Root, William, and Emerson 2000; William 2000), these are not an appropriate comparison to the Folsom site assemblage—at least with regard to most of the analyses undertaken here. However, information from such sites does occasionally prove relevant and useful. In addition, where possible I have utilized summary data from the Adair-Steadman (Tunnell and Johnson 2000), Cattle Guard (Jodry 1999a), and Hanson (Ingbar 1992) sites, as well as the Smithsonian Institution's Lindenmeier collection (Wilmsen and Roberts 1978). Finally, I have included information from Folsom point isolates assembled by Amick (1995), Judge (1973), and LeTourneau (2000).

Not all of these sources provide the same data, metric or otherwise, let alone in a format that matches that used here. Indeed, most only provide measurements of point length, width, and thickness; only a few sources provide data on basal width, fluting metrics, or the extent of grinding (table 8.6). These caveats notwithstanding, the data can be quite useful and constitute the "comparative sample" referred to throughout the remainder of this chapter.

Patterns in Lithic Raw Material Procurement

The assignment of stone to particular sources is based largely on macroscopic identification and, in part, on examination under ultraviolet light. There is, of course, considerable elemental variation within and between stone sources, so the source assignments here should always be viewed with a healthy skepticism. Non- or minimally destructive methods for more precisely assigning stone to source are badly needed.

That said, there were at least five separate sources of stone used in the manufacture of the projectile points from the Folsom site (fig. 8.6, tables 8.4 and 8.7). Alibates agatized dolomite and Tecovas jasper from the Texas Panhandle numerically dominate this assemblage, which is not the usual pattern for Folsom sites on the southern Plains (Wyckof 1999:43). Other stone types, possibly from sources in Colorado and Oklahoma, occur less frequently.

Projectile points made of Alibates are the most common in the Folsom site assemblage (n = 10; Sn 2, Sn4, Sn 6, Sn 12, Sn 14, Sn 15, Sn 17, Sn 23, Sn 25, Sn 28). The primary source of this stone is along a relatively short stretch of the Canadian River Valley north of the present city of Amarillo (Banks 1990:91; Holliday and Welty 1981; LeTourneau 2000; Wyckoff 1993). Although spatially discrete (~50 km²), the chert-bearing units within the dolomite are massive, averaging some 4 m in thickness (Wyckoff 1993:36). Alibates is also available in secondary pebble and cobble form as much as 600 km down the Canadian river, though it constitutes there only a very small percentage of the gravel train (~1%), and clast sizes are generally <6.5 cm in maximum dimension, although larger clasts are occasionally found (Banks 1990:91–92; Kraft 1997; LeTourneau 2000; Wyckoff 1993:51–53). The secondary gravel deposits are unlikely to have been the source of the Alibates used at the Folsom site, which occurs in relative abundance in consistently large "package" sizes of uniformly high quality, attributes not common in pickings of the random material carried in a gravel train (Meltzer 1984). The closest Alibates outcrop source is approximately 265 km southeast of Folsom, as the crow flies.

There are eight points (Sn 5, Sn 8, Sn 9, Sn 10, Sn 16, Sn 22, Sn 24, Sn 27) identified as Tecovas jasper. Like Alibates, Tecovas jasper is available in relatively discrete outcrops, but these are scattered over a larger area of the Texas Panhandle (Banks 1990:92; Holliday and Welty 1981; LeTourneau 2000:435–436; Mallouf 1989). Significant outcrops occur along the Canadian River on the western edge of the High Plains in eastern Oldham County and western Potter County (Mallouf 1989), in the vicinity of the Alibates source area (Banks 1990), and on the eastern edge of the High Plains escarpment near Quitaque, Texas (Banks 1990; Frederick and Ringstaff 1994; Hofman, Todd, and Collins 1991). Tecovas cobbles and pebbles also constitute a small portion of the gravel of the Canadian River (Banks 1984:72; LeTourneau 2000:436). The closest and farthest sources of Tecovas to Folsom are 190 and 375 km, respectively, both on a southeast bearing from the site.

There is a possibility that the stone identified as Tecovas from the Texas Panhandle was actually quarried much closer to Folsom. The Dockum Group, which includes the Tecovas formation, extends into New Mexico and includes

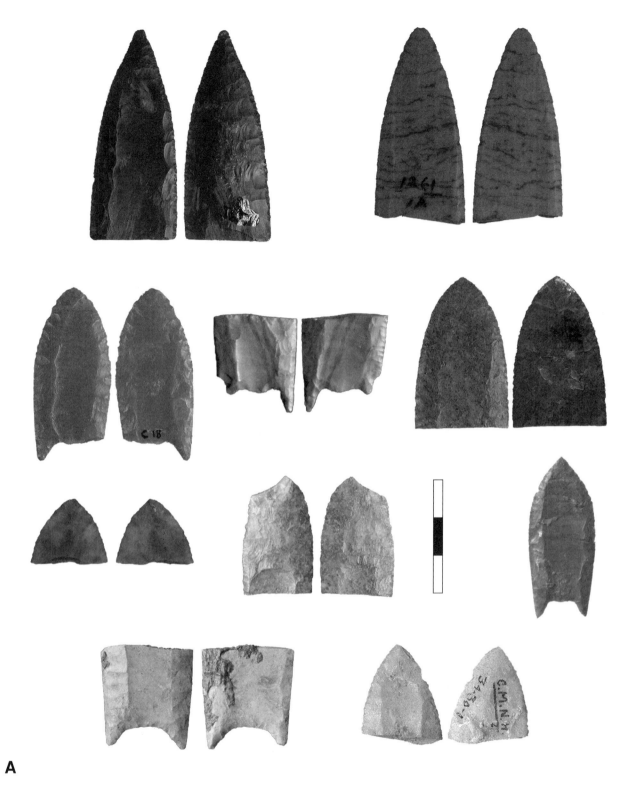

A

FIGURE 8.4 Projectile points from the Folsom site. Photographs and corresponding line drawings of the points (facing page) are to scale. Specimen numbers correspond to Table 8.2. Solid circles alongside edges in tine drawings indicate extent of edge grinding, if present. A. Specimen numbers 1–13. Note that the photograph and line drawing of Sn 3 are of a cast. The actual specimen is shown in Fig. 1.4. Sn 9 could not be photographed on both faces. (*Continued*)

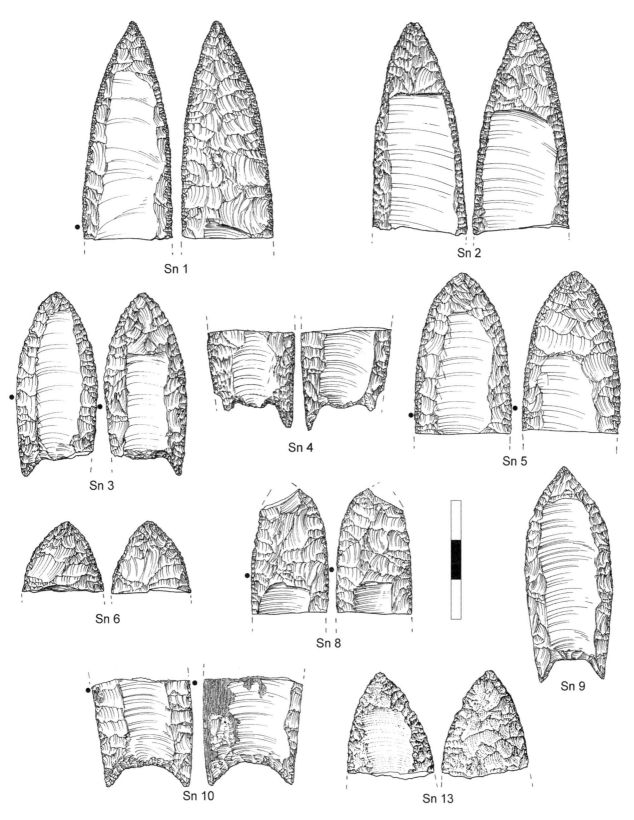

Sn 1

Sn 2

Sn 3

Sn 4

Sn 5

Sn 6

Sn 8

Sn 9

Sn 10

Sn 13

FIGURE 8.4 (*Continued*)

FIGURE 8.4 B. Specimen numbers 14–28. Sn 24 could not be photographed on both faces. (Photos by D. J. Meltzer; composite of photographs and line drawings by K. Monigal).

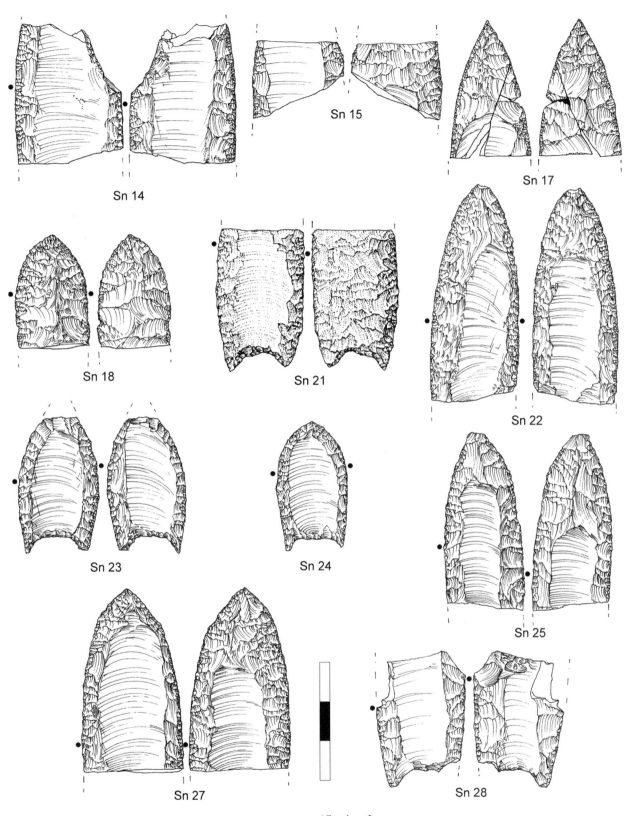

Sn 14

Sn 15

Sn 17

Sn 18

Sn 21

Sn 22

Sn 23

Sn 24

Sn 25

Sn 27

Sn 28

FIGURE 8.4 (*Continued*)

TABLE 8.3

Code	Measurement
1. MxLe	Maximum length of the complete point, taken from the tip to the base (or ears, if present). On point fragments, measurement is taken along the longitudinal axis of the remaining section.
2. MxWd	Maximum width, taken at the widest portion of the blade or base. On point fragments, measurement is at widest portion perpendicular to the longitudinal axis.
3. MxWB	Distance from the base to the spot where maximum width measured. Cannot be measured if base is missing.
4. BaWd	Base width, measured across the base of the point or between outer edges of point ears, if present.
5. MxTk	Maximum thickness, taken at the thickest portion of the blade or base.
6. MxTB	Distance from base to point of maximum thickness. Distance from the base to the spot where maximum thickness measured. Cannot be measured if base is missing.
7. BsCo	Depth of basal concavity, measured perpendicular to a chord drawn between the corners of the point.
8. FlTk	Thickness of blade within flute scars, at a point midway up the flute.
9. NFlO	Number of flutes, obverse face.
10. FlLO	Flute length, obverse face, measured from base of flute to its farthest extent.
11. FlWO	Flute width, obverse face, measured at a point midway up the flute.
12. NFlR	Number of flutes, reverse face.
13. FlLR	Flute length, reverse face, measured from base of flute to its farthest extent.
14. FlWR	Flute width, reverse face, measured at a point midway up the flute.
15. LGrn	Extent of grinding, left edge, measured from the base of the point or corner ear up the edge.
16. RGrn	Extent of grinding, right edge, measured from the base of the point or corner ear up the edge.

as a lateral equivalent the Baldy Hill formation (Baldwin and Muehlberger 1959; Banks 1990:93; LeTourneau 2000: 436; Lucas, Hunt, and Hayden 1987:fig. 13). That formation occurs as a very small and discrete outcrop at Baldy Hill along the Cimmaron drainage ~55 km east of the Folsom site. This formation contains nodules of red and yellow jasper, quartzite, and some silicified wood (chapter 3). Pieces of the jasper found at Baldy Hill tend to occur in a thin, small lens of relatively poorer quality than that which can be obtained at Tecovas outcrops (Baldwin and Muehlberger 1959:37; LeTourneau 2000:437; Lucas at al. 1987:115; Meltzer, unpublished field observations). Given this, it seems unlikely, though not impossible, that this was a source of the stone used at Folsom (also LeTourneau 2000:341). In general, the use of such look-alike sources in Folsom assemblages is rare (LeTourneau 1998b:78).

Two of the projectile points from Folsom (Sn 3, Sn 18) are apparently made of chert from the White River Group (Hoard et al. 1993:698), which outcrops across portions of western Nebraska into northeastern Colorado and southeastern Wyoming and, also, southwestern South Dakota (Greiser 1983; Hoard et al. 1992, 1993; Jodry 1999a; LeTourneau 2000). The best known of the primary sources within that region occurs at Flattop Butte, a small isolated mesa northwest of Sterling, Colorado. Flattop Butte is the closest of the White River Group cherts to Folsom, but is still ~450 km distant. Although the Flattop Butte cherts are macroscopically similar to other silicates in the White River Group, they are distinctive in their elemental composition (Hoard et al. 1993).

Because of the destructive nature of presently available analytical tests it is not possible to determine whether the stone for the two Folsom points was obtained from Flattop or another source within the White River Group. They certainly match up well macroscopically with the Flattop source, including the reddish inclusions that sometimes mark this source (Hoard et al. 1993:700). Thus, these points are tentatively assigned to Flattop (also LeTourneau 2000:341). If the identification is in error, it errs on the side of caution, for this is the closest of the White River Group sources and will not inflate estimates of the apparent distances the stone was moved.

Hofman, Todd, and Collins (1991:303) previously identified Sn 3 (figs. 1.4 and 8.4) as being made of Edwards Formation rather than Flattop chert (see also Hofman 1999b: 402; Jodry 1999a:table 48). Edwards chert outcrops over a large area of central Texas that extends toward the southeastern margin of the Southern High Plains (Banks 1990; Frederick and Ringstaff 1994; LeTourneau 2000:figs. 5.6, 5.7). If the stone for this point was obtained there, this suggests a very different direction and scale of movement, since the closest Edwards chert source is >550 km southeast of Folsom. Hofman, Todd, and Collins (1991:303) had explicitly considered the possibility of this particular specimen

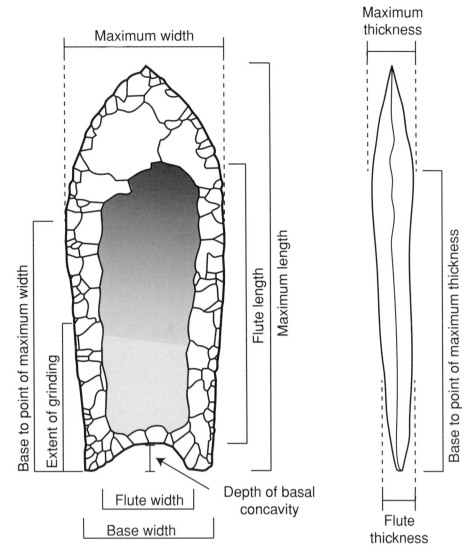

FIGURE 8.5 Schematic of a Folsom fluted projectile point showing the location of attributes measured in this study. Corresponding measurements on the reverse face (flute length and flute width), edge grinding on the right side, and mass are not shown.

being Flattop but rejected it: "Whereas the white light evaluation of this projectile point's raw material type was equivocal, the ultraviolet light test indicates that it is almost certainly not Flattop and very likely is Edwards [chert]." At the time of their examination, however, the point was on display within a protective fiberglass cover (Hofman, personal communication, 2002). LeTourneau was subsequently able to examine the specimen more thoroughly outside its case, and concluded that it was made of Flattop chert (LeTourneau, personal communication, 2001, 2000:341). Hofman (personal communication, 2002) accepts that revision, and I accept that source identification as well.

Two projectile points from Folsom (Sn 1, Sn 11) are made of high-quality petrified palm wood. They are quite comparable to material available in the Dawson Formation, which outcrops in the Black Forest area east of the Front Range

between Colorado Springs and Denver (Banks 1990; Hofman, Todd, and Collins 1991:302; Jodry 1999a:88–91, 333). Secondary cobble sources of this material occur as far east as Kit Carson, Colorado (Banks, personal communication, 1999), as well as in the upper San Luis Valley (Jodry 1999a:91). Given the size and quality of at least one of the petrified wood specimens at Folsom (Sn 1; the other is under glass at the AMNH and could not be examined closely), they most likely came from an outcrop source, in which case the stone was moved a distance of at least 240 km, perhaps more.

Finally, two point fragments (Sn 13, Sn 21) are made of a white quartzite. High-quality, knappable quartzite occurs throughout the Dakota (Early–Late Cretaceous), Lytle (Early Cretaceous; formerly Purgatoire Formation [Kues and Lucas 1987:169–170]), and Morrison (late Jurassic) formations, all of which are found in this region of the Plains and adjacent

TABLE 8.4

Metric Data on Attributes of Folsom Site Fluted Points

Specimen No.	Museum No.	Source Data	1. MxLe	2. MxWd	3. MxWB	4. BaWd	5. MxTk	6. MxTB	7. BsCo	8. FITk	9. NFIO	10. FILO	11. FIWO	12. NFIR	13. FILR	14. FIWR	15. LGm	16. RGm
1	DMNS1391/3	Orig	(56.07)	23.50	CM	CM	4.75	CM	CM	4.10	1	(40.78)	15.06	1	(3.44)	13.47	(3.19)	IND
2	DMNS1261/1A	Orig	(52.96)	24.90	CM	CM	4.71	CM	CM	2.97	1	(34.73)	17.66	1	(28.97)	18.96	IND	IND
3	DMNS1262/1A	Cast	45.43	21.32	23.60	18.88	3.77	30.76	5.12	2.83	1	35.46	12.18	1	26.42	10.72	20.17	(11.31)
4	Missing	Cast	(25.76)	(23.13)	(24.90)	(20.59)	3.66	(20.78)	5.60	2.67	1	(18.96)	12.38	1	(20.18)	13.24	IND	IND
5	DMNS1391/2	Orig	(40.13)	(25.77)	CM	CM	4.22	CM	CM	3.19	1	(30.17)	13.85	1	(18.84)	14.89	(4.57)	(6.35)
6	DMNS1263/1A	Orig	(17.71)	(21.37)	CM	CM	(3.44)	CM	CM	IND	1	(1.79)	8.51	IND	IND	IND	IND	IND
7	AMNH A																	
8	AMNH20.2.5871	Orig	(31.39)	19.99	CM	CM	3.59	CM	CM	2.29	1	(5.86)	12.77	1	(7.05)	9.17	(9.97)	(10.99)
9	AMNH20.2.5865	Cast	56.20	24.15	23.84	19.17	3.91	26.74	5.14	3.08	1	44.37	14.00	1	43.03	12.21	IND	IND
10	AMNH20.2.5867	Orig	(27.52)	(25.67)	(25.22)	22.43	3.93	(26.70)	5.61	3.89	1	(22.09)	12.23	1	(21.75)	15.83	(25.74)	(26.44)
11	AMNH20.2.5866	Cast	(35.51)	22.44	(20.57)	(17.12)	3.78	(28.02)	4.12	2.83	1	(23.26)	12.39	1	(29.90)	11.69	IND	IND
12	AMNH20.2.5868	Cast	(30.06)	20.89	(11.61)	19.82	3.65	(23.59)	4.49	2.16	1	(24.79)	12.39	1	(22.66)	16.18	IND	IND
13	UM 34-30-1 (t)	Orig	(26.62)	(22.69)	CM	CM	(5.2)	CM	See 21	(3.06)	1	(15.58)	(16.15)	1	(13.43)	(6.99)	IND	IND
14	AMNH20.2.5872	Orig	(35.31)	28.22	CM	CM	3.87	CM	CM	3.25	1	(32.63)	19.59	1	(32.46)	15.33	(19.68)	(11.42)
15	DMNS1391/1	Orig	(19.57)	(24.42)	CM	CM	(3.49)	CM	CM	2.50	1	(12.99)	(14.28)	1	(3.36)	(10.99)	IND	IND
16	AMNH J	Cast	(34.99)	21.68	(24.24)	18.4	3.73	(21.4)	3.98	2.99	1	24.29	9.92	1	29.45	13.39	IND	IND
17	AMNH20.2.5869	Orig	(35.03)	(21.57)	CM	CM	3.37	CM	CM	IND	1	(10.73)	(15.73)	IND	IND	IND	(13.65)	IND
18	AMNH20.2.5870	Orig	(28.94)	18.95	CM	CM	4.30	CM	CM	IND	1	(1.13)	(5.13)	IND	IND	IND	IND	(13.27)
19	Missing																	
20	Missing																	
21	UM 34-30-1 (b)	Orig	(36.20)	22.83	18.10	17.72	4.19	(9.92)	4.06	3.96	1	(31.86)	14.73	ABS	ABS	ABS	32.4	29.02
22	CAVO-115	Orig	(56.42)	23.24	(6.07)	CM	4.38	(13.29)	CM	3.31	1	(36.21)	13.08	1	(36.47)	13.81	(20.24)	(18.18)
23	CAVO-116	Orig	35.05	21.49	18.62	17.48	4.01	23.04	3.97	3.24	1	25.11	14.67	1	26.56	12.22	18.03	21.48
24	B. Burchard	Orig	32.32	18.74	19.79	16.24	4.59	17.26	2.95	3.22	1	26.09	13.18	1	29.76	12.88	19.78	20.26
25	A. Brown	Orig	(44.56)	21.42	(8.32)	CM	3.96	(11.12)	CM	3.45	1	(31.09)	9.43	1	(22.48)	13.28	(14.35)	(9.50)
26	Missing																	
27	D. Brown	Orig	(47.53)	26.77	(8.68)	(26.21)	4.52	(8.93)	CM	3.35	1	(37.36)	17.15	1	(25.24)	14.13	(7.06)	(6.60)
28	A2006.1	Cast	(31.08)	(24.18)	(20.12)	(20.32)	(5.16)	(14.15)	3.46	4.07	1	(28.14)	13.33	1	(24.02)	8.82	(20.2)	(21.97)

NOTE: Variable codes follow table 8.3. All measurements are in millimeters. Values in parentheses are measures of extant portions of broken specimens; a value of 0 entered under MxWB indicates that the widest portion remaining is at the base of the broken specimen (but that was not the base of the original point). Variables 1–7 in principle are measurable on all points but may not be in specific cases where landmarks (such as the base) are missing. CM (cannot measure) designates those instances. Variables 8–16 pertain to fluting and edge grinding, attributes that may or may not be present on a specimen. IND (indeterminate) designates those instances where, because a specimen was inaccessible or broken, it could not be determined whether fluting or grinding was present to be measured. ABS (absent) in those same columns indicates that the attribute is definitely absent. Orig, original.

TABLE 8.5
Nonmetric Data on Folsom Site Fluted Points

Specimen No.	Museum No.	Material	Reworked	Impact Fracture	Basal Grinding	Breakage Class	Comments
1	DMNH 1391/3	Black Forest	No	No	Indet.	2b	
2	DMNH 1261/1A	Alibates	No	No	Indet.	2c	
3	DMNH 1262/1A	Flattop	Yes	No	Yes	1	Point on display; ear broken off
4	Missing	Alibates	Indet.	Indet.	Indet.	4 (4b?)	Data from cast; cannot gauge extent of grinding
5	DMNH 1391/2	Tecovas	No	No	Indet.	2b	
6	DMNH 1263/1A	Alibates	No	No	Indet.	2c	
7	AMNH A	Indet.	Indet.	Indet.	Indet.	Indet.	
8	AMNH 20.2.5871	Tecovas	Yes	Yes	Indet.	3a	
9	AMNH 20.2.5865	Tecovas	No?	No	Indet.	1	Point on display; data from cast
10	AMNH 20.2.5867	Tecovas	Indet.	Indet.	Yes	4b	
11	AMNH 20.2.5866	Black Forest	Indet.	Yes	Indet.	4 (4b?)	Data from cast; cannot gauge extent of grinding
12	AMNH 20.2.5868	Alibates	Yes	Yes	Indet.	4 (4a?)	Point on display; data from cast; cannot gauge extent of grinding
13	UM 34-30-1 (tip)	Quartzite	No	No	Indet.	2c	
14	AMNH 20.2.5872	Alibates	No	Yes	Indet.	3a	
15	DMNH 1391/1	Alibates	Indet.	Indet.	Indet.	3c	
16	AMNH J (missing)	Tecovas	Yes	No?	Indet.	4 (4a?)	Data from cast and photograph; cannot gauge extent of grinding
17	AMNH 20.2.5869	Alibates	No	No	Indet.	2c	
18	AMNH 20.2.5870	Flattop	Yes	No	Indet.	2b	
19	Missing	Indet.	Indet.	Indet.	Indet.	Indet.	
20	Missing	Indet.	Indet.	Indet.	Indet.	Indet.	
21	UM 34-30-1 (base)	Quartzite	Indet.	No	Yes	4b	
22	CAVO-115	Tecovas	No	No	Indet.	2a	
23	CAVO-116	Alibates?	Yes	No	Yes	1	
24	B. Burchard	Tecovas	Yes	No	Yes	1	
25	A. Brown	Alibates	Yes (v. slight)	No	Indet.	2a	
26	Missing	Indet.	Indet.	Indet.	Indet.	Indet.	
27	D. Brown	Tecovas	Yes (v. slight)	No	Indet.	2b	
28	DMNH A2006.1	Alibates	Indet.	Yes	Yes	4b	

NOTE: Indet., indeterminate; v., very.

Rocky Mountains (Baldwin and Muehlberger 1959:56; Banks 1990:94; Jodry 1999a:97; Kues and Lucas 1987; LeTourneau 2000:446–447; Mateer 1987:223). The Dakota Formation, believed by Banks (1990:89) to be an oft-used source, outcrops ~40 km south of Folsom (Scott and Pillmore 1993) but is also quite extensive downstream along the Dry Cimarron Valley, beginning just 10 km east of the Folsom site and extending over the next ~100 km into the Oklahoma panhandle (Barnes 1984; Scott and Pillmore 1993). Unfortunately, that region has not been thoroughly or systematically explored for knappable stone, and where high-quality quartzite might exist within that broad swath is unclear (chapter 3).

TABLE 8.6

Projectile Point Metric and Nonmetric Data Available from Other Folsom-Age Sites Available for Comparative Analysis

Site	1. MxLe	2. MxWd	5. MxTk	4. BaWd	7. BsCo	8. FITk	Flute length[a]	Flute width[a]	Grinding[a]	Reworking	Impact fracture	Breakage	Stone
Adair-Steadman				Sum		Sum							
BwD—Bouldurian	Ind	Ind	Ind	Ind					Ind	Ind	Ind	Ind	Ind
BwD—Broilo		Ind	Ind	Ind	Ind				+/−				Ind
BwD—Hester	Ind	Ind	Ind				Ind		Ind	Ind		Ind	Ind
BwD—Mitchell		Sum	Sum	Sum	Sum								Sum
Cattle Guard	Sum	Sum	Sum	Sum	Sum		Sum	Sum				Sum	Sum
Cooper	Ind	Ind	Ind			Ind				Ind		Ind	Ind
Elida	Ind	Ind	Ind							Some		Ind	Sum
Folsom	Ind	Ind	Ind	Ind	Ind	Ind	Ind	Ind	Ind	Ind	Ind	Ind	Ind
Hanson													Sum
Lake Theo	Ind	Ind	Ind	Ind	Ind							Ind	Ind
Lindenmeier—Gantt	Ind	Ind	Ind	Ind			Ind	Ind		Ind	Ind	Ind	
Lindenmeier—W & R	Sum	Sum	Sum	Sum		Sum	Sum	Sum					
Lipscomb	Ind	Ind	Ind	Ind								Ind	Ind
Lubbock Lake	Ind	Ind	Ind	Ind									Ind
Shifting Sands	Ind	Ind	Ind	Ind		Ind			Ind			Ind	
Waugh	Ind	Ind	Ind	Ind								Ind	
Amick 1995	Sum		Sum	Sum									
Judge 1973		Sum	Sum	Sum	Sum	Sum							
LeTourneau 2000											Sum		

[a] No face (obverse/reverse) or edge (left/right) specified.

NOTE: Sites are from the "Comparative Sample"; see text. Ind, data available on individual specimens; Sum, only summary statistics available.

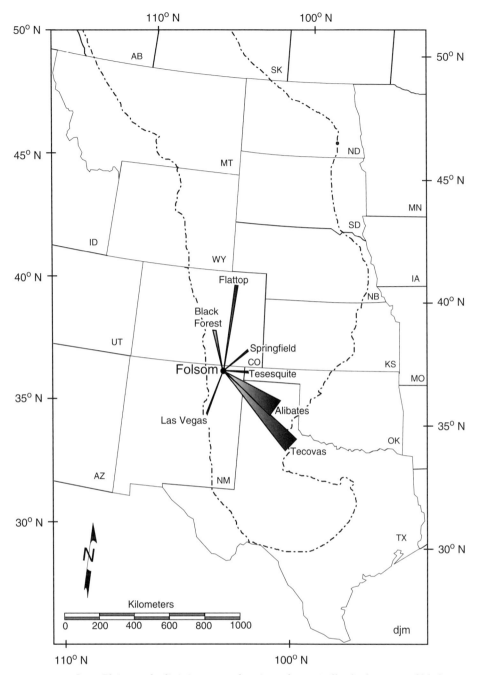

FIGURE 8.6 Great Plains and adjoining areas showing schematically the location of likely and/or possible outcrop sources of stone used at Folsom. The width of the arrows is proportional to the relative contribution of that stone source to the assemblage at Folsom.

That said, one high-quality quartzite source occurs along Tesesquite Creek, just southwest of Kenton, Oklahoma (Meltzer, unpublished field observations), due east of and just over 100 km distant from Folsom. That source area is broadly mapped as Morrison Formation (Barnes 1984) but may be Lytle Formation or perhaps Dakota. As Banks (1990:94) ruefully admits, "Clear distinctions have yet to be made" among these several formations. Quartzite of comparable quality to Tesesquite is also reported from the area around Springfield, Colorado (Dakota), northeast of and 130 km distant from

Folsom, and near Las Vegas, New Mexico (Morrison), which is 180 km southwest of Folsom (Banks 1990:94; Jodry 1999a:97).

Tesesquite Creek quartzite is macroscopically similar to the stone used at Folsom, but that does not mean that it was acquired at this particular source or preclude the possibility that it came from one of the other sources noted or a source yet unknown. The Tesesquite source would have been "en route" to a group coming north by northwest out of the Texas Panhandle and moving up the Dry Cimarron River (chapter 3). Of course, one cannot eliminate either the Springfield or

TABLE 8.7

Approximate Straight-Line Bearing and Distance from the Folsom Site to Likely and/or Possible Outcrop Sources of Stone
Used and Discarded at the Site

Source	Formation	Bearing (to Nearest 5°)	Distance (km)	Comments
Alibates (TX)	Quartermaster	125°	265	Distance to closest Alibates source area
Baldy Peak (NM)	Dockum/Santa Rosa	80°	55	Distance to Baldy Hill, NM
Black Forest (CO)	Dawson	350°	240–280	Distances to Calhan and Elizabeth (CO) source areas
Flattop (CO)	White River Group/Chadron	10°	450	Distance to Flattop Mesa, CO
Las Vegas (NM)	Morrison	215°	175	
Springfield (CO)	Dakota	65°	140	
Tecovas (TX)	Dockum/Tecovas	135°	190–375	Distances to Rotten Hill and Quitaque (TX) source areas
Tesesquite (OK)	Morrison?	90°	100	Distance to source area east of Kenton, OK

SOURCE: Data from Banks (1990); Jodry (1999a); Meltzer (field notes on source locations).

the Las Vegas source on the grounds that they are not so auspiciously located.

All of the stone discarded at Folsom came from sources east of the Rocky Mountains (fig. 8.6), which is in keeping with a general pattern seen in other Folsom age sites across the Southern Plains (Hofman 1999b:387; Wyckoff 1999). Folsom has a different complement of stone than that used in the Folsom-age assemblages in the San Luis Valley, as well as in the Albuquerque and Tularosa basins, which include raw material acquired from a number of the Texas sources but are often dominated by sources that outcrop west of the spine of the Rocky Mountains (Amick 1995; Jodry 1999a:fig. 23).

The Folsom site is also one of the few of this age in this part of the Plains and southern Rocky Mountains lacking Edwards Formation chert (cf. Amick 1995:415, 1999b; Bement 1999:75, 97, 115; Buchanan 2002; Hofman 1991, 1999b:table 4; Jodry 1999a:table 48; Stanford 1999:303; Wyckoff 1999). This is true not just of sites east of Folsom and thus closer to the Edwards source, but also of sites west of Folsom. In the San Luis Valley, for example, ~750 km from the closest Edwards source, Edwards chert comprises 3% to 46% of the site assemblages (Jodry 1999a:table 10, 1999b:75, 78).

There is no obvious explanation for the dearth of Edwards chert at Folsom. It might simply be that this particular group had not visited the Edwards source recently—or ever. Or it could be a result of sampling: If points of Edwards chert occurred at a low frequency within the tool assemblage, it might not appear in the relatively small sample available from the site. Among the sites in the comparative sample, there is a statistically significant relationship between sample size, measured by the total number of artifacts, and raw material richness, the number of different stone sources in the

assemblage. That relationship is significant but not particularly robust ($r^2 = 0.397$, $P = 0.007$), owing to several substantial outliers (fig. 8.7): Hanson and Reddin have significantly more stone sources represented than would be expected given their assemblage size (residuals = 1.53 and 1.17, respectively), while Shifting Sands has far fewer stone types than would be expected given a sample of this size (residual = −2.48). Moreover, comparably sized assemblages even farther from the Edwards source, at Linger ($n = 18$) and Zapata ($n = 28$), each yielded Edwards chert (Jodry 1999a:table 10).

Hofman (1999b:398–399) sees a "dramatic change" in raw material movement on the Southern Plains, marked by a dropoff in the incidence of Edwards chert at 400 km from the source. If that is so, it explains the absence of Edwards from Folsom, since it may be "beyond the limits of primary or dominant distribution of Edwards material" (Hofman 1999b:398). However, other studies of raw material distribution have not been able to substantiate that dropoff pattern (Amick 1995; LeTourneau 2000:228–229).

The high-quality stone used to make the Folsom site projectile points is in keeping with the general pattern seen in Folsom sites elsewhere (Amick 1999b:181; Hofman 1999b:387, 406; Hofman, Todd, and Collins 1991; Ingbar 1992; Jodry 1999a, 1999b; LeTourneau 2000). The pattern is presumably related to the technological demands of point production played against the needs of highly mobile hunter-gatherers, facing the unavoidable geographic fact that across much of the Plains high-quality stone is scarce and inevitably distant from the localities in which it was put to use and ultimately discarded (Amick 1999b; Bamforth 2002; Banks 1990; Buchanan 2002; Hofman 1991, 1999b, Hofman, Todd, and Collins 1991; Ingbar 1992; Ingbar and Hofman 1999:100; LeTourneau 2000:411; Sellet 2004).

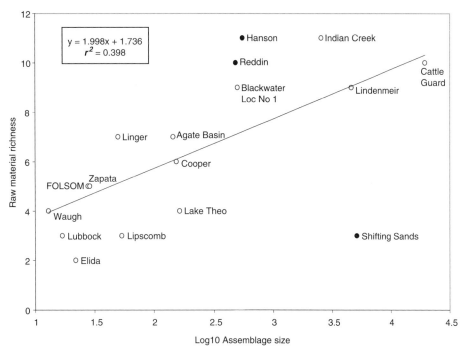

FIGURE 8.7 Least squares regression plot of raw material richness against Log10 assemblage size in projectile points from Folsom and the sites in the comparative sample. Sites with solid circles are more than 1 standard deviation from the regression line.

THE MECHANICS OF MOVEMENT

Taking source assignments at face value, the diverse raw material used at Folsom testifies to stone movement across distances of up to 450 km, and from widely divergent points on the compass (table 8.7). There are several possible processes that could account for this same pattern, identified as A–C in figure 8.8. This complement of stone could represent the convergence or aggregation on this spot of previously dispersed hunter-gatherers from areas north and south of the site, who carried with them stone they had directly acquired at outcrop sources (fig. 8.8A) (e.g., Jodry 1999a:262); or it could be the product of long-distance exchange of stone among hunter-gatherer groups that took place elsewhere on the landscape, and from which point one of the groups subsequently came to Folsom (fig. 8.8B); or this could be the stone supply of a single wide-ranging group who had visited all of these outcrops and collected stone at each (fig. 8.8C) (G. Jones et al. 2003; Kelly 1992; Meltzer 1989b).

Each possibility—and they are highly simplified in figure 8.8 for sake of discussion—bespeaks different social processes, stone acquisition patterns, and mechanics of movement. In the first and last cases the distance and direction to stone sources can provide information on the mobility of the several groups (A) or of one group (C). However, the estimates of the minimal distances moved will necessarily be lower in Λ than in C since the distances from individual outcrops to the discard point will always be less than the distance from outcrop to outcrop to outcrop—ultimately to the discard point. In case B, exchange could have occurred multiple times at various spots on the landscape, with all

the material discarded at the site by the one group carrying it, thereby rendering inferences of their mobility based on distance from outcrop to site virtually meaningless. Those distances would measure the movement only of stone on the landscape, and not of the group that discarded it.

How, then, to determine what the diverse pattern of stone at Folsom represents? Solutions to problems of equifinality, as this one is, often require suites of evidence, not just the data that prompted the problem (that is, the diverse pattern of stone types at Folsom). Evidence bearing on this question might be found in stylistic variation in projectile points, the relative diversity of stone sources by tool class, patterns of use and reworking, and/or independent evidence of aggregation. I therefore return to this question at the end of the chapter, after the available data and evidence have been presented.

Morphology and Morphometrics of Folsom Projectile Points

All projectile points from the Folsom site were finished; no point manufacturing is evident. Edge grinding, the final stage in point production (Frison and Bradley 1980:51; Tunnell 1977:151), was present on all specimens that preserved a portion of the base. Nor did the assemblage contain any evidence of point production: There are no manufacturing failures, channel flakes, preforms, or unfinished points present (e.g., Tunnell and Johnson 2000). Again, this likely reflects where excavations occurred, the kill area, and does not preclude the possibility that point manufacture or perhaps refurbishment

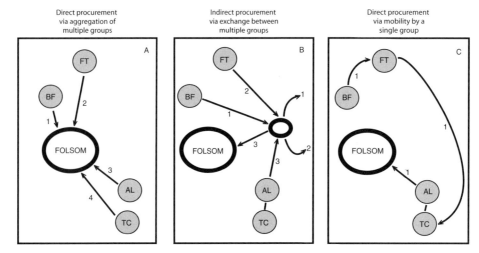

Direct procurement via aggregation of multiple groups

Indirect procurement via exchange between multiple groups

Direct procurement via mobility by a single group

FIGURE 8.8 Three models for the movement of stone across the landscape to the Folsom site. Arrows denote pathways of movement; numbers 1, 2, and 3 denote individual groups. AL, Alibates; BF, Black Forest; TC, Tecovas.

occurred nearby and outside the kill, as is the case elsewhere (Hofman, Amick, and Rose 1990; Jodry 1999a).

Only four of the points from Folsom (~17% of the total) are complete: Sn 3, Sn 9, Sn 23, and Sn 24 (table 8.1; Sn 9 broke during excavation). The remainder of the points occurs as tips, midsections, and bases. The points broke either in use or afterward, the 1920s excavations being especially hard on this assemblage (chapter 4). Specifics of the breakage patterns and processes are discussed in more detail below. Table 8.8 provides summary descriptive statistics for each of the metric variables, which were calculated using the complete points and those broken specimens in which it was possible to get a complete measurement of that particular attribute.

In all points from the Folsom site, save one, both faces were fluted. The exception is the quartzite base (Sn 21), which is made of a material that, as Roberts (1935:17) observed, might have been singly fluted owing to the difficulty of working this material (also Tunnell and Johnson 2000:11). And in all cases there is but a single flute per face; this is the usual pattern in Folsom (Amick 1999a; Wilmsen and Roberts 1978:111–112). Multiple fluting, which is common among earlier as well as later Paleoindian fluted point assemblages, including Clovis and Crowfield (Deller and Ellis 1984), is rare in Folsom points (Boldurian 1990:91; Frison and Bradley 1980:fig. 36; Tunnell and Jounson 2000:12). Amick, however, has recently reported, based on a study of channel flakes from Lone Butte (NM), that multiple fluting may occur at higher frequencies than previously supposed (Amick 2002; see also Ahler, Frison, and McGonigal 2002).

The descriptive statistics (table 8.8) indicate that an "average" point from the Folsom site is just over 42 mm in length (MxLe), is 22 mm in width (MxWd), and tapers to a slightly narrower base (BaWd; 18.8 mm), the tapering reduced drag and enhanced repeated penetration and removal of the point (Crabtree 1966:7; Ahler and Geib 2000:801). These are very thin points, averaging just over 4 mm in maximum thickness (MxTk) and approaching 3.2 mm thickness within the flute scar (FlTk). Because all of these are finished points, all are ground, and the grinding (LGrn and RGrn) generally extends ~23 mm up the edges from the base. The fluting (FlLR and FlLO) extends farther, on average ~31 mm.

In basic size and shape dimensions, the Folsom site points are well within the quantitative range of other Folsom projectile points (e.g., Amick 1995; Jodry 1999a; Judge 1973; Tunnell and Johnson 2000; Wilmsen and Roberts 1978). A series of t-tests run between the Folsom site specimens and those in the comparative sample, using only the complete points from each, reveals that the sample means differ only in measures of flute length and flute thickness ($t = 2.317$, $p = 0.025$, and $t = 4.344$, $p = 0.000$, respectively). Given the small number of complete points from Folsom ($n = 4$), and the fact that three of those points were reworked, thus reducing the flute length, those differences may not be especially significant.

The complete projectile points do not have a well-defined "center of gravity"; where they are widest (MxWB) is not where they are thickest (MxTB) (the averages are 21.5 and 24.5 mm, respectively). It is perhaps not surprising that the thickest spot on these points was beyond its widest spot and closer to the tip (cf. Ahler and Geib 2000:806); this was likely a vestige of manufacturing, whereby "excess stone" was left close to the tip to "lessen the shock received by the preform" during fluting (Crabtree 1966:6–7; also Root, William, and Emerson 2000:266). The distances from the base to the position of maximum width and thickness do not correlate with the flute length or the extent of edge grinding. The differences among these measures may have implications for understanding hafting and reworking, as discussed below.

Of course, one should not place too much interpretive weight on average values, which necessarily mask variation

TABLE 8.8
Summary Descriptive Statistics on Folsom Site Fluted Points

Variable	N	Mean	SD	Skew	Kurtosis	CV
1. Maximum length (*MxLe*)	4	42.25	10.88	0.72	−1.52	25.75
2. Maximum width (*MxWd*)	16	22.53	2.59	0.66	0.32	11.51
3. Distance from base to point of maximum width (*MxWB*)	4	21.46	2.65	−0.16	−5.01	12.36
4. Base width (*BaWd*)	8	18.77	1.85	0.93	1.70	9.86
5. Maximum thickness (*MxTk*)	20	4.04	0.40	0.36	−0.83	9.78
6. Distance from base to point of maximum thickness (*MxTB*)	4	24.45	5.74	−0.39	−0.30	23.46
7. Depth of basal concavity (*BsCo*)	11	4.41	0.87	0.00	−0.90	19.67
8. Thickness of blade within flute scars (*FlTk*)	20	3.17	0.55	0.09	−0.40	17.39
10. Flute length, obverse face (*FlLO*)	5	31.06	8.70	1.14	−0.20	27.99
11. Flute width, obverse face (*FlWO*)	20	13.43	2.67	0.42	0.68	19.89
13. Flute length, reverse face (*FlLR*)	5	31.04	6.88	1.95	4.00	22.16
14. Flute width, reverse face (*FlWR*)	18	13.35	2.46	0.16	0.71	18.47
15. Extent of grinding, left edge (*LGrn*)	3	23.40	7.84	1.64	—	33.5
16. Extent of grinding, right edge (*RGrn*)	4	22.73	4.23	1.89	3.57	18.63

NOTE: Variables from table 8.3; data from table 8.4. CV, coefficient of variation.

within the assemblages. That variation is not detectable in measures of standard deviation either (Judge 1973:171), or at least not in those cases where there is a linear and positive relationship between the mean and the standard deviation (Eerkens and Bettinger 2001). Such a relationship exists in these data. Measures with smaller means (e.g., maximum thickness) naturally have smaller standard deviations than those with larger means (e.g., maximum length) and, thus, appear to be more standardized (Judge 1973:171). Coefficients of variation (CVs) are a more reliable measure in this regard, as CV values scale the standard deviation to the mean (Eerkens and Bettinger 2001:498).

Table 8.8 provides CV values for the Folsom projectile point assemblage. As can be observed in these data, the attributes that vary more (e.g., CV ≥20) are length-related, including maximum length (MxLe), maximum thickness to base (MxTB), flute length (FlL O/R), and extent of left edge grinding (LGrn). That fluting length proves to be among the most variable of measures is hardly surprising; the length of a flute flake cannot be as well controlled as, say, the point width. That one of the measures of edge grinding is highly variable may be an artifact of this particular sample; of the three specimens comprising this small sample, two are heavily reworked.

In fact, three of the four complete points from Folsom were reworked. Assuming for the moment that reworking took place while the point was still embedded within the haft and thus affected only the exposed portion of the blade (Odell 1994:54), one would anticipate that point length would vary more than point width (Collins 1999b). In the larger comparative sample one can see that pattern (table 8.9): including all points, complete and broken, yields the greatest variation in maximum length (MxLe CV = 41.82), but progressively restricting the sample to complete but reworked points (MxLe CV = 29.44), and then only com-

plete and not reworked points (MxLe CV = 21.61), steadily reduces the variation along this dimension. Yet, even among the latter category, the CV values for various length measures are not low in either an absolute sense—virtually all are >20—or a relative sense, especially compared to measure of width dimensions.

In contrast, the least variable dimensions of the Folsom site and comparative sample projectile points are those related to width and thickness. At Folsom, CV values for basal width (BaWd) and maximum thickness (MxTk) are both <10. CV values in this range are close to but not within the "minimum error [variation] attainable" in artifacts produced manually and without the use of external rulers (Eerkens and Bettinger 2001:496–497). The variables of maximum width (MxWd) and maximum width to base (MxWB) have slightly higher CV values, while fluting thickness (FlTk) is higher still but <20 (table 8.8). Common to all of these attributes is that they might bear on (or are constrained by) the size of the haft within which the point was set. A similar outcome occurs in the larger comparative sample (table 8.9). All of this mirrors results first recorded by Judge (1973:261–264) and subsequently found to be characteristic of other Folsom and even other Paleoindian projectile points (e.g., Amick 1995; Buchanan 2002; Jodry 1999a; Meltzer and Bever 1995; Tunnell and Johnson 2000:36).

The consistently low CV values for width and thickness were argued by Judge to be a result of sizing these tools to fit socketed hafts, rather than vice versa (Judge 1973:265; also Crabtree 1966:7; Odell 1994:54). This argument, in turn, implied that socketed hafts, possibly of bone or wood, were more costly to manufacture than the points themselves (Judge 1973:175–176, 264–265; also Bamforth and Bleed 1997; Bleed 1986; Guthrie 1983; Keeley 1982; Meltzer

TABLE 8.9
Differences in Coefficient of Variation (CV) of Maximum Length, Width, and Thickness

	n	Mean	SD	CV
(a) Results from the Comparative Sample				
Maximum Length				
All points; includes broken specimens	432	28.0	11.71	41.82
Complete points; includes reworked specimens	116	36.2	10.66	29.44
Complete points; only reworked specimens	81	34.0	8.78	25.82
Complete points; only nonreworked specimens	10	52.6	11.37	21.61
Maximum Width				
All points; includes broken specimens	459	19.79	4.11	20.77
Complete points; includes reworked specimens	115	19.43	3.03	15.60
Complete points; only reworked specimens	80	19.22	2.63	13.70
Complete points; only nonreworked specimens	10	22.37	1.91	8.54
Maximum Thickness				
All points; includes broken specimens	251	3.54	.851	24.04
Complete points; includes reworked specimens	68	3.48	.840	24.14
Complete points; only reworked specimens	37	3.52	.649	18.44
Complete points; only nonreworked specimens	8	4.07	.730	17.94
(b) Results from Other Studies				
Judge (1973:table 8.6)				
Basal width	33	19.42	1.125	5.79
Maximum width	33	21.51	1.492	6.93
Maximum thickness	33	3.83	.481	12.54
Flute thickness	33	2.43	.442	18.15
Amick (1995:table 8.7)				
Maximum length	64	32.8	10.4	31.71
Basal width	295	18.5	2.4	12.97
Maximum thickness	521	3.7	0.7	18.92

NOTE: In (b), complete points only; presumably includes reworked and nonreworked specimens.

and Bever 1995; Shott 1986). Fluting in order to thin the point was obviously a part of the hafting process, although as Judge argues, it was not the most critical part of that process, as the presence of unfluted Folsom points indicates (Judge 1973:171–172, 175; also Ahler and Geib 2000:802; Amick 1995; Root, William, and Emerson 2000:269).

If, in fact, reducing the width of the point was critical to fitting it into a socketed haft, then one should see a consistency in the attributes that might have enabled the incremental reduction of width—namely, edge grinding (Jodry 1999a: 193; Titmus and Woods 1991). Judge (1973:263) found that the degree of edge grinding as measured on an ordinal scale of light versus heavy, and base width recorded as above or below the mean, was more closely associated in Clovis points than in Folsom points. This implied that in Folsom points the width of the base was equally controlled by other actions of the flint knapper, notably, by applying fine pres-

sure retouch to the edges (also Bamforth and Bleed 1997: 130; Crabtree 1966:5; Frison and Bradley 1980; Root, William, and Emerson 2000:266–267). There is nothing in the Folsom site sample to suggest that this was not the case.

However, another way to approach the matter is to examine the extent of edge grinding: If grinding was also or even exclusively used to fit a point to a haft, its extent ought to be consistent on both sides of the point, assuming that the haft and its binding lay perpendicular across the axis of the point. This appears to be the case. In the Folsom site assemblage there is a significant correlation between the extent of grinding along the left and that along the right edges ($r^2 = 0.89$).

To my earlier comment on variation in flute length, I would add here there is only a slight correlation of flute length on the two faces of a point ($r^2 = 0.47$), and flute length is not correlated with the extent of grinding (r^2 values range from 0.0002 to 0.34, and none are significant).

These results, of course, are not surprising, since it is far simpler to control grinding length than fluting length (Crabtree 1966:8; Flenniken 1978; Whittaker 1994). In most newly minted Folsom points, the flute routinely extends nearly the full length of the blade (Frison and Bradley 1980; William 2000; Wilmsen and Roberts 1978), well beyond the ground portion of the base and perhaps the elements of the haft (Ahler and Geib 2000, 2002). As the blade is reduced through reworking, the fluting and grinding length measures approach one another.

A NOTE ON FOLSOM POINT STYLES

Folsom points across a wide area can be astonishingly alike in form and technology (e.g., Morrow and Morrow 1999), testimony perhaps to the wide-ranging mobility of small groups, the brief period of time they were on the landscape, and the relative lack of cultural drift (also Jodry 1999a). Yet, it is also the case that not all Folsom fluted points look alike, even within a single assemblage, as Figgins well appreciated early on. But teasing stylistic variability out of a Paleoindian assemblage is no easy task (but see Bamforth 1991; Roosa 1977; Roosa and Deller 1982), let alone in an assemblage as small as this one and with as high an incidence of reworking. Much of the morphological variation apparent in point blades recovered from a kill site will be attributable to the use and life history of the point (Amick 1995; Hester 1972; Hofman 1992). Bases were protected in their hafts, but the technological demands of hafting apparently resulted in a standardization that limited the possibility of adding stylistic features in this portion of the point. Finally, identifying variation that might be referable to stylistic differences that are geographically distinctive or historically meaningful is a matter best approached at the regional scale rather than at a particular site (Hester 1972; Hofman 1992; cf. Roosa 1977). But even at that scale, efforts to identify stylistic attributes and patterns in Folsom points have been rare, mostly focused on the presence or absence of fluting (Agogino 1969; Amick 1995; Hofman 1992; Judge 1970; Wendorf and Krieger 1959; Wendorf et al. 1955; Wormington 1957). Nonetheless, such an effort can potentially help gauge the possibility multiple groups using stylistically distinct points contributed to a particular assemblage, perhaps by aggregating at that place (Hofman 1994).

At the Folsom site there are a number of points that are only slightly reworked (or not reworked at all) that preserve enough of their original form and technology to provide a possible glimpse into intra-assemblage variation that is not obviously a by-product of differences in raw material, technology, or reworking. In all, there are six points that fall into possible stylistic pairs. Sn 1 and Sn 2 (fig. 2.10 and 8.4A), found in 1926 in the northwest corner of the site, are relatively thick (maximum width to maximum thickness ratio [W:T] of 5.11), with wide and relatively deep flute scars, very fine marginal retouch, and blades that approach

a triangular shape. Sn 22 and Sn 25 (fig. 8.4B) are thinner and narrower overall (W:T ratio of 5.35), with thinner and shallower flute scars, less fine marginal retouch, and parallel-sided blades. Sn 5 and Sn 27 (fig. 8.4A and B) are slightly less parallel in blade form (though not as triangular as Sn 1 and Sn 2) and are the widest and thinnest of the sets (W:T ratio of 6.01), with relatively deep broad and deep flute scars, and have fine marginal retouch. Sn 1 and Sn 2 are alike despite one being made of Alibates and the other of Black Forest chert. Likewise, Sn 22 and Sn 25 are made of Tecovas and Alibates, respectively, while Sn 5 and Sn 27 are both made of Tecovas. Unfortunately, it is not known where the latter two pairs were within the bonebed.

On the assumption that there were several knappers among the hunters at Folsom, and that each had a particular style of fashioning projectile points (Bamforth 1991), these three pairs of points may have each been made by an individual, but not necessarily the same individual. Since Sn 1 and Sn 2 were each recovered in the same small corner of the site, I might further speculate that the hunter using these points stood atop the high bedrock wall where the paleotributary and paleovalley come together—assuming, that is, that the animal(s) with which the points were associated dropped close to where they were speared. Beyond that walk out a rickety limb, I would not venture to stray.

Folsom Point Hafting

The morphometric patterns beg the question of how Folsom points were hafted. The longstanding notion, of course, is a *fixed-haft* model, in which points were set firmly into socketed bone or wood foreshafts or shafts (e.g., Crabtree 1966; Judge 1973; reviewed by Ahler and Geib 2000:801–802). As illustrated, for example, by Crabtree (1966:fig. 12a), a newly minted point of ~70 to 75 mm in length (see Boldurian and Cotter 1999; Collins 1999b; Bradley 1991; Frison and Bradley 1980) would be inserted some 20 to 25 mm into a socket. To tighten the fit and give tensile strength to the point, small shims or splints of bone or wood might then be inserted as well; these could extend down the flute faces. A sinew binding would then be tightly wrapped around the socket and base of the point. The binding did not and need not extend fully up the face of the point, since the point was firmly anchored at its base. Thus, the shims could continue beyond the extent of the grinding. In any case, the depth of the socket and the length of the haft were essentially unchanging, and the points remained anchored within the haft over the course of its usable life.

Beyond the bound portion of the haft, of course, the edges of the blade were exposed. Over time and as a result of use, impact damage, or other attrition, the blade would be resharpened and reworked in anticipation of the next use (Hofman 1992). Reworking is assumed to have taken place while the point was still embedded within the haft. Once reworking had widened the front angle and reduced the blade to where it barely extended beyond the binding—say, when the specimen was about 30 mm to 35 mm in length—the remaining slug would be discarded (also Jodry 1999a:186), unless, of course, it never reached that size, having earlier broken beyond repair. Such is the obvious liability to a fixed-haft. Points that are basally anchored can transfer the energy of a blow received "head-on" through the haft and into the shaft, and may or may not break (but see the discussion of end shock below). But bending fractures received at an angle would almost certainly snap a fixed haft point, likely at the haft/no haft boundary. Depending on where that boundary was, and the nature of the blow, a point in a fixed haft could be rendered unusable in an instant.

Recognizing that liability, Ahler and Geib (2000, 2002) recently proposed an intriguing alternative model of hafting. They envision the process as having been akin to the forward-advancing mechanism of a modern *utility knife*. In the utility-knife model, newly minted fluted points would be set in a long, sliding, friction haft. They would be firmly anchored, but not at the base. Rather, they would be held tight along their faces and edges by a single split piece of bone or wood (or two separate pieces) that rested within the flute scar on each face and extended nearly the length of the point, leaving only a small, thin, relatively acutely angled tip exposed beyond the termination of the haft. The haft, as indicated by their illustrations (Ahler and Geib 2000:fig. 7, Ahler and Geib 2002:figs. 20.1, 20.2), would swathe at least ~60% and perhaps as much as 80% of the point, depending on the length of the specimen. This technique would minimize the vulnerable portion of the point, and thus when it broke as a result of use, damage, and so forth, far less stone would be lost (Ahler and Geib 2000:810–811, 2002:375). And when it broke, the haft binding would be loosened and the point slid forward to reexpose a tip, which would then be sharpened into the desired form, thickness, and front angle, then reanchored within the haft (Ahler and Geib 2000:811, fig. 4). That process would be repeated until the point was reduced to a slug that was too short, too dull, and too thick to penetrate effectively (Ahler and Geib 2000:fig. 7). The obvious virtue of this proposed hafting mechanism is that it would prolong the use-life of these points (Ahler and Geib 2000:806, 2002:379).

There are some data from the Folsom assemblage that bear on these hypothesized hafting techniques. Consider the first points found at Folsom (Sn 1 and Sn 2; fig. 2.10): These are two of the longest points recovered, and though both are missing their bases, they are still more than 50 mm in length, which places them among the longest of the 432 Folsom points in the comparative sample. Neither of these points is reworked. By the utility-knife model, at least ~60% of each would have been buried in the sinew-bound portion of the haft (see illustration in Ahler and Geib 2000:fig. 7), implying that the edge grinding on each should extend up most of their margins and that striations should occur on their faces where they were bound by foreshafts.

Although no striation data are available on these points, edge grinding data are (table 8.4). One of the points has only 3 mm of ground edge; the other is not edge ground. In fact,

both specimens are widest where they broke, suggesting that any trace of grinding would be farther toward the base (on the assumption that binding generally occurs below the point of maximum width, so as not to hinder penetration). Moreover, neither is a preform; both show the fine postfluting marginal retouch of finished points (Frison and Bradley 1980), and insofar as their archaeological context can be discerned, both were in the kill area, embedded in bison carcasses. As noted, one refit to a sliver found alongside a plaster-jacketed bison rib, and the other came from near the base of a vertebra (Figgins 1927; appendix B: July 14, 1926).

The utility-knife model would be correct if one were willing to make the assumption that these 50-mm-long broken points represent only the ~40% of the blade that was exposed beyond the binding. But if that was so, it would make the original unbroken specimens at least ~125 mm in length, longer than almost every known Folsom preform or point (Collins 1999b:26; also Boldurian and Cotter 1999:108; Bradley 1991; William 2000:173). Yet, if these two points were set in a fixed haft, that in turn would suggest that as much as 50 mm of their blades extended well beyond the edge-ground region, which would contradict any rules of conservation of raw material and render the points highly vulnerable to breakage—unless, of course, other elements of the haft (the shims) extended beyond the edge-ground region as well.

But these are only two projectile points from the site, they are broken, and there are no striation data available. Thus, they may not provide a viable test of a hafting model, though if, in fact, a disproportionate segment of these points extended beyond their hafts, they may hint at how much stone was available to these hunters (of which, more below).

Although a comprehensive test of these hafting models is beyond the scope of this study, the data assembled here do provide one critical measure by which the two can be compared and evaluated: for a key difference between the utility-knife model and the fixed-haft model is that the former predicts that edge grinding will scale with overall point length, while in the latter it will not.[4] This is so because the bond in the utility-knife model is maintained by facial friction along the flute faces, "accentuated by binding around the lateral haft element margins" (Ahler and Geib 2002:376).

Examining the extent of grinding is one of the five tests of the utility-knife model Ahler and Geib (2000) propose.[5] The data for it can be derived from the relative lengths of the blade and haft elements (Ahler and Geib 2000:813). The *haft element* is defined by them as that portion of the point—generally behind the spot of maximum width and its extent marked by lateral grinding—that provided purchase for hafting (Ahler and Geib 2000:809, 814). Usually tapered for deeper penetration, it is "shaped by steep and abrupt retouch and is intentionally dulled" (Ahler and Geib 2000:809). Like Ahler and Geib (2000) and others (e.g., Kay 1996; Jodry 1999a), I use the extent of edge grinding to mark the hafted area.[6]

The *blade element* is the part that extends distally beyond the haft element (Ahler and Geib 2000:813). Ahler and Geib suggest, and I concur, that two results should obtain if their utility-knife model is correct: (1) haft element length measurements should be much more variable than blade element length measurements in finished points; and (2) haft element length should be highly correlated with total point length in complete points, and blade element length should be significantly less highly correlated with total point length in complete points (Ahler and Geib 2000:814). In contrast, if these points were in fixed hafts, then haft element length should be less variable, and vary independently of overall point length.

The former can be tested by comparing CVs for these two length measures with the expectation that the CV for the haft element will be significantly greater than the CV for blade element length. The latter can be tested using correlation.

Unfortunately, edge grinding, the measure critical to determining haft element length and, by subtraction from maximum length, blade element length, is not customarily recorded in analysis of projectile points (table 8.6). Thus, I was only able to muster a sample of $n = 104$ for which edge grinding was recorded. Of those, only 41 specimens were complete, and for obvious reasons this analysis is best conducted with complete points.[7] These results are reported in table 8.10.

The first observation to make on these data is that, as anticipated by Ahler and Geib (2000), blade length is indeed shorter than haft element length. However, that by itself does not support the utility-knife model since at least half of the points in this sample ($n = 20$) were reworked and, as expected, have shorter blades. Moreover, the resulting CVs are just the opposite of what the utility-knife model predicts: CV values are higher for blade element length than for haft length. The difference in those CV values is statistically significant ($p = 0.0237$), using the significance test of Zar (1999:141–145). Thus, the data do not support the prediction of the utility-knife model that haft element length measurements should be much more variable than blade element length measurements. They, do, however, support the corresponding expectation of the fixed-haft model.

Table 8.10 also provides correlation coefficients between the blade element and haft element lengths and the maximum length. Again, the results are the opposite of those predicted by the utility-knife model, which anticipates that haft element length will be more highly correlated with total point length than blade element length. Although both are correlated with maximum length, the correlation is stronger for blade elements ($r^2 = 0.694$) than haft elements ($r^2 = 0.493$; both are significant at the 0.01 level).

Ahler and Geib (2000:814) suggest that this test should be carried out with a large sample from many contexts in order to "even out" variation due to the "vagaries of fracture and situational constraints." Their point is well taken. One suspects, given the results obtained here, that subsequent findings may not differ appreciably. In any case, a fuller exploration of such matters awaits.

In the meantime, let me return to the matter of standardization. One of the results evident in the data in tables 8.8 to 8.10 is the remarkably small amount of variation in

TABLE 8.10

Patterns in Blade Element vs. Haft Element Lengths (mm); from the Comparative Sample

(a) Variation in Blade Element and Haft Element Lengths

	Blade Element Length	Haft Element Length
Mean	18.72	20.64
SD	7.33	5.70
Count	41	41
CV	**39.16**[a]	**27.59**[a]

(b) Correlation Coefficients (r^2) for Blade Element and Haft Element and Maximum Lengths

	Maximum Length	Blade Element Length
Blade element length	0.694 (P = 0.000)	
Haft element length	0.493 (P = 0.000)	0.036 (P = 0.234)

[a]Boldfaced values indicate coefficient of variation (CV) difference between blade element and haft element lengths significant ($p = 0.0237$).

haft dimensions in Folsom points, supporting Judge's (1973) conclusion that points were made to fit into a haft of fixed dimensions (also Jodry 1999a:193; Keeley 1982). By extension, the haft appears to have been of fixed size, and not a function of the overall length of the point. If a sufficient amount of the blade remained after a point broke, it was resharpened and readied for use. That many Folsom points, including some of the specimens from the type site, show slight shouldering of the blade just beyond the haft/no-haft boundary supports the idea that reworking took place while the point was anchored in a fixed haft. If the blade broke beyond repair, it would have been removed from the haft, and a replacement point slotted into place.

Projectile Point Life Histories

Over the course of their use-lives and through the span of time and space separating the locality where the stone was acquired, and where artifacts were ultimately discarded or lost, projectile points underwent resharpening and reworking. This was done to maintain their utility for as long as possible against the inevitable attrition caused by use (Ahler and Geib 2000; Hofman 1992; Wyckoff 1999:55). Reworking would presumably come to an end when the blade of the point was close to the nub of the haft, at which point the remaining slug would be recycled into another tool or discarded. At Folsom the reworked complete points average 33.7 mm in length. Jodry found virtually the same size for discarded slugs at Cattle Guard, while the mean length for nonreworked complete points was a substantially longer ~52 mm (Jodry 1999a:186; Wilmsen and Roberts 1978).

Projectile point reworking commonly occurred on the blade and was marked by reduced length, slight shouldering of the blade above the haft area, deviations in blade shape including increased asymmetry, changes in flaking technologies such as retouch that took place after use/breakage or invasions into original flake scars, and changes in the front angle and leading edge sharpness of the point, both being more obtuse in reworked forms (Ahler and Geib 2000:805; Boldurian and Cotter 1999:106; Bradley and Frison 1996:45; Collins 1999b:26; Hofman 1992; Wheat 1979:77–78; Wilmsen and Roberts 1978:108–109). Reworking was directed at maintaining these specimens as projectile points, and primarily involved a reduction in length, rather than width or thickness (unlike, say, Dalton Paleoindian points, in which blade resharpening came in from the sides of the blades [Goodyear 1974]).

Pursuing this matter further, points that are longer will, all other things being equal, allow greater recovery from breakage, as there should be more material to work with after the point breaks (Ahler and Geib 2000:806). Therefore, one expects longer and thinner points (higher length:thickness [L:T] ratios) to have a lower incidence of reworking than points that are shorter and thicker (lower L:T ratios). This is indeed a statistically significant difference ($G^2 = 9.781$, $P = 0.002$), using the complete points ($n = 41$) from the comparative sample (table 8.11). Freeman-Tukey indicate that reworked points are significantly underrepresented in longer and thinner points (L:T ratios >12), while points without reworking occur at significantly higher frequencies than expected in that same category; in contrast, nonreworked points are significantly fewer among shorter and thicker points (L:T ratios <12).

There are, of course, instances where reworking occurred on the base (Meltzer, Mann, and LaBelle 2005). It is reasonable to assume that this occurred when a point snapped close to the base and well within the haft, thus ensuring that the broken blade segment was not lost. The intact blade, once released from the haft, could then be rebased to form-fit back into the haft. The fitting might include beveling the base or thinning the corners. I have only observed a few rebased points, and none were refluted. Given the customary near-full-length extent of Folsom fluting, such would have been unnecessary (Collins 1999b:26; cf. Boldurian 1990:93, 97; Boulurian and Cotter 1999:106). Basal reworking tends to be rarer than reworking of the blade (Collins 1999b; Hofman 1992:212; Wheat 1979; Wilmsen and Roberts 1978:108; Wyckoff 1999:46).

Resharpening and reworking, which took place in advance of the kill at Folsom, are clearly evident on 7 of the 18 points for which data are available (table 8.5; Sn 3, Sn 8, Sn 12, Sn 16, Sn 18, Sn 23, Sn 24). Two more points (Sn 25, Sn 27) display only the slightest traces of reworking and, for this reason, are not included with the others. Most of the heavily reworked specimens were close to or at the end of their use-lives judging by their relative L:W:T (W = width) ratios and the short amount of usable blade still present

TABLE 8.11

Association of Length:Width Ratio vs. Reworking, in Complete Points from Folsom and the Comparative Sample

	Reworked	Not Reworked	Total
Length:thickness ratio			
>12	3 (−1.44)	5 (1.99)	8
<12	30 (0.69)	3 (−1.44)	33
Total	33	8	41

NOTE: ($G^2 = 9.781$, $p = 0.002$): Freeman-Tukey deviates in parentheses. Significance at $p = 0.05 \pm 0.979$. Boldfaced values significant.

beyond the extent of edge grinding. Thus, nearly 39% of the observable points from Folsom were reworked slugs, likely to be discharged from the arsenal of weaponry of these hunters. All of the reworking of the Folsom site points took place on their tips; none of these specimens was rebased.

That almost 39% of the points from the Folsom site were heavily reworked is not inconsistent, percentage-wise, with evidence from other localities and may be on the low side, at least according to data of Hofman (1992:figs. 6.9, 6.10). Of course, given the number of points at Folsom for which reworking could not be determined ($n = 10$), that percentage may be inaccurate. In any case, in the larger comparative sample, which includes the Folsom data, an average of 27% of the points were reworked.

It has become customary to insert reworking tallies into Hofman's Retooling Index (Hofman 1992:fig. 6.9), in which the percentage of reworked tips is plotted against the mean length of complete points (see Bement 1999b:fig. 5; Bouldurian and Cotter 1999:107; Buchanan 2002; Jodry 1999a:fig. 55). Hofman proposed that a negative relationship will obtain between the two, such that the higher the percentage of reworked points, the lower the mean length, on the reasonable assumption that length will inevitably be lower in an assemblage with a high percentage of reworked points. Thus, on its face the Retooling Index only states a mechanical relationship that was already known.

But can the position of one site relative to others on the Retooling Index reveal significant intersite differences in the nature and amount of reworking, or the number of kill/retooling events, as is intended? Perhaps not, for several reasons. First, percentage data are being used on the Y axis, and percentages are inherently unstable measures in small samples: 2/5 and 40/100 are both 40%, but if one additional specimen is added or missed in the sample, the percentage of the former will change significantly, and the latter will not. Second, because the Retooling Index does not identify whether shorter points are reworked and longer ones are not, we run the risk of concluding that a relationship exists when it may not. It is conceivable that smaller points were not reworked if, for example, they broke in manufacture but were still pressed into service. Third, one cannot gauge the statistical significance of the position and differences among sites plotted on the Retooling Index graph, when samples of very different sizes are being compared, and percentages are used. In effect, the Retooling Index is not a useful measure of the incidence or degree of assemblage reworking and, thus, is not telling us what we think it is telling us.

How, then, can we gain a better understanding of the incidence of reworking and determine whether the differences one sees in the amount of reworked points in different sites are significant? To probe this question, I gathered data on reworking from a dozen sites including Folsom. These data are given in table 8.12. In anticipation of the statistical tests necessary to the task, I compiled frequency counts of reworked points rather than percentages.[8]

In exploring the data in table 8.12 a significant fact quickly emerges: There is a strong correlation between the frequency of reworked specimens and the sample size. This relationship obtains no matter whether one correlates the number of reworked points (1) with the total number of points in the assemblage ($r^2 = 0.918$, $p = 0.000$), (2) with the complete points + tip segments ($r^2 = 0.934$, $p = 0.000$); or (3) with only the complete points ($r^2 = 0.452$, $p = 0.016$). For a variety of reasons, it is most appropriate statistically to use complete points + tip segments, since one can generally assume that a single point has only one tip, and thus a complete point and a point tip must have come from different specimens. The exception to this assumption would occur in rare instances where a tip snapped off at a site, the point was reworked, and then it too was discarded at that same locality.[9]

The least squares regression plot of reworked points versus complete points + tip segments is shown in figure 8.9. This plot reveals, all other things being equal, that at any given time a consistent portion of any Folsom assemblage was comprised of reworked points. Given the sites involved, this relationship seems to occur regardless of how far from the stone source that group might be or how many kill-butchering-retooling episodes they had experienced.

However, and granting that the number of reworked points is a function of sample size, it is nonetheless possible to identify assemblages of projectile points that have a greater or lesser amount of reworking than one might expect. This can be accomplished by identifying statistically significant outliers on the regression line as measured by standardized residuals, which have a mean of 0 and a standard deviation of 1. Of the sites plotted in figure 8.9, three are significant outliers: Lindenmeier and Lubbock have more reworked points than would be expected (standardized residuals = 1.227 and 1.058, respectively), while Shifting Sands has fewer reworked points than would be expected (standardized residual = −2.187). Arguably, this regression plot is also sensitive to the vagaries of small samples, but when the same data are examined using contingency table analyses, which are more stable with small sample sizes, the same three sites prove to be significant outliers. That there are fewer reworked points than expected at Shifting Sands, a conclusion also reached by Hofman (1992:212), may in part

be a function of the fact that I tallied only the Folsom points from this site and did not include the Midland points, wherein there may have been a higher incidence of reworking.[10] In any case, note that the data from the Folsom site fall squarely on the regression line.

That the incidence of reworking is a function of sample size does not mean that variables such as the distance to the source mediated by the number of kill-butchery-retooling episodes (Hofman 1992) are irrelevant. Rather, it means that we must take into account sample size effects before comparing assemblages. Moreover, detecting the effects of source distance and retooling episodes should not be done solely with projectile points and independent of the other artifacts in the toolkit since, as Hofman (1992:203, 208) argues, reworking is perhaps best viewed at the assemblage level, and should include analysis of the relative size, occurrence, raw material richness, and use patterns of bifacial cores, flake blanks, and fluted and unfluted projectile points (also Bamforth 2002). It is important to add that sample size effects can also lurk in tool class richness and diversity, so differences in the relative amounts of cores versus flakes, etc., need to address that possibility. There are insufficient data from Folsom to pursue this question.

Let me, instead, turn to the relationship between reworking and raw material type, for it has been suggested that in an assemblage of projectile points made of stone acquired at a variety of sources, reworking should be more common in points made of stone from sources that were acquired earliest and that arrived on-site late in their use-lives (Bouldurian and Cotter 1999:106; Hofman 1992:208; Ingbar 1992:173; Ingbar and Hofman 1999:103). This hypothesis can be explored by examining whether or not there is a statistically significant association between the occurrence of reworking and the types of lithic raw material (table 8.13a). Assuming that the difference in the degree of reworking might prove significant, the data are sorted by heavy reworking, slight reworking, and no reworking. Contingency table analysis of the Folsom site data in table 8.13a yields a nonsignificant likelihood ratio chi-square statistic ($G^2 = 7.526$, $p = 0.481$). However, Freeman-Tukey analysis of the same data shows a subtle trend of a scarcity of nonreworked points made of Flattop chert; this does not necessarily imply that specimens of Flattop chert show a statistically significant inclination to be reworked, although the data trend in that direction.

One should not place too much emphasis on this result, however, given the inescapable fact that all the points used at Folsom were acquired at distant sources, and many of the cells in the contingency table have expected values of <5 (combining the heavily and slight reworked specimens does not solve the problem of low expected values). There may be only slight and insignificant differences in the frequency of reworking among so many exotic sources. Not surprisingly, a nonsignificant result also obtains when examining the association of reworking and lithic raw material at other sites where all the raw materials are exotic, such as Cooper (table 8.13b) (Bement 1999a, 1999b).

TABLE 8.12
Counts of Assemblage Size (Various Measures) and Reworking

(a) Data

Site	All Points	CO + Tip Only	CO Only	Reworked	Source(s)
Blackwater Draw	149	32	29	18	Boldurian (1981), Hester (1972)
Cattle Guard	211	60	11	34/42[a]	Jodry (1999a:191)
Cooper	34	25	15	13	Bement (1999b:table 8.1)
Elida	22	8	6	6	DJM assessment from figures in Hester (1962), Warnica (1961)
Folsom	28	14	4	9	DJM data (includes slightly reworked specimens)
LakeTheo	15	4	3	3	Buchanan (2002)
Lindenmeier	185	48	34	37	DJM assessment from figures in Gantt (2002)
Linger	21	11	5	4	DJM assessment from figures in Hurst (1943), Jodry (1999b:fig 72)
Lipscomb	30	12	9	6	DJM assessment from figures in Schultz (1943)
Lubbock	9	6	6	6	Johnson (1987)
Shifting Sands	17	17	8	2	Hofman et al. (1991)
Waugh	2	2	1	1	Hofman (1995)

(b) Correlation Matrix for Data in Part *a*

	CO + Tip	CO Only	Reworked
All points	0.952332	0.750492	0.958443 ($r^2 = 0.919$)
CO + tip segments	1	0.691851	0.966368 ($r^2 = 0.934$)
CO only		1	0.672749 ($r^2 = 0.452$)
Reworked			1

[a]Jodry (1999a:191) puts the occurrence of reworked tips at 70%. If that is 70% of just the point tips from the site, then the count is 34 (0.70·49). If, however, that is 70% of all points on which reworking might be discernible (i.e., all complete points and all tip segments), then the actual count would be 42 (0.70·60). Fortunately, the statistical consequences of using one or the other are not great, so I use the latter ($n = 42$).

What results obtain in assemblages in which both exotic and local sources were used? Unfortunately, only two have the requisite data for the test, Lake Theo (Buchanan 2002) and Lipscomb (Schultz 1943), and both of those yield nonsignificant likelihood ratio chi-square values (G^2), indicating once again that there is no statistically significant association between reworking and lithic raw material (cf. Hofman 1992:217). This occurs despite the fact that only the points made of the most distantly acquired stone (Edwards chert) exhibit reworking (tables 8.13c and 8.13d) (Buchanan 2002:141–142; cf. Root, William, and Emerson 2000:fig. 103c). In the case of Lipscomb, the Freeman-Tukey deviates indicate that Alibates points show significantly less reworking than would be expected by chance. The lack of an overall statistically significant difference may be due to the small sample sizes of the other stone types, which of course might itself be evidence of differential use of the stone.

However, even when the data from Lake Theo and Lipscomb are combined by partitioning raw material into three distance-to-source categories (within 50, 50–150, and more than 150 km), there is still not a statistically significant association between raw material and source distance (table 8.13e). Clearly, more data are needed to test the matter, and as Hofman (1992:208) rightly anticipated, the evidence might better emerge when incorporating the entire assemblage, as there may be reason to anticipate differences in raw material use and retooling between projectile points and other tool classes (Bamforth 2002; Bement 1999b:119; Buchanan 2002:130–131; Hofman 1999a, 1999b; Ingbar 1992:182; Sellet 2004; William 2000), a difference Amick (1999b:181) suggests might signal gender differences.

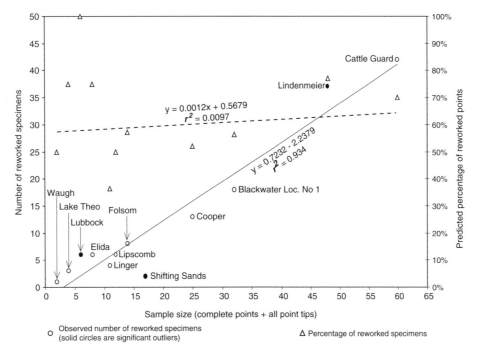

FIGURE 8.9 Least squares regression plot of frequency of reworked specimens against the frequency of complete points plus point tip segments from the comparative sample *(circles)*. Predicted percentage of reworked points against the frequency of complete points plus point tip segments *(triangles)*. Sites with solid circles are more than 1 standard deviation from the regression line. (Data from Table 8.12.)

Once at the Folsom site, both reworked and nonreworked points were pressed into service as weapons. Virtually all broke as a consequence. The scarcity of complete points at Folsom is not inconsistent with other Folsom-age sites; on average just over 21% of the points in the comparative sample are complete (also LeTourneau 2000:107), compared to 17% at Folsom. Folsom has a lower percentage of complete forms than virtually all others in the comparative sample, save Cattle Guard (5.29% [data from Jodry 1999a:fig. 50]).

Like reworked forms, the number of complete points is correlated with sample size, measured here simply as the total number of points in the assemblage: $r^2 = 0.563$, $p = 0.005$. That sample size is a strong predictor of the number of complete forms is also not surprising.

There is also a strong association between reworking and completeness, both in the Folsom sample and in the larger comparative sample (table 8.14). Specifically, whole points show significantly higher frequencies of reworking than would be expected by the null hypothesis that reworking is independent of completeness. One might argue that this result is misleading, since the category of broken points includes point midsections and bases, for which evidence of reworking cannot be obtained. However, the same pattern holds true when the sample includes only complete points and point tips.

That complete points have a significantly higher incidence of reworking is likely a consequence of the structural relationship between exposed blade length and vulnerability to breakage. That is, points that have been reworked will

have proportionately less of the blade exposed beyond the end of the haft, which reduces the chances later in a point's use-life that it will be leveraged and broken.

Patterns and Processes of Breakage

That the Folsom site projectile point assemblage had suffered considerable damage was apparent to early observers (Brown 1928b; Roberts 1935), as mentioned earlier. Brown and others attributed the scarcity of complete points either to the process of fluting, which possibly weakened and made the points vulnerable to breakage (Renaud 1931:12–13; Roberts 1935:17), or to the purposeful recovery of the "unbroken arrows" by hunters after the kill (Brown 1928b:4; also Hofman 1999a:126–128).

But did the removal of flutes so thin the points that "they became extremely fragile," and does that account for "the rarity of perfect [complete] specimens" (Roberts 1935:17)? Writing on the subject 40 years later, Crabtree did not think so. He believed that Folsom points were so well designed, despite the flute, that they were "one of the strongest of all projectile points," even compared to unfluted lanceolates. In his view, "no stemmed points made of comparable weight or size. . .would have had equal resistance to breakage" (Crabtree 1966:7; also Ahler and Geib 2000:802). Roberts' and Crabtree's views are not necessarily incompatible, for Crabtree was thinking about the entire projectile point/haft ensemble, and Roberts was not.

TABLE 8.13
Association of Lithic Raw Material and Reworking

(a) Folsom Site

	Alibates	Black Forest	Flattop	Quartzite	Tecovas	Unknown	Total
Heavily reworked	2	0	2	0	3	0	7
	(−0.39)	(−0.63)	(1.07)	(−0.63)	(0.43)		
Slightly reworked	1	0	0	0	1		2
	(0.34)	(−0.21)	(−0.39)	(−0.21)	(0.46)		
Not reworked	4	1	0	1	2	0	8
	(0.47)	(0.72)	(−1.18)	(0.72)	(−0.36)		
IND	3	1	0	1	2	4	11
Total	10	2	2	2	8	4	28

(b) Cooper Site

	Alibates	Edwards	Niobrara	Owl Creek	Total
Reworked	5	4	1	3	13
	(−0.14)	(0.25)	(0.72)	(−0.01)	
Not reworked	7	4	0	4	15
	(0.31)	(−0.02)	(−0.77)	(0.24)	
Total	12	8	1	7	28

(c) Lake Theo Site

	Alibates	Edwards	Tecovas	Total
Reworked	0	3	0	3
	(−0.36)	(0.50)	(−0.65)	
Not reworked	1	8	2	11
	(0.38)	(−0.14)	(0.45)	
Total	1	11	2	14

(d) Lipscomb Site

	Alibates	Edwards	Total
Reworked	0	6	6
	(−1.27)	(0.53)	
Not reworked	4	13	17
	(0.65)	(−0.21)	
Total	4	19	23

(e) Lake Theo and Lipscomb Sites Combined

	Distance from Source			
	0–50 km	50–150 km	>150 km	Total
Reworked	0	0	9	9
	(−0.72)	(−1.42)	(0.67)	
Not reworked	2	5	21	28
	(0.49)	(0.67)	(−0.31)	
Total	2	5	30	37

SOURCE: Data from (b) Bement (1999b:115); (c) Buchanan (2002); (d) Hofman and Todd (1990) and Schultz (1943); (e) parts c and d combined.

NOTE: Freeman-Tukey Deviates in Parentheses. Folsom site: $G^2 = 7.526$, $p = 0.481$, $p = .05$ significance level $= \pm 1.43$; calculations did not involve "IND" or "unknown" cells. Cooper site: $G^2 = 1.722$, $p = 0.632$; no significant values at 0.05 level. Lake Theo site: $G^2 = 1.657$, $p = 0.437$; no significant values at 0.05 level. Lipscomb site: $G^2 = 2.703$, $p = 0.100$; significance at $p = 0.05$ level shown ± 0.9799815; significant cell in bold. Lake Theo and Lipscomb sites combined: $G^2 = 4.403$, $p = 0.111$; significance at $p = 0.05$ level shown ± 1.131585; significant cell in bold.

Whether fluted or unfluted points were more susceptible to breakage can be tested by a detailed comparison of breakage patterns between the two; experimental data on the matter would also be of value. That is more than I can or wish to do here, so for now this question must remain unanswered. However, it is useful to explore breakage within Folsom fluted points, for as Ahler and Geib (2000:805) put it, "There is no avoiding the inevitable—they ultimately break." How and why they break are of some interest. Let me first address some of the causes of projectile point breakage and then examine breakage within the Folsom site assemblage.

A projectile point, when used as a weapon—and as Renaud (1934:4) observed early on, not all of them were (also Ahler 1970)—may fracture as a consequence of various actions. The most obvious of these is the impact fracture that sometimes occurs when stone meets bone, particularly the skeleton of an animal as massive as a bison (Frison 1991:177). Impact fractures are manifest in various ways. The most prevalent and readily recognized is the shatter of point tips and edges, which might include fractures that resemble deliberate burination, tip crushing/comminution, and, most dramatically, the presence of "reverse flute scars," flakes driven backward from the tip of the point toward the base (Ahler 1970, 1992; Bergman and Newcomer 1983; Dockall 1997; Frison 1987; Frison and Stanford 1982; Odell and Cowan 1986). These "longitudinal macrofractures" are the result of cone- or bending-initiation with propagation along one surface (Dockall 1997:325; Frison 1987:261–262; Wheat 1979). Impact fractures are common in Folsom-age bison kills and, for that matter, in bison kills of later periods as well (e.g., Bement 1999a; Bradley 1982; Bradley and Frison 1987; Frison 1974; Frison and Bradley 1980; Frison and Stanford 1982; Root 2000; Wheat 1979). Point impacts can also be detected on the struck bone, as shown experimentally by Frison (1991:149) and archaeologically by Bement (1999b:84–85).

Under certain circumstances, the compression waves of an impact might flow through the object, reach the base, and reverse direction, then intersect incoming longitudinal waves and fracture the point. The tip may or may not suffer damage. The result of this "back pressure" or end shock is commonly a lateral snap deep within the haft, perpendicular to the longitudinal axis of the point (e.g., Frison and Stanford 1982:fig. 2.60; see also Bradley and Frison 1987:271; Frison 1974:90–91, 1987; Frison and Bradley 1980:55; Judge 1973:265–266; Odell and Cowan 1986; Wheat 1979:89). End shock can occur within Folsom points but it may be rare: Projectiles as thin as Folsom points would have to receive the impact blow straight on the tip, that is, parallel to the longitudinal axis in profile and plan. If the impact came in at an angle, the point would more likely snap laterally at the haft/no-haft boundary (T. Baker, personal communication, 2003).

The latter raises a problem of equifinality, for lateral snaps can also result from actions other than impact. Consider the case of a projectile point that is firmly hafted and is levered up and down while the point is momentarily wedged into a solid object, the muscle and bone of a bison, say. In that

TABLE 8.14

Association of Projectile Point Breakage and Reworking

	Reworked	Not reworked	Total
Folsom Site			
Complete	3	1	4
	(1.31)	**(−1.07)**	
Broken	4	15	19
	(−0.68)	(0.53)	
Total	7	16	23
Comparative Sample			
Complete	81	10	91
	(4.05)	**(−6.55)**	
Broken	15	73	88
	(−5.90)	**(4.33)**	
Total	96	83	179

NOTE: Freeman-Tukey deviates in parenthesis; significant cells in bold. Folsom Site: $G^2 = 4.212$, $p = 0.40$; significance at $p = 0.05$ level shown ± 0.9799815. Comparative sample: $G^2 = 103.815$, $p = 0.000$; significance at $p = 0.05$ level shown ± 0.9799815.

instance, the tensile strength would most likely be overcome and the bend break would occur at the juncture between the two anchored portions of the point, the boundary between the base locked in the haft and the tip locked into the carcass (Titmus and Woods 1991:200). Lateral snaps can also occur in manufacture or from postdepositional taphonomic effects such as trampling (Bergman and Newcomer 1983; Dockall 1997; Odell and Cowan 1986; Shea 1988). Although all of these processes can produce lateral snaps, the manufacturing failures and postdepositional breaks should be relatively easy to spot, either by the absence of edge grinding, in the case of manufacturing failures, or by breakage that occurs randomly with respect to the haft/no-haft boundary, in the case of trampling.

Distinguishing lateral snaps caused by impact as opposed to bend breaks is more difficult. Where the lateral snap occurs deep within the haft area and closer to the base, is associated with the loss of the ears and corners of a point, or with longitudinal splits, or is accompanied by reverse flute scars, impact is the likely cause (also Frison and Stanford 1982:105–106). But if the lateral snap occurs at the haft/no-haft boundary, and is not associated with impact flutes, the cause could be either an impact or a bend break.

One of the end goals to understanding how a point breaks is ascertaining whether Folsom points were thrust or thrown, and if thrown, whether by hand or with the use of an atlatl, matters about which there is little consensus (e.g., Ahler and Geib 2000:804; Crabtree 1966:7; Frison 1991; Judge 1973: 157–158). The means by which the point was delivered, and here I only consider hand-thrown versus thrust, as the presence of atlatls in Paleoindian times is unresolved, speaks to the larger issue of hunting risk. Frison's observations on the matter are especially relevant here:

> Most wild animals that realize they are trapped often have a tendency to go berserk for a time and charge blindly into the restraints, whether they are dirt walls or logs in a corral fence. An infuriated animal the size of a mature bison can cause an almost incredible amount of destruction in a very short time. It probably would have been a wise move to spear or dart the animals from outside the trap, at least until the larger and more dangerous ones were no longer a serious threat. Once inside the trap, to dispatch the wounded animals, the thrusting spear would have been the ideal weapon in a close and direct confrontation situation. A cornered animal is likely to charge the hunter, in which case it is extremely vulnerable especially to a [thrust] spear. . . . (Frison 1991:167–170; also Binford 1997).

Apropos of this point, Binford's (1997) survey of the ethnographic literature documented that hand-thrown spears are unreliable shock weapons with short effective ranges, say, just ~6 to 8 m. Thrust spears are more effective, but they require game to be brought within close range, and for this reason, tactical aids, such as game drives, trapping, cornering, or otherwise disadvantaging the animals, are critical to reducing the risks of the hunt (Binford 1997; Churchill 1993).

Knowing whether the point was thrust or thrown might also shed light on the question of loss and discard (Wheat 1979). A thrust point is more readily tracked in the melee of a kill, assuming the hunter keeps a tight grip on the shaft and at least the hafted portion of the point remains intact; a thrown point leaves the hands of the hunter and stands a greater chance of being lost, especially if it breaks off inside the animal (Frison 1991:170; Hofman 1999a:124).

But distinguishing a point thrust from one thrown may not be a straightforward matter, since both can produce impact fractures (Dockall 1997:328; Frison 1991:177; Frison and Stanford 1982:105). Although there is some suggestive evidence that the degree of impact damage is greater from the higher-velocity impact of thrown spears (Ahler 1970:86, 106; Frison 1991:177), other work has found that velocity made no appreciable difference (Bergman and Newcomer 1983:243; Odell and Cowan 1986:204). Moreover, the impact force of a thrust spear on a charging bison could well equal the terminal velocity of a thrown spear. Ultimately, it may only be possible to distinguish a thrust from a thrown spear in those cases where the lateral snap is not associated with additional evidence of impact, and is arguably the result of a bend break—one that would imply someone had a firm grip when the point broke. This would occur, for example, when a point had been thrust into an animal and the shaft levered up and down for deeper penetration.

All of this suggests that knowing where lateral snaps occurred relative to the haft/no-haft boundary is of interest. Examination of a number of complete and reworked Folsom fluted points, including specimens from the type site, as well as from Shifting Sands (courtesy of Richard Rose), revealed that shouldering on reworked points generally began 7 mm to 8 mm beyond the farthest extent of the edge grinding. This 7 to 8 mm might reflect the distance the sinew extended beyond the grinding, or perhaps limits on how close a knapper could get to the haft area given the size of the flaking tool being used, or perhaps the desire on the part of the knapper to avoid having resharpening flakes terminate under the haft, or some combination thereof (T. Baker, personal communication, 2002; Hofman, personal communication, 2002). For analytical purposes, I assume that the area within ±7.5 mm of the ground/not ground edge marks the spot within which the haft/no-haft boundary occurred. On that assumption, a breakage classification scheme was developed (fig. 8.10).

This classification is a much-evolved version of a scheme created earlier for Clovis fluted points (Meltzer 1987, Meltzer and Bever 1995).[11] It has two levels. Recognizing that most of the Folsom literature records only whether a point is complete or occurs as a tip, midsection, or base, the first level divides points into these four categories, to ensure comparability. The second and more specific level subdivides the latter three groups (tips, midsections, bases) according to where the break occurred relative to the position of the haft/no-haft line or, more properly, to ±7.5 mm of the maximum extent of edge grinding (fig. 8.10).

Following earlier arguments, I hypothesize that points in classes 2a, 3a, and 4a broke from impact and end shock. My suspicion, and it can be little more than that without additional evidence, is that class 3b also resulted from impact and end shock. Point bases, midsections, or tips falling into classes 2b and/or 4b, lateral snaps close to the haft/no-haft boundary, could have resulted from oblique-angle impacts or bend breaks. Midsections in classes 3b, 3d, and 3e are equally ambiguous as to cause. In those instances where the haft/no-haft boundary is simply missing (classes 2c, 3c, 3f, 4c), the proximity of the break to that boundary cannot be determined, and refitted pieces are not available, I cannot surmise a cause.

Folsom site impact and breakage data are given in table 8.5 and figure 8.10. As shown by these data, nine specimens fall into breakage classes 2b and 4b, which I suggest are derived from oblique impacts or bend breaks. Six of those specimens have breaks suspected to result from impact and/or end shock, though perhaps not all of them broke in that way, as the ears are present on all point bases.

More broadly, tip (10) and base (7) portions are represented at about equal frequencies. This pattern contradicts the long-held notion (e.g., Hester 1962; Judge 1973:264; Roberts 1936:20) that kill sites ought to be marked by a preponderance of tips, and camps by a preponderance of bases, a notion based on the logic that broken points would only be removed from their hafts once the hunters were in camp (Keeley 1982). In fact, these results support Hofman's (1999a) more nuanced argument that frequencies of broken pieces will depend on a variety of factors, including the need for retipping, the use of the points as knives, the nature of the haft, whether tips were removed from the kill with meat packages, etc. (Hofman 1999a:124–125). One might further speculate that where point tips but not the conjoining bases are found with a skeleton, it marks the use of a thrust spear (assuming that the shaft was in the grip of the hunter when the point broke inside the carcass, and was retrieved); where a skeleton yields whole points or broken but conjoining tip and base segments, perhaps the spear was thrown and lost when a foreshaft broke off inside the carcass. Regardless, there is no necessary association expected between point tips and kills, as opposed to point bases and camps, and there is none at Folsom, where all specimens came from the kill area (see also Amick 1999a:2; Jodry 1999a:273).

It is also the case that the Folsom site differs significantly from sites in the comparative sample with regard to the frequency of breakage. As shown in table 8.15, at the Folsom site tips occur more often and midsections less often than would be expected by chance. The low incidence of midsections at Folsom may well be a consequence of 1920s field methods, which missed small fragments; most of the tips recovered are large (mean length = 38.46 mm) and thus more visible archaeologically. The higher incidence of bases and complete points ($n = 11$) marks the number of hafting components potentially lost.

It was observed earlier that there was not a significant association between lithic raw material and the incidence of reworking (table 8.13), suggesting that attrition was independent of stone type. One can pursue that inquiry further by examining the association between the five lithic raw material types and the four breakage classes. Doing so does not produce a statistically significant co-occurrence ($G^2 = 6.521$, df = 12, $p = 0.888$). In fact, when the assemblage of projectile points from Folsom is partitioned by raw material, and metric variables are grouped by Alibates and Tecovas, the two types for which there were sufficient sample sizes (>5 points), there is no statistical difference in breakage by raw material type (Meltzer, Todd, and Holliday 2002:28). This is not necessarily the expected pattern (e.g., Buchanan 2002:141).

Impact fractures are present on five (~21%) of the Folsom site points for which data are available (table 8.5; Sn 8, Sn 11, Sn 12, Sn 14, Sn 28). Two of those points (Sn 8, Sn 14) are midsections that also appear to have corresponding end shock. This sample is too small to test the prediction, made above, that points with impact fractures might co-occur with end shock and/or fall into certain breakage classes (namely, 2a, 3a, and 4a). It will be difficult to amass a larger sample of Folsom points or any set of hafted Paleoindian lanceolates, since the occurrence of impact fractures is not widely recorded in the literature. Impact fracture tallies are available only for two of sites in the comparative sample; at these sites they occur at a frequency comparable to that seen at Folsom (17.5% at Blackwater Draw, ~24% at Linger).

Loss and Discard

That Brown and Roberts raised the issue of the scarcity of complete points at Folsom suggests that they believed the

TABLE 8.15

Comparison of Breakage Frequencies between Folsom Site and Sites in the Comparative Sample

Breakage Class	Comparative Sample	Folsom	Total
Complete	86	4	90
	(0.00)	(0.23)	
Tips	107	10	117
	(−0.46)	**(1.94)**	
Midsections	231	3	234
	(0.47)	**(−2.60)**	
Bases	126	7	133
	(−0.11)	(0.65)	
Total	550	24	574

NOTE: $G^2 = 11.374$, $p = 0.010$; Freeman-Tukey deviates in parentheses; significance at $p = 0.05$ level shown ± 1.200227; significant cells shown in bold.

number of projectile points recovered there was less than one might expect for a kill of more than 30 bison—hence their suggestion that Folsom points broke readily or were carefully collected afterward. But though they may have suspected that there were more points used at the kill, they had no other sites with which to compare the Folsom tallies.

There are now sufficient data from Folsom and non-Folsom Paleoindian sites to explore the relationship between the number of animals killed (MNI) and the number of projectile points at a site. Data of Hofman (1999a: table 1) yield a positive correlation between the number of points and the number of animals ($r^2 = 0.66$). However, among just Folsom-age sites ($n = 9$), a weaker and non-significant relationship obtains ($r^2 = 0.368$, $p = 0.083$).[12] That relationship is strongly influenced by a distant outlier, the Cattle Guard site, with 211 points and an MNI of 49 bison (Jodry 1999a:177, 1999b:80). Removing Cattle Guard results in a significant correlation ($r^2 = 0.635$, $p = 0.018$). Obviously, the sample at hand is not altogether adequate to address this matter, and thus it is too early to conclude that the number of points recovered from the Folsom site is what would be expected for a bison kill of this size.

Even if one could so conclude, let me hasten to add that this does not mean that the sample from Folsom or any other kill site is equivalent to the original number of weapons originally used in the hunt. How many were brought to bear is not known, nor likely ever can be. It can only be surmised that there were at least two dozen or so, for the simple and obvious reason that the number of recovered points at any given site can be a function of additional factors besides the number of bison killed, among them the taphonomic history of the site, the size and portion of the area excavated (cf. Hofman 1999a), and, of course, whether the complete or still useable projectile points were systematically recovered and removed from the kill and, if so, how. If hunters salvaged their weapons, the sample of points from

1. Complete

2. Tips (distal end present; proximal end broken)

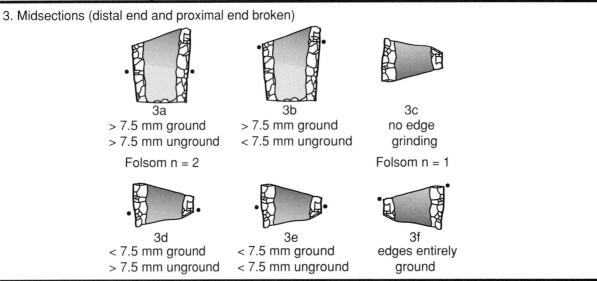

1

Folsom n = 4

2a
> 7.5 mm edge
grinding present

Folsom n = 2

2b
< 7.5 mm edge
grinding present

Folsom n = 4

2c
no edge grinding

Folsom n = 4

3. Midsections (distal end and proximal end broken)

3a
> 7.5 mm ground
> 7.5 mm unground

Folsom n = 2

3b
> 7.5 mm ground
< 7.5 mm unground

3c
no edge
grinding

Folsom n = 1

3d
< 7.5 mm ground
> 7.5 mm unground

3e
< 7.5 mm ground
< 7.5 mm unground

3f
edges entirely
ground

4. Bases (distal end broken; proximal end present)

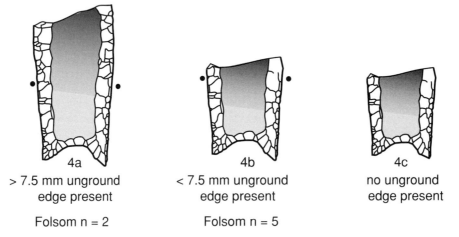

4a
> 7.5 mm unground
edge present

Folsom n = 2

4b
< 7.5 mm unground
edge present

Folsom n = 5

4c
no unground
edge present

FIGURE 8.10 Breakage classification scheme for Folsom fluted projectile points, showing the number of specimens from the Folsom site in each breakage class.

this or any other site would inevitably underestimate to an unknown degree the number of weapons originally in use.

It is difficult to conceive of a test that would detect whether spent projectile points were sought and recovered by the hunters at this locality. That said, it has nonetheless long been assumed that because of the "cost" in time and raw material of their manufacture, Folsom points would not have been casually discarded (Crabtree 1966:8). The corollary to that assumption is that points still long enough to be serviceable would have been salvaged and reworked. This, in turn, suggests that serviceable points that were not salvaged must have been lost (Hofman 1999a; also Buchanan 2002:141). Fully a third of the points and point fragments at Folsom were >40 mm in length and, thus, presumably long enough to reuse. Is it possible to determine whether these were lost and, therefore, provide a backdoor approach to the question of whether the hunters sought to recover their points from the kill?

Hofman (1999a:124) has argued that the proportion of "lost points," which he equates with the number of archaeologically recovered complete points, will be positively correlated with the density of carcasses at the kill, calculated as MNI divided by total bonebed area. Sites with few carcasses, widely dispersed and thoroughly butchered, are thought more likely to have been successfully picked over by hunters in search of still-serviceable weapons and butchering tools, especially if the weather was cool and the occupation extended as, perhaps, at Folsom (Hofman 1999a:128). Recovery chances would have been especially favorable for butchering tools, which start out in the hands and not deeply embedded in the animal. However, Hofman also suggests that widely dispersed carcasses might reflect a "less concentrated and perhaps less accessible herd," which in turn might have required the hunters to use atlatls; and that would increase their chances of losing their projectile points since the weapons left their possession when hurled at the animals (Hofman 1999a:128). These two lines of reasoning are not altogether contradictory, though they might appear so, since in one case a low density of carcasses means greater recovery of one class of tools yet lesser recovery of another. But as Hofman also observes, there are other variables involved in whether a projectile point was recovered by the hunters, including the intensity of carcass processing, the duration of occupation, and site surface conditions (Hofman 1999a:124).

But staying with the matter of carcass density, Hofman believes that there were, all other things being equal, relatively higher losses of projectile points at Folsom compared to Lipscomb, which he attributes to the wider dispersal of the bison carcasses at Folsom, at the very least, ~1 animal per 10 m^2 (chapter 7), again based on the reasoning that the animals "had room to move" and hunters had to use atlatls and easily lost track of their points. Although it seems unlikely that the carcasses at Folsom were moved significantly from the spot where the animal collapsed (chapters 5 and 7), where the animal collapsed may not be the spot where it was speared. Hence, one cannot assume that the relative dispersal of carcasses indicates that animals were bunched or dispersed when

attacked, or that their archaeological position reveals whether atlatls or thrusting spears were used to bring them down. Nor does it seem reasonable to argue that hunters who used thrusting spears necessarily had a better chance of recovering their weapons, unless the points did not break off inside the animal and the hunters kept a tight grip throughout.

In fact, the recovery of projectile points by hunters after a kill might be quite independent of whether spears were thrust or thrown. More likely, recovery or loss/abandonment has as much or more to do with where in the animal the projectile points were embedded, the degree of butchering, whether the point-bearing parts of the animal were removed from the kill area and further processed, and perhaps whether there was a need to recover the artifacts.

Unfortunately, the position within the skeleton is known with certainty for only three projectile points from Folsom, all of which were adjacent to ribs. Two of these lack bases (Figgins 1927a:figs. 3, 4) and may have been detached from their hafts and, thus, perhaps invisible when butchering began. However, these two are among the largest specimens in the assemblage, and certainly could have been reworked into usable points. That these points were not recovered may indicate that the rib areas associated with these points were not processed.

Even if they were, there are circumstances in which serviceable points could have been salvaged but were not; that is, the points were not lost but rather were abandoned. The latter would likely occur when groups had a sufficient stone supply. In such cases, it would not be necessary or worthwhile to spend the time or effort to locate spent points, or invest the time in cleaning or reworking them for reuse, time that could otherwise be put to other pursuits (Bamforth and Bleed 1997:127–128).

In contrast, groups low on stone may have made considerable efforts to search for broken pieces to see whether they might be serviceable, even if the points were not readily visible. That the large points were not salvaged by the Folsom hunters may say more about the actual and anticipated supply of stone than about the means by which the points were delivered, or the dispersal of the carcasses, which in this instance seemingly would have allowed for the recovery of useable pieces of stone. Although it seems reasonable to suggest that complete and/or still usable projectile points should be recovered and removed by hunters after a kill, there seems to be little hard evidence for that having occurred at Folsom.

Other Tools from the Folsom Site

Four other tools are part of the site assemblage; a fifth was claimed to be, but the claim is highly suspect.[13] Unfortunately, no contextual information is available on two of the specimens, and their current whereabouts are unknown (also Hofman 1999a:126). Recall that Roberts (1939:534, 1940:59) described them as a "nondescript flake knife, and . . . a generalized type of scraper," and that they might be the specimens that appear in a 35-mm slide labeled "Folsom," provided by Fred Wendorf (fig. 8.3). If so,

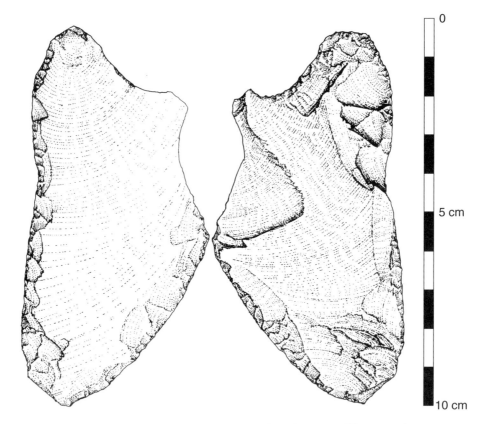

0

5 cm

10 cm

FIGURE 8.11 Quartzite knife found by Ele Baker at the Folsom site. (Illustration by K. Monigal.)

both of the missing artifacts appear to be expedient flake tools, which might have been struck from bifaces or remnants of formal tools that had reached the end of their use-lives. Both appear to be too large to have been struck from any of the projectile points found on site.

The Baker specimen (fig. 8.11) is the largest of any artifact found at the site (maximum length = 95.55 mm). It was initially identified as a "side-scraper" (Reed 1940:4) but has bifacial use-wear indicative of having functioned primarily as a knife. Its rounded edges from grinding and use, and lack of evidence for damage from hitting bone, raise the possibility that this tool was used as a skinning knife. Such large quartzite tools are not uncommon in Folsom-age sites, perhaps because of their ability to hold an edge (Frison 1991:324; Frison and Bradley 1980:113; Hofman, Amick, and Rose 1990; Jodry 1999a:109).

The Baker knife is made of gray quartzite but appears to be a different variety of quartzite from that used to make the quartzite projectile points. Given how little is known of the range of this material in different source areas, this may not be the case. As mentioned earlier, it is not possible to pinpoint the stone source from which one or both of these quartzites may have come.

Nor is it possible to determine with precision where on-site the quartzite knife was found, save that it came from the North Bank, at a depth of what may be ~3.9 m below the surface (fig. 8.2). That would put the specimen in the paleovalley,

which provides a clue to where the associated camp/processing area might have been located: for if this specimen was part of the butchering toolkit, it might suggest that at least some carcass processing took place in the paleovalley.

A flake tool (fig. 8.12) made of Black Forest chert was found in 1999 on the surface just below (north of) the 1920s back-dirt berm. This small, expedient tool (maximum length = 27.76 mm), made on a biface thinning flake, has minor edge damage from use and slight retouch is present along one edge, while a burination blow is present along the opposite edge.

Finally, one of the projectile point bases from the site shows evidence of use (Sn 10). Its broken lateral edge was put into service as a scraper in an expedient manner, causing slight and irregular edge damage along that edge. Unfortunately, only a cast of this particular point survives.

Save for the broken projectile point reused as a scraper, none of the other tools shows evidence of formal preparation and manufacture; both appear to have been made and used in an expedient manner, dulled or broken in use, and then been discarded. This is in keeping with the pattern seen in other Folsom kill/initial butchering localities (e.g., Jodry 1999a). Formal, complex, and substantially modified tools tend to characterize more intensive processing or habitation localities (Amick 1999a:3–4; Jodry 1999a).

The scarcity of tools, formal or otherwise, may reflect the distribution of carcasses, which made them easier to process and the tools easier to recover (Hofman 1999a), or, more

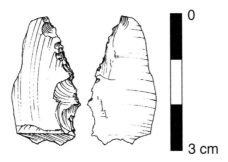

FIGURE 8.12 Black Forest chert flake tool found at the Folsom site in 1999. (Illustration by K. Monigal.)

simply, a sampling phenomenon: how and what portions of the site have been excavated.

Little can be said of the lithic raw material used to make these tools, and perhaps the most telling observation is a negative one: The high-quality stone from Alibates and Tecovas, of which the majority of the projectile points were made, was not used in the production of these flake tools (cf. Buchanan 2002). From this, one might infer that artifacts from more distant (time/space) sources that had already reached the end of their effective use-lives were, as the need arose, reworked and pressed into service as flake tools (also Amick 1999b:181; Bamforth 2002; Ingbar 1992). There is some hint of that in the flake tool made of Black Forest chert. Fully testing that hypothesis will require examining lithic debitage, but that is comprised of just a small amount of microdebitage ($n = 25$), mostly the by-product of tool use, which is not especially helpful in this regard (Meltzer, Todd, and Holliday 2002:28).

Finally, and as a matter of record, during the 1999 season a subangular cobble of rhyolite or dacite, 63 mm in maximum length, was found in a 1×1-m test unit in the F8 block, ~60 m south/southeast of the bonebed and deep within *f2* sediments. The *f2* in this area of the site is massive and relatively clean, with occasional thin lens of small gravel and fine but subtle sedimentary structures (laminations), along with flecks of charcoal, one of which, located ~27 cm above the cobble, yielded an age of $10,510 \pm 50$ [14]C yr B.P. (CAMS-74649). A cobble of this size is out of place in this sort of sedimentary context. Yet, close examination of the specimen reveals no apparent human modification. Although under suspicion as a possible manuport, Occam's razor suggests that its presence there is natural.

Summary: The Folsom Artifact Assemblage

The artifacts brought to and used at Folsom included at least 28 projectile points and 4 flake tools. Dominated as it is by projectile points, it is typical of an assemblage from a kill/butchering area. Most of the artifacts were recovered from the deeper portions of the site, and all from the paleo-tributary, save for two of the projectile points and the Baker quartzite knife. How many artifacts were missed during the

original investigations or since is unknown. It is perhaps telling, however, that it has been at least a decade since a stray projectile point emerged from the backdirt (Sn 28), and that was two decades after the previous known find (Sn 27). That just two points were found over the last three decades, in contrast to the half-dozen that emerged from the backdirt in the two decades following excavations, suggests that the supply of "backdirt points" may be nearly exhausted. Given the recent efforts to slow erosion and protect the site, it may be some time before any more artifacts are found.

In terms of their morphology, these are Folsom points—What else could they be?—but as was apparent in the 1930s (appendix E), and is even more evident today, there is considerable morphometric variation on that theme. The variability is primarily the result of reworking and resharpening over the use-lives of these points, such that exhausted and discarded forms appear to be significantly different from, and much smaller than, newly fashioned points (Amick 1995; Hofman 1992). It is not the case that either reworking or breakage correlates with raw material type, as one might expect of the points that had been carried the longest (Hofman 1992). However, that expectation holds true in other sites in the comparative sample.

Variation in the Folsom points is primarily manifest in blade size and shape. In contrast, there is a strong degree of standardization in the width and thickness of point bases, which were evidently being fashioned to fit hafts, and not vice versa (Judge 1973). Those hafts, as best can be determined by the available data, were not of the utility-knife form proposed by Ahler and Geib (2000) but were apparently fixed hafts in which the lower ~20 mm of the points was firmly anchored (based on the data in tables 8.7 and 8.9). But by being hafted in that manner, as Ahler and Geib (2000, 2002) rightly observe, the blades were exposed and highly vulnerable, and indeed this is an assemblage that was considerably damaged in this momentary encounter. This was partly because of the mass, agility, and speed of the prey (Frison 1991), but also because some of these points were, arguably, thrown rather than thrust; in either case several use-life–ending impact fractures were produced.

Yet despite the beating the toolkit absorbed, it does not appear as though these hunters were disadvantaged as a result. Left behind in the carcasses were a number of points that could have readily been refurbished and reused—along with, of course, the many shattered pieces and a few complete specimens worn down to little more than slugs.

Viewed from a different analytical angle, it is also apparent that the incidence of reworking and the number of complete (or broken) points in a given assemblage are related to the sample size, as, for that matter, is raw material richness (e.g., figs. 8.7 and 8.9). Sample size effects in these measures are hardly unexpected (G. Jones, Grayson, and Beck 1983; also, papers in Leonard and Jones 1989). Ironically, in places where similar scalar relationships were presumed to occur, such as between the number of points and the number of bison, or between the number of points

TABLE 8.16
Archaeological Correlates of Different Models for the Movement of Stone across the Landscape

	Stylistic Variation	Patterns in Lithic Raw Material	Evidence of Aggregation
1. Stone brought by multiple groups	Significant variation in point styles	Occurs in comparable amounts; no expectations of differential reworking	Features (hearths) and range of activities
2. Stone brought by a single group who acquired it via exchange	Significant variation in point styles	Occurs in varying amounts; expect less reworking of rare items	Not applicable
3. Stone brought by a single group who acquired it at the source	No significant variation in point styles	Occurs in varying amounts; expect reworking proportional to raw material frequency	Not applicable

SOURCE: Based in part on Meltzer (1989b:22–30).

and the size/density of the kill area (e.g., Hofman 1999a), they did not occur.

The observation that the products of artifact procurement, maintenance, and use are correlated with sample size does not mean that the patterns here lack interpretive value. Through other analytical avenues, such as the examination of residuals, it is possible to assess whether sites or assemblages are patterning in ways that are meaningful archaeologically.

In all, the frequencies of reworking and breakage seen at Folsom are not appreciably different from those seen in other sites of this period, testimony perhaps to the fact that at any given moment a group of Folsom hunter-gatherers was carrying with them toolkits in varying states of repair, ranging as here from newly minted points (e.g., Sn 1 and Sn 2) to points that had become effectively obsolete but had at least one more good use in them (e.g., Sn 23 and Sn 24). Such strategies and tactics of curation and use were likely a common response to the logistical challenge of being highly mobile hunters on a landscape in which prey and stone sources were widely dispersed.

As a part of that, tools made of very different and widely scattered lithic raw material sources were included in this toolkit (fig. 8.8). At least five separate stone types are in the assemblage, the majority of which (n = 18) were made from stone acquired from the Alibates and Tecovas outcrops in the Texas panhandle, a minimum of 200 km southeast of the site (table 8.7). A minority (n = 4) were apparently made of Flattop and Black Forest chert; the former were obtained at least 450 km north of the site (table 8.7). No stone from sources in the Front Range or west of the Rocky Mountains is present in the assemblage. Perhaps more curious, no Edwards chert from Texas was used, this despite its appearance in sites farther west. All this is assuming the accuracy of the source identifications, granting the absence of evidence from other tool classes, and the small size of the assemblage, which might not include rarer raw material types.

At present there is no evidence that tool replacement or point manufacture took place at Folsom, or that such was necessarily a pressing matter, given the large size of some of the points left behind. Assuming that this is not merely a vagary of sampling only the bonebed, Folsom was not a "gearing-up" locality (chapter 1). And because the several stone types occur at uneven frequencies, with (admittedly only subtle) hints of uneven degrees of reworking, and the assemblage is not dominated by a single stone type, it seems reasonable to conclude that this was an assemblage "on the move." It was one in which tools were used, broken, and discarded at very specific moments—such as this bison kill—but were being replaced more gradually over time and space as stone sources were encountered (Sellet 2004).

Although all the stone came from distant sources, it appears (following Hofman 1992) that the amount of time elapsed since those sources were visited was not great, and the intervening kill/retooling episodes were few. But how did it get to Folsom? It is time to return to that question.

The data allow no definitive answers, though certain possibilities can be assigned lower probabilities, using the archaeological correlates identified in table 8.16 (after Hofman 1994; Meltzer 1989b). The first possibility (fig. 8.8A) held that the stone was procured directly at outcrop sources by different groups, who then converged on this site. Evidence for such an aggregation ought to include some variation in projectile point styles; comparable numbers of each stone type, since each group would in principle have been carrying a full complement of weaponry; and independent archaeological evidence of aggregation. Although there are, indeed, stylistic and technological hints that some of the points were made by different individuals, the stylistic variation is not beyond what might be expected of different knappers within the same group (Roosa 1977). More telling, perhaps, is the disparity in the frequency of stone types, the absence of independent archaeological evidence of aggregation, and the lack of paleoecological evidence of resources that might make this area a destination (Hofman 1994; Root, William, and Emerson 2000). This possibility seems an unlikely explanation for the observed patterns.

The second possibility (fig. 8.8B) stipulated that a single group had acquired multiple stone types from other groups via exchange. Here one might also expect variation in projectile point styles ("exotic styles"), based on the presumption that exchange usually involved finished items, rather than unfinished blocks of stone (for the reasoning behind this presumption, see Meltzer [1989b:22–30]; also Hester and Grady [1977] and Hofman [1992]). In this instance, a disparity in the frequency of stone types would not be unexpected, but there might be differences in the incidence of reworking, assuming that the rarer gifted items were preferentially maintained with an assemblage (Meltzer 1989; also Ellis 1989; Weissner 1984). Although assemblage-wide there was no relationship between raw material and reworking, the rare items made of Flattop chert showed more rather than less reworking (table 8.13a).

The final possibility (fig. 8.8C) held that a single well-traveled group acquired the stone directly from multiple outcrops. Variation in projectile point styles would not be expected, but a disparity in the frequency of stone types would be. More important, given the distances in space and time that must have been involved, were this stone collected by a single group and used over time in proportion to its frequency in the toolkit, there ought to be evidence of greater incidence of reworking of the rarer items. These expectations are met in this assemblage.

On balance, the third possibility seems to be the most likely explanation for the pattern of raw materials at Folsom, though with the repeated disclaimer that this result is based strictly on the projectile points, which may give a biased view of procurement strategies relative to other tools that may have been part of the assemblage. Still, were the stone used in those other tools known, it might not necessarily change the result arrived at here.

Assuming, then, that this assemblage was acquired by a highly mobile group, the apparent source locations provide a measure of the geographic territory utilized (e.g., G. Jones et al. 2003; Root 2000). The straight-line distances involved in moves between these sources were considerable but surely minimize the actual distances traveled by this group, and cannot provide even a rough measure of seasonal, annual, and lifetime ranges, which among contemporary hunter-gatherers can be greater by an order of magnitude than the distances traveled in a single season or year (Amick 1995, 1996; Binford 1983:110,115; Kelly 1995:table 4-1). Nor, of course, are we in a position to do more than speculate on how long raw material from a source could last in a toolkit and, hence, whether the toolkit here represents a single seasonal round or something longer.

Given the numerical dominance of stone from the Texas Panhandle, it would appear that their most recent resupply had been in that area. If that is so, this group had much earlier been to the Colorado sources; the two points of Flattop chert highlight the considerable span of time/space for which items could be carried in a toolkit. As discussed in chapter 3, had these hunters followed the river and stream drainages out of the panhandle area (fig. 3.2), they could have easily come through the Folsom site en route to crossing the Clayton-Raton volcanic fields via Trinchera Pass, passing the Dakota and Morrison formation quartzite sources along the way.

Notes

1. With the subsequent discovery of other Paleoindian projectile points, the differences between the Folsom and the Lone Wolf Creek specimens came into sharper relief for Figgins. By 1935, Figgins (1935:5) recognized that the Lone Wolf Creek points, although more like Folsom than Yuma points, were nonetheless not members of either type.

2. To clarify: that number is derived by taking the total number of points and point fragments (28) and subtracting the 2 now-missing fragments (Sn 7 and Sn 20), which may or may not conjoin.

3. Including Cooper in the comparative sample raises a potential analytical complication. Although Bement (1999a, 1999b) argues that there were three separate bison kills at the Cooper site, LaBelle (2000) examined the Cooper site data and suggested the alternative possibility that there was but a single bison-killing episode at the site (see also Carter and Bement 2003). Resolution of this issue cannot be accomplished here, but I raise the issue because there is merit in LaBelle's suggestion, for it does help explain some curious aspects of the site, as, for example, why three groups on three separate occasions purportedly separated by several years ended up at the same time of year killing nearly identical numbers of bison at the same small and inconspicuous spot on the landscape. Moreover, it presents something of an analytical quandary, for as Bement himself observed (1999b:115–118), the Cooper data can look different when viewed separately as opposed to in the aggregate. Take, for example, the relationship between the MNI and the number of projectile points. It matters not whether Cooper is treated as a single kill (of, say, 29 bison) or as multiple kills, regarding the question of the correlation between MNI and the *total* number of points (r^2 values are 0.674 and 0.757, respectively). However, when examining the correlation between the MNI and the number of complete points, the results differ substantially if Cooper is a single kill, as opposed to three separate kills (r^2 values are 0.295 and 0.899, respectively). That a far more robust correlation results from treating Cooper as separate kills is merely a statistical artifact of having an additional three data points, all of which have very tightly scaled numbers of bison and points. But since the tight scaling may be a statistical "artifact," they may not be meaningful. Ultimately, an analytical decision must be made, and in this instance I follow Occam's razor and treat Cooper as a single bison kill.

4. Another key difference between the two is the presence of basal grinding, which ought to occur in points embedded in a basally abutted fixed haft, to reduce damage on impact—and is not anticipated in the utility-knife model (Ahler and Geib 2000, 2002). Although all five of the Folsom site points that preserve the base are ground, the matter cannot be determined for the other 23 points (table 8.5), rendering these data suggestive but ultimately insufficient as a test of the two models.

5. Although Ahler and Geib (2000) offer five archaeological tests in support of their model, they do not provide a side-by-side comparison of it against the fixed-haft model. Such a systematic comparison would be desirable (Ahler, personal communication, 2004), especially since (in my opinion) the outcome of several of their proposed tests could equally support a fixed-haft model. But acceptance of my opinion must be tempered by the fact that I have not systematically and fully compared the two models either. Ultimately, the matter of Folsom hafting deserves further exploration and testing.

6. It might be useful to make a distinction between the hafted area and the bound area. Ahler and Geib illustrate the two as being separate (e.g., Ahler and Geib 2000:figs. 5, 7) but state elsewhere that "the forward limit of the haft is marked by the extent of lateral margin dulling" (814). Yet, if the foreshaft did extend beyond the sinew-wrapped basal section, as they illustrate and as would have made their utility-knife arrangement quite effective, then points ought to break in the tip area, well beyond the terminus of grinding. This is a matter that can be readily tested with assemblages where broken points can be refit, which unfortunately are not available here. An alternative approach to defining the hafted region might be through high-power magnification, to see whether abrasion or striations occur on the flake faces or arrises that might be attributable to a tight binding (Kay 1996:329; Kay, in Jodry 1999a:187). However, these binding traces can vary with the "hafting approach" and will depend on the mechanical forces exerted on the haft element during the use-life of the tool (Kay 1996:329). So far, too few systematic analyses have been done and too few data exist to use this approach in any general study.

7. The points in this analysis are from Blackwater Locality 1, Folsom, Hot Tubb, and Shifting Sands, as well as unpublished data on points recorded from the central Plains provided by Jason LaBelle. As Ahler rightly notes (personal communication, 2004), this analysis could also be conducted on point bases or tips, as long as such fragments included enough of the edge to clearly indicate the limits of grinding. Unfortunately, such is not always clear in the published data on point fragments.

8. It would have been preferable to use the numerical tallies from which Hofman's (1992) percentages were originally derived to ensure comparability. However, frequencies of reworked specimens were not provided, nor could they be derived from the graphs.

9. Using the total number of points as the measure of sample size is problematic since it includes all fragments and, thus, violates the assumption of item independence: bases and point tips are tallied separately but may, in fact, be two parts of the same specimen. This could potentially inflate the total number of points and, thus, reduce the incidence of reworking. Using only complete points as the measure of overall sample size might skew the results if the count of reworked specimens is derived not just from the complete specimens, but also from broken point tips (thus inflating the relative number of reworked specimens).

10. One can argue both sides of the question of whether to include or exclude Midland points in an analysis of Folsom point assemblages, especially those on the Southern High Plains where Midland points often co-occur with Folsom; this includes the Midland type site (Wendorf et al. 1955), as well as Blackwater Locality No. 1, Shifting Sands, and many others (LeTourneau 2000:121–131). Although Midland and Folsom points are clearly related in some cultural historical sense, the latter are nonetheless largely a Southern High Plains phenomenon, suggesting that there is more here than just the possibility that these two points represent two portions of a use-life process (Amick 1995; Hofman 1992). If they did, Midland points ought to occur in other stone-poor areas of the Plains, but they do not; and they ought not to occur in stone-rich areas, but they do (see also Collins 1999b:26). Moreover, the relatively uncommon unfluted Folsom points that occur on the Central and Northern Plains (e.g., Root, William, and Emerson 2000) do not appear (to me at least) to be very similar to Midland points. That there is variation in projectile points during this period comes as no surprise; nor does the possibility that this variation could result from several possible causes. In some cases, raw material shortage may have played a role; in other areas, not. That these points have distinct temporal/spatial boundaries suggests that style plays a role. In any case, for analytical purposes and because I am including assemblages from outside the Southern High Plains where Midland points sensu stricto do not occur, I do not include them in this analysis. The "Midland Problem" remains to be solved.

11. Jodry (1999a:fig. 50) independently developed a somewhat similar classification for the Folsom materials at Cattle Guard, and though it highlights edge grinding, it did not use this variable systematically in the classification.

12. The sites comprising this sample are Cattle Guard, Cooper, Folsom, Kincaid, Lake Theo, Linger, Lipscomb, Lubbock Lake, and Waugh.

13. A unifacial scraper, possibly made of Washington Pass (Narbona Pass) chalcedony, surfaced during Anderson's work at the site in the 1970s and was subsequently discussed in the literature (Hofman 1999a:126). However, Anderson strongly suspects that the piece was planted. She reports:

> An individual who occasionally volunteered on my field crew was always bragging about the sites he had found and showing up with unusual artifacts and "rare" historic items, such as an historic Southern Plains culture feather headdress. He once gave me an Alibates Clovis point and offered to take me to the site where he found it. Well, the point looked strange to me—the proportions were wrong and there was no basal grinding—(later identified by both Vance Haynes and Joe Ben Wheat as a "McCormick" point), and the stratigraphy at the "site" was all wrong to be anything but recent. While at the Folsom site, this individual pointed out the scraper in place—in place in fresh cow muck and mud at the bottom edge of the arroyo, well below the cultural horizon. There was no "old" impression of the scraper in the muck, but one so fresh that it oozed water and had no well-defined impression. While he "fessed up" to the McCormick point, he never would admit to planting the scraper. He always displayed a cocky, "got you," attitude and was no longer welcome to work with us. (Anderson, personal communication, May 1, 2001)

Under the circumstances, there is no reason to consider this piece any further.

NINE

Folsom

From Prehistory to History

DAVID J. MELTZER

At the outset of this book, I detailed a series of questions that guided the recent field, laboratory, and historical investigations at Folsom. I expressed the caveat that not all of them would necessarily be answered or were even answerable. That has proven to be true. Nonetheless, in the course of asking and attempting to answer those questions, much has been learned of the Folsom site, in terms of both what occurred there in Late Glacial times and the events there in the 1920s. Let me return to those questions, as a framework for summarizing the archaeology, geology, history, and paleoecology of Folsom, doing so in a slightly different order than the first time around.

Answered and Unanswered Questions

By the fall of 1928, after three years of excavations on-site, matters stood thus: The Colorado and American museums had removed much of the bonebed on the South Bank of Wild Horse Arroyo—an area of roughly 270 m²—and extended their excavations a short distance into the North Bank (figs. 4.2 and 4.13) The bonebed had yielded at least 14 Folsom fluted projectile points. Ultimately that number would be doubled, as more points appeared during laboratory preparation of the plaster-jacketed skeletal remains (e.g., Figgins to Brown, October 23, 1929, VP/AMNH) and, over the next half-century, on the backdirt from the unscreened earth of the excavations. These newly recognized points represented a previously unknown bison hunting complex.

Altogether between 1926 and 1928, the site produced several thousand skeletal elements of a new species of bison, *Bison taylori,* later synonymized with *B. antiquus.* Very few articulated skeletons were recovered, though it was tacitly assumed that clusters of bone elements represented single animals. The highest density of those remains was in the small area of the Colorado Museum excavations and around

the western edge of the South Bank excavations, where the remains were "more or less mixed."

Brown thought at least 30 animals had been killed here, including "male, female, and yearlings . . . [all] killed at the same time," the event occurring some 15,000 to 20,000 years ago, at the close of the Pleistocene (Brown 1929; Brown to Hay, January 10, 1929, VP/AMNH). Wissler (e.g., 1910), well familiar with Plains bison hunting techniques, supposed there must have been some means of holding the animals and considered the possibility that the bison had been ambushed at a water hole. That squared with the suspicion that the kill occurred in summer, as Brown and others reasoned that the winters in the area were too harsh for bison and humans, a fact seemingly confirmed by the scarcity of archaeological remains in the region.

Many of the Folsom site's archaeological details remained to be worked out after the 1926–1928 seasons, but few of these were ever tended to. Nonetheless, by providing the critical evidence necessary to resolve the human antiquity controversy, these investigations had proved their worth.

Folsom in Historical Context

WHY AT FOLSOM, BUT NOT AT ANY OF THE OTHER SITES PREVIOUSLY CHAMPIONED AS PLEISTOCENE IN AGE?

The simplest reason is that Folsom was a Pleistocene-aged site, and the evidence was undeniable. But there is more to it than that, since Folsom was not the only Pleistocene-age site in play at the time. There were at least two others on the table that are now known to be of comparable antiquity: Lone Wolf Creek and 12 Mile Creek. Lone Wolf Creek, of course, was being vigorously advocated by Cook (1927a) and Figgins (1927a) at the very same time and even in their Folsom announcement. But neither 12 Mile Creek nor Lone Wolf Creek had been found under compelling circumstances, nor had they been examined carefully while the remains were still in situ. Cook got to Lone Wolf Creek (fig. 2.3) as soon as he could, but by then the excavation was over and all the bison remains, possibly just one or a few animals, were removed. By 1927 the 12 Mile Creek site was merely an historical footnote, its excavator (Samuel Williston) having long since passed from the scene and, in any case, never having made a strong claim for the site having any great antiquity.

In those preradiocarbon days, ascertaining the antiquity of a site required firmly associating human artifacts or remains with geological markers of the Pleistocene, such as an extinct fauna. At Folsom, a kill site that contained over 30 skeletons of an extinct bison species, whatever that species turned out to be, and with almost as many projectile points embedded in various of those skeletal remains, it was possible for several waves of visitors over a two-year period to examine the evidence for themselves. By 1927, such visits had an important role and a long history in archaeology. They were conceived at a time when determining the age of a site had to be done in the field, by examining its geology, stratigraphy, context, and associated remains. Successful visits had taken place at Trenton, Lansing, and Vero and shown that these sites were not demonstrably Pleistocene in age. Folsom was that old, and that was confirmed by the site visit in September 1927, a visit that followed directly on the advice of Aleš Hrdlička.

I am under no illusion that it was historically inevitable that the breakthrough in the decades-long human antiquity controversy would occur at Folsom, only that a site like Folsom was needed to break the Pleistocene barrier. Arguably, it did not even have to be a kill site. However, demonstrating a Pleistocene human presence in those decades unquestionably demanded a site in which the evidence of human antiquity was reliable and utterly unimpeachable. Projectile points embedded in the ribs of an extinct species of bison (fig. 1.4) easily met that requirement. Other kill sites could have; a non-kill site might have. But the long history of dispute shows just how difficult it was for non-kill sites to clear this high bar, and indeed none did.

Thus it was that because of a long train of events, set in motion by a thunderstorm in the late summer of 1908, that Folsom ultimately became the site that broke the Pleistocene barrier. Replay the tape of history, put a few more bison and a few less clumsy hands to work at Lone Wolf Creek in 1924, and then send out telegrams, and perhaps Folsom would have been relegated to being merely one of many Pleistocene sites, and not the first of all.

WHY WAS CREDIT FOR RESOLVING THE HUMAN ANTIQUITY CONTROVERSY GIVEN TO OTHERS, NOT TO COOK AND FIGGINS?

Resolving this question requires unpacking two separate issues: *discovery and resolution*. To be sure, the Folsom discovery and the resolution of the human antiquity controversy coincided in time, and one assuredly led to the other.

But they were not the same event. The discovery in 1926 that this was an archaeological site, and the subsequent finding of a point in situ in 1927, provided the essential data. Taking a measure of those data, turning them into meaningful and reliable evidence, assessing why Folsom was different from Snake Creek, Lone Wolf Creek, and Frederick, and determining Folsom's larger meaning for American archaeology involved and indeed required a different set of discussions by a separate group from those who made the discovery. The reason, as detailed earlier, is that controversy in science is resolved when a core set of elite scientists, those whose opinions count and who have assessed the evidence in hand, reaches a consensus on its meaning. It matters little what opinions or pronouncements might be offered by those who are not part of that elite core (fig. 2.15).

Although neither Cook nor Figgins appreciated the fact, it is quite clear that resolution of scientific controversy works largely because of, and not in spite of, the fact that such inequality exists. It is only the elite scientists who can bring about closure (Rudwick 1985:428). We know it would hardly matter if Cook claimed to have resolved the human antiquity dispute, because he *had* claimed to have resolved the human antiquity dispute, and his claims had had no influence whatsoever.

In sharp contrast, Kidder wrote and spoke only briefly on the matter, while Hrdlička said virtually nothing at all. Yet Kidder's few words and Hrdlička's deafening silence—it was by then well known what Hrdlička's reaction was to sites and claims with which he disagreed—resonated loudly across the landscape and profoundly influenced the archaeological community. What they said, or didn't say, legitimized the study of a Pleistocene human presence in the Americas.

WHAT MADE FOLSOM SO IMPORTANT TO AMERICAN ARCHAEOLOGY?

When resolution of the human antiquity controversy was achieved at Folsom, the American archaeological landscape was forever altered. This is so because Folsom was more than just another site, and the human antiquity controversy was about more than simply when prehistory began in America. The debate over human antiquity reached deep into the conceptual core of the discipline—it was all about time.

In a midtwentieth century retrospective, anthropologist A. L. Kroeber (1952:191) expressed his bewilderment that earlier archaeologists "as experienced as Holmes . . . saw their archaeological pasts as completely flat." For them, human history on this continent had no major epochs of cultural change; much of the depth of America's past could be plumbed merely by following the traces of modern Native American tribes into the past using the Direct Historical Approach. As Kidder (1936:145) described it, the "effort was directed toward identification of ancient sites with modern tribes." It was an approach that demanded continuity between present and past. Significant variation in archaeological materials was thought to occur across space, not time (e.g., Holmes 1919).

Folsom exploded the belief that American prehistory was so shallow and, with it, much of the theoretical underpinning and methodological tools on which American archaeology had been relying for decades. Ethnohistoric homology necessarily gave way to ethnographic analogy, but use of analogy had to become more sophisticated than it had been, given the great temporal divide between the historic period and the world of Pleistocene hunter-gathers. Culture history emerged in the wake of Folsom, and the gap it created between the Late Pleistocene and the Late Prehistoric. Folsom was a relief to Kidder, in providing the "chronological elbowroom" to stave off the diffusionists' claims that American civilization simply hadn't enough time to develop independently, but it also presented the challenge of filling in the details of what transpired in the intervening millennia.

When scientific controversy is resolved, debate dies down, and research turns to working out the details of the new paradigm of understanding (Oldroyd 1990:345). That happened here. Over the next several decades, archaeologists contented themselves with working out the details of the Paleoindian occupation, in the process finding dozens of sites, which in turn established a pattern and set of expectations of North American Paleoindian sites and adaptations. The question of human antiquity in the Americas was set aside, essentially because it was believed the answer was in hand (Bryan 1941).

But in the early 1950s a new cycle of claims and controversy over human antiquity began, with Alex Krieger's (1953:238–239) dire warning that a Hrdličkian "dogmatism" was creeping back into American archaeology, one that this time "allowed" an antiquity on this continent of only 10,000 to 15,000 years. He thought there was a still earlier human presence. That next round of controversy, which began in earnest in the early 1960s (e.g., Krieger 1964; Leakey, Simpson, and Clements 1968), would prove to be strikingly similar to the one that began a century earlier at Trenton and that was finally resolved at Folsom in 1927 (Meltzer 2005, 2006).

The Paleoindian Occupation at Folsom: Some Conclusions

ARE THERE INTACT ARCHAEOLOGICAL DEPOSITS REMAINING AT THE FOLSOM SITE?

Left behind at the close of the excavations in early October of 1928 were a 34-ft vertical wall along the North Bank and a large and open excavation area on the South Bank, ringed by a berm of backdirt, centered around an earthen datum pillar. None of the excavation was backfilled, but then that seemed unnecessary, as it was believed that there was little of the bone quarry left.

Still, there were hints that this was not altogether the case. When we began excavations in 1997, intact portions of the bison bonebed were located relatively quickly on the South Bank, just where the archival material hinted it might be (chapter 4), testimony to the practical payoff that occasionally rewards historical research. From 1997 to 1999 we, in turn, excavated 19 m^2 of the bonebed on the South Bank, excavated another ~18 m^2 more of test pits on bone-bearing units elsewhere on the South and North banks, and put in another ~60 m^2 of test units and slit trenches (fig. 4.13). In total, an area approaching 375 m^2 has been excavated at the Folsom site, and a still larger portion has been probed and mapped by coring and augering.

Yet, despite the extensive excavations, and the erosion that has occurred there since the 1920s, there are still intact Paleoindian deposits at Folsom. More of the bonebed extends to the north and west of our M17 block, and apparently intact (albeit possibly reworked) Late Glacial sediments in the paleovalley are present beneath the deep overburden of the North Bank. The North Bank is something of a wild card in all this: Erosion has been extensive there since 1928, and though an angle of repose has been reached, the bank itself is still slowly slipping downward. Efforts have been put in place to check that process, so as to preserve what might remain beneath it.

Although more of the site remains, after 1999 no further excavations were conducted by us at Folsom. This was by design. Over the course of three seasons of fieldwork, and with the analysis of the collections from the original investigations, we had gained a much better understanding of Folsom's structure and contents. Continued excavation in the remaining portions of the bonebed would almost certainly have led to the recovery of additional faunal remains and perhaps artifacts as well, and would surely enhance our understanding of the details of the bonebed. But the odds were against our making any discoveries in the bonebed that would radically alter our understanding of the site. The incremental gains that might be made were not enough to justify the cost to this important archaeological resource. The Folsom site deserves to be preserved and protected for future study, so that in another 70 years new methods and techniques can be brought to bear, and our work, in turn, can be evaluated by a later generation of archaeologists.

Admittedly, I might have reached a different conclusion about continuing excavations after 1999 had we located a camp.

IS THERE AN ASSOCIATED CAMP/HABITATION AREA AT FOLSOM?

There is good reason to suspect that there was once a camp with the bonebed, if only because of the several days it must have taken to butcher and process 32 large bison (chapter 7). Such a camp would likely have been nearby, to protect the kill from scavengers. It might possibly have provided a more representative sample of the formal and non-formal tools of stone and bone used in butchering, meat and hide preparation, tool production, or other activities, as recovered at many Folsom sites (e.g., Hofman, Amick, and Rose 1990; Jodry 1999a; Root, William, and Emerson 2000; William 2000), all of which are lacking in the projectile point–dominated assemblage from the Folsom bonebed.

Unfortunately, none of the cores, augers, or test units turned up evidence of a camp, and we looked both close to and well away from the bonebed, particularly in areas where a camp might have been expected to occur, such as within the paleotributary itself or atop the valley wall flanking it. The search also involved survey in the uplands surrounding the site, including the interfluves between Wild Horse Arroyo and the adjacent drainages, but this effort also failed to turn up traces of any other Folsom-age presence. Our results were consistent with those of Carl Schwachheim and Gerhardt Laves in the 1920s and of Adrienne Anderson's survey in the 1970s. The situation there is in sharp contrast to Folsom-period occupations in other areas.

Although none of our efforts yielded traces of an associated camp, one of the test units provided geological evidence that may help explain its absence. Charcoal recovered from the surface of the Smoky Hill Shale that formed the valley wall of the paleotributary west of the bonebed yielded a radiocarbon age of 9,820 ± 40 ^{14}C yr B.P. The implication of that age is clear enough: The bedrock surface of the valley wall was swept clean in Late Glacial times, along with any Folsom-age archaeological remains that might have been deposited on that surface. Thus, the same geomorphic episode that carried shingle shale (f3) across the surface of the Folsom bonebed may have also carried away traces of a Folsom camp. No artifacts were seen mixed into the f3 above the bonebed.

Of course, this assumes that a camp would have been situated on the flat uplands of the valley wall above the bonebed, near where the bison carcasses were most densely packed. Although a reasonable assumption, this also would have put the camp on a high and exposed surface. This was a fall kill, and if it was a cold fall, as such can be, the hunters may have chosen to stay down in the paleovalley or within the paleotributary, where they would have had more protection from exposure to the wind and elements.

If that is so, then a camp could still be present and deeply buried or otherwise inconspicuous; we did not exhaust the possible areas in which remains of a camp might occur. Or it may have simply been washed away by later cycles of erosion and arroyo cutting, of which there were several.

WHAT IS THE GEOLOGICAL HISTORY AND CONTEXT OF THE FOLSOM SITE?

The Folsom site extends across two very different geomorphic settings: a paleovalley, the ancestral channel of Wild Horse Arroyo, and its adjoining paleotributary. These drainages were incised into the Smoky Hill Shale bedrock

and had begun to fill during Late Glacial times. The initial in-filling (stratum f1) took place perhaps as early as 12,350 ^{14}C yr B.P. and was episodic and rapid. It produced a substantial debris flow fan of silt and shingle shale that jutted into the paleovalley from the mouth of the paleotributary.

The main portion of the bison bonebed lies in the central portion of the paleotributary just above where it joins the paleovalley. The bison carcasses were dropped on an aggrading surface of loess or loess-derived fine-grained silt (stratum f2) that began to accumulate as early as 11,500 ^{14}C yr B.P., continued until ~10,100 ^{14}C yr B.P., and quickly and almost completely covered the carcasses after the kill. The lack of a well-defined stratigraphic surface or contact on which the bones were resting, along with the evidence for only weak soil development within the f2, testifies to the nearly continuous nature of deposition throughout this time.

For approximately 200 to 300 radiocarbon years around the Pleistocene-Holocene boundary (10,000 ^{14}C yr B.P.), the geomorphic regime switched from deposition to erosion. In the paleotributary, the top of stratum f2 was slightly eroded, then rapidly buried beginning around 9,800 ^{14}C yr B.P. by episodic sheet flows of poorly sorted and angular shingle shale (stratum f3) coming off the paleotributary walls and piling up in thicknesses of as much as 30 cm.

In the paleovalley, the underlying climatic and geomorphic controls and timing were the same, but the geological consequences were very different. There the upper portion of the f2 was heavily eroded, and stratum f3 appears as a thick (>1-m), complexly laminated deposit, containing multiple lenses of rounded and size-sorted gravels and secondary carbonate nodules, interspersed by fine-grained silts. The f3 here testifies to significant low gradient fluvial action, as opposed to the colluvial processes that dominated at this time in the paleotributary.

There is, in both the paleotributary and the paleovalley, stratigraphic evidence of multiple erosional and depositional cycles throughout the Holocene, the most significant of which appear in the Early and Middle Holocene, with erosional hiatuses between the f3 and the deposition of m1 and between the m1 and the deposition of m2. The earlier of those two cycles is not as well constrained chronologically as the latter. The m1/m2 hiatus, which falls within the Middle Holocene, is marked by the deep downcutting of the McJunkin channel, which snaked between the valley bedrock walls at sharp angles to the earlier paleovalley. Valley incision is widespread across the Great Plains and Mountain West during this period, the result of the Altithermal drought (Albanese and Frison 1995; Artz 1995; Mandel 1992, 1995; Meltzer 1999). The McJunkin channel began to fill, producing unit m2, around 4,800 ^{14}C yr B.P.

Units m1 and m2 are thick, laminated in portions of the sections, and generally fine-grained, likely having been deposited in relatively quiet fluvial and paludal settings. There are nonetheless occasional gravel lenses, evincing higher-energy fluvial pulses. These units comprise much of the surface deposits in the area, save where unit *w*, a Late Holocene facies of *m2* that possibly marks another cut-and-fill episode in the valley, rests unconformably on that surface.

Although the occasional thick lens of cobbles and coarse gravels in unit *w* hints at changes in the geomorphic regime in the valley, the Late Holocene in the Folsom area is generally a time of landscape stability, evidenced by pedogenic development within units *m1* and *m2*.

WHAT IS THE AGE OF THE FOLSOM BISON KILL?

Previous work had put the age of the bonebed at ,~10,890 ^{14}C yr B.P. (Haynes et al. 1992), although that estimate came with the caveat that the half-dozen charcoal samples from the North Bank that yielded this average age might not be anthropogenic. This is almost certainly true. There is, as Wissler and others observed 70 years earlier, a considerable amount of natural background charcoal in the deposits at Folsom. We found only one small cluster of charcoal in the sidewalls of the M17 block that is arguably cultural in origin, but it was Middle Holocene in age and obviously unrelated to the Paleoindian occupation.

Recognizing the nonanthropogenic source of the charcoal, and the added complication of dating charcoal samples from the paleovalley that are not likely to be in primary context, we concentrated our efforts on dating organic remains from the paleotributary. Not surprisingly, the charcoal there returned a range of ages that stretched over more than a radiocarbon millennium, from 11,500 to 10,100 ^{14}C yr B.P. However, a half-dozen radiocarbon ages on bison bone and teeth from both the paleotributary and the paleovalley, and thus from different animals, produced a much tighter range (fig. 5.12). The average of these ages clusters tightly at 10,500 ^{14}C yr B.P., and this provides the best estimate of the age of the Folsom bison kill.

This recent suite of bone radiocarbon ages overlap at 1σ (calibrated) with a bison bone collagen age run from the site three decades ago (Hassan 1975), but was long suspected to be unreliable (Haynes et al. 1992). The recent bone radiocarbon ages suggest otherwise. More intriguing, the radiocarbon ages from the Folsom bison overlap at 1σ (calibrated) with the radiocarbon age on the Archuleta Creek *B. antiquus*. The partial excavation of that skeleton did not yield any cultural remains, so it is not known whether this animal escaped the melee of the kill at Folsom and died on the floor of this creek 4 km away. But its antiquity certainly allows that possibility.

WHAT WAS THE CLIMATE AT THE TME OF THE FOLSOM SITE OCCUPATION?

At 10,500 ^{14}C yr B.P., the Folsom site occupation falls squarely within the Younger Dryas Chronozone (YDC). The ecological effects of this millennium-long cold snap were not as pronounced in this corner of temperate North America as they were elsewhere. Glacial ice in the nearby Sangre de Christo Mountains did not show any appreciable YDC re-advance (Richmond 1986), as it did at higher elevations farther north in the Rocky Mountains. Temperatures were certainly cooler than at present; nearly a dozen species of snails found in the bonebed presently live at higher elevations than the site, and their δ^{18}O records hint at lower YDC temperatures, especially during the summer. But judging from the records assembled there, it was not cold enough for long enough to lower temperatures beyond the thermal capacity of the snails or to produce a downslope shift of tundra, alpine, or subalpine plant species.

Like temperature, precipitation was lower in Late Glacial times, as is apparent in the presence of land snails but no aquatic forms; the δ^{18}O and δ^{15}N records from snails and bison; and the thick drape of loess or loess-derived fine-grained sediment that filled the paleotributary prior to and during the occupation of the site. Climate models suggest the possibility of diminished winter precipitation at this time, and the δ^{18}O values in snails and the relatively high percentage of C_4 grasses hint at increased summer precipitation (Fall 1997; also Kutzbach 1987; R. Thompson et al. 1993). The data suggest but cannot resolve whether there was a seasonal shift in precipitation that accompanied the net decrease in annual precipitation (fig. 6.16).

Even though conditions in the Folsom area were relatively dry during Late Glacial times (cf. Haynes 1991), the region did not experience the harsher episodic drought that marks this period on the High Plains (Holliday 2000a). Trees were present on the Folsom landscape, and judging by the presence of cattail, *Ephedra,* and several aquatic taxa in the Bellisle and bonebed pollen, small lakes and freshwater ponds were scattered about the landscape. But in the paleotributary, there was neither a flowing stream nor standing water nor evidence of aquatic plants or snails.

Within a few centuries of the bison kill, there were several significant changes in climate and environment. These included an increase in temperature and in the length of the growing season, and an increase in precipitation, marked by changing snail δ^{18}O values above the bonebed, and possibly tied to an increase in the intensity of the summer monsoon (also Mayer et al. 2005). With these climatic changes, the YDC in the Folsom area came to a close.

These changes were manifest by erosion that began sometime after ~10,200 ^{14}C yr B.P. at both the site and Bellisle Lake—and, indeed, across much of the area (Mann 2004) and well beyond (Mayer et al. 2005). At the site itself, erosion of the top of the *f2* was followed by deposition of the *f3* (ca. 9,800 ^{14}C yr B.P.), which in the paleovalley displays clear evidence of fluvial action, and by deposition of Zone B-2 at Bellisle Lake (ca. 8200 ^{14}C yr B.P.), which for long periods had standing water and around which forests were beginning to expand. However, this and later periods of the Holocene continued to experience episodic warming and drying, which produced complicated depositional and erosional sequences throughout this region (Mann 2004).

WHAT BIOTIC RESOURCES WERE AVAILABLE TO HUNTER-GATHERERS AT THE TIME OF THE FOLSOM SITE OCCUPATION?

The pollen and macrofossil records indicate that a variety of tree species were present on the Late Glacial landscape, including spruce, pine, juniper, and oak, all of which occur in the area today. Yet, this was not a heavily forested area but, instead, an open parkland, probably more open than at present, in which small gallery forests occurred along drainages, and copses of trees grew in moist and protected areas on the sides of Johnson Mesa.

The ground cover in between was primarily grass, with a smaller component of forbs and shrubs, judging by the isotopic and pollen evidence. The grassland was dominated by C_4 species (figs. 6.12 and 6.14), which covered a greater portion of the landscape than at present. Their dominance is likely a result of several factors, including lowered atmospheric CO_2 in Late Glacial times, which favored C_4 over C_3 growth (Ehleringer, Cerling, and Helliker 1997; Koch, Diffenbaugh, and Hoppe 2004), and perhaps a seasonal shift in precipitation (fig. 6.16).

Although this is an ecotonal area today (fig. 3.11) and likely was in the Late Glacial, at present it is not an area that produces abundant nut or fruit-bearing trees or edible plants, at least ones that provide sufficiently rich caloric returns on which hunter-gatherers could rely, especially during the critical months of winter and early spring (chapter 3). No evidence emerged in the course of this investigation of any additional or significant plant food resources occurring there during the cooler and drier times of the Paleoindian occupation.

Theoretical models (Binford 2001) and the archaeological record agree: This was historically the kind of environment where hunting was the dominant subsistence pursuit. But the light scatter of projectile points over the landscape suggests that hunting did not occur very often. And in Late Glacial times, there is no evidence that other game beyond bison were being exploited at the Folsom site, or that such were especially abundant on the landscape. The striking lack of carnivore damage in the bonebed indirectly testifies to the character of the environment, for despite the availability and accessibility of bison carcasses through the winter and spring that followed the kill, which are precisely the times when one would expect the most extensive use of carcasses by nonhuman predators and scavengers, this large pile of unearned resources of meat and marrow was not extensively exploited. All this suggests that prey numbers were sufficiently low or unpredictable that the area did not support a large resident population of carnivores or scavengers.

Folsom was good bison habitat during the summer months when C_4 grasses covered the landscape. There is no evidence in the $\delta^{13}C$ values of the Folsom bison of any significant contribution of cool-season C_3 grasses to their diet, which would be expected were bison herds overwintering there. One possible hypothesis for why bison did not inhabit the area year-round is that cooler Late Glacial springtime temperatures kept snow on the ground longer, which in turn delayed the emergence of the early-season C_3 grasses on which bison depend.

DID FOLSOM GROUPS OCCUPY PROTECTED FOOTHILLS AND INTERMONTANE BASINS DURING THE COLD SEASON?

It does not appear that humans were overwintering in the Folsom area in Late Glacial times either. Without a sufficient supply of winter game, in the absence of other critical resources that might attract and support a residential group, and assuming that snowfall in the YDC could be as heavy as it is today, this would not have been a place to linger through the cold months. Nor is there is any archaeological evidence that the Folsom hunters did: They killed these bison in the fall and abandoned the site very soon thereafter. There is no hint of any winter features or activities, such as structures, meat caches, intensive processing of skeletal elements for fat and bone grease, storage facilities, or the exploitation of additional food resources that one might expect of long-term encampment, and that one sees in other such sites (e.g., Binford 1993; Frison 1982a, 1982b; Hill 1994, 2001; Todd 1991).

This does not falsify Amick's (1996) hypothesis that Folsom groups spent their winters off the Plains and in more protected settings such as this; it only means that overwintering did not occur in this particular area. This evidence does not and cannot negate the possibility that such a strategy was practiced by Folsom groups in other protected foothill settings or intermontane basins, and indeed, investigations in several such places, notably the Gunnison Basin, Middle Park, and San Luis Valley of Colorado, suggest that might be the case (Andrews 2004; Jodry 1999a, 1999b; Kornfeld and Frison 2000; Kornfeld et al. 1999).

WHAT MIGHT BE INFERRED OF THE TACTICS AND STRATEGIES FOR BISON HUNTING AT THE FOLSOM SITE?

The paleotributary was flanked by walls 3 m to 4 m high and had a knickpoint at its head (fig. 5.7). There were still higher and steeper walls in the paleovalley, perhaps as much as 8 m high, though no apparent knickpoint. Or so it appeared from the mapping of exposed sections of the valley wall of Wild Horse Arroyo, and was glimpsed through the various remote sensing methods used to map the deeply buried contact between the consolidated and the unconsolidated Smoky Hill Shale. Of course, this bedrock contact is not the surface on which the Paleoindian kill occurred, since by the time the kill took place that surface had been draped by *f1* and *f2* sediments. Nonetheless, given the age and depth of the *f2*, the morphology of the upper surface of the Smoky Hill Shale, the manner in which the Smoky Hill Shale erodes, and the stability of its slopes as gauged by the presently exposed valley wall, the bedrock surface as mapped likely provides a passable rendition of the shape of the landscape at the time of the kill.

That being the case, it is reasonable to conclude that the topography of both the paleotributary and the paleovalley could have been used to advantage in trapping or otherwise hindering the movements of large game such as bison. Whether the topography actually was used that way is another matter, although there is no evidence for any other features—artificial or natural barriers, a corral, etc.—that might have served the tactical purposes of the hunters (Frison 1991). Nor is there evidence of bison bone on the upland surfaces above the paleotributary and paleovalley, although that absence could be a result of subsequent geomorphic processes that, as is now clear, swept much of that surface clean.

It is not known whether the paleotributary or the paleovalley was the focal point of the kill; bison remains have been found in both settings. Overall, a great deal more bison skeletal material has been found in the paleotributary, and in a manner suggesting that these remains were found close to where the animals dropped. Taken at face value, the unquantified observations of the original investigators of the relative density of bison remains across the site suggest that a great many of those animals were bunched up and dropped and/or butchered on the west side of the paleotributary, close to where it joined the paleovalley.

Perhaps, as hypothesized, the animals found in the paleovalley were those few that managed to scramble down and out of the trap presented by the paleotributary, only to be killed by hunters waiting in the main valley. Unfortunately, so little is known of the archaeological context of the bison remains in the paleovalley that I cannot eliminate the prospect that animals were dispatched in this area as well, and that at one time there was a large bonebed there, the traces of which have since been erased by erosion or deeply buried by Holocene fill. Nonetheless, I can eliminate the possibility that the bone in the paleovalley was washed in from the paleotributary, based on the simple fact that bison remains in the paleovalley have been recovered well upstream of its juncture with the paleotributary.

An alternative scenario, of course, is that the kill took place in the paleotributary and that carcass parts were moved by the hunters to a secondary butchering and processing area on the floor of the paleovalley, perhaps to afford better protection from the elements. But if that were the case, the complete cranium found during the 1972 salvage excavations by Trinidad State makes little sense: Why haul a massive bison skull into that area?

Strategically, all evidence points to this being an encounter kill rather than an instance of ambush or intercept hunting. Although the regional vegetation could support bison herds in summer, their abundance and predictability and those of other prey must have been relatively low, again given the apparent scarcity of carnivores. Nor were there features there that would lend themselves to ambush tactics; although this was a relatively dry period and water holes might have been a prime attractor for game, there was no

standing water at the site itself. The site might have been a suitable spot in which to trap and dispatch a herd, but not necessarily one where hunters would expect to intercept one. Finally, the thorough nature of the butchering and processing and the manner in which the carcasses were segmented (fig. 7.12), suggest that it was done by a nonresidential group of mobile hunters who were en route elsewhere.

But how many hunters? This is an interesting question, but most of us shy away from guessing at the numbers that may have been involved in such an activity (but see Wheat 1972:123), and for good reason: There are virtually no data that permit well-grounded hypotheses on the matter, and no means of testing such hypotheses if there were (Hofman 1994:357). One can speculate, of course, using the number of recovered projectile points to estimate the number of points originally used at the site and, from that, the number of hunters represented by that tally. But the many undemonstrated assumptions that would need to be made at each step, such as about the relationship between the number of points used in prehistoric times and the number recovered archaeologically, or the number of points per hunter, would render the result highly suspect. Similarly, one could estimate the total weight of meat and bone from the kill, how much was transported off-site, and, by making certain assumptions about how much an individual hunter could carry, produce a guess at the total number of hunters involved in transport and thus, perhaps, the kill. But this effort too relies on a string of assumptions about the representativeness of the sample; carcass weights; what was transported as opposed to disappeared for taphonomic reasons; that transport occurred, and the parts were not simply consumed in a nearby but undiscovered camp; whether all hands and dogs participated in transport; and the like (Cannon 2003; Hofman 1994; Speth 1983; Wheat 1972). In the end, conjuring the number of hunters at the kill would be misleading.

Those disclaimers aside, the Folsom kill is not easily explicable as the work of one or a few hunters. Unlike, say, a bison jump, in which relatively small number of hunters could stampede and kill a large number of animals (Frison 1991; Verbicky-Todd 1984), the paleotributary and paleovalley in which the Folsom kill took place would have required more hands to accomplish the task. To effectively trap and kill 32 animals required individuals to maneuver the animals up the paleovalley and into the paleotributary, others to be stationed on the perimeter walls surrounding the kill area, and still others to help "bottle" up and kill the herd once it was trapped (e.g., Frison 1991:170; Verbicky-Todd 1984). The Archuleta bison, if future testing turns up associated artifacts, may shed light on how securely that bottle was capped (Meltzer, Mann, and LaBelle 2004).

Given those requirements, the labor-intensive demands of the butchering that followed (Frison 1991:300–302), the thoroughness with which the bison carcasses were processed, and the amount of meat and bone transported off-site, this must have been a communal effort, in the strict sense of involving a cooperative group. What that translates to in absolute numbers is anyone's guess, although the archaeo-logical evidence at least sheds some light on the question of whether those hunters had previously been dispersed across the landscape, then aggregated at Folsom to make the kill (below).

IS THERE EVIDENCE FOR MORE THAN ONE BISON KILL AT FOLSOM?

There is further, albeit indirect, testimony to this kill having been the result of an encounter rather than an ambush or intercept; it appears to have been a one-time event. The stratigraphic evidence, including bone backplots (fig. 7.7), points to there being but a single horizon on which the bison remains are resting. There is no sign of differential weathering of the bone or later trampling of the remains to support the possibility that bison entered the archaeo-logical record at different times over several seasons or years. The ages of the animals, derived from tooth eruption and wear patterns, bespeak a tight set of age cohorts. And finally, the ages of the bison bone from various areas of the site are virtually identical within the resolution of calibrated radiocarbon ages. All this is quite in keeping with the general Folsom pattern, in which particular spots on the landscapes—save, for example, stone outcrops—are not used redundantly (Frison, Haynes, and Larson 1996; Kelly and Todd 1988; LaBelle 2004; LaBelle et al. 2003; Stanford 1999).

WHAT WAS THE NATURE OF THE BUTCHERING AND PROCESS-ING OF THE BISON AT THE FOLSOM SITE?

The Folsom hunters may have been traveling quickly, but they were not necessarily traveling lightly, given the evidence for extensive dismemberment of the carcasses, much of which apparently took place near where or where the animals were dropped.

The full complement of tools used in that process is not known; only that several flake tools and a quartzite skinning knife were part of the toolkit. One of the broken points was evidently used as well. Other stones tools may have been, but these were either transported off-site or left elsewhere on-site and not recovered. Cutmarks on bone are present (fig. 7.14), though rare. There is no evidence that broken bones were pressed into service as tools.

The hunters targeted and then pieced into transportable packages a disproportionate number of the high-utility elements, such as ribs, vertebrae, and upper limbs. Left behind were low-utility elements (Wheat 1972), such as lower limbs, mandibles (though not until the tongues had been removed), and crania, many of which were found on the west side of the site. Crania were statistically underrepresented, which may be more a consequence of erosion and weathering than of human use, for none of the recovered skulls had been broken to remove the brains. The attrition of the crania notwithstanding, in general there was no density-mediated destruction of skeletal parts.

Although this processing pattern fits a general model of light or "gourmet butchering" (Ewers 1955:160; Todd 1987b, 1991; Wheat 1972:100; Wissler 1910:41–42), there was not a strict focus on the highest-utility elements (fig. 7.12). Some lower-utility elements were also removed, perhaps as a by-product of their attachment to other elements: riders, in Binford's (1978b) terms, among them the caudal (tail) vertebrae that, as Brown figured, probably "followed the hide" (cf. Wheat 1972). Other lower-utility elements were likely removed because they obstructed access to higher-utility elements but, once removed, were nonetheless transported off-site (Barlow and Metcalfe 1996; Cannon 2003).

In the fall of the year and after a long summer of feeding, bison cows are at their peak of fat stores (Ewers 1955:152; Frison 1982b, 1991; Jodry 1999b; Todd 1991:fig. 11.5; Verbicky-Todd 1984:51). This is, of course, the time of the year when the kill took place at Folsom, and this was dominantly a cow-calf herd, thus presenting a prime opportunity for hunters to target fat reserves, which were a critical resource. Yet, the Folsom hunters did not take full advantage of this opportunity. To be sure, they removed tongues and back hump ribs, but there is no evidence of bone impact fractures created in search of marrow, or of the highly destructive processing of bone for grease.

High–marrow content long bones (e.g., Brink 1997: 270–271; Emerson 1993:152; Hill 2001:230) are underrepresented in the bonebed assemblage, presumably from having been transported off-site. Whether they were subsequently processed for their fat and grease cannot be determined. The hypothesis that Paleoindian groups may have been "fat indifferent" (Todd 1991:218) is not weakened by the data at Folsom. This evidence does not, however, help explain why Paleoindian groups, in contrast to Late Prehistoric and Historic bison hunters, were "fat indifferent." Nor does it shed light on whether that indifference is attributable to Late Glacial environmental conditions (e.g., Todd 1991:232; Jodry 1999b:328), except, perhaps, indirectly.

The paleoecological evidence from Folsom and across the Southwest and Great Plains indicates a much greater C_4 grass cover in Late Glacial times (e.g., Connin, Betancourt, and Quade 1998; Ehleringer, Cerling, and Helliker 1997; Koch, Diffenbaugh, and Hoppe 2004). Given bison preferences for C_4 species like buffalo grass and blue grama (Peden 1976:228), these animals could indeed have been fatter for longer periods of the year across more of their range. Thus, Paleoindian hunters need not have maximized fat returns at each opportunity or during particular seasons. Of course, if the underlying cause of greater C_4 grass cover in Late Glacial times is lowered atmospheric CO_2, then this hypothesis fails to explain why Late Prehistoric and Historic bison hunters were also "fat indifferent," since atmospheric CO_2 amounts were not significantly higher in pre-Industrial times than they were in the Late Glacial (Raynaud et al. 2000). Recent bison should have been just as fat and abundant as their Pleistocene ancestors, yet they were still being far more systematically and thoroughly exploited by hunter-gatherers.

But then the difference in the degree of bison carcass processing between Paleoindian and the Late Prehistoric times may have less to do with bison ecology, C_4 grass cover, or differences in Late Glacial and Holocene seasonality, than with hunter-gatherer ecology and demography. Paleoindian populations were small and spread thinly on a vast landscape. By Late Prehistoric times, human populations on the Plains were much larger and were living on a relatively more densely packed landscape. The respective resource demands were very different. Under the historical circumstances of the later period, when bison products were not just a critical subsistence resource but also a valuable exchange commodity (e.g., Spielmann 1983), more thorough use of carcasses would be expected.

WHAT IS THE TAPHONOMIC HISTORY OF THE FOLSOM BISON BONEBED?

After the butchering and the processing was complete, the bison bones lay on the surface long enough for soft tissue to decay and for weathering of bone surfaces and root etching of the undersurfaces to begin. However, they were not exposed for very much longer, for those surfaces are in excellent condition, the bones were hardly gnawed by carnivores or rodents, and none suffered obvious dry-bone breaks from trampling or weathering. Given the season of the death (fall), active decomposition may not have begun until the following spring or early summer. At a bare minimum, one summer's surface exposure of defleshed bones would have been required for the weathering to have begun. Overall, exposure may have lasted no longer than, say, a few years time—long enough to allow for slight weathering and root etching and some natural disarticulation of the skeleton.

During the time the bone was exposed in the paleotributary (and probably the paleovalley, though we have little evidence on the point, save the bison cranium recovered in 1972), it was not moved to any significant degree by fluvial action. A few of the smaller and more mobile elements may have been repositioned, but otherwise the whole was covered by a gentle fill of fine-grained $f2$ silts that buried the bonebed to a depth of at least 20 cm, enough to cover even the largest crania.

The original burial depth may have been deeper still, but erosion beginning around ~10,200 ^{14}C yr B.P. began stripping off that silt mantle. Before the erosion cut into the top of the bonebed in the paleotributary, however, deposition of the $f3$ shingle shale commenced, set off by increased precipitation, as noted above. At least two and possibly more pulses of shingle debris washed off the valley wall and across the paleotributary, yet very little of it came to rest directly on bison bone (fig. 7.6). Those layers of shingle shale effectively armored the bonebed in the paleotributary, preventing access by burrowing animals, putting it well out

of reach of the root zone, and apparently supporting the 1 m to 3 m of overburden that would ultimately accumulate here, thus protecting the bonebed from crushing and preserving elements close to where they had been discarded during butchering and processing. The bison bone in the paleotributary remained protected until Wild Horse Arroyo began downcutting and laterally incising the northern edge of the paleotributary following the flood of 1908.

The Late Glacial climate change that triggered deposition of the $f3$ shingle shale had very different consequences for the bison remains in the paleovalley, which were also buried in fine-grained $f2$ sediment, but which show clear evidence of significant postdepositional transport and tumbling. The upper portion of unit $f2$ in this setting was heavily eroded and exposed, and bison bone elements were plucked from primary context and carried down-valley, coming to rest at high angles (as much as 79°) in and among lens of gravel. The bones occur not as clusters of carcass parts, but as individual elements, with broken or missing segments, edges, or ends. During still later Holocene arroyo cutting in the paleovalley, particularly in the Middle Holocene, some of the bison bone from the Folsom kill was almost certainly reexposed and then lost to erosion. It is possible that not all the bison remains discarded in the paleovalley suffered that fate, but finding such under the 5- to 6-m-thick overburden has proven difficult.

DID FOLSOM HUNTER-GATHERERS EXPLOIT FAUNAL RESOURCES OTHER THAN BISON?

Skeletal remains of five species of animals besides bison apparently came out of the site during the original investigations (Hay and Cook 1930), and three additional species were recovered during our excavations. Elements from another 16 species have, over the last seven decades, migrated into one of the Folsom drawers at the Denver Museum, but fortunately these taxonomic interlopers are easily spotted. Disregarding these, the eight taxa known to have come from the site do little to support the suspicion that this Folsom group exploited animals other than bison.

The majority of those animals are rodents and, as was recognized early on, were likely to have been later intrusions in the bison bonebed (Cook to Hay, February 11, 1930, OPH/SIA; Hay and Cook 1930:35). Knowing where and how they were found would bolster that case, but contemporary field techniques and records being what they were, little more can be said. Two species of burrowing rodents were recovered in our excavations from below the level of the bonebed and in a single excavation unit on the south side of the M17 block, where bison remains were relatively rare. These occur as nearly completely skeletons and show no evidence of processing for food.

Deer shaft fragments were reported from the original excavations (Hay and Cook 1930) and could possibly represent game exploited by Folsom groups. Where those remains were found was not recorded, nor are their present whereabouts known. The femur shaft fragment we recovered in our excavations on the North Bank was not obviously that of a deer but more likely of an antilocaprid, an animal that occurs in other Folsom sites (Frison 1982a; Hill 1994, 2001; Wilmsen and Roberts 1984).

There is sufficient taxonomic ambiguity about this specimen and the supposed deer fragments recovered in the 1920s that they could be from the same individual. The specimen we recovered came from the same general area in which those earlier finds of deer remains were reportedly made, based on a directionally challenged map prepared by Fred Howarth. The specimen we excavated came from stratum *f3* and was directly radiocarbon dated to 9,270 ± 50 ^{14}C yr B.P. Although it is unlikely that I can ever prove the point, it is entirely possible that this femur fragment came from the same animal whose remains were found in the 1920s.

If so, it resolves in the negative the question of whether deer or antelope were exploited by hunters at the Folsom site. But then, given that Folsom appears to have been a short-lived hunting foray on the part of an otherwise mobile group, and not a site in which there was longer-term residence or overwintering, such is hardly surprising. This evidence cannot decide the larger question of the diet breadth of Folsom groups, or whether they were bison-hunting specialists (Hofman and Todd 2001; LaBelle, Seebach, and Andrews 2003).

WHERE OR HOW DID THE FOLSOM HUNTERS PROCURE THE STONE FOR THEIR TOOLKITS?

The projectile point assemblage readily affirms the truism that Folsom groups had a penchant for using high-quality exotic stone in their tool production (Frison 1991; Hofman 1992; Stanford 1999). Insofar as those sources can be reliably identified, the four chert and jasper types came from outcrops at minimum distances of 190 to 450 km (fig. 8.6). The quartzite used in the manufacture of the two projectile points and the Baker knife cannot be identified to a particular formation, or even to the same formation, but given the overall distribution of quartzite in the region, it could not have been obtained any closer than 10 to 40 km away.

The remainder of the stone was acquired on the High Plains from sources in northeastern Colorado to the panhandle of Texas, well north and well south and southeast of the Folsom site. In broad strokes this resembles the patterns seen in other Plains Folsom sites (Hofman 1990; Stanford 1999), but that by itself does not resolve the underlying cause of the pattern, or whether it reflects large-scale territories or, perhaps, long-distance mating and small-scale exchange for the purpose of maintaining alliances (e.g., MacDonald 1998). That said, none of the projectile points in this assemblage fit a profile of exchanged items (Meltzer 1989b; also Hofman 1992).

The diversity of lithic sources present here does, however, broadly conform to the expectation of an aggregation event, insofar as groups converging on this spot from different regions would carry weaponry fashioned of stone from those areas (Hofman 1994:352). Of course, that same pattern would also result from procurement at the source by the same, highly mobile group. Were this an aggregation site, one would also expect to see differences in technology and style corresponding with raw material, but the stylistic contrasts detected in a few pairs of projectile points (e.g., fig. 2.10) at Folsom do not an aggregation make. Besides, those differences in point style cross-cut raw material types. Although that pattern could conceivably represent unfinished pieces of stone from different areas exchanged at this spot and then fashioned into stylistically distinct forms (Meltzer 1989b), the absence of evidence of point manufacture lends little support to this possibility.

More broadly, there is no evidence in the archaeological record at Folsom to indicate a large aggregation, and apparently little in the environment to attract or support one; as Hofman (1994:351–352) points out, aggregations tend to occur where critical resources such as wood, stone, backup food resources, and water can be counted on occurring, and where there are good odds of making a large kill (also Kelly 1995:219–221; Root 2000). The Folsom area does not meet these criteria. There is nothing in the archaeology or paleoecology of Folsom to support the hypothesis that the kill was made in the course of an aggregation by previously dispersed hunters.

Assuming that this pattern of stone use resulted from direct procurement, it testifies to a settlement course that covered a straight-line distance of ~700 km, the minimum north-to-south distance between the two most distant sources, Flattop and Tecovas. Since these groups were surely not moving in straight lines across the landscape, the actual distances involved were undoubtedly much greater. The oblique triangle formed by just three of these points on the landscape—the Flattop and Tecovas outcrops and the Folsom site—has an area of ~90,000 km². A foraging territory of this size is hardly excessive by either Folsom or Paleoindian standards or those of other mobile hunter-gatherers (e.g., Amick 1996; Binford 1983; Hester and Grady 1977; Hofman 1991; G. Jones et al. 2003; Kelly 1995:table 4-1). Indeed, the overall foraging territory may have been much larger.

What is unusual, at least relative to stone use seen in other Folsom sites on the Great Plains and in the southern Rocky Mountains (Amick 1999b; Hofman 1991, 1999b; Jodry 1999a, 199b), is that their course seemingly did not include a provisioning stop at Edwards chert outcrops in central Texas. It might have, of course, but any evidence of it was not left at Folsom. Furthermore, Folsom includes no stone from any chert sources west of the spine of the Rocky Mountains, some of which were no farther from the site than Flattop and which were in heavy use in the nearby San Luis Valley and Albuquerque Basin (Amick 1995; Jodry 1999a:fig. 23).

The "lithic conveyance zone" (Jones et. al. 2003:32) at Folsom is one that extended north and south on the Great Plains, and may represent a territory distinct from that used by Folsom groups living farther west in the intermountain valleys and basins.

All other things being equal, and they rarely are, the relative frequency or diversity (evenness) of stone sources recovered in the Folsom bonebed ought to reflect the diversity of stone brought to the site by the hunters, assuming that hunters did not prefer one high-quality chert over another and that one stone type did not break more often or easily than another. The former seems unlikely, and the latter is not supported by the data here. Furthermore, the diversity of stone as well as the degree of reworking represented in a toolkit should, in some general sense, represent the pattern of acquisition, such that the most recently acquired stone will also be the most abundant (G. Jones et al. 2003), while the fewer, highly curated items provide a distant echo of sources long since visited.

Granting these assumptions, the group at Folsom can be placed earliest on the Piedmont and Plains of Colorado, though whether earlier in the Black Forest area or Flattop cannot be determined. From there they tracked south, reaching the Alibates/Tecovas region, whence they moved back to the northwest into the Folsom area, the latter segment being a course easily taken by merely following the regional drainages (fig. 3.2). Or so it appears, based on their projectile points. Were there a larger complement of artifacts from the site, including less formal tools made of other (local?) stone types, a different inference might be reached.

HOW DID THE FOLSOM HUNTER-GATHERERS ORGANIZE THEIR TECHNOLOGY?

Mobility on this scale presents the logistical challenge of ensuring that sufficient stone in the appropriate packages—whether, for example, as finished projectile points or, perhaps, as large bifaces that could be fashioned into a variety of tool forms (Kelly 1988)—was available for use, whenever and wherever the need arose (Ingbar and Hofman 1999). In this particular instance, the need arose hundreds of kilometers from where the stone was obtained.

It appears from the richness and diversity of the lithic raw material that this group had been provisioning their toolkit over time, as stone resources were encountered in the course of movement across space. All of this points toward gradual replacement rather than gearing-up, with different lithic raw materials incorporated into the toolkit serially but discarded together in this episode of use (Ingbar 1992; Sellet 1999, 2004).

Although the Folsom group had resupplied most recently at the Alibates and Tecovas outcrops, there are few hints of the time and energy they invested in the process, save that they did not take the opportunity while there to do a wholesale replacement of the toolkit they had with them. They continued to carry fluted points and perhaps other artifacts made of Black Forest and Flattop chert. The Folsom evidence conforms to Sellet's (2004) prediction that gradual replacement is the provisioning pattern of groups on the move during the warm season. It says nothing, however, of the possibility that a substantial gearing-up effort had occurred the previous winter. Still, the fact that fluted points made of Black Forest and Flattop chert were still available by the time the group reached Folsom after a stopover in the Texas Panhandle speaks to a planning depth associated with a gearing-up event.

There is no evidence in the Folsom assemblage of biface reduction, on-site point production and fluting, or refurbishment of the toolkit after the kill. Taking the next step and arguing that this negative evidence is positive testimony that the group had an ample supply of stone, or that all the artifacts came in as finished fluted points and not as bifacial cores or some other form of tool, skates dangerously close to a tautological abyss. Even if that plunge is avoided, the absence-of-evidence argument would be equally vulnerable to the empirical challenge that this assemblage came from a bonebed, which is not where point manufacture and retooling would be expected to occur. Those activities could still have taken place at Folsom, in an undiscovered camp beyond the kill area, as is the case elsewhere (Hofman, Amick and Rose 1990; Jodry 1999a).

That caveat duly made, perhaps some indirect measure of available stone can be gained. The assemblage contains no pseudofluted or unfluted forms (e.g., Hofman 1992), or projectile points that ended their use-lives in the bonebed as other tools, but it does contain a considerable amount of still-usable stone in the form of broken but relatively large projectile points. If stone was in short supply, this group was not making much of an effort to conserve it.

WHAT TACTICS DID FOLSOM HUNTER-GATHERERS USE TO MAINTAIN THEIR TOOLKITS?

Although there is little evidence that this group was dangerously low on stone, they were not squandering it either. Nearly 39% of the points had been resharpened and reused and, by the time they were left at Folsom, had little usable blade edge left. Analysis using the comparative sample showed that despite widely varying sample sizes, most Folsom assemblages have a comparable percentage of reworked points, a relationship that seems to occur independent of distances to stone sources. At the risk of generalizing beyond what these data will bear, I think it reasonable to hypothesize that in any given projectile point assemblage from a Folsom site located away from a stone source and provisioned by a strategy of gradual replacement, there will be a constant, if not relatively predictable, amount of reworked points (given the relationship between reworking and sample size). If that proves to be the case, the interesting assemblages may be those where reworking occurs more or less often than would be expected by their sample size (fig. 8.9).

One tactic for stone conservation and weaponry maintenance that they were apparently not using at Folsom, however, was a utility-knife model of hafting. However much that hafting arrangement would, in principle, have prolonged the use-life of these points (Ahler and Geib 2000), the evidence presented here from both Folsom and the comparative sample fails to support that model. These points were basally anchored in fixed hafts.

IN WHAT MANNER DO THE POINTS FROM FOLSOM VARY, AND IS THAT VARIATION A RESULT OF RAW MATERIAL, TECHNOLOGY, FUNCTION, OR STYLE?

Their mode of hafting was apparent in the morphometric analysis of the assemblage, which revealed that the least variable dimensions in these points— and this is true of other Folsom and Paleoindian assemblages, as Judge (1973) first observed decades ago—are those related to the point width and thickness, especially of the basal portion. The obvious conclusion, which Judge (1973) and others since (Amick 1995; Buchanan 2002; Odell 1994; Tunnell and Johnson 2000) have drawn, is that the demands of hafting trumped those of knapping, and points were made to fit socketed hafts, not vice versa.

The haft area is the portion of the point that is best protected from the exigencies of use and, thus, one of the areas of the point where stylistic attributes might be preserved. But because this is also the portion of the point for which there are very narrow limits within which the form can vary and still fit effectively into a set of socketed hafts (Judge 1973; Odell 1994), the chances of incorporating stylistic features is correspondingly lessened. The slight differences in base forms in this assemblage did not appear to reflect stylistic variability. Nonetheless, half a dozen of the points in which a large part of the original and unreworked blade is present do show variation (by pairs) in form and technology (fig. 2.10).

That some stylistic variability was present in this assemblage is not unexpected or unusual for an assemblage from a single site, produced by one or a few members of a group. Whether they were the most skilled flintknappers of the group, those Bamforth (1991:312) argues would have been responsible for point production prior to a communal hunt, is impossible to say. The only relevant observation on this matter is that the technical skill on display did vary somewhat. Of course, skill levels vary among members of any human group; still, none of these points were poorly made or the work of obvious novices.

The primary dimension of morphological variability relates to the use and use-lives of these points. All of the projectile points from Folsom are finished specimens, jettisoned from the toolkit at various stages in their use-lives—a few still large enough to be resharpened and reused, others worn down to slugs of ~35 mm. That size is quite comparable to that of projectile points discarded at other sites (e.g., Jodry 1999a; Wilmsen and Roberts 1978), suggesting that this length may represent a discard threshold beyond which points are too small to be effective. Because of the variation in the degree of use and reworking, length-related dimensions vary significantly in these points, as is true of virtually all other Paleoindian assemblages (cf. Goodyear 1974).

WHAT IS REVEALED IN THE BREAKAGE AND DISCARD PATTERNS OF THE ARTIFACTS?

Unlike the more gradual attrition that takes place in cutting and scraping tools, projectile points can be rendered useless almost instantaneously. That happened at Folsom. Of the 24 points for which data are available, only 4 were complete. Nearly a third of the broken specimens showed evidence of impact, and just under half were snapped laterally (bend breaks), possibly from impact. Large numbers of broken points are common in Folsom sites and, like reworking, are related to assemblage size. Granting that an unknown number of unbroken or repairable specimens were carried off-site, what does the breakage or loss of over two dozen weapons represent?

Sellet's detailed nodule analysis of the production of fluted points in the winter camp at Agate Basin indicated that a minimum of 38 Folsom points were manufactured there (Sellet 2004). It is impossible to put this number into a meaningful comparative context, given that a host of critical variables, such as the number of hunters and knappers that were at Agate Basin, the number of Folsom points that were already in their inventory, or the size of the excavation area and the representativeness of the sample, are unknown and mostly unknowable. But taking that number simply at face value, the 28 points lost to use at Folsom would equate to nearly 75% of the inventory acquired during the seasonal gearing-up episode at Agate Basin.

By that measure, the breakage or loss of 28 points at Folsom was a substantial hit to the inventory. It was brought about, Hofman (1999a:128) suspects, because the area at Folsom within which the bison were trapped was so open that the animals were less concentrated, were less accessible, and thus had to be killed from a distance by atlatl darts, resulting in a "higher loss of projectile points per animal." Of course, the very openness of the Folsom bonebed would also have allowed, as Hofman (1999a:128–129) also suggests, a thorough butchering of the carcasses, which should have enabled the recovery of most of the reusable projectile points (Hofman 1999a:128–129), so perhaps the loss was not as significant as it appears.

A couple of observations are relevant here. First, impact fractures are certainly present in this assemblage, but these could have resulted from spears either thrown or thrust. Indeed, there is good reason to suspect that the hunters at Folsom did use both, given the tactical advantages of each (Binford 1997; Churchill 1993; Frison 1991). When the bison were initially trapped in the paleotributary or paleo-valley, hunters atop the valley walls could have thrown their spears from a safe distance. After the most aggressive and dangerous animals were killed, the hunters could then move in and thrust their spears into the animals still alive (Frison 1991:170).

Second, at least eight of the points and fragments from the bonebed are >40 mm in length, and four of those are >50 mm in length, all well above the apparent discard threshold and, thus, in principle still usable. Six of those eight points lack bases, and if one accepts the speculation that broken tips result from thrust spears (chapter 8), then in those cases there was a hunter with a firm hold on the shaft when the point broke off, who presumably knew precisely where the point entered the bison. That these large and useable fragments were not salvaged for reuse may thus have no bearing on the question of whether atlatls were used at the Folsom site.

In the end, the recovery or loss of projectile points may have less to do with the openness of the bonebed, and is more likely a function of where in the animal the points were embedded, the degree of butchering relative to those areas of the carcass, whether the point-bearing parts of the animal were removed from the kill area and later recovered, and whether there was even a need to recover the artifacts. At Folsom, the stone supply available for use after the kill must have been deemed sufficient that the hunters did not have to search through the "mass of meat and gore" (Wheat 1979:95) to salvage usable and recyclable pieces. They could just pack up and leave, and evidently, they did.

Coda

The kill at Folsom was made by a group of wide-ranging hunters who, perhaps, had been following the northwesterly course of rivers and streams away from the Alibates and Tecovas outcrops on the Southern High Plains (fig. 3.2) where they recently reprovisioned their toolkit. They may have been aiming for a pass through the range of high mesas and volcanoes that extend eastward from the Sangre de Christo Mountains (fig. 3.1). A herd of bison was spotted, plans were made, and the animals were maneuvered and killed using features of the landscape to advantage. The bison were butchered, and much, but by no means all, of the nutritional value of the animals was exploited. Bones and carcass segments were prepared for transport, gear was collected, and the hunters left before winter set in.

The Folsom site is in many ways an accident of history, in its creation, preservation, and discovery. A nonresidential foraging group, en route elsewhere, held over here for a few days (a week?) in Late Glacial times. Although successful at the hunt, beyond the food these 32 bison provided for the ensuing weeks or month(s), the long-term consequences of the kill for this group were likely not especially significant. Their stay was brief. There is no evidence that they returned to this spot, though they or other groups may have moved through the area en route to Trinchera Pass on later occasions.

Within a few years all traces of the kill were buried in silt; within a few centuries it was armored by shingle shale; within a few millennia it was buried beneath thick pond and fluvial sediments. There the bonebed remained, essentially intact, until a town-breaking flood over 10,000 years later brought an astute cowboy out to check his cattle and fence lines, and to spot something unusual in the bottom of the arroyo, near where he broke wild horses.

The irony, of course, is that, however fleeting, ordinary, and even inconsequential this episode may have been in the lives of the Folsom hunter-gatherers who killed those bison some 10,500 years ago, their actions would have a profound and lasting impact on American archaeology.

Field Procedures and Protocols

David J. Meltzer

Because of the historic importance and remaining archaeological potential of the Folsom site, it is appropriate to devote a few pages to some of the more mundane but critical details of our field techniques, including our mapping and coding systems. Doing so provides a permanent record for those who might contemplate work on the site or our collections in the future, and will help readers better assess the reliability of or any biases in our data. Readers interested more in what we found, and less in how we found or recorded it, are certainly under no obligation to read through this otherwise prosaic business.

As the bulk of the archaeologically recovered remains came from the 1998 and 1999 seasons, the descriptions that follow focus on the recovery techniques and methods used during those seasons. Where relevant (or different), the procedures used in 1997 are noted and described.

The Folsom Grid System

Horizontal and vertical control during our work at the Folsom site used a metric grid set to a permanent datum established on the site in March 1970 by Adrienne Anderson. This point is coded in all our work and maps as Datum AA. In addition, a series of four concrete datum points were placed by us on-site, in order to avoid the problem of shifting or disappearing datums—which, in fact, happened to our impermanent datum points put on the site in 1997.[1] Having learned our lesson, the four concrete datum points were established in 1998. In addition, temporary subdatums were established close to particular excavation areas to facilitate use of the instruments. The coordinates of the five permanent datum points on the Folsom site are noted in table A.1. All of these datum points are on the South Bank, except Datum E, which is on the North Bank.

Datum AA was not the origin point of our grid, as we anticipated at the outset that our testing and excavations would occur south and north of that point. Making it the origin (i.e., North 0, East 0) would create a gridded nightmare of units both south and north, east and west. So that all conceivable grid coordinates would be northings and eastings from a distant origin point, we arbitrarily assigned Datum AA the grid position North 1,000 meters East 1,000 meters, Z (elevation) 100 meters. These coordinates are not the same as the coordinates assigned that point by Anderson, who gave this datum an arbitrary elevation value of 100 *feet*. Each excavation unit was designated using the coordinates of its southwest corner, e.g., North 1033 East 998 (N1033 E998).

As a hedge against the mislabeling of grid unit bags or field forms, and the confusion and loss of information that inevitably follow, there was superimposed over the 1-m site grid an additional numbering system. In this system a 5 × 5-m grid was superimposed over the 1-m grid; each 5 × 5-m block—25 individual units—was assigned a unique alphanumeric code, based on the grid lines that intersected at its southwest corner. The relevant portion of the block system is illustrated in fig. A.1; as can be seen in this example, the corner of the 5 × 5-m block marked by the intersection of N1030 E995 is the M17 block.

To separately identify the 25 individual 1 × 1-m units within each 5 × 5-m block, a unique number is assigned to each of the 1-m squares. The numbering begins with 1 in the southwest corner of the 5 × 5-m block, increases west to east, then repeats on the next line to the north. Unit 1 within a 5 × 5-m block is always in the southwest corner; Unit 25 is always in the northeast corner, as shown in fig. A.1. For example, the 1-m square with the grid coordinates N1033 E998 also has the alphanumeric designation M17-19.

Grid Coordinates of Permanent Datum and Temporary Subdatum Markers Established at the Folsom Site

Datum	Northing (N)	Easting (E)	Elevation (Z)	Notes
AA (Adrienne Anderson datum)	1000.000	1000.000	100.000	This is the first permanent datum on site, set in 1970 by Adrienne Anderson, on the eastern edge of the site clearing of the South Bank. Grid values for it were arbitrarily set by us, and are not the same as those used by Anderson. The marker is rebar set in concrete, located off the southeast corner of the 1928 backdirt pile; elevation at top of rebar.
B	969.9993	999.9978	102.3597	South set, placed in southeast corner of the site clearing (South Bank), 30.0005 m south of AA, on bearing 180°00'30". Concrete with aluminum cap. Grid point and elevation at dimple in cap. Set June 24, 1998.
C	1000.0006	974.9996	100.5153	West set, placed in western side of site clearing (South Bank), 25.0005 m west of AA, on bearing 270°00'05". Concrete with gutter spike. Grid point and elevation at dimple in spike. Set June 24, 1998.
D	1033.9997	984.0019	101.2957	Set on the western wing of the 1928 backdirt berm, in the northwest corner of the site clearing (South Bank), 37.5757 m northwest of AA, on bearing 334°48'20". Concrete with aluminum cap. Grid point and elevation at dimple in cap. Set June 24, 1998.
E	1076.0007	1038.0007	98.6759	Set on the North bank, on edge of clearing between 1926–1927 and 1928 camps. 84.9712 m north of AA, on bearing 26°33'55". Concrete with aluminum cap. Grid point and elevation at dimple in cap. Set June 24, 1998.
Subdatum A	1025.850	1004.093	98.339	Temporary spike in the 1928 "pillar" area, established in 1998 for setup and instrument use close to M15 and M17 blocks. While the spike stayed in the ground, its position shifted slightly over the seasons, and its precise position was reestablished for use each season.
Subdatum B	1042.818	1003.746	97.456	Temporary spike in the 1926 excavation area, for mapping L23 block.

In turn, each mapped specimen or sample within each unit was given a unique, sequential number within each 1 × 1-m excavation unit, and these were recorded on the containers in which the specimen or sample was collected, and on the excavation data sheets and unit level forms. Thus, a specific item recovered in N1033 E998 might be designated M17-19-123, the last number being its unique specimen number.

Therefore, a single excavation unit has a unique combination of a block designation (M17), a number within that block (19), and a number for that square (N1033 E998), while each individual specimen has an additional unique and sequential number. The purpose of this redundant system was to reduce the inevitable pile of mislabeled and unidentifiable specimens, bags, samples, etc. As the road to hell is paved with good intentions, in a

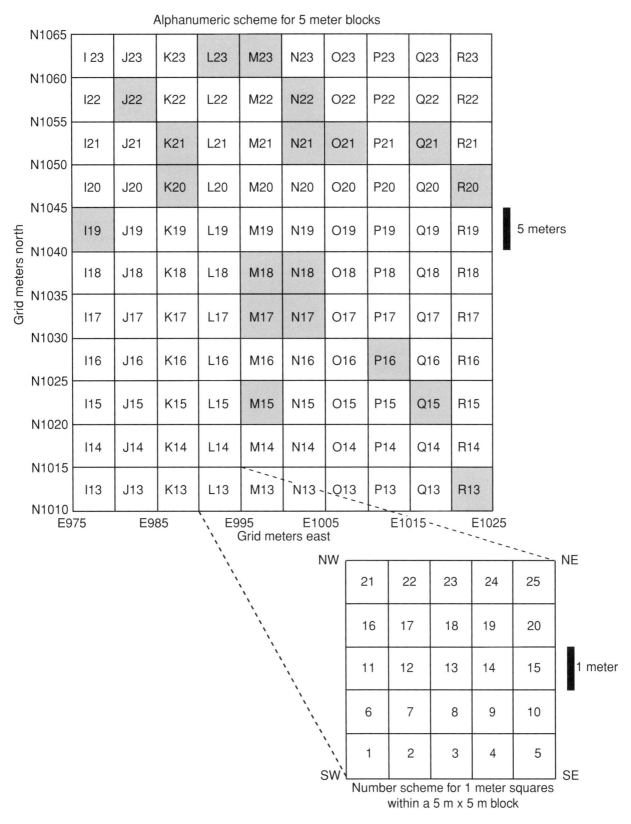

FIGURE A.1 Plan map of the alphanumeric 5 meter × 5 meter block system in use at the Folsom site in 1997–1999, and the plan map of the numbering scheme for 1 × 1 meter units within each 5 × 5 meter block. Shaded blocks are ones in which excavation occurred In 1997–1999. The test excavation in the F8 block, well to the south, is not shown. In the case of M17, most of the 5m × 5m block was excavated; in most of the other blocks, only 1 meter square was excavated.

project of this size one can never wholly eliminate those occasional glitches. Nonetheless, by having multiple identification numbers for each unit and item, which were also recorded on multiple forms, we stood a fair chance of assigning wayward items to their proper place.

Excavation Levels

A uniform and arbitrary level system was established across the site, such that each 5 cm of absolute elevation had one—and only one—level designation, regardless of the unit in which it occurred or where it occurred. Thus, Level 146 in one excavation unit was at precisely the same absolute elevation as Level 146 in all other excavation units. This is not to imply that all archaeological materials were at the same elevation across the site; they were not. Rather, this was done to maintain comparability among the many units that were excavated, and so that one could instantly know the elevation of excavations in one unit relative to any other, simply by comparing the level number. Level numbers increased as elevation decreased: The top of Datum AA, at elevation 100.00, is the base of Level 99. The highest bison bone first encountered in the M17 block was in Level 144 (97.799-97.750 elevation)—in contrast, the highest bison bone on the North Bank in the L23 block was much lower, in Level 168 (96.599-96.550).

Mapping

In order to establish as precise a fix as possible on the actual location of the Folsom site, a high-precision Trimble GPS was brought to the field in 1999 by Dr. Richard Reanier, and readings were taken at 30-sec intervals over a period of 1 hr each at Datum B and Datum C. Unfortunately, Datum AA is beneath a tree and satellite coverage is reduced. The process was then repeated at a section corner atop nearby Preston's Point, on Johnson Mesa. These readings were subsequently differentially corrected using data from the USGS base station in Amarillo. To gain ground control on the relative position of the Preston's Point GPS station with the site datum points, the EDM/Total Station was set up on Datum B, while the triple-prism target was placed on Preston's Point, at the same position as the GPS setup. Multiple shots were taken of the target and then averaged, the distance from instrument to target being 1,603 m. The averaged position puts Preston's Point on the site grid, and this substantially aids in refining the GPS calculations and mapping efforts. The calculated datum UTM positions are on file.

All horizontal and vertical control on-site was maintained by constant mapping using EDM/Total Stations: These were TOPCON and SOKKIA instruments in 1997 and SOKKIA instruments thereafter. Computer data loggers were used from midway through the 1998 field season and thereafter. The EDM/Total Stations were on-site and in use every day of excavations. All core and auger holes, geological features such as bedrock exposures and cut-and-fill sequences, excavation corners and levels, profile points, faunal remains, charcoal and gastropod samples, and routine unit samples were mapped with the EDM/Total Station. The coordinates of all mapped points and items from the excavation area were recorded both by hand on the unit data sheets and with the data logger. EDM/Total Station shots of points and features taken outside of the excavation area were recorded in field books or with the data logger.

A Brief Digression on Piece-Plotting

Piece-plotting seems such a noble enterprise that one feels uncomfortable, even heretical, thinking that perhaps there are situations where it just is not worth all the time and effort it consumes. I am not ready to go so far as O'Connell (1995) on the matter, and abandon piece-plotting skeletal elements in a bonebed, although I confess I find his arguments increasingly compelling.

Nor, as a general principle, am I ready to abandon piece-plotting of charcoal samples, recognizing the vital importance of assessing association and context. But like all general principles, one should never apply it mindlessly, because then it becomes less a principle and more a pointless ritual. Given that most if not all of the charcoal in the bonebed at Folsom was natural and not anthropogenic, its precise position with respect to the bones bordered on the immaterial, save that both charcoal and bone came from the same stratigraphic unit and were mostly of the same geological age (see chapter 5). But because one can never know that in advance, as we did not, and given the critical role of radiocarbon dating, piece-plotting all charcoal is justifiable, if not the safest approach to take.

But henceforth I will no longer piece-plot individual snails. Those we could actually see in the ground were naturally the largest of the species present at the site, but the many hundreds we piece-plotted did not include all or even most of the individuals of those species. Just how many snails we missed in the field, not only of the largest species, but also of the total snail fauna, only became apparent after James Theler carefully screened bulk sediment samples from Folsom through very fine-mesh sieves in his laboratory at UWLC (Chapter 6). He recovered large numbers of snails and snail species, as well as snail eggs and slug parts, many remains so tiny one could hardly hope to spot them in the dirt under the glare of the hot sun, and naturally, we had not. Attempting to recover and piece-plot individual gastropods in the field is not an efficient use of time; gathering of bulk samples for laboratory processing is (chapter 6).

Surface Survey

In order to gain a sense of the local geology and archaeology of the Folsom site area, in 1997 we conducted a brief surface survey of the areas surrounding the site. The survey began in the immediate vicinity of the site itself and involved extensive but unsystematic traverses of the uplands above and surrounding the Folsom locality. The survey then expanded outward from the site, and ultimately covered the tributary arroyo that enters Wild Horse Arroyo several hundred meters downstream from the site, the interfluves between the Dry Cimarron River and Wild Horse Arroyo, the uplands immediately south and west of the site, and between Wild Horse Arroyo and Cherry Creek, and the uplands immediately north and east of the site. We also surveyed the eastern edge and parts of the talus slope of Johnson Mesa; the open valley south of the Dry Cimarron River and south of the site, including several of the prominent hills that rise from the valley floor and provide good views of the surrounding region; and a lengthy section along Archuleta Creek.

We also attempted to relocate two of the three caves (fumarols) reported by A. B. Anderson (1975) on the edge of Johnson Mesa; specifically, we sought caves 29CX26 and 29CX97, which were located relatively close to one another, one of which (29CX97) was excavated in 1928 by parties from the American Museum as they were finishing their work at the Folsom site (chapter 4). This was part of an effort to assess the size, geological context, and potential contents of caves/shelters in this area. After considerable searching under very difficult conditions—the area skirting the edge of the Mesa is thickly vegetated by an oak and locust thicket—the two caves were found. Both proved to be quite small, devoid of any archaeological material, save for a historic period tin can in 29CX97, and difficult to see. As that oak/locust thicket drapes the slopes of Johnson Mesa, and neither Anderson (1975) nor longtime rancher in the Folsom area, Fred Owensby (personal communication, 1997), report any more cave or shelter sites, it was decided not to undertake a more extensive survey along the edges of Johnson Mesa.

In the survey of the creeks and canyon, attention was focused on features of possible archaeological or geological interest. Charcoal samples were collected in 1997 by C. V. Haynes, who joined us for part of that season's work, from the tributary arroyo downstream from the site. In Archuleta Canyon, the exposures were somewhat more limited, and no archaeological materials were in evidence—even at a site (29CX7) reported by Anderson (1975). Nor were there any soil, stratigraphic, or paleoenvironmental features of interest. Mann's discovery of the Archuleta *Bison antiguus* remains occurred several years later after erosion of the canyon walls.

In general, the 1997 survey was somewhat hampered by the poor surface visibility in the heavily vegetated valleys but better visibility on the open uplands. Virtually no archaeological material was found; over the area covered and the dozens of person-days devoted to the survey,

fewer than five flakes (and very small ones, at that) were found.

Excavation and Recording Procedures

All slit trenches and test excavations in 1997 were done by hand, using shovels and trowels. In those test units in which fine-scale and fragile remains are encountered, dental and bamboo picks were used (the latter when bone was found). All excavated material from intact sediments was passed through ⅛-in. (3.175-mm)-mesh screen. Sediment from the 1920s backdirt areas and units was also screened. Sediment from the shallow slit trenches was not systematically screened. Excavation in the test units was done using arbitrary levels, within the natural stratigraphic units.

The 1997 excavations showed that the deposits above the bonebed were sterile, and in 1998 and 1999, all overburden in the excavation blocks was removed with shovels but not screened down to stratum *f3*, the easily spotted but very difficult to shovel shingle shale lens that covers the bonebed. It was a relatively simple matter to know when it was time to stop shoveling overburden and prepare to come down into the bonebed. No heavy equipment was used at any time on the site, though on occasion during the removal of the overburden and later during backfilling, it was sorely wished for by all hands.

Once in the bonebed sediments, excavation in 1998 and 1999 took place in 1 × 1-m units, but separately within those units in four (50 × 50-cm) quadrants. Unit excavation was performed with trowels and other hand tools, until bone was encountered. When bone was encountered, bamboo and other finer and nonmetal excavation tools were used to minimize excavation damage. All the sediment excavated from each 5-cm level was bagged separately according to the 50 × 50-cm quadrangle from which it came. Sediment bags were tied, tagged, and then transported by flatbed truck to a water-screening operation set up on the Dry Cimarron River approximately 1 mile from the site. There, the sediment was washed through a nested pair of ⅛-in. (3.175-mm)-mesh and 1/16-in. (1.5875-mm)-mesh screens.

The screens were then set out on a nearby rack for drying, after which the recovered material was given a cursory picking for any small flakes, intact shell remains, or other pieces of interest, which were placed in vials or film canisters. All the material caught in those screens—including shingle shale—was separately bagged and transported back to the laboratory at SMU, where, over a three-year period, the contents of each bag were systematically examined. Any small flakes, bone fragments, gastropods, etc., were removed from the matrix, while the shingle shale was weighed (in grams), and its volume measured (by use of graduated cylinders; in cubic centimeters). This effort was particularly important in the analysis of the impact of the *f3* shingle shale on the bonebed (chapter 7).

Because excavation was in arbitrary 5-cm levels within the *f2* sediment, there was a minimum provenience designation for all samples and screened materials of at least $50 \times 50 \times 5$ cm. In practice, however, provenience was much finer: Items of archaeological interest, such as bone, charcoal, and snails, were piece-plotted in three dimensions with the EDM/Total Station to within 1 mm. There was a varying size cutoff for different kinds of materials with regard to the piece-plotting and other treatment. Bone fragments smaller than 1 cm were not piece-plotted, unless they were diagnostic or of particular interest. The position of all larger skeletal elements was mapped precisely, occasionally with multiple EDM shots (depending on the size of the element). All charcoal samples and snails recovered in situ were collected and mapped with the EDM, despite often being much smaller than 1 cm.

After being mapped and photographed in place, larger and/or more fragile skeletal elements were carefully wrapped in light plaster casts (using Johnson & Johnson *Ready-Wraps*), while relatively smaller or sturdier pieces of bone were collected and wrapped in aluminum foil. Charcoal found in excavation areas was plotted, separately collected, and put in aluminum foil, small vials, or film canisters and then in labeled bags.

For each unit, data sheets recorded detailed information on the specimens recovered from units; these data sheets were modeled after those developed by L. C. Todd and D. Rapson for use on bison bonebeds on the Central and Northern Plains, in order to allow more direct comparison of the recovered materials from these and other sites. Before collecting any item from a unit, a number of observations and pieces of information were made and recorded. These included information on unit, level, stratum, provenience points on the item, the item's orientation and inclination, three-dimensional coordinates, and the kind of material, i.e., bone, charcoal, etc. This was followed by a series of more specific attributes recorded for each item, tailored to the kind of material, and then how the piece was packaged for removal, whether wrapped in foil, in plaster, placed directly in a plastic bag, etc., who did the work, when, EDM shot number, and comments, if any. If the item was bone, insofar as possible the described attributes included the species, the skeletal element, and the portion and segment present; data were also recorded on its relative degree of fusion, breakage, or burning; whether the element was in articular position or conjoined with other elements; its maximum length; and, as noted already, its orientation and inclination. The latter records are especially useful for assessing the taphonomic history of skeletal elements, and data that can only be gathered in the field.

In addition, in every 1×1-m excavation unit, a set of sediment samples was routinely taken in every 5-cm level, generally in the southwest quadrant of the unit for consistency's sake. If this was not possible, samples were taken from another quadrant. In 1998, three samples were taken per unit per level; in 1999, the number was reduced to two. The samples were placed in cloth *Hubco* soil sample bags, primarily the 3.5×5-in. size, but larger bags were used on occasion if supplies ran out. These samples were collected for later extraction and analysis of sediments, pollen, phytoliths, gastropods (snails), and beetles. The position of each sample was mapped with the EDM/Total Station, and all samples were separately numbered within the unit and recorded on the excavation data form.

For each level within each unit, a level form was completed, which contained information on the dates of excavation of that level, the beginning and ending depths of the level (target and actual), a sketch map of the floor, comments on the stratum and items recovered, checklists for samples, photographs, map numbers, and so on.

The above procedure was used in the bonebed areas. A slightly different procedure was used in excavating the test units in possible camp areas. All 1×1-m test units were dug in 10-cm levels, using shovels and trowels, and sediment from all levels was dry-screened through ¼-in. (6.35-mm) mesh. Level forms were completed for each 1×1-m test unit, which recorded starting and ending elevations, numbers assigned to any mapped items in that level, samples collected, photographs taken, and general comments and observations on the level.

In addition to the multiple forms and maps, both color and black-and-white photographs were taken for documentation, as well as, in 1998, some video footage. Finally, throughout the project daily field notes were kept by the principal investigator.

Closing Up

As virtually all of the excavated sediment had been removed from the site, primarily through the water screen washing but also as samples, a considerable amount of the original fill was no longer available for backfilling. Moreover, because it was anticipated that some of the excavation areas might be reopened at a later date, by us or others, it was desirable to backfill the units in such a way that our prior excavations would be relatively easy to spot and easy to uncover.

For these reasons, an off-site source of fill was sought, and found in a sand and gravel quarry on the slopes of nearby Johnson Mesa; this mixture of well-sorted sand and well-rounded quartz and quartzite gravels and cobbles is identified as a remnant of the Ogallala Formation (chapter 3). The sands there are a distinctive reddish color—unlike any of the deposits in or near the Folsom site. But because the deposit also contains considerable gravels, including small chert pebbles (which are relatively rare), it was decided to screen the material that was destined to immediately cover the bonebed. The fill was initially screened through 1/4-in. mesh, then through 1/2-in.

mesh when the wet sand continued to clog the finer-mesh screen. In 1998, the gravel removed from the screening of the sands was used to fill the deepest of the excavation units in the M15 block, as there are no plans or evident need to return to these units.

The screened sand was used for the lowest layers of fill in M17 closest to the bonebed. In addition, an effort was made at the quarry itself to gather fill from sections that were dominantly sand. If any future work is done in the M17 block, which seems unnecessary, given that all of the excavated units were taken down to sterile deposits, the fill in the units will be easy to spot and care should be taken not to confuse natural chert pebbles with artifacts—here or elsewhere on-site.

In preparation for backfilling, the floors and lower walls of the excavation units in the M17 block were covered with black landscape cloth, and the screened sands were carefully spread over the top. This continued until all of the area was well covered in the clean sand, and then unscreened sand and gravel was thrown in on top, followed by the backdirt from the upper layers, mostly McJunkin Formation clays.

There are no plans or evident need to return to the test units on the uplands, and because the supplies of the imported sand were limited, none was used in the backfilling of these units. Instead, the floors of the units were covered in plastic, then backfilled with backdirt originally from the block excavations. Aluminum soda cans were also tossed in with the backfill.

Note

1. In the course of reestablishing the grid system on the site at the start of the 1998 field season, it was discovered that the datum points established at the site in 1997, two iron rebars set in the ground at 90° angles to Datum AA, had shifted over the winter. Trying to re-establish their present position also revealed that the 1997 grid was at a declination of 6.84° off magnetic north and, thus, was not at True North as supposed. Rather than maintain this anomalous orientation, it was decided to take the opportunity to reestablish the grid, aligning it on magnetic north as of June 1998. Obviously this position will shift in the years to come, which suggests that a preferable alternative might have been to set the grid to true north with the correct declination. Unfortunately, we did not have the current declination on hand in the field. Reestablishing the grid required recalculating the position of the ~3,900 mapping shots taken in the summer of 1997; a formula for doing so was developed and all mapping points from 1997 were so translated.

The Folsom Diary of Carl Schwachheim

Edited with annotations by David J. Meltzer

More so than any other individual, Carl Schwachheim (1878–1930) a Raton, New Mexico, blacksmith and skilled amateur naturalist, was the main thread running throughout the original fieldwork at the Folsom site. It was Schwachheim, alone among the many individuals McJunkin told about the bones, who bothered to go to the Crowfoot Ranch to examine the remains in place. It was Schwachheim who, with Fred Howarth, took the initiative to have the bison remains examined at the Denver Museum of Natural History. And, of course, it was Schwachheim who was the primary and often sole excavator at the site from 1926 to 1927, and part of the crew in 1928.

For 11 years, from May 6, 1918, through July 4, 1929, Carl Schwachheim kept a diary. It is filled with thoughtful but often rather terse observations on natural history, flora, and fauna. The diary covers the period Schwachheim worked at Folsom, and as the significance of the site emerged, more of the entries are devoted to comments about the skeletal remains and artifacts being uncovered. Because Schwachheim sketched many of the artifacts, it is possible to identify which points were found when, where, and with what bison skeletal remains (chapter 8). Where possible, the specific projectile points Schwachheim refers to in his diary are identified below, using the American Museum designation, as discussed in chapter 8 and shown in table 8.2. One can bemoan the lack of detail recorded by these early field-workers, but Schwachheim's diary at least helps fill in some of the critical details of the excavation (chapter 4), representing as it does a day-by-day account of the discoveries and events at the site.

All those parts that refer to the work at Folsom site are reprinted here, including comments on the weather, to help provide a fuller understanding of the conditions of the fieldwork. I also append annotations explaining the significance of particular events, the identity of visitors,

and photographs of the diary from Two landmark days. For the sake of clarity, I have corrected misspellings, added punctuation, and adjusted capitalization, but I meddled as little as possible, so as to give the full flavor of the original text.

The diary, as well as the photographs and files of Carl Schwachheim, were kindly made available by Emily Burch Hughes and Tom Burch of Raton, the children of Tillie Burch, Carl Schwachheim's sister. Fred Owensby provided help with the identification of the local ranchers mentioned in the diary. The first reference to the Folsom site occurs with a terse note in December of 1922, and it is there that this annotated diary begins.

Background Entries

December 10, 1922 [Sunday; page 34]

Went to Folsom and out to the Crowfoot Ranch looking for a fossil skeleton. Found the bones in arroyo ½ mile north of ranch & dugout nearly a sack full which look like buffalo & Elk. We only got a few near the surface. They are about 10 ft down in the ground.
Note: Accompanying Schwachheim that day were Fred Howarth and James Campbell, who took the photograph of Howarth sitting on the floor of Wild Horse Arroyo, pointing to bone (fig. 2.8). Schwachheim's entry makes no mention of artifacts: only that he was there looking for a fossil skeleton.

July 1, 1923 [Sunday; pages 38–39]

Near Folsom. Yucca, poppymallow, & mariposa lilies in bloom. The bones were of the American bison. Young larks, doves, & chipmunks nearly grown. Wheat and oats about 3 inches high on Johnson Mesa. Many butterflies of different kinds.
Note: It is unclear which bones he is referring to in this entry, but since he is in the Folsom area, it is possible he is referring to the

Folsom site. Who might have made the identification in question is not known. The Folsom bison bones were not identified at the Denver Museum until 1926; see the next several entries.

November 28, 1924 [Friday; page 57]

Crossed the top of Johnson Mesa to Folsom & Oak Canon here. We found several kinds of fossil shells in the limestone. Would like to investigate more here. Saw a sulphur butterfly. Saw a large flock of quail. Fred [Howarth] got 2 and a rabbit. Found 2 scrapers & a fine arrow point. Would like to have a week in this Canon. Indian grave on point between Oak and Cimarron Canon. They say there is trout here but I don't believe there is.

January 25, 1926 [Monday; page 73]

[In Denver] Took in both museums also the public library & capitol. They are sure fine. We met Mr. [Jesse] Figgins director of the [*entry ends here*]
Note: The background for why Schwachheim and Howarth took the bones to the Colorado Museum is provided in chapter 2.

March 7, 1926 [Sunday; page 74]

Went with Mr. Figgins & Mr. [Harold] Cook to look at the fossils on the Crowfoot Ranch. They say they are worth while. Then we visited the lava beds east of Capulin Mt.
Note: This is the visit that was memorialized in the photographs, of Figgins and Howarth and of Cook and Howarth, eating a picnic lunch at the site (fig. 1.3). It appears that figure 3.4A, one of several photographs of Schwachheim standing on the North Bank, may also have been taken this day.

The 1926 Field Season at Folsom

Note: Schwachheim arrived at the Crowfoot Ranch in early May of 1926 to begin fieldwork. He worked steadily through May and June, but little mention is made of anything related to the work at the site until June, and as can be seen in those entries below, the information is minimal. Entries during this period are instead devoted to observations of birds, flowering plants, and animals seen or collected.

May 2, 1926 [Sunday; page 77]

Out to the crowfoot ranch near Folsom.

June 4, 1926 [Friday; page 80]

[Harold] Cook left at 10 o'clock.
Note: There is no earlier mention of when Cook arrived on-site.

June 18, 1926 [Friday; page 81]

Cook came yesterday & he worked on the skull yesterday & today. He left this evening. The skull looks like the water buffalo and it is very large. The horns at the base being about 15 inches. Found a larks nest with 5 eggs in it.

June 30, 1926 [Wednesday; page 83]

Had about 25 or 30 visitors to see the bones. Mrs. Owen sent me a dandy lunch. Mr. Owen's mother was with the visitors.
Note: The Ben Owen family at that time lived at the Pitchfork (now Hereford Park) Ranch, which adjoined the Crowfoot Ranch.

July 9, 1926 [Friday; page 83]

The young ground sparrows have left the nest & I have it. I found a rock full of fossils which look like they may be some kind of fish. We have some of the bones packed & will pack the skull as soon as it gets a little dryer. It is 33 inches to tip of horn cores. Length of head 26½ inches. Eye to eye 15 inches. Horn core at base 13½ inches. Length of horn cores 12 inches. Length of lower jaw 18½ inches. Mariposa lilies and poppy mallow is in full. The mariposa has the 3-cornered seed pod on the highest of the 2 stems. I counted in our tent around the lamp 15 different kinds of small moths some of them very beautifully marked.

July 14, 1926 [Wednesday; pages 84–85]

Found part of a broken spear or large arrow head near the base of the fifth spine taken out. It is about 2 inches long & is of a dark amber colored agate & of a very fine workmanship. It is broken off nearly square & we may find the rest of it. I sure hope we do. It is a question which skeleton it was in, but from the position of them it must have been in the skeleton of the smaller one & just inside the cavity of the body near the back. It was found 8½ feet beneath the surface with an oak tree growing directly over it 6 inches in diameter, showing it to have been there a great length of time [*fig. B.1*]
Note: This is the first record of a projectile point from the site: Schwachheim provides a sketch, which makes it possible to identify the point as DMNH 1391/3(fig. 2.10).

July 27, 1926 [Tuesday; page 85]

The large spiders which live in holes in the ground have young. The young hang on to the old one all over her body. The Indian paintbrush has seed now. We found another skull. It is very badly crushed & broken like others had walked on it while it was in the water.

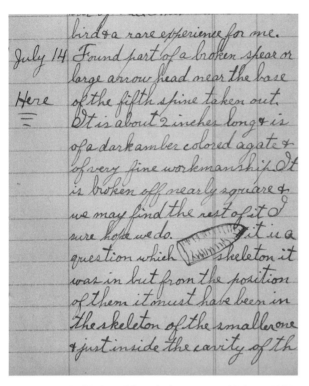

bird & a rare experience for me.
July 14 Found part of a broken spear or
large arrow head near the base
Here of the fifth spine taken out.
It is about 2 inches long & is
of a dark amber colored agate &
of very fine workmanship. It
is broken off nearly square &
we may find the rest of it. I
sure hope we do. It is a
question which skeleton it
was in but from the position
of them it must have been in
the skeleton of the smaller one
& just inside the cavity of th

FIGURE B.1 Carl Schwachheim's diary entry of July 14, 1926, which records the Folsom fluted point found on site. (Photo by D. J. Meltzer, courtesy of Carl Schwachheim Collection.)

August 22, 1926 [Sunday; page 87]

Size of second skull Tip to tip 42 inches. Length of head 27 inches. Between the eyes 18 inches, horn cores at the base 14 inches, between the horns 17 inches, point of jaw to angle 18 inches. About 9 feet below the surface. Mr. Orr and wife from the museum spent the day in camp. They are on their vacation on a motorcycle.

August 24, 1926 [Tuesday; page 88]

Shipped the second skull & another box of bones today.

September 6, 1926 [Monday; page 88]

Got a mastodon or mammoth tooth from Mrs. Honey. Think it came from north of their house.
Note: Minnie Honey lived on a nearby ranch, between the site and the town of Folsom.

October 12, 1926 [Tuesday; page 90]

Mr. Gripe, Morrow, & Ryan of Folsom were in camp. Mr. Gripe says near his place on dry cimarron are some very good fossil tracks. Also near there are Indian pictures he says.
Note: Raymond Morrow was the father of the late Jane Morrow Owensby, whose husband Fred Owensby is the longtime archaeological steward at the Folsom site. This entry documents the fact

that Schwachheim was still at Folsom in mid-October 1926. He apparently stayed in the area for at least the next two weeks. He did go down the river to the Gripe's place, and then on into Kenton, Oklahoma, where he saw mammoth bones in the Bank window. Those bones came "from the north somewhere."

The 1927 Field Season at Folsom

Note: Schwachheim was in Frederick, Oklahoma, from February 24, until about March 5, 1927. He was subsequently sent to Vernon, Texas, on April 10, 1927, apparently to examine some bison remains that had just been found. He arrived at the Crowfoot Ranch on May 30, 1927.

May 30, 1927 [Monday; page 103]

Crowfoot Ranch 1927. Came over here this morning.

June 3, 1927 [Friday; page 103]

The driest time in 25 years no green grass or flowers yet. Sick about 4 days.

June 10, 1927 [Friday; page 104]

First rain just a little shower hardly wet the ground.

June 12, 1927 [Sunday; page 104]

Rained nearly all night and is still at it 9 o'clock. Was up with cramps nearly all night but feel better now.

June 13, 1927 [Monday; page 104]

Had a bad time all day yesterday with chills and fever but feel better tonight but very sore. Rained nearly all day. Received a letter from Fred [Howarth] this evening the first from any one since I came.

June 14, 1927 [Tuesday; page 104]

Cold and foggy until noon then cleared so I worked this afternoon but I am very sore and weak yet.

June 15, 1927 [Wednesday; page 104]

Rained nearly all day today.

June 16, 1927 [Thursday; page 105]

Worked today but it was pretty muddy this morning.

June 19, 1927 [Sunday; page 105]

Shorty [Floyd Blair] came today at noon.
Note: Floyd "Shorty" Blair was on the field staff of the Colorado Museum of Natural History.

July 17, 1927 [Sunday; page 107]

Bob arrived at noon today.
Note: The Colorado Museum of Natural History Annual Reports make no mention of who "Bob" is—and Schwachheim does not provide a last name.

July 18, 1927 [Monday; page 107]

Fred did not show up today so I turned the big lizard out.

July 24, 1927 [Sunday; pages 107–108]

Sunday Bob, Shorty and I went to Capulin. Were caught in a thunder shower on top. It was a grand sight. Rained and thundered for nearly an hour with lightening flashes on every side.

July 29, 1927 [Friday; page 108]

Wildfire in bloom on west end of big shale bank also on next one west of the big one.

August 18, 1927 [Thursday; page 109]

It rains everyday now.

August 29, 1927 [Monday; pages 109–110]

I found an arrow point this morning it is of a clear colored agate or jasper. It is not exposed the full length but it is hollow on the sides and looks something like this. The point was near a rib in the matrix. One barb is broken off. Since noon Mr. [Floyd] Blair found another not in place but in the loose dirt; is much the same shape 1 inch wide at break and ¾ at base. Shaped like this but more of it. Made of dark red flint. These are not of the notched points but are the link between the spear or lance and the notched points of a later day. Sent a letter to the boss [Figgins] today [*fig. B.2*].
Note: This entry refers to point DMNH 1262/1A (Sn 3), the specimen that Brown, Kidder, and Frank Roberts would all witness in situ, which for many decades was on display with the associated bison ribs at the Museum. Schwachheim's letter to Figgins triggered the wave of telegrams for a site visit (chapter 2). The whereabouts of the second point are not known.

September 4, 1927 [Sunday; page 111]

Mr. Figgins, Mr. [Barnum] Brown were here and took pictures of the arrow point in place. Then came Fred [Howarth], Jim, and Mr. and Mrs. [Frank] H.H. Roberts of the Smithsonian. Mr. Brown is with the American Museum of Natural History.

September 6, 1927 [Tuesday; page 111]

Mr. Roberts and wife were back and took pictures of the arrow in place. His address is Frank H.H. Roberts of the Bureau of American Ethnology Smithsonian Institution Washington D.C.

September 8, 1927 [Thursday; page 117]

Roberts and A.V. Kidder came back and took some pictures. Kidder is working for the Phillips Academy Andover Massachusetts under W.K. Moorhead [*sic*]. H. Cook and wife were here today.
Note: In late October Schwachheim subsequently visited Pecos Pueblo, likely a result of his meeting Kidder at Folsom.

September 23, 1927 [Friday; page 112]

Sent in first shipment of bones. 11 boxes.

September 29, 1927 [Thursday; page 112]

Shorty will leave tomorrow & I will ship the [camp] outfit. Then I will go into Raton Sunday with Lud [Shoemaker].
Note: Lud Shoemaker lived at the Crowfoot Ranch.

September 30, 1927 [Friday; page 113]

Sent in last shipment of bones, camp equipment & arrow point today. Shorty left also.
Home [to Raton].

The 1928 Field Season at Folsom

Note: In 1928, Schwachheim was hired on to the American Museum of Natural History's field crew. His diary for that year begins with moving out to set up the field camp at Folsom.

June 3, 1928 [Sunday; page 123]

Rained all afternoon. Glenn [Streeter] & I got stuck with the truck down to the axle 1 mile from camp.
Note: Streeter was on the staff of the American Museum.

June 5, 1928 [Tuesday; page 123]

Got stuck 3 times before getting to camp but everything is in camp now.

June 6, 1928 [Wednesday; page 123]

Harold [Cook] brought Mr. [Peter] Kaisen and son [Ernest Kaisen] out. Got back to camp at 9 o'clock tonight.

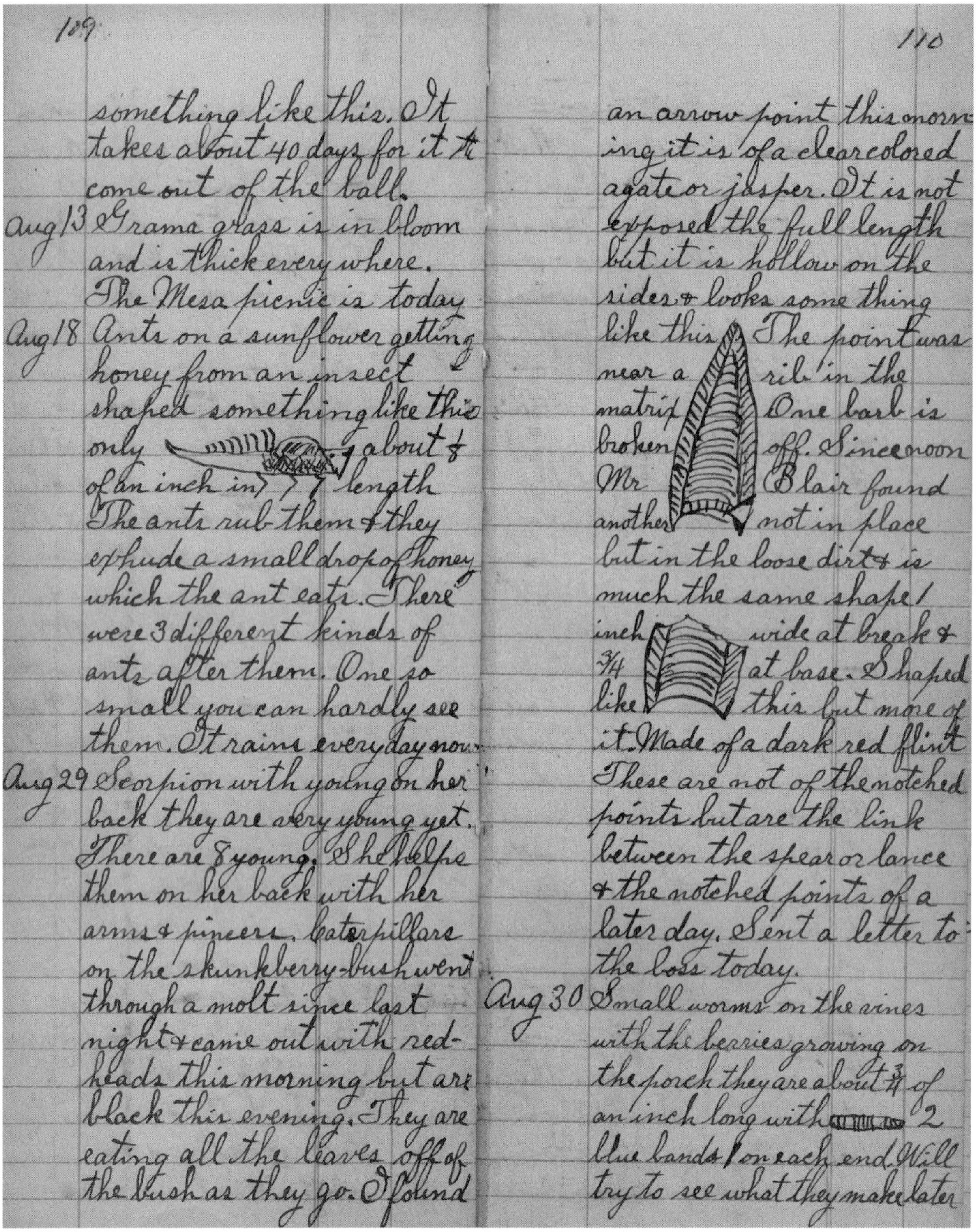

something like this. It
takes about 40 days for it to
come out of the ball.

Aug 13 Grama grass is in bloom
and is thick every where.
The Mesa picnic is today

Aug 18 Ants on a sunflower getting
honey from an insect
shaped something like this
only ⟨sketch⟩ about ⅛
of an inch in ⟨sketch⟩ length
The ants rub them & they
exhude a small drop of honey
which the ant eats. There
were 3 different kinds of
ants after them. One so
small you can hardly see
them. It rains every day now

Aug 29 Scorpion with young on her
back they are very young yet.
There are 8 young. She helps
them on her back with her
arms & pincers. Caterpillars
on the skunkberry bush went
through a molt since last
night & came out with red-
heads this morning but are
black this evening. They are
eating all the leaves off of
the bush as they go. I found

an arrow point this morn-
ing it is of a clear colored
agate or jasper. It is not
exposed the full length
but it is hollow on the
sides & looks some thing
like this ⟨sketch⟩ The point was
near a ⟨sketch⟩ rib in the
matrix ⟨sketch⟩ One barb is
broken ⟨sketch⟩ off. Since noon
Mr ⟨sketch⟩ Blair found
another ⟨sketch⟩ not in place
but in the loose dirt & is
much the same shape 1
inch ⟨sketch⟩ wide at break &
¾ ⟨sketch⟩ at base. Shaped
like ⟨sketch⟩ this but more of
it. Made of a dark red flint
These are not of the notched
points but are the link
between the spear or lance
& the notched points of a
later day. Sent a letter to
the loss today.

Aug 30 Small worms on the vines
with the berries growing on
the porch they are about ⅜ of
an inch long with ⟨sketch⟩ 2
blue bands on each end. Will
try to see what they make later

FIGURE B.2 Carl Schwachheim's diary entry of August 29, 1927, which records the first Folsom fluted point found in situ. (Photo by D. J. Meltzer, courtesy of Carl Schwachheim Collection.)

Note: Cook described Peter Kaisen was "one of the oldest and best of the American Museum field collectors." Brown had put Kaisen in charge of the fieldwork at Folsom that season, and Cook was along at the request of the AMNH "to lay out the work to be done, and see that things were started off properly" (Cook, Unpublished Report on Activities of the Department of Paleontology, CMNH, for the year 1928; HJC/AGFO).

June 12, 1928 [Tuesday; page 123]

[Lud] Shoemaker [Crowfoot Ranch] branded 200 head today. Mr. A[lfred] E. Jenks of the University of Minnesota & Jim were in camp.

Note: Jenks was an archaeologist and ethnologist who was quite enthralled by Folsom and, for a time, sought to excavate at the site himself (chapter 4). He would later become involved in the Minnesota Brown's Valley human skeleton (Jenks 1937).

June 14, 1928 [Thursday; pages 123–124]

Found a very crude a metate in the ditch about 200 feet below the quarry today.

Note: This is likely the same small tributary where Howarth collected the charcoal Cook later submitted to Willard Libby, ostensibly as a sample from the Folsom site (chapter 5).

June 17, 1928 [Sunday; page 124]

Mr. Kaisen, Glenn, and I went up to the rim-rock [the edge of Johnson Mesa]. Wild peas and several kinds of lupine are in bloom now. Saw the black & white swifts flying along the mesa.

June 18, 1928 [Monday; page 124]

[Gerhardt] Laves found a cave above the road as you come down the hill from the mesa.

Note: Gerhardt Laves, a graduate student at the University of Chicago, was out that summer at the behest of Clark Wissler, Curator of Anthropology at the American Museum. Wissler appointed Laves to survey the area around the site, in the hopes of locating the "camp of the Folsom hunters." This was likely Cave No. 1 (see chapter 4).

June 23, 1928 [Saturday; page 124]

Mr. [Barnum] Brown, Jim, and Fred [Howarth] came out today.

June 25, 1928 [Monday; page 124]

Glenn found a piece of broken arrow point something like this: *[CS sketch AMNH point A Sn 7]*

June 27, 1928 [Wednesday; page 124]

Mr. Brown found a broken point today it is shaped like all the others. Found 9 feet 1½ inches below the surface. *[CS sketch AMNH point B Sn 8]*

June 28, 1928 [Thursday; pages 124–125]

Left camp for Gripes ranch at 7:05. Arrived at 10:00. Dinner at windmill in Peacock Canon. Dug out cave & I found 2 broken arrowpoints modern. In camp 3 miles up Allen canon at 8 o'clock fine place to camp.

June 29, 1928 [Friday; page 125]

Climbed up to look at 2 walled caves opposite Mr. Gripes place north. The lower one looks like Mexican work while the high one looks like Indian. They are both walled with stone. Sure some climb up there. Down to Mr. McCuiston's ranch [*downriver on the Dry Cimarron*] where Brown took picture of some of the chimneys. Back to camp at 8. Gripes 47 miles camp to Folsom by Floyds 12 miles. Morrows to Folsom 5½ miles.

July 1, 1928 [Sunday; page 125]

Owen's, Shoemaker, Budds [*Johnson Mesa residents*] were visitors in camp today.

July 2, 1928 [Monday; page 125]

[Barnum] Brown left today. Glenn took him to Raton [New Mexico].

Note: Brown's stint at Folsom—9 days—was limited, owing to his other field commitments that season.

July 4, 1928 [Wednesday; page 125]

Slept late. We could not get a chicken so will have no big feed. It turned out better than it looked as they had company at Shoemakers & made ice-cream and gave us some & some chicken sandwiches so we fared pretty good.

July 7, 1928 [Saturday; page 126]

We have 3 skeletons nearly out, but are short all the skulls. It hailed this afternoon and many of the stones were the size of hen's eggs. They went through Mr. K's [Kaisen's] tent like it was paper.

July 8, 1928 [Sunday; page 126]

Ralph Cally, Van Dyne and wives, Henry Floyd and family, also Mr. and Mrs. Justin of Trinidad were visitors today.

July 13, 1928 [Friday; page 126]

Earnie [Kaisen] struck another arrow point of jasper. It is identical in shape and chipping with the rest. He broke the point off but the rest is in the bank in place, not yet exposed. [CS sketch AMNH point C Sn 9]

July 16, 1928 [Monday; page 127]

Mr. Kaisen found a broken point on the dump. It seems to be a little wider than the others, but the same shape as all the others. It was broken off about 1¼ inches from the base & has lime [calcium carbonate] formed on it. [CS sketch AMNH point D Sn 10]

July 17, 1928 [Tuesday; page 127]

I found another point today. It looks like agate and is about 2 inches long with the point broken off. It looks something like this. In shape it is just like the others & the workmanship is very fine. [CS sketch AMNH point E Sn 11]

July 19, 1928 [Thursday; page 127]

Earnie's point [AMNH Point C Sn 9] is about 6 inches from the bones & is 2 ³/₈ inches in length. It is complete only where he broke it with the pick.

July 22, 1928 [Sunday; pages 127–128]

Bob Landberg and [Nelson] Vaughan came up from their camp on the way to Raton. I went with them & we got back at 6:30 had supper and they left for their camp at 9. [Barnum] Brown got in at 10 o'clock.
Note: Robert Landberg and Nelson Vaughan both excavated fossils for the Museum, Vaughan at the Lone Wolf Creek site, in Colorado City, Texas (chapter 2). That summer, the two were collecting a fossil bison locality near Raton reported by Fred Howarth (Cook, Unpublished Report on Activities of the Department of Paleontology, CMNH, for the year 1928; HJC/AGFO).

July 23, 1928 [Monday; page 128]

We opened up everything this morning & while doing so I found another point which the rain had washed out. It has been struck by a pick or tool & the point broken off. It is of nearly clear agate and the same shape as the others. Only about ½ is exposed. I hope the base is there. [CS sketch AMNH point F Sn 12]
Note: By then "three arrows had been discovered in an undisturbed position," and that afternoon or evening Brown sent out his series of telegrams inviting "prominent scientists" to view the artifacts in place (Brown 1928; replies received are in Field Correspondence, 1927–1932, VP/AMNH). See chapter 4.

July 26, 1928 [Thursday; page 128]

Helped Laves with his cave but found only some scrap bones, part of a broken metate, also parts of 2 broken manos & some chips.

July 27, 1928 [Friday; page 128]

I found the point of a broken point, something like this [CS sketch AMNH point G Sn 13] & this evening, Mr. Kaisen found the widest one of them all. It is of jasper & the very finest of work. He struck it & broke it into 5 pieces. It is shaped about like this [CS sketch AMNH point H Sn 14] & it is sure fine stone.

July 28, 1928 [Saturday; page 129]

Glenn and Mr. Brown went to Bear Canon to look for Indian graves, but found none. They brought back Mr. [Neil] Judd of Washington D.C.
Note: Neil Judd was a curator in the Department of Anthropology, United States National Museum, Smithsonian Institution. As discussed in chapter 2, he was there in 1927 when Figgins brought in one of the first Folsom points to show Holmes and Hrdlicka, and he (Judd) had offended Figgins by suggesting that such points were not that uncommon—and that he had found them at Pueblo Bonito. Judd took that to mean that the points were not necessarily very old. Figgins afterward wrote that he thought Judd rather ignorant (Figgins to Hay, July 1, 1927, OPH/SIA). As it happens, Judd was right: he had found a fluted point at Pueblo Bonito—it had evidently been picked up as a curio in prehistoric times [D. Stanford, personal communication, 2004].

July 30, 1928 [Monday; page 129]

Mr. [Frank] Roberts & wife, Harold Cook, Mr. Meade, & 2 other of the [American] museum trustees were company here yesterday. Mr. Judd is going with Roberts over to his work 20 miles west of Pagosa Springs on Pedra River. Mr. Kaisen found another fragment of an arrow point. It is of white agate & a very fine stone. [CS sketch AMNH point I Sn 15]

July 31, 1928 [Tuesday; page 129]

Mr. [Kirk] Bryan, teacher of Geology at Harvard, I think, arrived this morning & will be here a few days I think.
Note: As discussed in chapter 2, Bryan visited the site at the behest of Alexander Wetmore, Director of the Smithsonian Institution. He traveled to the site with Roberts and Judd, and along the way they took a series of photographs, along with ones at the site (e.g, figs. 1.5 and 2.14).

August 1, 1928 [Wednesday; page 129]

The Robertses, Judd and Bryan left this evening for Roberts' camp. 8 miles down the river from chimney there are some fossil fish.

August 2, 1928 [Thursday; page 129]

Mr. [Clark] Wissler arrived this morning. Fred and Emma [Howarth] brought him out.
Note: As discussed in chapter 4, Wissler had sent Laves out to survey the Folsom region for the camp that might have been associated with the bison kill.

August 7, 1928 [Tuesday; pages 129–130]

Mr. Brown left for Colo[rado]. Mr. & Mrs. [Earl] Morris of Aztec N.M. came in this afternoon & will be here for a few days.
Note: Earl Morris was the National Park Service archaeologist at Aztec Ruin.

August 9, 1928 [Thursday; page 130]

Laves went down to the Riley camp in oak canyon to dig out a cave in sandstone. He will be there a week or two. Ernie's arrow is in Slab No. 15.

August 10, 1928 [Friday; page 130]

Mr. Wissler left this morning for Texas. Tall larkspur is in bloom now. . . .
Note: This same day Schwachheim wrote to Brown, "We have found to date 9 broken points. Oh! Yes, one was a fine one, but Ernie struck it with a pick breaking it . . ." (Schwachheim to Brown, August 10, 1928, VP/AMNH).

August 13, 1928 [Monday; page 130]

Boy Scouts from Clayton were here today. The Mullen is in bloom.

August 14, 1928 [Tuesday; page 130]

Laves left for Raton tonight at 10 minutes to 7. He is going from there to Taos.

August 20, 1928 [Monday; page 130]

Received a broken spearhead from Frances but no letter today.

August 28, 1928 [Tuesday; page 131]

Ernie found the base of another arrow today; this one is No. 10 for the season and the lowest one found 11'-9". It is the only one found so far on the north side & is of an agate or jasper red with a few clear streaks. [*CS sketch AMNH point J Sn 16*]

August 29, 1928 [Wednesday; page 131]

Ernie found another point this morning; it is the same material as his last one & nearly fits on it. This is No. 11.

He struck it with a pick & broke it in 3 pieces. [*CS sketch AMNH point K Sn 17*]

August 31, 1928 [Friday; page 131]

All bones out of north side of bank are plastered.

September 3, 1928 Labor Day [Monday; page 131]

Mr. and Mrs. [Charles] Berkey of N.J., Mr. Majors, Miss Johnson and another lady were here today.
Note: Charles P. Berkey was a Columbia University geologist; Brown ran into him in Denver after leaving the Folsom site that August and asked him to go down to the site "to study the geological situation" (Brown to Kaisen, August 27, 1928, AMNH VP Archives).

September 4, 1928 [Tuesday; page 131]

Mr. [Childs] Frick & wife were here today.
Note: Childs Frick was a Trustee of the American Museum of Natural History and a benefactor of the vertebrate paleontology program.

September 7, 1928 [Friday; page 131]

Harold Cook & wife came in today.

September 10, 1928 [Tuesday; page 131]

The German geologist [A. Penck] and the Archeologist Mr. [Fay Cooper] Cole and wife were here today [*see fig. 2.13*].
Note: Albrecht Penck was a geologist and geographer, then at the University of Berlin. Fay Cooper Cole (1881–1961), was a University of Chicago archaeologist whose expertise was in eastern North America but who took an active interest in early material. J. B. Griffin reports that in the years after Folsom, Cole often sent students out west looking for Paleoindian material (Griffin, personal communication, 1980). Both Penck and Cole had come to Folsom in response to the telegrams Brown sent out in mid-July. Cole had been unable to get away earlier: "University work and commitments prevent visit to Folsom before first week September. If expedition still there will come then. Hope Nelson or Wissler can see material while still in place" (Cole to Brown, July 25, 1928, VP/AMNH).

September 12, 1928 [Thursday; page 131]

Finished digging in the quarry today.

September 14, 1928 [Saturday; page 131]

Ernie left for Lawrence Kansas to start school there.

September 17, 1928 [Monday; page 132]

Took in 7 boxes of bones then worked in the cave. We met Harold [Cook], his wife, and Miss Wilson in Folsom.

September 21, 1928 [Friday; page 132]

Too wet to do much until noon then took in the last of the bones.

September 28, 1928 [Friday; page 132]

Mr. Brown came today. Met us in Folsom & came out. We have been working about 7 days in the cave.

October 1, 1928 [Monday; page 132]

Mr. Brown & Mr. Kaisen left for Raton on their way to N.Y. The black & white moths have been flying north all day.

October 2, 1928 [Tuesday; page 132]

Glenn and I broke camp & packed. Got away at 1 o'clock. Into Raton at 4.

October 3, 1928 [Wednesday; page 132]

Glenn pulled out at 10.

[Carl Schwachheim passed away two years later.]

Historical Archaeology of the Folsom Site

David J. Meltzer and Donald A. Dorward

Carl Schwachheim's Diary (appendix B) provides valuable insight into the nature of the work and the ebb and flow of activities over the several field seasons. However, there is also a material record of the activities that took place at Folsom from 1926 to 1928, since field crews for both the Colorado and the American museums camped on-site. In 1997 we conducted a metal detector survey in the areas of the field camps, in order to supplement the written records, see what further insight might be gained of the original investigations, their methods and tools, and the excavators, and perhaps a get glimpse into what camp life and field conditions were like in the late 1920s.

Schwachheim's camp was set up in the same place in 1926 and 1927, and relocating it proved to be quite simple, as the crew tents appear in many of the photographs taken at the time. The camp was on the North Bank of Wild Horse Arroyo, immediately across from where the Colorado Museum excavations were taking place. In both years there were at least two tents on-site, one for Schwachheim, the other presumably for other crew. In 1926, that included Frank Figgins, with occasional visits from Harold Cook; in 1927, Floyd Blair joined him on-site (fig. C.1), as did a young man named "Bob" (see appendix B).

The 1928 camp was larger than the previous years' and included some half-dozen tents. It was not immediately adjacent to Wild Horse Arroyo, but its position could be fixed with the help of a photograph taken from a high point downvalley (fig. C.2). Once the general area of the tents was located, a brief ground check confirmed that this was the right spot, as there were small nuggets of plaster of paris on the surface—remnants of the specimen preparation that took place in camp (fig. 4.6). The 1928 camp was ~70 m northeast of the excavation area and well above the arroyo, tucked away behind a small grove of trees.

Methods

These decades-old camps are now, of course, part of the archaeological record of the site, and in order to assess their nature and extent, a metal detector survey was conducted in each of the camp areas, using a Tescro Royal Sabre metal detector. The metal detector survey proceeded in ~1-m-wide, east-to-west transects across each area, continuing until no artifacts had been detected for several meters. The area of the 1926–1927 camp was vegetated, but not to the degree that it significantly hampered the ability of the surveyor to go several meters into the surrounding woodland.

However, the survey of the 1928 camp area was constrained by the denser brush and woods around its perimeter. Also complicating the survey in the 1928 camp was the near-surface presence of volcanic rocks, which occasionally had sufficient metallic components to produce a response signal. It was nonetheless possible to separate these signals from those of metal artifacts.

Whenever the instrument signaled the presence of metal, the surface was cleared of its leaf and grass litter and the artifact was exposed. The instrument was able to detect metal objects as much as 10 cm below the surface; however, most items were recovered within 4cm to 6 cm of the surface in the 1926–1927 camp and within 2 cm of the surface in the 1928 camp. Once exposed, artifacts were bagged, but then left in place for subsequent mapping. Nearly 60 items were located in the 1926–1927 camp area, and some 70 items in the 1928 camp area. Of those artifacts, roughly one-fourth, notably those with diagnostic value, were collected. These remains are listed in table C.1.

FIGURE C.1 Carl Schwachheim (right) and Floyd Blair in camp, 1927. (Photo courtesy of Carl Schwachheim Collection.)

FIGURE C.2 View west up Wild Horse Arroyo, toward Johnson Mesa, 1928. Note the road toward the site, and the 1928 field camp in the upper right. (Photo courtesy of American Museum of Natural History.)

TABLE C.1
Historic Artifacts Mapped in the Areas of the 1926–1927 and 1928
Field Camps

Item	Number
(a) 1926–1927 Field Camp	
Barbed wire	2
Button	1
Cans (canned meat, ham, lard, syrup, tea, biscuits?)	10
Flashlight spring	1
Heater parts	3
Nails	12
Pepsodent tube	1
Ring (scissor handle?)	1
Rivets	2
Staples	1
Whisk broom handle	1
Wire	3
(b) 1928 Field Camp	
Cans	4
Chain link	1
Coin (1915 half-dollar)	1
Glass bottle (catsup?)	1
Miscellaneous metal fragments	2
Nails	33
Shell casing (22 caliber)	1
Shellac can lid	1
Small tin disk, possibly a bottle seal	1
Table knife	1
Tin cup handle	1
Wire (heavy- and light-gauge bailing wire)	23

The 1926–1927 Camps

Examination of the spatial distribution of those remains (fig. C.3) reveals a clearly defined *drop zone* immediately around where the tent was placed and an equally well-defined *toss zone* in a small ravine that ran at an angle north of the tent.

Fourteen cans, all food related, were found in the toss zone. Most of these were small, three-piece cans that would hold 12 oz to 15 oz of meat. However, there were also cans that held lard, tea, syrup, and a canned ham. This last item is of particular interest, since this product was not developed until 1926 by the Hormel Company.

The amount of debris from food preparation and eating seems insubstantial and much less than would be expected given the extended periods for which Schwachheim and others lived on-site, which lasted from early May to early October in 1926 and from late May to late September in 1927 (appendix B). It is possible that these remains underestimate the amount of food in camp,

since cardboard boxes, bread wrappers, butcher paper, and cotton, burlap, or paper sacks would not likely survive. And, of course, Schwachheim or animal scavengers could have hauled his garbage off-site.

Still, the distance to the nearest grocer, the sparseness of food remains, the absence of evidence for a significant amount of cooking and eating—though we did recover a table knife and tin cup handle—suggests that there was little eating in camp, except snacking or meals made on odd occasions (fig. C.4). Although his diary is silent on the point, it appears that many of Schwachheim's meals were taken at the Crowfoot Ranch a mile or so away (Howarth to Figgins, July 5, 1926, JDF/DMNS).

In camp, Schwachheim did attend to personal matters—like brushing his teeth. His brand was Pepsodent, available in metallic tubes since the turn of the century (fig. C.5). Other items found in the survey included some of the excavation tools, including a whisk broom handle, and miscellaneous parts such as a flashlight

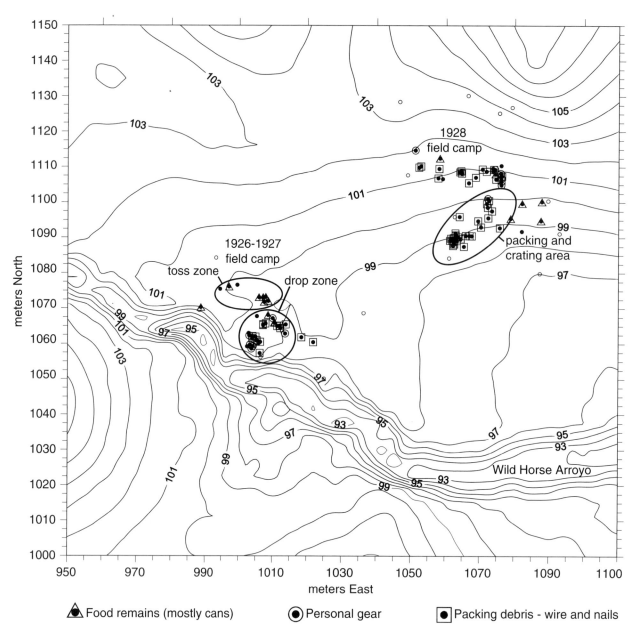

FIGURE C.3 Map of the distribution of historic artifacts from the 1926–1927 and 1928 field camps. The 1926 and 1927 tents occupied the open space in the midst of the "drop zone," and there is a shallow ravine in the area of the "toss zone." The "packing and crating" area in the 1928 field camp contained a large number of nails and segments of bailing wire.

spring and staples. There were also fragments of a kerosene or coal heater, of the sort common in household use from the late nineteenth century until after World War II. In rural areas, it was available for sale from the Sears, Roebuck and Company catalog. Given the weather Schwachheim experienced in late spring or early fall, this item would have come in handy for heating the tents.

Rounding out the inventory were a couple of sections of antique barbed wire, which by the 1920s was considered harmful to cattle and was probably long out of produc-

tion. These sections may predate the excavations and be unrelated to the field camp. There were also some two dozen nails, all nearly the same size, likely used in crating plaster-jacketed bone specimens or in tent frame construction.

The 1928 Camp

Judging by contemporary field photos (e.g. fig. 4.6), there were at least five tents on-site in 1928 (fig. C.6). As in previous seasons, the crews likely took their meals at the

FIGURE C.4 Floyd Blair cooking in camp, 1927. The kettle over the fire was left on site, and was recycled as a datum cover by later field crews. (Photo courtesy of American Museum of Natural History.)

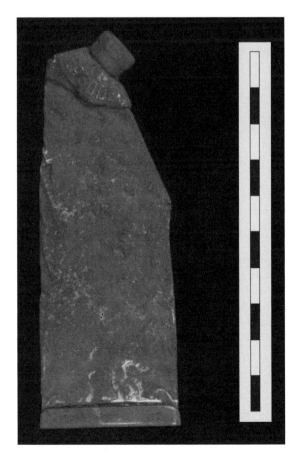

FIGURE C.5 Schwachheim's Pepsodent toothpaste tube recovered in the 1926–1927 field camp "drop zone." (Photo by D. J. Meltzer.)

FIGURE C.6 Camp life, 1928. (Photo courtesy of Carl Schwachheim Collection.)

Crowfoot ranch house, so there is little food debris in this area: just four food cans were recorded in the survey—less than from the prior year's camp. But as in years past, they appear to have occasionally eaten in camp. A complete, 10-sided, clear catsup bottle was found, marked on the bottom "PACKED BY CALPACK CORP," with a five-pointed star: California Packing Corporation was formed in 1916 with the merger of several California canneries, which would ultimately (1967) become Del Monte.

Occurring in abundance on the margin of the 1928 camp was the debris from the preparation of the bison bones for shipment to New York. The archives record that 8,300 lb of plastered bone was shipped to New York in 35 wooden crates (chapter 4). Our metal detector survey in the area where the photograph in fig. 4.6 was taken located scores of nails and cut sections of bailing wire, suggesting that the shipping crates were built on-site (fig. C.3). They were substantial crates, judging by the size (16D) of some of the nails.

The only artifact of significance was a 1915 half-dollar, found in the woods behind the tents. It likely fell out of a pair of pants, probably for the obvious reason, and almost certainly the person who lost it was very unhappy about it. In 1928, 50 cents was a considerable sum.

Sediment Mineralogy and Bone Preservation

Todd A. Surovell

This appendix reports the results of Fourier-transform infrared spectroscopic (FTIR) analysis of sediment mineralogy from the Folsom site and its implications for bone preservation and distributional analysis. The study was initiated in 1998 when a diffuse bonebed was uncovered in the M17 block (N1030-N1034), but the M15 (N1020-N1024) block only 10 m to the south contained only a very few fragments of bone. Although the presence of bone in one part of the site indicated conditions conducive to the preservation of bone, the scarcity of bone in another implied nothing about preservation since bone may never have been there in quantity or, alternatively, was once there and had since been lost to diagenesis, namely, mineral dissolution. Therefore, some independent means of assessing variable preservation conditions must be undertaken to aid in the interpretation of osteological distributions.

FTIR analysis of sediment mineralogy is one method for addressing site geochemistry and bone preservation. The basic premise is that the sediments that bury bones have undergone the same diagenetic processes as the bones themselves. Sediment mineralogy can then be used as a proxy for site geochemistry and preservation conditions. This method was initially developed and applied at a number of cave sites in the Mediterranean region, namely, Kebara, Hayonim, and Tabun caves in Israel and Theopetra Cave in Greece (Karkanas et al. 1999, 2000; Stiner et al. 2001; Weiner, Goldberg, and Bar-Yosef 1993; Weiner et al. 1995). In these sites, the mineralogical composition of sediments was found to correlate with bone preservation conditions. A series of authigenic phosphate minerals occurs in diagenetic chains of transformation, and identification of the dominant mineral species present in particular portions of these caves allowed researchers to estimate the relative degree of diagenesis affecting the site across space. It was found that sediments in areas conducive to bone preserva-

tion contained calcite and/or dahllite (carbonated hydroxyapatite), but in areas where bone had been lost, these minerals were absent, being replaced by less soluble phosphates, or in cases of extreme diagenesis, only relatively insoluble silicates remained (Karkanas et al. 1999, 2000; Schiegl et al. 1996; Stiner et al. 2001; Weiner, Goldberg, and Bar-Yosef et al. 1993; Weiner et al. 1995). Therefore, bone is expected to be preserved in portions of sites where sedimentary minerals of equal or lesser solubility than bone mineral itself are found. If such minerals are absent, it can be inferred that they have been lost to dissolution by groundwater, as would any bone that might have been present. Of course, this research is only possible for sites where the mineralogy of sediments is conducive to such an analysis. Ideally, mineral assemblages should span the solubility level of bone or, minimally, contain minerals of lower solubility than bone hydroxyapatite.

The present study is the first to apply this method to an open-air context. The technique is perhaps most effective in cave settings since phosphate-charged ground waters (created by interaction of water with bat and bird guano) create diverse mineralogies that not only span the solubility level of bone hydroxyapatite, but also contain sedimentary apatites. Also, the aforementioned sites have considerable time depth, generally exceeding 40,000 years, during which time, diverse authigenic minerals are formed in quantities detectable by infrared spectroscopy. Sediment mineralogy of the Folsom site, however, is relatively simple, being comprised almost entirely of clays, quartz, and calcite. Fortunately, the presence, absence, and relative concentration of calcite, which is less soluble than bone apatite, can serve as a rough indicator of bone preservation conditions. Generally, if calcite is present, bone should be present as well. If calcite is absent, relatively acidic conditions are implied, which decreases the

Sediment Mineralogy Data from FTIR Analyses of Folsom Site Sediments

Sample No.	Provenience			FTIR Data		
	Northing	Easting	Elevation	Calcite	Silicates	Calcite:Silicates
1	1031.657	997.572	98.898	0	0.174	0
2	1031.638	997.578	98.808	0	0.211	0
3	1031.630	997.595	98.642	0	0.113	0
4	1031.653	997.596	98.594	0	0.141	0
5	1031.646	997.616	98.523	0	0.310	0
6	1031.653	997.618	98.397	0.012	0.286	0.041
7	1031.663	997.624	98.305	0	0.217	0
8	1031.665	997.636	98.207	0.022	0.134	0.162
9	1031.655	997.648	98.120	0.034	0.199	0.172
10	1031.659	997.655	98.029	0.017	0.122	0.141
11	1031.663	997.681	97.948	0.090	0.296	0.304
12	1031.620	998.036	97.857	0.072	0.086	0.835
13	1031.619	998.028	97.753	0.125	0.181	0.691
21	1034.250	997.997	98.105	0.135	0.243	0.556
22	1034.238	998.022	98.054	0.136	0.195	0.697
23	1034.230	998.026	98.011	0.202	0.248	0.815
24	1034.250	998.022	97.947	0.291	0.233	1.249
25	1034.263	998.029	97.897	0.198	0.187	1.059
26	1034.283	998.027	97.859	0.383	0.244	1.570
27	1034.272	998.022	97.813	0.390	0.351	1.111
28	1034.261	998.025	97.764	0.541	0.355	1.524
29	1034.261	998.027	97.710	0.263	0.294	0.895
30	1034.254	998.027	97.653	0.139	0.196	0.709
31	1034.260	998.029	97.608	0.135	0.228	0.592
32	1034.250	998.030	97.550	0.186	0.322	0.578
33	1034.260	998.030	97.480	0.119	0.292	0.408
34	1034.260	998.030	97.430	0.281	0.331	0.849
100	1024.250	998.250	97.325	0.090	0.122	0.741
101	1024.250	998.250	97.275	0.149	0.214	0.696
102	1024.250	998.250	97.825	0.093	0.135	0.689
103	1024.250	998.250	97.525	0.119	0.222	0.537
104	1024.250	998.250	97.775	0.085	0.120	0.705
105	1024.250	998.250	97.375	0.164	0.130	1.262
106	1024.250	998.250	97.975	0.109	0.130	0.838
107	1024.250	998.250	97.575	0.130	0.100	1.300
108	1024.250	998.250	97.875	0.071	0.110	0.646
109	1024.250	998.250	97.925	0.061	0.098	0.619
110	1024.250	998.250	97.625	0.099	0.091	1.091
111	1024.250	998.250	97.475	0.135	0.166	0.813
112	1024.250	998.250	97.725	0.108	0.140	0.771
113	1024.250	998.250	97.675	0.126	0.127	0.992
114	1024.250	998.250	97.425	0.125	0.111	1.126
115	1033.250	999.250	98.025	0.174	0.217	0.802
116	1033.250	999.250	97.975	0.090	0.157	0.573
117	1033.250	999.250	97.925	0.159	0.176	0.903
118	1033.250	999.250	97.875	0.150	0.153	0.980
119	1033.250	999.250	97.825	0.180	0.195	0.923
120	1033.250	999.250	97.775	0.251	0.248	1.012
121	1033.250	999.250	97.725	0.211	0.263	0.802

Sample No.	Provenience			FTIR Data		
	Northing	Easting	Elevation	Calcite	Silicates	Calcite:Silicates
122	1033.250	999.250	97.675	0.150	0.169	0.888
123	1033.250	999.250	97.625	0.110	0.166	0.663
124	1033.250	999.250	97.575	0.124	0.206	0.602
125	1033.250	999.250	97.525	0.199	0.223	0.892
126	1033.250	999.250	97.475	0.177	0.168	1.054
127	1033.250	999.250	97.425	0.176	0.159	1.107
200	1030.213	997.559	98.653	0.006	0.327	0.018
201	1030.200	997.568	98.548	0	0.243	0.001
202	1030.199	997.582	98.468	0	0.177	0
203	1030.174	997.589	98.361	0.022	0.274	0.078
204	1030.174	997.589	98.261	0.033	0.320	0.104
205	1030.174	997.589	98.181	0.043	0.338	0.128
206	1030.174	997.589	98.121	0.043	0.218	0.197
207	1030.174	997.589	98.001	0.059	0.338	0.173
208	1030.174	997.589	97.910	0.087	0.215	0.404
209	1030.196	998.014	97.748	0.107	0.161	0.665
210	1030.196	998.014	97.658	0.174	0.157	1.108
211	1030.196	998.014	97.568	0.204	0.222	0.919
212	1030.196	998.014	97.468	0.183	0.222	0.824
213	1030.196	998.014	97.368	0.140	0.197	0.711
214	1028.508	997.870	98.488	0	0.437	0
215	1028.508	997.870	98.398	0.034	0.272	0.125
216	1028.508	997.870	98.288	0.042	0.285	0.147
217	1028.508	997.870	98.208	0.049	0.259	0.190
218	1028.508	997.870	98.108	0.025	0.414	0.060
219	1028.508	997.870	98.018	0	0.253	0
220	1028.508	997.870	97.938	0.027	0.282	0.095
221	1028.508	997.870	97.858	0.050	0.212	0.235
222	1028.508	997.870	97.768	0.135	0.190	0.711
223	1028.508	997.870	97.678	0.173	0.264	0.655
224	1028.508	997.870	97.598	0.107	0.182	0.588
225	1028.508	997.870	97.518	0.101	0.183	0.552
226	1028.508	997.870	97.418	0.126	0.181	0.696
227	1028.508	997.870	97.338	0.152	0.220	0.691
228	1028.508	997.870	97.248	0.026	0.181	0.146
229	1028.508	997.870	97.178	0.129	0.282	0.457
230	1028.508	997.870	97.088	0.052	0.110	0.472
231	1028.508	997.870	97.018	0.139	0.256	0.543
232	1028.508	997.870	96.918	0.107	0.158	0.677
233	1028.508	997.870	96.858	0.103	0.197	0.523
234	1028.508	997.870	96.788	0.088	0.157	0.557
235	1028.508	997.870	96.698	0.092	0.137	0.672
236	1026.800	997.825	98.536	0	0.161	0
237	1026.800	997.825	98.456	0.031	0.257	0.120
238	1026.800	997.825	98.396	0.024	0.195	0.123
239	1026.800	997.825	98.306	0.093	0.206	0.450
240	1026.800	997.825	98.236	0.231	0.368	0.628
241	1026.800	997.825	98.146	0.124	0.133	0.932
242	1026.800	997.825	98.086	0.262	0.133	1.970

(*continued*)

Sample No.	Provenience			FTIR Data		
	Northing	Easting	Elevation	Calcite	Silicates	Calcite:Silicates
243	1026.800	997.825	98.016	0.088	0.106	0.833
244	1026.800	997.825	97.956	0.171	0.176	0.972
245	1026.800	997.825	97.856	0.232	0.210	1.105
246	1026.800	997.825	97.786	0.120	0.192	0.625
247	1026.800	997.825	97.706	0.195	0.201	0.970
248	1026.800	997.825	97.616	0.088	0.206	0.427
249	1026.800	997.825	97.526	0.221	0.320	0.691
250	1026.800	997.825	97.436	0.110	0.136	0.809
251	1026.800	997.825	97.356	0.111	0.218	0.509
252	1026.800	997.825	97.276	0.052	0.233	0.221
253	1026.800	997.825	97.186	0	0.438	0
254	1026.800	997.825	97.096	0.078	0.267	0.292
255	1026.800	997.825	96.996	0.061	0.151	0.401
256	1026.800	997.825	96.926	0.091	0.207	0.442
257	1026.800	997.825	96.866	0.087	0.171	0.506
258	1026.800	997.825	96.746	0.088	0.206	0.427

likelihood of bone preservation. Therefore, quantities of calcite relative to silicates can serve as a rough index of bone preservation conditions, assuming equivalent relative concentrations as a starting point for each sample. This study is also complicated by the presence of potential secondary carbonates. For example, if acidic conditions at one point in time removed any sedimentary carbonates and bone present, followed by relatively alkaline conditions during which time secondary carbonates are deposited, this situation could create the illusion that geochemical conditions through time were conducive to bone preservation. Therefore, it is critical that some consideration of the source of carbonates be made.

Methods

During the 1998 and 1999 field seasons, a total of 114 sediment samples were taken from seven columns (table D.1). Three columns were taken from the west excavation profiles of the M17 and M15 excavation blocks (sample nos. 1–13, 21–34, and 200–213), two columns were taken from auger holes in the unexcavated portion of the transect (sample nos. 214–235 and 236–258; N1024-N1029), and two columns were taken as aliquots of sediment samples removed during excavation from units M15-24 and M17-20 (sample nos. 100–127; N1024 E998, and N1033 E 999). Samples were taken every 10 cm from profiles, approximately every 8.5 cm in auger holes, and every 5 cm from

excavation. The samples form a north-south transect across the site between approximately N1024.25 and N1034.25 and approximately E997.5 and E999.25. Samples were also taken of the Smoky Hill shale bedrock from the south bank of Wild Horse Arroyo adjacent to the site.

All samples were prepared by the KBr pelleting technique (Fridmann 1967; Smith 1997) Sample preparation began by removing two or three small peds from the sample. These were pulverized into a fine powder and homogenized in an agate mortar and pestle. Each sample was reduced to a few hundred micrograms by repeated removal of excess of material with a lab wipe and regrinding. Approximately 50 mg of KBr was added and mixed by light grinding. This mixture was then pressed into a 7-mm-diameter pellet using a Spectratech Mini-Press (International Crystal Laboratories). Infrared analyses were performed at the Department of Anthropology, University of Arizona, with a Midac Corp. Prospect-IR FTIR spectrometer. Instrument settings were as follows: scans = 32; resolution = 4 cm^{-1}; gain = auto; range = 4,000–400 cm^{-1}; and mode = absorbance. The empty sample chamber was used as the background spectrum.

Spectral analyses were performed with Grams 386 software (Galactic Industries Corp). Semiquantitative estimates of calcite concentration are based on peak-height ratios on absorbance spectra using the 876-cm^{-1} carbonate peak and the 797- to 799-cm^{-1} silicate peak, which is present in both quartz and clays (fig. D.1). All peak heights are

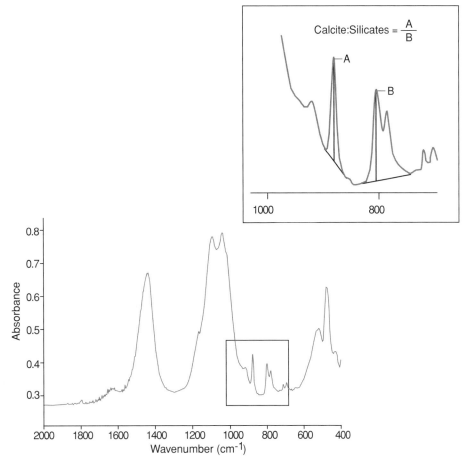

FIGURE D.1 The derivation of the calcite:silicates index on a typical FTIR absorbance spectrum of Folsom site sediments. The index uses the ratio of peak heights of the 876 cm-1 carbonate and the 797–799 cm-1 silicate peak, present both in quartz and clays. Both heights are measured above the baselines as shown

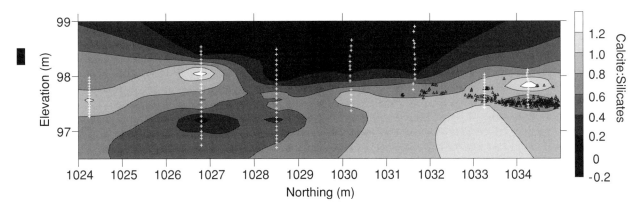

FIGURE D.2 Extrapolated calcite:silicates values for the sampling transect (E997.5-999.25). Locations of samples are shown as white crosses, and the piece-plotted bones in proximity of the transect are shown as black triangles. Darker areas of the map show low calcite concentrations and imply poor bone preservation. Lighter areas show greater calcite concentrations and imply better bone preservation. Plot made with SURFER v. 7.0 (Golden Software) using a radial basis function extrapolation with an anisotropy setting of 2.4 to account for uneven sampling intervals with respect to northing and elevation.

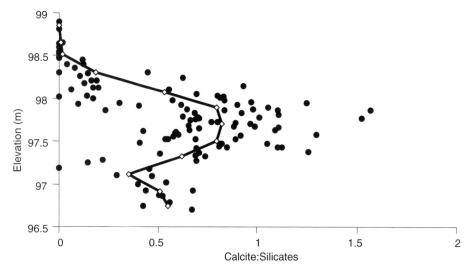

FIGURE D.3 Carbonate:silicates values by elevation for all sediment samples. The line depicts mean values for arbitrary 20 cm-elevation intervals.

measured above the baselines shown in fig. D.1. While this measure should correlate well with the actual mass ratios of these minerals, it cannot be directly translated into mass ratios without empirical calibration. Consequently, in this study, the calcite:silicates index is only used as a relative measure.

Results

As mentioned above, all samples, including bedrock, were comprised of varying amounts of calcite, clay, and quartz (table D.1). In fact, site sediments and bedrock appear to be indistinguishable, at least with respect to the mineral species present. Occasional clasts of basalt were encountered in sediment samples, but they generally did not contribute significantly to sediment spectra. Clays from the site compare favorably with a Ca-montmorillonite standard.

Calcite concentrations varied dramatically within the site. A number of samples from the McJunkin formation (stratum *m2*) lacked calcite entirely, evidencing the relatively acidic conditions that existed during its formation in Late Holocene times (~4000 B.P.). Some McJunkin samples did contain small amounts of calcite, generally in fine

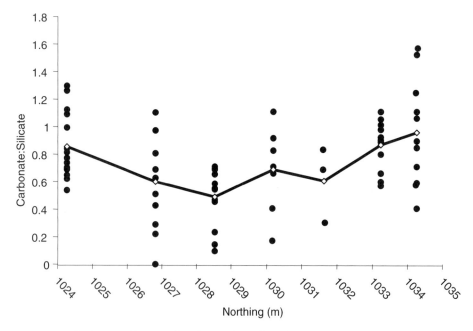

FIGURE D.4 Carbonate:silicates values by northing for sediment samples along elevation 97–98 m (approximate elevation of bone bed). The line depicts mean values for each sample column.

stringers, suggesting secondary deposition. All samples within the bone-bearing *f2* contained some calcite, with the highest concentrations occurring within colluvial debris flows with significant input of bedrock debris. Calcite concentrations generally increased beneath the lower contact of *m2*, reaching maximum levels at elevations between 98 and 97.5 m (figs. D.2 and D.3). Calcite concentrations beneath 97.5 m drop off to intermediate levels. Across the sampling transect, calcite concentrations tend to be higher on its northern and southern extremes and lower toward its center (figs. D.2 and D.4). Although some variability in calcite concentrations across space is certainly a depositional phenomenon, much of the patterning is likely explained by leaching that occurred in the late Holocene during the formation of the *m2* stratum. The upper and central portions of the transect seem to have experienced the greatest leaching of carbonates, while the lower, northern, and southern extremes may have been somewhat buffered from leaching, or little impacted, due simply to their greater distance from acidic groundwater at that time.

Discussion

This patterning has clear implications for spatial variation in bone preservation. Foremost, all areas of the sampled transect within stratum *f2* appear to have retained conditions favorable to bone preservation throughout the site's history. The scarcity of bone in the M15 block, which contains relatively high calcite concentrations, therefore, can truly be considered a scarcity of bone deposition as opposed to being a product of poor preservation conditions. Bone preservation, however, should be poor in the center of the transect as opposed to its northern and southern extremes (figs D.2 and D.3). In fact, as one moves southward from N1033, the condition of bone declines, providing support that some leaching of bone apatite has occurred. The central portion of the transect was likely most impacted by late Holocene geochemistry since the bonebed rises toward stratum *m2*, and stratum *m2* cuts more deeply into late Pleistocene sediments in this portion of the site. It is further predicted that the unexcavated portion of the intervening transect (N1025–N1029) would likely have the most poorly preserved bone.

Defining Folsom

Theme and Variations

David J. Meltzer

Like the flurry of taxonomic debate that followed the discovery of the Folsom bison (chapter 7), the discovery of Folsom points sparked a decade-long discussion over matters of projectile point typology. However, the two differed in several significant ways. For one, questions of bison taxonomy involved the repeated examination of the Folsom site specimens; the projectile points from Folsom did not figure so prominently in the typological debate or its resolution. In the debate over point typology, less was riding on the outcome—at least at the site itself. Knowing the number of bison taxa present at Folsom potentially had a bearing on whether one or more kills took place (chapter 7); cleanly defining the Folsom point type had no such interpretive consequences. Finally, and as noted in chapter 8, the typological trouble over the Folsom point type did not begin at this site, but only several years later, as additional forms of Paleoindian projectile points were reported from the western Plains and other portions of North America (Howard 1935; Renaud 1931, 1932, 1934; Roberts 1935, 1937, 1939; Shetrone 1936).

This burgeoning sample of points of broadly similar form and technology were presumably of similar age, but enough variation was apparent to suggest that they might not be of identical antiquity. Moreover, it was quickly becoming clear that the "Folsom" label was inadequate to the task of categorizing all these forms, especially the lanceolate points from the "Yuma District" of the High Plains of eastern Colorado that lacked fluting (e.g., Cook 1931b; Renaud 1931, 1934). Once the variation on the generalized lanceolate theme had become fully apparent, debate began over just what constituted a Folsom point and how it differed from but was historically related to unfluted lanceolates ("Yuma" points), as well as the so-called "Generalized Folsom" or "Folsom-like" points (Holliday and Anderson 1993). The latter, which were larger and less carefully made fluted forms, had, by the

mid-1930s, been found at Clovis and Dent and across much of eastern North America (e.g., Cotter 1937; Figgins 1934, 1935; Howard 1935:105–123, 1943; Krieger 1947; Renaud 1931, 1932, 1934; Roberts 1935:8–9, 1937:161, 1939, 1940:56; Shetrone 1936; Wormington 1948).

By the mid-1930s, Folsom had become a type everyone believed they could recognize but about which few could agree on a definition (LeTourneau 1998a:57–59), while Yuma had become a "catch-all" category that contained a wide variety of dissimilar unfluted forms that included, it seemed to Roberts, just about anything that was not a "true Folsom or a barbed and tanged arrowhead of the recent Plains type" (Roberts 1936:21; Wormington 1948:3, 8). This wasn't entirely polemic on Roberts's part: even some Folsom and "Generalized Folsom" points fell into Renaud's Yuma types (Howard 1935:110; also Figgins 1935:2; Roberts 1936:21, 1940:61–62).

The effort to resolve what was variously termed the "Folsom Problem," "Folsom-Yuma Problem," or "Yuma Problem" (e.g., Howard 1935:110; Roberts 1936:21, 1940:61), that is, to tease apart the typology and age of Folsom points and their historical and chronological relationship with other large lanceolates, fluted and unfluted, lasted through much of the 1930s and into the 1940s (Wormington [1948] provides a contemporary summary; see also LeTourneau [1998a], who takes a different approach than used here). It is useful to explore that discussion, albeit briefly and selectively, for, as noted in chapter 8, that debate helped clarify (or not) what was and was not Folsom, and because understanding that debate helps explain why, ironically, there emerged no definition of the Folsom type point from the Folsom type site.

If the Folsom-Yuma Problem had a root cause, it was in the work of E.B. Renaud, this despite the fact that his early discussions of Folsom and Paleoindian projectile points were characterized by flashes of remarkable insight.

Moreover, his classification scheme at least began from a clear and explicit theoretical foundation: Renaud knew what he should do and wanted to do. Unfortunately, his attempt to define the types and subtypes of Folsom and Yuma points, including those points from Folsom itself (e.g., Renaud 1931, 1932, 1934), in no small measure led to the confusing terminological morass into which these types (especially Yuma) quickly sank.

Complicating the Folsom-Yuma Problem, leaving aside strictly typological matters, was that it was no easy task to resolve the stratigraphic and archaeological relationship among these various point forms. It was generally believed that all possessed comparable antiquity; they were clearly different from points used by more recent archaeological groups. Yet, with perhaps one exception (Cook 1931b), Yuma points were known only from surface collections, primarily from Dust Bowl blowouts in eastern Colorado, and there was no stratigraphic information on sequence or contemporaneity. In a few instances, Yuma points were alleged to be associated with extinct fauna, notably bison and mammoth (e.g., Barbour and Schultz 1932; Bell and Van Royen 1934; Cook 1931b). However, the association proved to be little more than that both points and bones were found in the same blowouts, along with material of far more recent origin, making any associations suspect (Figgins 1934:2–3, 1935:3–4; Renaud 1931:16; Roberts 1935:9; Wormington 1948:3–5).

Renaud's efforts at developing a "descriptive terminology" and "basic typology" of Folsom and Yuma points began in the early 1930s, after examining the material from Folsom and some 50 points in private collections in eastern Colorado, primarily those of the Andersen family (LaBelle 2002). His initial classification (Renaud 1931) had three main elements: *point shape* or overall morphology, whether parallel-sided, tapered, triangular, or stemmed; *base geometry*, whether straight, concave, concave but "squarish," or convex; and *flaking of the blade*, the degree and frequency of pressure retouch flakes: and *flaking of the base*, basal thinning and fluting, which he believed to be historically related. Renaud spoke of each of these elements and their variants as separate *types*: an individual point had both an overall shape "type" and a base "type" and flaking "type." For the most part these individual "types" did not overlap, but too many did: a point could be stemmed, but also have a tapered shape. Others were somewhat ill defined, as, for example, his attributes of blade and base flaking patterns, which invoked categories such as "fine," "abrupt," "extensive," and "irregular," supplemented by metric data from his sample on flake spacing along the blade and length up from the base (Renaud 1931:7–12).

Although he seemed to understand that his "types" were actually attributes of objects that, when intersected, could be used as the basis for defining distinct classes (e.g., Lyman, O'Brien, and Dunnell 1997:6), he did not explicitly treat the matter in this "top-down" manner. Nor

would that have been entirely possible, since the variants of his attributes ("types") were not always mutually exclusive and nonoverlapping. Instead, he took a "bottom-up" approach and examined individual points, as, for example, the points from Folsom, and then observed generally where they fell according to his separate blade and base types. Thus, Folsom points were shape type 2 (tapers toward base), and base type C (concave, but with a concavity that "is squarish or irregular, not rounded or notched as in type B"), with a longitudinal groove being "the most typical and constant" characteristic of the point (Renaud 1931:14).

It was not a bad start, all things considered, but because his classification scheme was object-driven and more a description of extensionally defined types, rather than being based on intensionally defined classes (*sensu* Dunnell 1971; see Lyman, O"Brien, and Dunnell 1997:7; also LeTourneau 1998a:58), he hit the inevitable wall of how to deal with objects that did not fit cleanly into one of his "types" (elements), and there were many such objects. Yet, instead of retooling the attributes of his classification, he simply created additional ad hoc "types" or, more correctly, attributes, using variables that were neither exclusive from or consistent with his existing ones. In some cases, this produced new, single-item types (Figgins 1935:2). Thus, his blade shape type 4 was created solely to "take care of a couple of points from the Yuma district," and was defined by stem form rather than overall shape, as his shape types 1–3 had been defined (Renaud 1931:8). The result of all this was a classification in which a particular object could fall into one or more types, with Yuma points spread across the typological terrain: not surprising, given the great range of variation in the points he was assigning to the Yuma category.

Although identifying Yuma point types proved to be a classificatory challenge, doing so at least highlighted for Renaud just how distinct Folsom was, and revealed to him a historical relationship between Yuma and Folsom. As he judged the matter, the essential attributes differentiating the Folsom type were the "long and broad lengthwise groove, more or less fine marginal retouching between the edges and the lateral ridges, [and] concave base with often large and sharp base points" (Renaud 1931:12). Yuma points might have some of those same attributes, but only points with all of those attributes could rightly be called Folsom and, thus, be assumed to represent the same people and culture and be of the same age.

After all, as he observed, Solutrean points from western Europe shared some of those same attributes such as fine marginal retouch (also Renaud 1934:11), but that did not make them Folsom points. To "promiscuously" apply a type label like Folsom to other lanceolates because they had one or more of those attributes would be "unscientific" and "lead to confusion and misunderstanding" (Renaud 1931:13, 15), as was clearly happening with his ill-defined Yuma points.

That Yuma points shared some of the morphological attributes of Folsom points was, nonetheless, of considerable interest, for in those similarities Renaud (1931:12) detected "the successive stages of typologic advance culminating in the typical and specialized Folsom point." Most significant in this regard was basal thinning, which he saw as a means of facilitating hafting. Renaud detected in his sample of Yuma points a range in the length of basal thinning flakes. Assuming that increase represented an improvement in knapping skill over time and thus was temporally sensitive, then fluting marked the climax of a process of technological evolution, the culmination of the "complete technical skill of a difficult art" (Renaud 1931:11). In turn, progressively longer flutes marked "the various steps of the evolution from first attempts to final exaggeration" (Renaud 1934:3).

Renaud believed that Yuma points revealed the "morphologically . . . earlier phases of a progressive evolution," a conclusion he pronounced "logical and acceptable on the basis of comparative typology" (Renaud 1931:11–12; also 1934:2). Still, he admitted that it remained "purely theoretical." There was "no concrete means to establish such a claim," lacking as he was geological or stratigraphic evidence of the greater antiquity of Yuma points (Figgins 1934:2; Renaud 1931:15). Nor could he otherwise demonstrate that points with shorter basal thinning flakes were older than points with longer ones. Harold Cook (1931b:103), who recorded many Yumas and had been over much of the same eastern Colorado ground, was unconvinced and, instead, considered Folsom and Yuma points as contemporaneous.

Although not a defining element of his point classification, Renaud (1931:15) recognized that a "reasonable fluctuation in proportions is, of course, expected according to the size of the pieces." On the basis of his initial examination of the points from Folsom, as well as others then available, he described a "typical Folsom point" as averaging

> 22 to 23 mm at maximum breadth, 16 to 17 mm [width] at the base, 4 mm in thickness, 10 to 11 flakes or retouches along the edges, although great variations are possible, and has a longitudinal flake, usually on both faces, on at least two-thirds of the length of the point and 13 to 14 mm wide. (Renaud 1931:15)

He observed variation in the range and average length of the points (Renaud 1931:14). But because so few of the points he examined were complete, he did not consider those values to be especially informative, save in showing that Folsom points were generally not as long as Yuma points and confirming his suspicion that fluting weakened the blade and made them vulnerable to breakage (Renaud 1931:13, 1934:9).

By 1934 Renaud (1934:6) had amassed a sample of a thousand points, of which 237 were Folsom and the remainder Yuma. As far as he was concerned, the classification he'd devised in 1931 was based on a collection so

representative and types "so stable that no correction was needed in subsequent years." To be sure, there was occasionally a need to designate a few new subtypes, marked by differences in the "length of the longitudinal groove," in order to take into account the newly discovered "generalized Folsoms" from Clovis and Dent (Renaud 1934:2). Such is the fate of a classification that is based on extensionally defined units, for the addition of new specimens almost inevitably requires a modification of the type definition (Lyman, O'Brien, and Dunnell 1997:8).

The classification, as it appeared in 1934, however, was not as similar to his initial classification as he claimed. Although he continued to refer to "blade" types, in fact his scheme had evolved into one where the initial broad dichotomy was drawn between unfluted Yuma ("blade" types 1–4, with a total of seven subtypes) and fluted Folsom forms ("blade" type 5, with three subtypes), for a total of 10 separate "blade" types; unfluted versus fluted as an essential dichotomy first emerges in Renaud (1932). Variants within Yuma were then identified by overall point morphology, while variants within Folsom were defined by the length of the flute: type 5a, extending about one-third the length of the point; type 5b, extending about one-half the length of the point; and type 5c, extending two-thirds or more of the length of the point. But because he was mixing dimensions in his classification, Folsom points and other fluted points could be just as easily classified into one of his several Yuma types, as Howard (1935:112) later observed. Renaud retained his base types and flaking patterns essentially as they were in 1931, but by 1934 base type had mostly become a secondary attribute, used to describe different forms found on Folsom and Yuma types, and the only flaking pattern he made use of was fluting. Hence, his Yuma types were shown as point outlines, and the Folsom types as outlines with roughly drawn flutes (Renaud 1934:plate 1).

Renaud (1934:8) explored how points fell out when classified by intersecting "blade" and base types, and as one might expect, the results were not altogether clean: virtually all 10 "blade" types were represented in each of the base types. Nonetheless, the frequency data convinced him there were certain patterns of association or lack thereof, such that virtually no Yuma points had base type C (concave and squarish), and no Folsom points had straight bases (Renaud 1934:8–9).

Perhaps even more explicitly than he had in 1931, Renaud emphasized the historical contiguity of basal flaking, for now he had the newly discovered "generalized Folsom" points to serve as a typological bridge between unfluted Yuma points and the fully fluted "true" Folsom points. Believing, as he did, that full-length Folsom fluting represented the final exaggerated development of a long evolving "industry," then clearly those "generalized Folsoms" were part of a series:

> It seems then that our subdivision of the general Folsom type into three sub-types on the basis of the development

of the longitudinal groove is substantiated by a systematic comparison of the three principal dimensions [length, width, thickness], in which we observed a very fair correlation, and, in every case, the true original Folsom type took its place at the end of the series as if actually it was the last refinement of a general class in morphologic evolution. (Renaud 1934:11)

Looking at point length in particular provided further confirmation that fluting was a process that reduced the life span of the point, for Renaud (1934:9) detected, across the three Folsom types (5a–5c) in his sample,

> a progressive shortening of the blade as we pass from 5a to 5b and especially to 5c, suggesting an evolution of length parallel in inverse direction to the growing of the longitudinal groove. The technical difficulty, the increasing danger of spoiling the piece, seems to have much reduced the length of the points. The correlation is one more argument in favor of the true Folsom type being evolved from the longer and simpler Yuma type through the progressive stages designated as 5a and 5b.

In 1934, Renaud (1934:5) blithely asserted that "the designation and use of the terms, Folsom and Yuma, has never been challenged publicly or in print by archaeologists knowing the accepted practice of naming artifact types, industries or cultures." He was mostly correct. But that changed within a matter of months of his saying so.

Figgins had been tracking Renaud's work and was plainly unhappy with it. Although he did not admit as much explicitly, he seemed particularly irked that Renaud gave pride of antiquity to Yuma points, and not Folsom points. Figgins (1934:2, 1935:6) saw the "technique" of Folsom and Yuma point production as completely different, and Folsom as a separate culture with its own "long line of evolutionary progression from cruder types," not merely the "final stage of an earlier [Yuma] type":

> Since there is no material similarity in form and detail between Folsom and Yuma artifacts, nor evidence of intergradation of their widely divergent characters, the ground upon which a relationship is proposed appears to be extremely insecure. . . . Their relative position is purely conjectural.

Indeed, if the criterion of "progressive improvement" was measured by the "quality of workmanship," then the more finely crafted Yuma points were superior and must *follow*, not precede, Folsom points in time (Figgins 1934:5–6, 1935:7).

As Figgins (1934:3) saw it, the reason Renaud and others (e.g., Bell and Van Royen 1934) were mistaken about a similarity and historical link between Yuma and Folsom was that they were unfamiliar with the "characters which distinguish Folsom artifacts." It was at this juncture that Figgins (1934:3) provided his own definition of Folsom (chapter 8) as having "wide spalls removed from the sides, beginning at the base and sometimes extending quite to the tip, producing a hollowed or 'fluted' effect. The bases

are concave, often to a depth of a quarter or an inch or more and thus forming ear-like backward projections. In practically all cases Folsom artifacts are widest forward of a point midway of their length, and their width sometimes equals, or exceeds, half of their length." On its face, Figgins's description of a Folsom point seems little different from Renaud's. But Renaud had insisted that all the identified attributes had to be present in order for a point to be given that type designation. Figgins (1935:2) allowed far more latitude in his concept of "type"—at least with regard to Folsom.

As noted in chapter 8, Figgins was not offering a type definition, just identifying characteristics with which to identify the form. He was clearly cognizant of the variation within Folsom points, even to the point of accepting the possibility that Folsom points need not be fluted.

But if there was such variability within Folsom points, then what constituted the type? The diagnostic characteristics Figgins (1934:3, 1935:4–5) thought most important were the "relative proportions of width to length, and the reduced width backwards of a point forward of the midsection," as well as the edge retouch and deeply concave bases. Fluting might or might not be present, although it was in many instances (Figgins 1935:2–3, 6). Identifying multiple attributes proved somewhat awkward as a type definition, for by them any relatively wide, tapered, even unfluted lanceolate could be considered Folsom.

Yet, despite the wide variability he was willing to accept within the type, Figgins (1934:4), with Renaud clearly in his sights, insisted that "not one of the . . . described characters [in Folsom points] are present in Yuma artifacts." And that was clear since he saw one attribute as being a necessary, sufficient, or universal feature of Yuma points, and that was their distinctive basal morphology (Figgins 1935:6). Folsom points simply "do not have squared bases, rarely paralleling edges, and never taper forward from a maximum basal width" (Figgins 1935:6). Perhaps. But given the wide range of morphological variability in Yuma points, Figgins's insistence (1935:5–6) that all Yumas were parallel sided, were widest at their base, or had square bases rang hollow.

Others tried to define what constituted a Folsom point, as opposed to a Yuma or "Generalized Folsom," on the heels of Renaud and Figgins, notably E. B. Howard and Roberts. Howard approached the problem from the vantage of his work at Clovis and, thus, had a contrasting set of objects to sort. Because of that, and his appreciation of the fact that some of the defining attributes of a "true" Folsom were obviously not restricted to just those points found at Folsom, he asked rhetorically whether "we are not justified in applying the name [Folsom] to any points found elsewhere which exhibit all of the same characteristics" (Howard 1935:109)?

To answer that question, he began with Renaud's (1934) classification and its several "Folsom" types but then quickly abandoned the effort when he realized that any point or group of like points could fall into one of several of Renaud's "types."[1] Such was the inevitable consequence of Renaud's mixing his classificatory criteria, identifying Yuma points by their shape and Folsom points by their fluting:

> Referring to [Renaud"s] classification we find, for example, that the commonest types—those to which we have referred as having a distribution over most of the United States and Canada and which we have called Folsom-like points, would be included under Renaud's type 5-a [Folsom type] and under types 2-a and 2-b [Yuma type] (if they had been shown grooved, as these forms very often are). (Howard 1935:110)

Presumably none of Renauds Yuma (shape-based) types were fluted, but many points that were fluted were so shaped. Moreover, if the points from Folsom were the standard by which the true Folsom type was to be defined (with their nearly fully fluted faces, fine secondary retouch, concave bases, and "rabbit ears" [Howard 1935:106]), then Renaud"s Folsom types 5a and 5b should not be called Folsom (Howard 1935:110–112).

Howard could easily see that the points he'd recovered at Clovis and like points elsewhere across the country were not true Folsom points, though they did have that "specialized groove." He supposed there had to be "some sort of relationship" between the forms. He was reluctant to say, however, what that relationship was (matters have not changed much [see Collins 1999b:31]), or even whether the more finely fluted points such as "true Folsoms" were earlier or later than his "generalized Folsoms" (Clovis), though he certainly leaned in that direction (Howard 1935: 121; Roberts 1938).

To account for these distinctive fluted forms, and to remove the complications of Renaud's sorting points by shape *or* fluting, Howard modified Renaud's classification. He created three broad groups, first by sorting points into fluted versus nonfluted forms, then by sorting the fluted group into two subgroups by shape, one of which was the "true Folsom" form, the other "Folsom-like" points. Yuma was reserved for unfluted forms, corresponding in part to Renaud's original types 1, 3, and 4 (Howard 1935:112 [Howard identified the corresponding Renaud types for heuristic purposes only]).

Howard had it relatively easy, seeking to differentiate "true Folsom" points, from "Folsom-like" (Clovis) forms. A more difficult task befell Roberts, whose work at Lindenmeier raised the questions of how to sort variation within "true Folsom" points. At first, and before he'd tackled the Lindenmeier assemblage, he had little difficulty identifying the attributes of a Folsom point:

> A true Folsom specimen is a thin leaf-shaped blade. The tip is slightly rounded and the broadest part of the blade . . . tends to occur between the tip and a line across the centre of the face. A typical feature is a long groove extending along each face about two-thirds of the length, which produced lateral ridges paralleling the edges of the blade. . . . The base is concave, often with long sharp base points. There normally is a more or less fine marginal retouching, a secondary removal of small flakes between the edges and the lateral ridge of the central groove. Another feature frequently observed is that of smoothed edges around the base and extending along the edges for about one-third the length of the blade. The usual material from which such objects were made was jasper, chert or chalcedony. (Roberts 1934:18)

Like Figgins and Howard, Roberts too had an archetype in mind. But he was pragmatic enough to realize that not all points met that ideal, least of all the specimen found in situ at Folsom in 1927 (Sn 3). Although he labeled this the type specimen, and the standard against which other points were deemed Folsom or not, Roberts (1935:5, 1939:533) could see that it was not "typical" in any meaningful sense. Few type points are. Indeed, he supposed it was one of *two* types of points present at Folsom and in the Lindenmeier assemblage (Roberts 1935:16).

The first form, which included the "type" specimen, had a slightly rounded tip and was somewhat "stubby," with the broadest part of the blade "between the tip and a line across the center of the face" just beyond the end of the haft area (Roberts 1935:fig. 2a, lines A–B). The second form, also present at Folsom but "rarely mentioned in discussions," was not stubby but rather "long and slender in outline" with "a tapering rather than a rounding tip" (Roberts 1935:16). At this juncture, Roberts believed the difference in blade shape was an intended by-product of manufacturing and was stylistically or typologically significant. Wormington (1949:26–27, 1957:34) similarly distinguished two forms of Folsom point but attributed the difference between the two types to use in "hunting different types of game." Neither evidently realized that the difference was merely a consequence of reworking, as Crabtree and others would later point out (e.g., Crabtree 1966:8; Tunnell 1977:143). Roberts maintained that formal distinction between the two types the following year (Roberts 1936:18), although it soon disappeared without comment from his subsequent papers on the topic (e.g., Roberts 1937, 1939, 1940).

Although he considered blade shape to be important in differentiating subtypes within Folsom, Roberts did not consider it diagnostic of the type itself. Nor, for that matter, were various of the other attributes of Folsom points, such as basal concavity and leaflike shape, singly or in combination, necessary and sufficient to identify a point as Folsom. Instead, Roberts considered fluting to be the "essential feature" of the type.

But whether a Folsom point required flutes on both faces or just one was debatable (Roberts 1935:17). Roberts knew that at least two points from Folsom, specimens Sn 1 and Sn 13/Sn 21, appeared to be fluted on just one side. However, Figgins's (1934) claims to the contrary, he argued that one of those (Sn 1) had "just a trace of the upper end of the [flute]" and thus it was actually fluted on *both* sides—the flute not extending further because of a flaw in the stone (Roberts 1935:17; also Tunnell and Johnson 2000:11). The other point did have just one flute, which Roberts attributed to the difficulty of working quartzite. In view of that evidence, he concluded that "a true Folsom point should be fluted on both sides, but an otherwise typical example may occasionally have the feature on only one side" (Roberts 1935:17).

Holding up fluting as the diagnostic attribute of a true Folsom was not without complications, for Roberts discovered at Lindenmeier one point that was "extremely thin [and] which would not have permitted the removal of such [flute] flakes" (illustrated in Roberts 1935:plates 5, 6). Although it had been found in the Folsom-age deposits, and in general outline and "style of chipping" it was plainly related to Folsom points by his definition of the type, Roberts (1935:18) could not accept it as a Folsom point.

However uncomfortable he was with unfluted Folsom points, Roberts (1936:21) was at least certain they were not Yuma points, even if there was "considerable confusion as to what constitutes [a Yuma] point." It was obvious something had to be done to "reach an agreement on what is meant by Yuma and that its use be restricted to something more specific than its present catch-all connotation" (Roberts 1936:21; see also Figgins 1935:2–3; Howard 1935; Wormington 1939).

Resolution of sorts would come following a systematic examination and partitioning into separate groups of a large sample of Yuma points by Holmes and Wormington, additional excavations, and through discussions among many of the principals at meetings in Philadelphia in 1937 (the "Early Man" Conference) and Santa Fe in 1941 (Holliday and Anderson 1993; Howard 1938, 1943; Wormington 1944:25–29, 1948:9–10). Emerging from those efforts was the realization that "Yuma" included points of many distinct forms and technology. In recognition of that, they were given new, site-based labels, a number of which, such as Eden, Plainview, San Jon, and Scottsbluff, are still in use today (Wormington 1948:10–11; cf. Wheat 1972). For a time in the early 1940s "Yuma" was kept alive as a suffix (e.g. "Eden Yuma"), but that practice was mostly a consequence of historical inertia and was soon dropped. Even after losing their common name, all of these forms, however much they might vary in their overall morphology, hafting geometry, or flaking, retained one inviolate trait: they were "never" fluted (Howard 1938:444; Wormington 1939, 1948).

Over this same time, the relative age of various Yuma points was becoming less ambiguous. By 1936, Roberts could report stratigraphic evidence at Lindenmeier indicating that Folsom points were earlier than Yuma (Eden) at the site, and if they overlapped, it "was at best only a late contemporaneity . . . with a later survival of the Yuma" (Roberts 1936:21–22; cf. Cotter 1939:153). By 1940, he would dismiss the claim of a greater antiquity for Yuma as "only conjectural typological seriation" (Roberts 1940:62; Wormington 1948). Radiocarbon evidence for the younger ages of "Yuma" points would become available in the 1950s from several sites, including Clovis, Cody, and San Jon (Roberts 1951:20–21; Sellards 1952:72–74).

All of this left the way clear to resolve the taxonomic, historical, stratigraphic, and chronological relationship among fluted points. Howard and Roberts had both perceived technological and typological differences between "true Folsoms" and those larger and less finely fluted and crafted "generalized Folsom" points from the Clovis site and other parts of North America, especially from eastern North America. Both, in fact, further suspected that, since "Generalized Folsom" points had a much wider distribution and based on the assumption that "cruder" fluting was earlier, they must be older than true Folsoms (Roberts 1939:544; also Howard 1935:119, 122; Roberts 1935:8, 1937:161–162). But that formal distinction was "frowned upon by the archaeological taxonomists" (Roberts 1938:534). Instead, at the 1937 "Early Man" Conference, the term Folsom was retained as a generic label for all fluted points. As it emerged from that meeting, a Folsom was simply a "leaf-shaped blade. It has a varying base; neither barbed nor stemmed. It is fluted on one or both sides, wholly or partially. It is pressure flaked from both sides" (Howard 1938:444). Roberts grumbled that "definite qualifiers" ought to be used, but even Cotter (1939:152), then in the midst of his work at Clovis, accepted this more generic label.

Soon thereafter, however, Cotter had sufficient evidence from Clovis to indicate that the "long and heavy" and only slightly fluted "Folsoms" were associated with mammoth remains in the speckled sand or Gray Sand, while in the overlying bluish clay or diatomite, smaller "Folsoms" of slighter design were found with bison (Cotter 1938:117; Haynes 1975; Sellards 1952:58). To be sure, that difference in point size might simply reflect different caliber weaponry being brought to bear on animals of very different body sizes (Roberts 1938:544), but as the chronological and stratigraphic relationship between "generalized Folsom" (Clovis) and Folsom points became increasingly clear-cut, and with the increasingly obvious geographic differences in their distributions, it was agreed in 1941 that there were sufficient differences between these fluted points to codify them as Clovis and Folsom types (Holliday and Anderson 1993; Howard 1943:227–229; LeTourneau 1998a:62). Wormington (1939:7) subsequently offered a detailed definition of the Folsom point type:

> Folsom points . . . are pressure flaked and of excellent workmanship. They have an average length of about two inches, are thin, more or less leaf-shaped with concave bases usually marked by ear-like projections. The one characteristic, however, which distinguished the Folsom

from all others is the removal of longitudinal flakes from either face. The removal of these flakes results in the formation of grooves or channels extending from one-third to almost the entire length of the point, which gives it a hollow-ground appearance when seen in cross-section. Rare examples have been found where the grooves have been formed only on one face but these are to be regarded as aberrant.

The subsequent editions of the volume in which this first appeared would repeat this definition nearly verbatim (Wormington 1944:7, 1949:21, 1957:27), adding to it only a couple of observations: namely, that "there is frequently a small central nipple in the basal concavity " and that "the lower edges and base normally bear evidence of grinding" (Wormington 1957:27). In all cases, however, the most important characteristic for her remained the presence of fluting (also Cotter 1939:152; Howard 1943:229). Of course, Wormington's definition was established long before points that were unfluted, but otherwise similar in all respects to Folsom points, were being reported, a process that began in the 1950s at the Midland site on the Southern High Plains (Wendorf and Krieger 1959; Wendorf et al. 1955).

Although Wormington (1939:7) followed the "accepted rule of naming an artifact type after the type station, that is, the first site or station in which it is discovered," unlike Roberts she did not specify a type specimen from the site. A few years later Howard (1943:229) remarked that by consensus there was a "typical" (not type) specimen, but he too was reluctant to ascribe it any special significance:

> We have come to think of the specimen in Pl. VIII, a, [*fig. 8.4A; Sn 9*] as being typical and it would serve no good purpose to try to change this view; but it is well to observe that there are several of the other varieties in the collection from Folsom, although most of them are broken specimens.

The varieties Howard had in mind were points that were widest above, below, or at their center—obviously, he appreciated the fact that breakage could obscure this measure.

More recently, Ingbar and Hofman (1999:99) identified one of the Lindenmeier points as "the common standard" (a trend perhaps begun by Crabtree 1966:3). But, of course, within any assemblage of Folsom points—and the ones from Folsom and Lindenmeier are no exceptions—there is considerable morphometric variability about a "typical" form, as a result of variation in manufacture, point reworking, raw material availability, the number of kill or retooling events, the temporal/spatial distance from the last or to the next quarry visit, etc. (Amick 1995:34; Hofman 1992:193; Ingbar and Hofman 1999; LeTourneau 1998a; Meltzer, Todd, and Holliday 2002; a matter anticipated by Figgins 1934:4).

In the end, and despite the lack of an agreed-upon type specimen, let alone one from the type site, many offered and most agreed on the definition of a Folsom point. For most it was essential that the point be fluted; for a few others, including Figgins, fluting was considered unnecessary. Lingering ambiguity in the definition did not matter, for in the 1940s one merely had to distinguish Folsom points from Clovis and Late Paleoindian forms like Yuma— and that was proving to be easy enough. Matters became more complicated with the appearance of Midland and unfluted Folsom points, which have bedeviled archaeologists ever since (Agogino 1969; Amick 1995; Hofman 1991; Howard 1943; Judge 1970; LeTourneau 1998a; Root, William, and Emerson 2000; Wormington 1957).

Yet, despite ambiguity in the definition of a Folsom point, or the lack of a type specimen against which one can compare forms (LeTourneau 1998a:66), Folsom points became and are still one of the most instantly recognizable of Paleoindian projectile points.

Note

1. Howard (1935:110) nonetheless came to Renaud's defense, pointing out that Renaud, "like everyone else who has made any attempt to unravel this 'Folsom problem,' has been feeling his way along, and therefore has had no compunction in discarding his earlier conclusion when necessary." He recognized that it was important to "sincerely acknowledge that our task has been made easier by the pioneer work of Professor Renaud. Our effort has been to contribute to what has already been done, not to detract" (Howard 1935:115). Having just dismantled Renaud's classification scheme, that was faint praise indeed.

REFERENCES CITED

A Note on Archival Sources

Unpublished archival materials are cited in this volume using the following abbreviations.

AH/NAA: Aleš Hrdlička Papers, National Anthropological Archives, Washington, DC

ANTH/AMNH: Department of Anthropology Archives, American Museum of Natural History, New York

BAE/NAA: Bureau of American Ethnology Papers, National Anthropological Archives, Washington, DC

BB/AMNH: Barnum Brown Papers, American Museum of Natural History, New York

DIR/DMNS: Papers of the Director, Denver Museum of Nature and Science, Denver, Colorado

EA/UA: Ernst Antevs Papers, Department of Geosciences, University of Arizona, Tucson

EHB/NSM: Erwin H. Barbour Papers, Nebraska State Museum, Lincoln

FB/APS: Franz Boas Papers, American Philosophical Society, Philadelphia, Pennsylvania

FHHR/NAA: Frank H. H. Roberts Papers, National Anthropological Archives, Washington, DC

HFO/AMNH: Henry Fairfield Osborn Papers, American Museum of Natural History, New York

HJC/AGFO: Harold J. Cook Papers, Agate Fossil Beds National Monument, Scottsbluff, Nebraska

HJC/AHC: Harold J. Cook Papers, American Heritage Center, Univ. of Wyoming, Laramie, Wyoming

JCM/CIW: John C. Merriam Papers, Carnegie Institution of Washington, Washington, DC

JCM/LC: John C. Merriam Papers, Library of Congress, Washington, DC

JDF/DMNH: Jesse D. Figgins Papers, Denver Museum of Nature and Science, Denver, Colorado

NJ/SIA: Neil Judd Papers, Smithsonian Institution Archives, Washington, DC

OPH/SIA: Oliver P. Hay Papers, Smithsonian Institution Archives, Washington, DC

USNM/SIA: United States National Museum, Permanent Administrative Files, Smithsonian Institution Archives, Washington, DC

WHH/SIA: William Henry Holmes Papers, Smithsonian Institution Archives, Washington, DC

Published Works

Abbott, C.C. 1877. On the discovery of supposed Paleolithic implements from the glacial drift in the Valley of the Delaware River, near Trenton, New Jersey. *Peabody Museum Annual Report* 10:30–43.

———. 1881. *Primitive industry*. Salem, MA: G. Bates.

———. 1889. Evidences of the antiquity of man in eastern North America. *Proceedings American Association for the Advancement of Science* 37:293–315.

———. 1892. Recent archaeological explorations in the Valley of the Delaware. *Univ. of Pennsylvania Series in Philology, Literature, and Archaeology* 2:1–30.

Adams, D.K., and Comrie, A.C. 1997. The North American monsoon. *Bulletin of the American Meteorological Society* 78, 2197–2213.

Adovasio, J., and Pedler, D. 1997. Monte Verde and the antiquity of humankind in the Americas. *Antiquity* 71:573–580.

Agogino, G. 1969. The Midland Complex: Is it valid? *American Anthropologist* 71:1117–1118.

———. 1971. The McJunkin controversy. *New Mexico* 49:41–47.

———. 1985. The amateur involvement in the discovery of the Folsom type site. *The Chesopiean* 23:2–4.

Aharon, P. 2003. Meltwater flooding events in the Gulf of Mexico revisited: Implications for rapid climate changes during the last deglaciation. *Paleooceanography* 18(4): 1079–1092.

Ahler, S. A. 1970. *Projectile point form and function at Rodgers Shelter, Missouri*. Missouri Archaeological Society Research Series no. 8.

———. 1992. Use-phase classification and manufacturing technology in Plains Village arrowpoints. In *Piecing together the past: applications of refitting studies in archaeology*, ed. J. L. Hofman and J. Enloe, 36–62. BAR International Series 578.

Ahler, S. A., and P. R. Geib. 2000. Why flute? Folsom point design and adaptation. *Journal of Archaeological Science* 27: 799–820.

———. 2002. Why the Folsom point was fluted: Implications from a particular technofunctional explanation. In *Folsom technology and lifeways*, ed. J. Clark and M. B. Collins, 371–390. Special Publication no. 4. Tulsa, OK: Department of Anthropology, Univ. of Tulsa.

Ahler, S. A., G. Frison, and M. McGonigal. 2002. Folsom and other Paleoindian artifacts in the Missouri River Valley, North Dakota. In *Folsom technology and lifeways*, ed. J. Clark and M. B. Collins, 69–112. Special Publication no. 4. Tulsa, OK: Department of Anthropology, Univ. of Tulsa.

Akerman, K., and J. Fagan. 1986. Fluting the Lindenmeier Folsom: A simple and economical solution to the problem, and its implications for other fluted point assemblages. *Lithic Technology* 15:1–6.

Albanese, J. 1978. Paleotopography and bison traps. *Plains Anthropologist Memoir* 14:58–62.

Albanese, J., and G. Frison. 1995. Cultural and landscape change during the Middle Holocene, Rocky Mountain area, Wyoming and Montana. In *Archaeological geology of the Archaic Period in North America*, ed. E. A. Bettis, 1–19. Geological Society of America Special Paper 297. Boulder, CO.

Aleinikoff, J., D. Muhs, R. Sauer, and C. Fanning. 1999. Late Quaternary loess in northeastern Colorado, II-Pb isotopic evidence for the variability of loess sources. *Geological Society of America Bulletin* 111:1876–1883.

Allen, B., and R. Anderson. 1993. Evidence from western North America for rapid shifts in climate during the last glacial maximum. *Science* 260:1920–1923.

Allen, B. D., and R. Anderson. 2000. A continuous, high–resolution record of late Pleistocene climate variability from the Estancia basin, New Mexico. *Geological Society of America Bulletin* 112:1444–1458.

Allen, B. L. 1959. A mineralogical study of soils developed on Tertiary and Recent lava flows in northeastern New Mexico. Unpublished Ph.D. diss., Michigan State Univ.

Alley, R. 2000. Ice-core evidence of abrupt climate changes. *Proceedings of the National Academy of Science* 97:1331–1334.

Alley, R., and P. Clark. 1999. The deglaciation of the northern hemisphere. *Annual Review of Earth and Planetary Science* 27:149–182.

Alsoszatai-Petheo, J. 1986. An alternative paradigm for the study of early man in the New World. In *New evidence for the Pleistocene peopling of the Americas*, ed. A. Bryan, 15–23. Orono, ME: Center for the Study of Early Man.

Ambrose, S. 1990. Preparation and characterization of bone and tooth collagen for stable carbon and nitrogen isotope analysis. *Journal of Archaeological Science* 17:431–451.

Ambrose, S. 1991. Effects of diet, climate, and physiology on nitrogen isotope abundances in terrestrial food webs. *Journal of Archaeological Science* 18:293–317.

Amick, D. 1994. Technological organization and the structure of inference in lithic analysis: An examination of Folsom hunt-ing behavior in the American Southwest. In *The organization of North American prehistoric chipped stone tool technologies*, P. Carr, ed., 9–34. International Monographs in Prehistory, Archaeological Series 7. Ann Arbor, MI.

———. 1995. Patterns of technological variation among Folsom and Midland projectile points in the American Southwest. *Plains Anthropologist* 40:23–38.

———. 1996. Regional patterns of Folsom mobility and land use in the American southwest. *World Archaeology* 27:411–426.

———. 1999a. New approaches to understanding Folsom lithic technology. In *Folsom lithic technology: Explorations in structure and variation*, ed. D. S. Amick, 1–11. International Monographs in Prehistory, Archaeological Series 12. Ann Arbor, MI.

———. 1999b. Raw material variation in Folsom stone tool assemblages and the division of labor in hunter-gatherer societies. In *Folsom lithic technology: Explorations in structure and variation*, ed. D. S. Amick, 169–187. International Monographs in Prehistory, Archaeological Series 12. Ann Arbor, MI.

———. 2000. Regional approaches with unbounded systems: The record of Folsom land use in New Mexico and west Texas. In *The archaeology of regional interaction*, M. Hegmon, ed., 119–147. , Boulder: Univ. Press of Colorado.

———. 2002. Lone Butte revisited: A Folsom hunting station in the Tularosa Basin of New Mexico. Paper presented at the 60th Plains Conference, Oklahoma City.

Amundson, R., O. Chadwick, C. Kendall, Y. Wang, and M. DeNiro. 1996. Isotopic evidence for shifts in atmospheric circulation patterns during the late Quaternary in mid-North America. *Geology* 24:23–26.

Anderson, A. B. 1975. "Least cost" strategy and limited activity site location, Upper Dry Cimarron River Valley, northeastern, New Mexico. Unpublished Ph.D. diss., Univ. of Colorado, Boulder.

Anderson, A. B., and C. V. Haynes. 1979. How old is Capulin Mountain? Correlation between Capulin Mountain volcanic flows and the Folsom type site, northeastern New Mexico. In *Proceedings of the First Conference on Scientific Research in the National Parks, Vol. II*, R. Linn, ed., 893–898. U. S. Department of the Interior, NPS Transactions and Proceedings, no. 5.

Anderson, G. W., T. Hilley, P. Martin, C. Neal, and R. Gomez. 1982. *Soil survey of Colfax County*. Washington, DC: Soil Conservation Service, United States Department of Agriculture.

Andrews, B. 2004. Field research in 2003 at the Mountaineer site (5GN2477. Gunnison County, Colorado. In *Quest Archaeological Research Fund: Summary of field and laboratory research, 2003–2004*, D. J. Meltzer, ed. Report on file. Dallas, TX: Department of Anthropology, Southern Methodist University.

Antevs, E. 1925. Swedish Late-Quaternary geochronologies. *The Geographical Review* 15:280–284.

———. 1931. Late Glacial correlations and ice recession in Manitoba. *Geological Survey of Canada, Memoir* 168:1–76.

———. 1935. The spread of aboriginal man to North America. *The Geographical Review* 25:302–309.

Armour, J., P. Fawcett, and J. Geissman. 2002. 15 k.y. paleoclimatic and glacial record from northern New Mexico. *Geology* 30:723–726.

Arnold, J. R., and W. F. Libby. 1950. *Radiocarbon dates (September 1, 1950)*. Chicago: Institute for Nuclear Studies, Univ. of Chicago.

Arnold, K. 1998. Semester summary of Folsom, New Mexico site 29CX1 snail population survey for unit N1024 E998. Unpublished report on file. Dallas, TX: Department of Anthropology, Southern Methodist Univ.

———. 1999. Microstratigraphic investigation employing gastropod samples for unit N1033 E998 of the Folsom site, 29CX1, New Mexico. Unpublished report on file, Department of Anthropology. Dallas, TX: Southern Methodist Univ.

Artz, J. 1995. Geological contexts of the Early and Middle Holocene archaeological record in North Dakota and adjoining areas of the Northern Plains. In *Archaeological geology of the Archaic Period in North America*, ed. E. A. Bettis, 67–86. Geological Society of America Special Paper 297. Boulder.

Aubele, J., and L. Crumpler. 2001. Raton-Clayton and Ocate Volcanic fields. In *Geology of the Llano Estacado*, 69–76. New Mexico Geological Society Guidebook, 52nd Annual Field Conference.

Baerreis, D. A. 1980. Habitat and climatic interpretation from terrestrial gastropods at the Cherokee Site. In *The Cherokee Excavations: Mid–Holocene Paleoecology and Human Adaptation in Northwest Iowa*, ed. D. C. Anderson and H. S. Semken, Jr., 101–122. New York: Academic Press.

Bailey, H. 1960. A method for determining the warmth and temperateness of climate. *Geografiska Annaler* 42:1–16.

Bailey, V.(1931. *Mammals of New Mexico*. North American Fauna no. 53, 412. Washington, DC: United States Department of Agriculture, Bureau of Biological Survey.

Balakrishnan, M. 2002. Stable carbon and oxygen isotopic composition of land snail aragonite shells: Modeling, measurements and paleoenvironmental interpretations. Unpublished Ph.D. diss., Southern Methodist Univ.

Balakrishnan, M., and C. Yapp. 2004. Flux balance models for the oxygen and carbon isotope compositions of land snail shells. *Geochimica et Cosmochimica Acta* 68:2007–2024.

Balakrishnan, M., C. Yapp, D. J. Meltzer, and J. Theler. 2005a. Paleoenvironment of the Folsom site ~10,500 ^{14}C years B.P. as inferred from the stable isotope composition of fossil land snail shells. *Quaternary Research* 63:31–44.

Balakrishnan, M., C. Yapp, J. Theler, B. Carter, and D. Wyckoff. 2005b. Environmental significance of ^{13}C/^{12}C and ^{18}O/^{16}O ratios of modern land snail shells from the southern Great Plains of North America. *Quaternary Research* 63:15–30.

Baldwin, B., and W.R. Muehlberger. 1959, Geologic studies of Union County, New Mexico. *State Bureau of Mines and Mineral Resources, Bulletin 63*. New Mexico Institute of Mining and Technology, Socorro.

Bamforth, D. 1985. The technological organization of Paleoindian small group bison hunting on the Llano Estacado. *Plains Anthropologist* 30:243–258.

———. 1988. *Ecology and human organization on the Great Plains*. New York: Plenum Press.

———. 1991. Flintknapping skill, communal hunting and projectile point typology. *Plains Anthropologist* 36:309–322.

———. 2002. High-tech foragers? Folsom and later Paleoindian technology on the Great Plains. *Journal of World Prehistory* 16:55–98.

Bamforth, D., and P. Bleed. 1997. Technology, flaked stone technology, and risk. *Archaeological Papers of the American Anthropological Association* 7:109–139.

Banks, L. 1990. *From Mountain Peaks to Alligator Stomachs: A Review of Lithic Sources in the Trans-Mississippi South, the Southern Plains, and Adjacent Southwest*. Oklahoma Anthropological Society, Memoir 4.

Barbour, E., and C. B. Schultz. 1932. The mounted skeleton of *Bison occidentalis*, and associated dart points. *Bulletin of the Nebraska State Museum* 32:263–270.

———. 1936. Palaeontologic and geologic consideration of early man in Nebraska. *Bulletin of the Nebraska State Museum* 45:431–449.

Barnes, V. 1984. *Geologic atlas of Texas: Dalhart sheet*. Bureau of Economic Geology, Austin, Texas.

Barnosky, C. W., P. M. Anderson, P. J. Bartlein, P. J. 1987. The northwestern U.S. during deglaciation; vegetational history and paleoclimatic implications. In *North America and Adjacent Oceans During the Last Deglaciation*, ed. W. F. Ruddiman, and H. E. Wright, 289–321. Boulder, CO: Geological Society of America.

Barry, R. G. 1983. Climatic environments of the Great Plains, past and present. *Transactions of the Nebraska Academy of Science* 11:45–55.

Bartlein, P. J., K. H.Anderson, P. M. Anderson, P. E. Edwards, C. J. Mock, R. S. Thompson, R. S. Webb, T. Webb, and C. Whitlock. 1998. Paleoclimate simulations for North America over the past 21,000 years: Features of the simulated climate and comparisons with paleoenvironmental data. *Quaternary Science Reviews* 17:549–585.

Barton, D. 1941. Father of the dinosaurs. *Natural History* 48: 308–312.

Bartram, L. 1993. Perspectives on skeletal part profiles and utility curves from eastern Kalahari ethnoarchaeology. In *From bones to behavior*, ed. J. Hudson, 115–137. Carbondale: Southern Illinois Univ. Press.

Beauchamp, W. 1897. Aboriginal chipped stone implements of New York: *New York State Museum Bulletin* 16.

Beck, J. W., D. A. Richards, R. L. Edwards, and D. Biddulph. 2001. Extremely large variations of atmospheric ^{14}C concentration during the last glacial period. *Science* 292: 2453–2458.

Behrensmeyer, A. K. 1978. Taphonomic and ecologic information from bone weathering. *Paleobiology* 4:150–162.

Bell, E., and W. van Royen. 1934. An evaluation of recent Nebraska finds sometimes attributed to the Pleistocene. *Wisconsin Archaeologist* 13:49–70.

Belovsky, G. E., and J. Slade. 1986. Time budgets of grassland herbivores: Body size similarities. *Oecologia* 70:53–62.

Bement, L. 1997. The Cooper site: a stratified Folsom bison kill in Oklahoma. *Plains Anthropologist Memoir 29*:85–100.

———. 1999a. View from a kill: The Cooper site Folsom lithic assemblages. In *Folsom lithic technology: explorations in structure and variation*, D. S. Amick, ed., 111–121. International Monographs in Prehistory, Archaeological Series 12. Ann Arbor, MI.

———. 1999b. *Bison hunting at Cooper site*. Univ. of Oklahoma Press. Norman.

Bennett, K. 2002. Programs for plotting pollen diagrams and analyzing pollen data, version 4.10. Available at: http://www.kv.geo.uu.edu.se/psimpoll.html.

Bequaert, J. C., and W. B. Miller. 1973. *The mollusks of the arid Southwest, with an Arizona check list*. Tucson: Univ. of Arizona Press.

Berger, J., and C. Cunningham. 1994. *Bison: Mating and conservation in small populations*. New York: Columbia Univ. Press.

Bergman, C., and M. Newcomer. 1983. Flint arrowhead breakage: Examples from Ksar Akil, Lebanon. *Journal of Field Archaeology* 10:238–243.

Betancourt, J.L. 1990. Late Quaternary biogeography of the Colorado Plateau. In *Packrat Middens*, ed. J.L. Betancourt, T. R. Van Devender, and P.S. Martin, 259–292. Tucson: Univ. of Arizona Press.

Binford, L. R. 1978a. Dimensional analysis of behavior and site structure: learning from an Eskimo hunting stand. *American Antiquity* 43:330–361.

———. 1978b. *Nunamiuit ethnoarchaeology*. New York: Academic Press.

———. 1979. Organization and formation processes: Looking at curated technologies. *Journal of Anthropological Research* 35:255–273.

———. 1980. Willow smoke and dogs' tails: Hunter-gatherer settlement systems and archaeological site formation. *American Antiquity* 45:4–20.

———. 1981. *Bones: ancient men and modern myths*. New York: Academic Press.

———. 1983. *In pursuit of the past*. New York: Thames and Hudson.

———. 1986. In pursuit of the future. In *American archaeology: Past and future. A celebration of the Society for American Archaeology*, ed. D.J. Meltzer, D.D. Fowler, and J.A. Sabloff, 459–479. Washington, DC: Smithsonian Institution Press.

———. 1993. Bones for stones: Considerations of analogues for features found on the central Russian Plains. In *From Kostenki to Clovis: Upper Paleolithic—Paleo–indian adaptations*, ed. O. Soffer and N. Praslov, 101–124. New York: Plenum Press.

———. 1997. Linking ethnographic information on man-bear interaction to European cave bear deposits. Paper presented at the Man and Bear International Conference, Auberives-en-Royans, Isere.

———. 2001. *Constructing frames of reference*. Berkeley: Univ. of California Press.

Binford, L. R. and J. A. Sabloff. 1982. Paradigms, systematics, and archaeology. *Journal of Anthropological Research* 38:137–153.

Birkeland, P. 1999. *Soils and geomorphology*. 3rd ed. New York: Oxford Univ. Press.

Blackmar, J. 2001. Regional variability in Clovis, Folsom, and Cody land use. *Plains Anthropologist* 46:65–94.

Bleed, P. 1986. The optimal design of hunting weapons: maintainability or reliability? *American Antiquity* 51:737–747.

Blumenschine, R., and C. Marean. 1993. A carnivore's view of archaeological bone assemblages. In *From bones to behavior*, ed. J. Hudson, 273–300. Carbondale: Southern Illinois Univ. Press.

Boldurian, A. 1981. An analysis of a Paleo-Indian lithic assemblage from Blackwater Draw Locality no. 1 in eastern New Mexico. Unpublished master's thesis, Eastern New Mexico Univ.

———. 1990. Lithic technology at the Mitchell locality of Blackwater Draw: A stratified Folsom site in eastern New Mexico. *Plains Anthropologist Memoir* 24.

Boldurian, A. and J.L. Cotter. 1999. *Clovis revisited: new perspectives on Paleoindian adaptations from Blackwater Draw, New Mexico*. The Univ. Museum, Univ. of Pennsylvania, Philadelphia.

Boldurian, A., and S. Hubinsky. 1994. Preforms in Folsom lithic technology: A view from Blackwater Draw, New Mexico. *Plains Anthropologist* 39:445–464.

Borchert, J. R. 1950. The climate of the central North American grasslands. *Annals of the Association of American Geographers* 40:1–39.

Bowen, B. M. 1996. Rainfall and climate variation over a sloping New Mexico Plateau during the North American Monsoon. *Journal of Climate* 9:3432–3442.

Bradley, B. 1982. Flaked stone technology and typology. In *The Agate Basin site: a record of the Paleoindian occupation of the northwestern High Plains*, G. Frison and D. Stanford, eds., 181–208. Academic Press. New York:

———. 1991. Flaked stone technology in the Northern High Plains. In *Prehistoric hunters on the High Plains*. 2nd ed., ed. G. Frison, 369–395. New York: Academic Press.

———. 1993. Paleo-Indian flaked stone technology in the North American High Plains. In *From Kostenki to Clovis: Upper Paleolithic—Paleo–indian adaptations*, ed. O. Soffer and N. Praslov, 251–262. New York: Plenum Press.

Bradley, B., and G. Frison. 1987. Projectile points and specialized bifaces from the Horner site. In *The Horner site: The type site of the Cody Cultural Complex*, ed. G. Frison and L. Todd, 199–231. New York: Academic Press.

———. 1996. Flaked-stone and worked-bone artifacts from the Mill Iron site. In *The Mill Iron site*, ed. G. Frison, 43–69. Albuquerque: Univ. of New Mexico Press.

Brink, J. 1997. Fat content in leg bones of *Bison bison*, and applications to archaeology. *Journal of Archaeological Science* 24:259–274.

Broecker, W. S., and J. Kennett. 1989. Routing of meltwater from the Laurentide Ice Sheet during the Young Dryas cold episode. *Nature* 341:318–323.

Broilo, F. 1971. An investigation of surface collected Clovis, Folsom, and Midland projectile points from Blackwater Draw and adjacent localities. Unpublished master's thesis, Eastern New Mexico Univ.

Brown, B. 1928a. Recent finds relating to prehistoric man in America. Discussion of "The origin and antiquity of man in America" by A. Hrdlička. *New York Academy of Medicine Bulletin* 4(7):824–828.

———. 1928b. The Folsom culture. Paper presented at the International Congress of Americanists.

———. 1929. Folsom culture and its age. *Geological Society of America Bulletin* 40:128–129.

———. 1932. The buffalo drive. *Natural History* 32:75–82.

———. 1936. The Folsom culture—An occurrence of prehistoric man with extinct animals near Folsom, New Mexico. *Report of the XVI Session, International Geological Congress* 2:813.

Bryan, A. 1986. Paleoamerican prehistory as seen from South America. In *New evidence for the Pleistocene peopling of the Americas*, ed. A. Bryan, 1–14. Center for the Study of Early Man, Orono.

Bryan, K. 1922. Erosion and sedimentation of the Papago country, Arizona, with a sketch of the geology. *United States Geological Survey Bulletin* 730:19–90.

———. 1925. Date of channel trenching (arroyo cutting) in the arid Southwest. *Science* 62:338–344.

———. 1929. Discussion of "Folsom culture and its age," by B. Brown. *Geological Society of America Bulletin* 40:128–129.

———. 1937. Geology of the Folsom deposits in New Mexico and Colorado. In *Early man*, ed. G.G. MacCurdy, 139–152. Philadelphia: Lippincott.

———. 1940. Erosion in the valleys of the Southwest. *New Mexico Quarterly* 10:227–232.

———. 1941. Geologic antiquity of man in America. *Science* 93:505–514.

Bryant, V., and R. Holloway. 1985. Late-Quaternary paleoenvironmental record of Texas: An overview of the pollen evidence. In *Pollen records of Late-Quaternary North American sed-*

iments, ed. V. Bryant and R. Holloway, 39–70. American Association of Stratigraphic Palynologists.

Bryant, V., R. G. Holloway, J. G. Jones, and D. L. Carlson. 1994. Pollen preservation in alkaline soils of the American Southwest. In *Sedimentation of organic particles*, ed. A. Traverse, 47–58. Cambridge: Cambridge Univ. Press.

Buchanan, B. 2002. Folsom lithic procurement, tool use, and replacement at the Lake Theo site, Texas. *Plains Anthropologist* 47:121–146.

Bull, W. B. 1991. *Geomorphic responses to climatic change*. New York: Oxford Univ. Press.

Busacca, A., J. Beget, H. Markewish, D. Muhs, N. Lancaster, and M. Sweeney. 2004. Eolian sediments. In *The Quaternary period in the United States*, ed. A. Gillespie, S. C. Porter, and B. Atwater, 275–309. New York: Elsevier Science.

Butzer, K. W. 1980. Holocene alluvial sequences: Problems of dating and correlation. In *Timescales in geomorphology, ed.* R. A. Cullingford, D. A. Davidson, and J. Lewin, 131–142. London: John Wiley and Sons.

Byerly, R., and D. J. Meltzer. 2005. Historic Period faunal remains from Mustang Springs, West Texas. *Plains Anthropologist.* 50:93–110.

Caire, W., J. Tyler, B. Glass, and M. Mares. 1989. *Mammals of Oklahoma*. Univ. of Oklahoma Press. Norman.

Calvin, E. M. 1987. A review of the volcanic history and stratigraphy of northeastern New Mexico, the Ocate and Raton-Clayton volcanic fields. In *Northeastern New Mexico*, 83–85. New Mexico Geological Society Guidebook, 38th Field Conference.

Cannon, M. D. 2003. A model of central place forager prey choice and an application to faunal remains from the Mimbres Valley, New Mexico. *Journal of Anthropological Archaeology* 22:1–25.

Cannon, M. D. and D. J. Meltzer. 2004. Early Paleoindian foraging: examining the faunal evidence for large mammal specialization and regional variability in prey choice. *Quaternary Science Reviews* 23(18/19):1955–1987.

Carter, B. J., and L. Bement. 2003. Geoarchaeology of the Cooper site, northwest Oklahoma: Evidence for multiple Folsom bison kills. *Geoarchaeology* 18:115–127.

Cerling, T., and J. Harris. 1999. Carbon isotope fractionation between diet and bioapatite in ungulate mammals and implications for ecological and paleoecological studies. *Oecologia* 120:347–363.

Chamberlin, R. 1917a. Interpretation of the formations containing human bones at Vero, Florida. *Journal of Geology* 25:25–39.

———. 1917b. Further studies at Vero, Florida. *Journal of Geology* 25:667–683.

Chamberlin, T. C., ed. 1892. Geology and archaeology mistaught. *The Dial* 13:303–306.

———. 1893a. Professor Wright and the geological survey. *The Dial* 14:7–9.

———. 1893b. The diversity of the glacial period. *American Journal of Science* 45:171–200.

———. 1902. The geologic relations of the human relics of Lansing, Kansas. *Journal of Geology* 10:745–779.

———. 1903. The criteria requisite for the reference of relics to a glacial age. *Journal of Geology* 11:64–85.

———. 1917. Symposium on the age and relations of the fossil human remains found at Vero, Florida. *Journal of Geology* 25:1–3.

———. 1919. Investigation versus propagandism. *Journal of Geology* 27:305–338.

Chisholm, B., J. Driver, S. Dube, and H. Schwarcz. 1986. Assessment of prehistoric bison foraging and movement patterns via stable-carbon isotope analysis. *Plains Anthropologist* 31:193–206.

Christenson, A. 1982. Maximizing clarity in economic terminology. *American Antiquity* 47:419–426.

Christopherson, R. J., and R. J. Hudson. 1978. Effects of temperature and wind on cattle and bison. *57th Annual Feeder's Day Report* 57:40–41.

Christopherson, R. J., R. J. Hudson, and R. J. Richmond. 1976. Feed intake, metabolism and thermal insulation of Bison, Yak, Scottish Highland and Hereford calves during winter. *55th Annual Feeder's Day Report* 55:51–52.

Churchill. S. 1993. Weapon technology, prey size selection, and hunting methods in modern hunter-gatherers: Implications for hunting in the Paleolithic and Mesolithic. In *Hunting and animal exploitation in the late Paleolithic and Mesolithic of Eurasia*, ed. G. Peterkin, H. Bricker, and P. Mellars. Archaeological Paper 4:11–24. Washington, DC: American Anthropological Association.

Clark, P., R. Alley, and D. Pollard. 1999. Northern hemisphere ice-sheet influences on global climate change. *Science* 286:1104–1111.

Clark, P. U., S. J. Marshall, G. K. C.Clarke, S. W. Hostetler, J. M. Licciardi, J. T. Teller. 2001. Freshwater forcing of abrupt climate change during the last deglaciation. *Science* 293:283–287.

Clark, P. U., J. Mitrovica, G. Milne, and M. Tamisiea. 2002. Sea-level fingerprinting as a direct test for the source of global Meltwater Pulse IA. *Science* 295:2438–2441.

Clarke, G., D. Leverington, J. Teller, and A. Dyke. 2003. Superlakes, megafloods, and abrupt climate change. *Science* 301:922–923.

Claypole, E. 1893a. Prof. G. F. Wright and his critics. *Popular Science Monthly* 42:764–781.

———. 1893b. Major Powell on "Are there evidences of man in the glacial gravels?" *Popular Science Monthly* 43:696–699.

Colbert, E. 1937. The Pleistocene mammals of North America and their relation to Eurasian forms. In *Early Man*, ed. G. G. MacCurdy, 173–184. Philadelphia: Lippincott.

Collatz, G., J. Berry, and J. Clark. 1998. Effects of climate and atmospheric CO_2 partial pressure on the global distribution of C_4 grasses: Present, past and future. *Oecologia* 114:441–454.

Collins, M. B. 1999a. *Clovis blade technology*. Austin: Univ. of Texas Press.

———. 1999b. Clovis and Folsom lithic technology on and near the Southern Plains: Similar ends, different means. In *Folsom lithic technology: Explorations in structure and variation*, ed. D. Amick, 12–38. International Monographs in Prehistory. Ann Arbor, MI.

Collins, R. F. 1949. Volcanic rocks of northeastern New Mexico. *Geological Society of America Bulletin* 60:1017–1040.

Connin, S., J. Betancourt, and J. Quade. 1998. Late Pleistocene C_4 plant dominance and summer rainfall in the southwestern United States from isotopic study of herbivore teeth. *Quaternary Research* 50:179–193.

Cook, H. J. 1925. Definite evidence of human artifacts in the American Pleistocene. *Science* 62:459–460.

———. 1926. The antiquity of man in America. *Scientific American* 137:334–336.

———. 1927a. New geological and palaeontological evidence bearing on the antiquity of mankind in America. *Natural History* 27:240–247.

———. 1927b. New trails of ancient men. *Scientific American* 138:114–117.

———. 1928a. Further evidence concerning man's antiquity at Frederick, Oklahoma. *Science* 67:371–373.

———. 1928b. Glacial age man in New Mexico. *Scientific American* 139:38–40.

———. 1931a. The antiquity of man as indicated at Frederick, Oklahoma: A reply. *Journal of the Washington Academy of Sciences* 21:161–167.

———. 1931b. More evidence of the "Folsom Culture" race. *Scientific American* 144:102–103.

———. 1947. Some background data on the original "Folsom Quarry" site: And of those connected with its discovery and development. Unpublished manuscript on file, Agate Fossil Beds National Monument, Scottsbluff, NE. February 27.

———. 1952. Early man in America. Unpublished manuscript on file. Agate Fossil Beds National Monument, Scottsbluff, NE. March 31.

———. 1968. *Tales of the 04 Ranch: Recollections of Harold J. Cook, 1887–1909*. Lincoln: Univ. of Nebraska Press.

Cormie, A., and H. Schwarcz. 1996. Effects of climate on deer bone $\delta^{15}N$ and $\delta^{13}C$: lack of precipitation effects on $\delta^{15}N$ for animals consuming low amounts of C_4 plants. *Geochimica et Cosmochimica Acta* 60:4161–4166.

Cotter, J. 1937. The occurrence of flints and extinct animals in pluvial deposits near Clovis, New Mexico, Pt. IV: Report on the excavations at the Gravel Pit in 1936. *Proceedings of the Philadelphia Academy of Natural Sciences* 89:1–16.

———. 1938. The occurrence of flints and extinct animals in pluvial deposits near Clovis, New Mexico, Part VI—Report on excavations in the Gravel Pit, 1937. *Proceedings of the Philadelphia Academy of Natural Sciences* 90:113–117.

———. 1939. A consideration of "Folsom and Yuma culture finds." *American Antiquity* 5:152–155.

Court, A. 1974. The climate of the coterminous United States. In *Climates of North America*, ed. R. Bryson and F. K. Hare, 193–343. Amsterdam: Elsevier.

Cowling, S., and M. Sykes. 1999. Physiological significance of low atmospheric CO_2 for plant-climate interactions. *Quaternary Research* 52:237–242.

Crabtree, D. 1966. A stoneworker's approach to analyzing and replicating the Lindenmeier Folsom. *Tebiwa* 9:3–39.

Cronin, J. G. 1964. A summary of the occurrence and development of ground water in the southern High Plains. *United States Geological Survey Water–Supply Paper* 1693.

Danforth, C. H. 1931. Report on Section H. The American Association for the Advancement of Science. *Science* 74:117–118.

Daniel, G. 1975. *A hundred and fifty years of archaeology*. Cambridge, MA: Harvard Univ. Press.

Darwin, F. 1898. *The life and letters of Charles Darwin*. New York: D. Appleton.

Daubenmire, R. 1985. The western limits of the range of the American bison. *Ecology* 66:622–624.

Davis, E. M. 1953. Recent data from two Paleo–Indian sites on Medicine Creek, Nebraska. *American Antiquity* 18:380–386.

Davis, W B. 1974. *The mammals of Texas*. Austin:Texas Parks and Wildlife.

Dawkins, W. B. 1883. Early man in America. *North American Review* 137:338–349.

Delaygue, G., T. F. Stocker, F. Joos, and G. Plattner. 2003. Simulation of atmospheric radiocarbon during abrupt oceanic circulation changes: Trying to reconcile models and reconstructions. *Quaternary Science Reviews* 22:1647–1658.

Deller, D. B., and C. Ellis. 1984. Crowfield: a preliminary report on a probable Paleoindian cremation in southwestern Ontario. *Archaeology of Eastern North America* 12:41–71.

DeMallie, R., ed. 2001. Plains. *Handbook of North American Indian*. Vol. 13. Washington, DC: Smithsonian Institution.

DeNiro, M., and C. Hastorf. 1985. Alteration of $^{15}N/^{14}N$ and $^{13}C/^{12}C$ ratios of plant matter during the initial stages of diagenesis: Studies utilizing archaeological specimens from Peru. *Geochimica et Cosmochimica Acta* 49:97–115.

Dick-Peddie, W. 1993. *New Mexico vegetation: Past, present and future*. Albuquerque: Univ. of New Mexico Press.

Dillehay, T. 1989. *Monte Verde: A late Pleistocene settlement in Chile, Vol. 1, Palaeoenvironment and site context*. Washington, DC: Smithsonian Institution Press.

———. 1997. *Monte Verde: a late Pleistocene settlement in Chile, Vol. 2, The archaeological context and interpretation*. Washington, DC: Smithsonian Institution Press.

Dillon, T. J., and A. L. Metcalf. 1997. Altitudinal distribution of land snails in some montane canyons in New Mexico. In *Land Snails of New Mexico*, ed. A. Metcalf and R. A. Smartt. Bulletin 10, 109–127. Albuquerque: New Mexico Museum of Natural History and Science.

Dixon, E. J., and R. Marlar. 1997. A new discovery at the Folsom type site. *Plains Anthropologist* 42:371–374.

Dockall, J. 1997. Wear traces and projectile impact: A review of the experimental and archaeological evidence. *Journal of Field Archaeology* 24:321–331.

Drake, R. J. 1975. Fossil nonmarine molluscs of the 1961–63 Llano Estacado paleoecology study. In *Late Pleistocene environments of the Southern High Plains*, ed. F. Wendorf and J. J. Hester, 201–245. Publication 9. Dallas, TX: Fort Burgwin Research Center, Southern Methodist Univ.

Driver, J. C. 1990. Meat in due season: The timing of communal hunts. In *Hunters of the recent past*, ed. L. B. Davis and B. O. K. Reeves, 11–33. London: Unwin Hyman.

Dungan, M., R. Thompson, J. Stormer, and J. O'Neill. 1989. Excursion 18B: Rio Grande rift volcanism: Northeastern Jemez zone, New Mexico. In *Field excursions to volcanic terranes in the western United States, Vol. 1, Southern Rocky Mountain region*, ed. C. Chapin and J. Zidek, 435–483. New Mexico Bureau of Mines and Mineral Resources, Memoir 46.

Dunnell, R. C. 1971. *Systematics in prehistory*. New York: Free Press.

Eerkens, J. W., and R. Bettinger. 2001. Techniques for assessing standardization in artifact assemblages: Can we scale material variability? *American Antiquity* 66:493–504.

Egeland, C. 2003. Carcass processing intensity and cutmark creation: An experimental approach. *Plains Anthropologist* 48:39–51.

Ehleringer, J. R. 1978. Implications of quantum yield differences in the distributions of C_3 and C_4 grasses. *Oecologia* 31:255–267.

Ehleringer, J. R., T. Cerling, and B. Helliker. 1997. C_4 photosynthesis, atmospheric CO_2, and climate. *Oecologia* 112:285–299.

Elias, S. 1996. Late Pleistocene and Holocene seasonal temperatures reconstructed from fossil beetle assemblages in the Rocky Mountains. *Quaternary Research* 46:311–318.

Ellis, C. J. 1989. The explanation of northeastern Paleoindian lithic procurement patterns. In *Eastern Paleo-indian lithic resource procurement and processing*, ed. C. Ellis and J. Lothrop, 139–164. Boulder, CO: Westview Press.

Elmore, F. H. 1976. *Shrubs and trees of the southwest uplands.* Tucson, AZ: Southwest Parks and Monuments Association.

Emerson, A. 1993. The role of body part utility in small-scale hunting under two strategies of carcass recovery. In *From bones to behavior*, ed. J. Hudson, 138–155. Carbondale: Southern Illinois Univ. Press.

Evans, J. 1860. On the occurrence of flint implements in undisturbed beds of gravel, sand, and clay. *Archaeologia* 38:280–307.

Evans, O. 1930. The antiquity of man as shown at Frederick, Oklahoma: A criticism. *Journal of the Washington Academy of Sciences* 20:475–479.

Ewers, J. 1955. The horse in Blackfoot Indian culture. *Bureau of American Ethnology Bulletin* 159.

Fagan, B. 1987. *The great journey.* London: Thames and Hudson.

Fall, P. L. 1987. Pollen taphonomy in a canyon stream. *Quaternary Research* 28:393–406.

———. 1992a. Pollen accumulation in a montane region of Colorado, USA: A comparison of moss polsters, atmospheric traps, and natural basins. *Review of Palaeobotany and Palynology* 72:169–197.

———. 1992b. Spatial patterns of atmospheric pollen dispersal in the Colorado Rocky Mountains, USA. *Review of Paleobotany and Palynology* 74:293–313.

———. 1997. Timberline fluctuations and late Quaternary paleoclimates in the southern Rocky Mountains, Colorado. *Geological Society of America Bulletin* 109:1306–1310.

Fawcett, W. 1987. Communal hunts, human aggregations, social variation, and climatic change: Bison utilization by prehistoric inhabitants of the Great Plains. Unpublished Ph.D. diss., Univ. of Massachusetts, Amherst.

Figgins, J. D. 1927a. The antiquity of man in America. *Natural History* 27:229–239.

———. 1927b. Report of the Director. *Annual Report of the Colorado Museum of Natural History for the Year 1926*, 9–19.

———. 1928. How long has man hunted. The light fossils shed on the question. *Outdoor Life* 61:18–19, 82.

———. 1933a. A further contribution to the antiquity of man in America. *Proceedings of the Colorado Museum Natural History* 12(2).

———. 1933b. The bison of the western area of the Mississippi Basin. *Proceedings of the Colorado Museum Natural History* 12(4).

———. 1934. Folsom and Yuma artifacts. *Proceedings of the Colorado Museum of Natural History* 13(2).

———. 1935. Folsom and Yuma artifacts. Part II. *Proceedings of the Colorado Museum of Natural History* 14(2).

Findley, J., A. Harris, D. Wilson, and C. Jones. 1975. *Mammals of New Mexico.* Albuquerque: Univ. of New Mexico Press.

Fiorillo, A. R. 1989. An experimental study of trampling: Implications for the fossil record. In *Bone Modification*, ed. R. Bonnichsen and M. Sorg, 61–72. Orono, ME: Center for the Study of the First Americans.

Fisher, J. 1992. Observations on the Late Pleistocene bone assemblage from the Lamb Spring Site, Colorado. In *Ice Age Hunters of the Rockies*, ed. D. J. Stanford and J. S. Day, 51–81. Niwot: Univ. Press of Colorado.

Fisher, T. 2003. Chronology of glacial Lake Agassiz meltwater routed to the Gulf of Mexico. *Quaternary Research* 59:271–276.

Fitzgerald, J., C. Meany, and D. Armstrong. 1994. *Mammals of Colorado.* Denver: Denver Museum of Natural History and Univ. Press of Colorado.

Flenniken, J. 1978. Reevaluation of the Lindenmeier Folsom: A replication experiment in lithic technology. *American Antiquity* 43:473–480.

Flores, R., and T. Cross. 1991. Cretaceous and Tertiary coals of the Rocky Mountains and Great Plains regions. In *Economic geology, U.S.*, ed. H. Gluskoter, D. Rice, and R. Taylor. Vol. P–2 of *The geology of North America*, 547–571. Boulder, CO: Geological Society of America.

Folsom, F. 1992. *Black cowboy: The life and legend of George McJunkin.* Niwot, CO: Roberts Rinehart.

Ford, J. A., and G. R. Willey. 1941. An interpretation of the prehistory of the eastern United States. *American Anthropologist* 43:325–363.

Forman, S., and P. Maat. 1990. Stratigraphic evidence for late Quaternary dune activity near Hudson on the piedmont of northern Colorado. *Geology* 18:745–748.

Forman, S., R. Ogelsby, and R. Webb. 2001. Temporal and spatial patterns of Holocene dune activity on the Great Plains of North America: Megadroughts and climate links. *Global and Planetary Change* 29:1–29.

Frederick, C. D., and C. Ringstaff. 1994. Lithic Resources at Fort Hood: Further Investigations. In *Archeological investigations on 571 prehistoric sites at Fort Hood, Bell and Coryell counties, Texas*, ed. W. N. Trierweiler. United States Army Fort Hood Archeological Resource Management Series Research Report no. 31, 125–181, appendixes.

Fredlund, G., and L. Tieszen. 1997. Phytolith and carbon isotope evidence for Late Quaternary vegetation and climate change in the southern Black Hills, South Dakota. *Quaternary Research* 47:206–217.

Frest, T. J., and R. S. Rhodes II. 1981. *Oreohelix strigosa cooperi* (Binney) in the Midwest Pleistocene. *The Nautilus* 95(2):47–55.

Fricke, H., and J. O'Neil. 1996. Inter- and intra-tooth variation in the oxygen isotope composition of mammalian tooth enamel phosphate: Implications for palaeoclimatological and palaeobiological research. *Palaeogeography, Palaeoclimatology, Palaeoecology* 126:91–99.

Fridmann, S. A. 1967. Pelleting techniques in infrared analysis—A review and evaluation. In *Progress in infrared spectroscopy, Vol. 3*, ed. H. A. Szymanski, 1–22. New York: Plenum Press.

Frison, G. C. 1974. Archaeology of the Casper site. In *The Casper site*, G. Frison, ed., 1–111. Academic Press. New York:

———. 1982a. Folsom components. In *The Agate Basin site: A record of the Paleoindian occupation of the northwestern High Plains*, ed. G. Frison and D. Stanford, 37–76. New York: Academic Press.

———. 1982b. Paleo-Indian winter subsistence strategies on the High Plains. In *Plains Indian studies: A collection of essays in honor of John C. Ewers and Waldo R. Wedel*, ed. D. H. Ubelaker and H. J. Viola, 193–201. Smithsonian Contributions to Anthropology no. 30. Washington, DC: Smithsonian Institution.

———. 1987. The tool assemblage, unfinished bifaces, and stone flaking material sources for the Horner site. In *The

Horner site: The type site of the Cody Cultural Complex, ed. G. Frison and L. Todd, 233–278. New York: Academic Press.

———. 1991. *Prehistoric hunters on the High Plains*. 2nd ed. New York: Academic Press.

———. 1998. The northwestern and northern Plains Archaic. In *Archaeology on the Great Plains*, ed. W. R. Wood, 140–172. Lawrence: Univ. Press of Kansas.

Frison, G. C., and R. Bonnichsen. 1996. The Pleistocene–Holocene transition on the Plains and Rocky Mountains of North America. In *Humans at the end of the Ice Age: The archaeology of the Pleistocene–Holocene transition*, ed. L. G. Straus, B. V. Erikson, J. M. Erlandson, and D. R. Yesner, 303–318. New York: Plenum Press.

Frison, G. C., and B. Bradley. 1980. *Folsom tools and technology at the Hanson site, Wyoming*. Albuquerque: Univ. of New Mexico Press.

Frison, G. C., and L. C. Todd. 1986. *The Colby site: Taphonomy and archaeology of a Clovis kill in northern Wyoming*. Albuquerque: Univ. of New Mexico Press.

Frison, G. C., C. V. Haynes, and M. L. Larson. 1996. Discussion and conclusions. In *The Mill Iron site*, ed. G. Frison, 205–216. Albuquerque: Univ. of New Mexico Press.

Frison, G. C., L. C. Todd, and B. Bradley. 1987. Summary and concluding remarks. In *The Horner site: The type site of the Cody cultural complex*, ed. G. C. Frison and L. C. Todd, 361–369. New York: Academic Press.

Frye, J. C., A. B. Leonard, and H. Glass. 1978. Late Cenozoic sediments, molluscan faunas, and clay minerals in northeastern New Mexico. New Mexico Bureau of Mines and Mineral Resources, Circular 160.

Gadbury, C., L. Todd, A. Jahren, and R. Amundson. 2000. Spatial and temporal variations in the isotopic composition of bison tooth enamel from the Early Holocene Hudson-Meng bone bed, Nebraska. *Palaeogeography, Palaeoclimatology, Palaeoecology* 157:79–93.

Gantt, E. 2002. The Claude C. and A. Lynn Coffin Lindenmeier collection: An innovative method for analysis of privately held artifact collections and new information on a Folsom campsite in northern Colorado. Unpublished master's thesis, Colorado State Univ.

Gibbs, G. 1862. Instructions for archaeological investigations in the United States. *Smithsonian Institution Annual Report for 1861*, 392–396.

Gilder, R. 1911. Scientific "inaccuracies" in reports against probability of geological antiquity of remains of Nebraska Loess man considered by its discoverer. *Records of the Past* 10: 157–169.

Gile, L. H. 1975. Causes of soil boundaries in an arid region. *Soil Science Society of America Proceedings* 39:316–330.

Gilmore, M. 1919. Uses of plants by the Indians of the Missouri River region. *Bureau of American Ethnology Bulletin* 33.

Gladwin, H. 1937. The significance of early cultures in Texas and southeastern Arizona. In *Early Man*, ed. G. G. MacCurdy, 133–138. Philadelphia: Lippincott.

Goddard, P. E. 1926. The antiquity of man in America. *Natural History* 26:257–259.

Goldberg, P., and T. Arpin. 1999. Micromorphological study of selected samples from the Folsom site. Report on file. Dallas, TX: Department of Anthropology, Southern Methodist Univ.

Goodfriend, G. 1992. The use of land snails in paleoenvironmental reconstruction. *Quaternary Science Reviews* 11: 665–685.

Goodfriend, G., and G. L. Ellis. 2002. Stable carbon and oxygen isotopic variations in modern Rabdotus land snail shells in the southern Great Plains, USA, and their relation to environment. *Geochimica et Cosmochimica Acta* 66:1987–2002.

Goodfriend, G., and M. Margaritz. 1987. Carbon and oxygen isotope composition of shell carbonate from desert land snails. *Earth and Planetary Science Letters* 86:377–388.

Goodfriend, G., M. Margaritz, and J. Gat. 1989. Stable isotope composition of land snail body water and its relation to environmental waters and shell carbonate. *Geochimica et Cosmochimica Acta* 53:3215–3221.

Goodyear, A. C. 1974. The Brand Site: A techno-functional study of a Dalton Site in Northeast Arkansas. *Arkansas Archeology Survey, Research Series*, no. 7.

Gosse, J., E. Evenson, J. Klein, B. Lawn, and R. Middleton. 1995. Precise cosmogenic ^{10}Be measurements in western North America: Support for a global Younger Dryas cooling event. *Geology* 23:877–880.

Gould, C. 1906. The geology and water resources of the eastern portion of the Panhandle of Texas. *United States Geological Survey, Water Supply and Irrigation Paper*, no. 154.

Graham, R. 1986. Plant-animal interactions and Pleistocene extinctions. In *Dynamics of extinction*, ed. D. Elliot, 131–154. New York: John Wiley.

Grayson, D. K. 1983. *The establishment of human antiquity*. New York: Academic Press.

———. 1990. The provision of time depth for paleoanthropology. In *Establishment of a geologic framework for paleoanthropology*, ed. L. Laporte. Special Paper 242:1–13. Boulder, CO: Geological Society of America.

Gregg, J. 1844. *Commerce of the prairies*. New York: Langley.

Gregory, W. K. 1927. *Hesperopithecus* apparently not an ape nor a man. *Science* 66:579–581.

Gregory, W. K., and M. Hellman. 1923a. Notes on the type of *Hesperopithecus haroldcookii* Osborn. *American Museum Novitates* no. 53.

———. 1923b. Further notes on the molars of *Hesperopithecus* and of *Pithecanthropus*. *American Museum of Natural History Bulletin* 48:509–530.

Greiser, S. 1983. A preliminary statement about quarrying activity at Flattop Mesa. *Southwestern Lore* 49:6–14.

———. 1985. Predictive models of hunter-gatherer subsistence and settlement strategies on the central High Plains. *Plains Anthropologist Memoir* 20.

Griffin, J. B. 1946. Cultural change and continuity in eastern United States archaeology. In *Man in northeastern North America*, ed. F. Johnson. Vol. 3 of *Papers of the Robert S. Peabody Foundation for Archaeology*, 37–95. Andover, MA: Phillips Academy.

Grimm, E. C. 1987. CONISS; a FORTRAN 77 program for stratigraphically constrained cluster analysis by the method of incremental sum of squares. *Computers and Geosciences* 13:13–25.

Grocke, D., H. Bocherens, and A. Mariotti. 1997. Annual rainfall and nitrogen isotope correlation in macropod collagen: Application as a paleoprecipitation indicator. *Earth and Planetary Science Letters* 153:279–285.

Gruber, J. W. 1965. Brixham Cave and the antiquity of man. In *Context and meaning in cultural anthropology*, ed. M. Spiro, 373–402. New York: Free Press.

Gryba, E. 1988. A stone age pressure method of Folsom fluting. *Plains Anthropologist* 33:53–66.

———. 1989. A mousetrap 10,000 years too late. *Plains Anthropologist* 34:65–68.

Gunnerson, D. 1974. *The Jicarilla Apaches: A study in survival.* DeKalb: Northern Illinois Univ. Press.

Gustavson, T., and R. Finley. 1985. Late Cenozoic geomorphic evolution of the Texas Panhandle and northeastern New Mexico. Bureau of Economic Geology Report of Investigations no. 148.

Gustavson, T., R. Baumgardner, C. Caran, V. Holliday, H. Mehnert, M. O'Neill, and C. Reeves. 1991. Quaternary geology of the Southern Great Plains and an adjacent segment of the Rolling Plains. In *Quaternary nonglacial geology; Coterminous U.S.,* ed. R. B. Morrison. Vol. K-2 in *The geology of North America,* 477–501. Boulder, CO: Geological Society of America.

Gutentag, E. D., F. J. Heimes, N. C. Krothe, R. R. Luckey, and J. B. Weeks. 1984. Geohydrology of the High Plains aquifer in parts of Colorado, Kansas, Nebraska, New Mexico, Oklahoma, South Dakota, Texas, and Wyoming. *United States Geological Survey Professional Paper* 1400-B.

Guthrie, R. D. 1983. Osseus projectile points: Biological considerations affecting raw material selection and design among Paleolithic and Paleoindian peoples. In *Animals and archaeology, Vol. 1, Hunters and their prey,* ed. J. Clutton–Brock and C. Grigson. BAR International Series no. 163. London.

———. 1990. *Frozen fauna of the Mammoth Steppe.* Chicago: Univ. of Chicago Press.

Guyer, G. W. 1988. Untitled news story, August 31, 1908, originally published in *La Epoca,* the Folsom weekly newspaper. In *Folsom 1888–1988. Then and now,* 31–32. Folsom, NM: Centennial Book Committee.

Hadjas, I., G. Bonani, P. Boden, D. Peteet, D. Mann. 1998. Cold reversal on Kodiak Island, Alaska, correlated with the European Younger Dryas by using variations of atmospheric 14C content. *Geology* 26:1047–1050.

Haley, J. E. 1936. *Charles Goodnight: Cowman and plainsman.* Boston: Houghton Mifflin.

Hall, S. A. 1985. Quaternary pollen analysis and vegetational history of the Southwest. In *Pollen records of Late-Quaternary North American sediments,* ed. V. Bryant and R. Holloway, 95–123. American Association of Stratigraphic Palynologists.

———. 1990. Pollen deposition and vegetation in the southern Rocky Mountains and southwest Plains, USA. *Grana* 29:47–61.

———. 1994. Modern pollen influx in tallgrass and shortgrass prairies, southern Great Plains, USA. *Grana* 33:321–326.

Hallowell, A. I. 1928. Proceedings of the American Anthropological Association for the year ending, December 1927. *American Anthropologist* 30:532–543.

Hanson, J. 1984. Bison ecology in the northern Plains and a reconstruction of bison patterns for the North Dakota region. *Plains Anthropologist* 29:93–113.

Harrison, B., and K. Killen. 1978. *Lake Theo: a stratified, early man bison butchering and camp site, Briscoe County, Texas.* Special Archaeological Report 1. Canyon, TX: Panhandle-Plains Historical Museum, Canyon.

Hassan, A. 1975. Geochemical and mineralogical studies on bone material and their implications for radiocarbon dating. Unpublished Ph.D. diss., Southern Methodist Univ.

Hawkes, K., and R. Bleige Bird. 2002. Showing off, handicap signaling, and the evolution of men's work. *Evolutionary Anthropology* 11:58–67.

Hawkes, K., J. F. O'Connell, and N. G. Blurton-Jones. 2001. Hadza meat sharing. *Evolution and Human Behavior* 22: 113–142.

Hay, O. P. 1918. Doctor Aleš Hrdlička and the Vero man. *Science* 47:459–462.

———. 1927. An account of three recent important finds of relics of man associated with remains of Pleistocene Mammalia. Paper presented at the Annual Meeting of the American Association for the Advancement of Sciences, Nashville, TN. Manuscript on file, Oliver Hay Papers, Smithsonian Institution Archives.

———. 1928. On the antiquity of relics of man at Frederick, Oklahoma. *Science* 67:442–444.

Hay, O. P., and H. J. Cook. 1928. Preliminary descriptions of fossil mammals recently discovered in Oklahoma, Texas and New Mexico. *Proceedings of the Colorado Museum Natural History* 8:33.

———. 1930. Fossil vertebrates collected near, or in association with, human artifacts at localities near Colorado, Texas; Frederick, Oklahoma; and Folsom, New Mexico. *Proceedings of the Colorado Museum Natural History* 9: 4–40.

Haynes, C. V. 1975. Pleistocene and recent stratigraphy. In *Late Pleistocene environments of the Southern High Plains,* ed. F. Wendorf and J. J. Hester, 57–96. Publication 9. Dallas, TX: Fort Burgwin Research Center, Southern Methodist Univ.

———. 1990. The Antevs-Bryan years and the legacy for Paleoindian geochronology. *In Establishment of a geologic framework for paleoanthropology,* ed. L. Laporte. Special Paper 242:55–68. Boulder, CO: Geological Society of America.

———. 1991. Geoarchaeological and paleohydrological evidence for a Clovis-age drought in North America. *Quaternary Research* 35:438–450.

———. 1995. Geochronology of paleoenvironmental change, Clovis type site, Blackwater Draw, New Mexico. *Geoarchaeology* 10:317–388.

———. 2003. Dating the Lindenmeier Folsom site, Colorado, U.S.A., before the radiocarbon revolution. *Geoarchaeology* 18: 161–174.

Haynes, C. V., A. Anderson, and F. Frazier. 1976. Geochronology of the Folsom site. Manuscript on file, Department of Anthropology, Univ. of Arizona, Tucson.

Haynes, C. V., R. Beukens, A. T. Jull, and O. K. Davis. 1992. New radiocarbon dates for some old Folsom sites: Accelerator technology. In *Ice age hunters of the Rockies,* ed. D. J. Stanford and J. S. Day, 83–100. Niwot: Univ. Press of Colorado.

Haynes, G. 1980. Prey bones and predators: Potential ecologic information from analysis of bone sites. *Ossa* 7:75–97.

Haynes, H. W. 1881. Their comparison with paleolithic implements from Europe. *Proceedings Boston Society of Natural History* 21:132– 137.

Heath, R. C. 1988. Hydrogeologic settings of regions. In *Hydrogeology,* ed. W. Back, J. S. Rosenhein, and P. R. Seaber, *The Geology of North America.* Volume O-2:15–23. Boulder, CO: Geological Society of America.

Heaton, T. 1999. Spatial, species, and temporal variations in the $^{13}C/^{12}C$ ratios of C_3 plants: Implications for palaeodiet studies. *Journal of Archaeological Science* 26:637–649.

Heaton, T., J. Vogel, G. von la Chevellerie, and G. Collett. 1986. Climatic influence on the isotopic composition of bone nitrogen. *Nature* 322:822–823.

Hedges, R. E., R. Stevens, and M. Richards. 2004. Bone as a stable isotope archive for local climatic information. *Quaternary Science Reviews* 23:959–965.

Henry, J. 1862. Report of the Secretary. *Smithsonian Institution Annual Report for 1861,* 13–48.

Hester, J. 1962. A Folsom lithic complex from the Elida site, Roosevelt County, New Mexico. *El Palacio* 69:92–113.

———. 1972. *Blackwater Locality no. 1: A stratified early man site in eastern New Mexico,* 164–180. Dallas, TX: Fort Burgwin Research Center.

Hester, J.J., and J. Grady. 1977. Paleoindian social patterns on the Llano Estacado. *Museum Journal* 17:78–96.

Hewett, J. 1971. The bookish black at Wild Horse arroyo. How the Folsom man came to light. *New Mexico* 49:20–24.

Hibbard, C.W., and D.W. Taylor. 1960. Two Late Pleistocene faunas from southwestern Kansas. In *Contributions from the Museum of Paleontology.* Vol. 16(1). Ann Arbor: Univ. of Michigan.

Hietala, H. 1989. Contemporaneity and occupational duration of the Kubbaniyan sites: An analysis and interpretation of the radiocarbon record. In *The prehistory of Wadi Kubbaniya,* ed. A. Close, ed., Vol. 2, 284–291. Dallas, TX: Southern Methodist Univ. Press.

Hill, M.E., and J.L. Hofman. 1997. The Waugh site: A Folsom-age bison bonebed in northwestern Oklahoma. *Plains Anthropologist* 42:63–83.

Hill, M.G. 1994. Subsistence strategies by Folsom hunters at Agate Basin, Wyoming: A taphonomic analysis of the bison and pronghorn assemblages. Unpublished master's thesis, Univ. of Wyoming, Laramie.

———. 2001. Paleoindian diet and subsistence behavior on the northwestern Great Plains of North America. Unpublished Ph.D. diss., Univ. of Wisconsin, Madison.

Hill, W.E. 1992. *Santa Fe Trail: Yesterday and today.* Caldwell, ID: Caxton Printers.

Hillerman, T. 1971. Or: How Folsom was saved to history. *New Mexico* 49:25–27.

———. 1973. *The Great Taos bank robbery and other Indian Country affairs.* Albuquerque: Univ. of New Mexico Press.

Hinsley, C. 1981. *Savages and scientists: The Smithsonian Institution and the development of American anthropology.* Washington, DC: Smithsonian Institution Press.

Hoard, R., S. Holen, M. Glascock, H. Neff, and J. Elam. 1992. Neutron activation analysis of stone from the Chadron formation and a Clovis site on the Great Plains. *Journal of Archaeological Science* 19:655–665.

Hoard, R., J. Bozell, S. Holen, M. Glascock, H. Neff, and J. Elam. 1993. Source determination of White River Group silicates from two archaeological sites in the Great Plains. *American Antiquity* 58:698–710.

Hoff, C.C. 1962. Some terrestrial Gastropoda from New Mexico. *Southwestern Naturalist* 7(1):51–63.

Hofman, J.L. 1989. Prehistoric culture history—Hunters and gatherers in the southern Great Plains. In *From Clovis to Comanchero: Archaeological overview of the southern Great Plains,* ed. J. Hofman et al., 25–60. Arkansas Archaeological Survey Research Series 35.

———. 1990. Salt Creek, recent evidence from the eastern Folsom margin in central Oklahoma. *Plains Anthropologist* 35:367–374.

———. 1991. Folsom land use: Projectile point variability as a key to mobility. In *Raw material economies among prehistoric hunter-gatherers,* ed. A. Montet–White and S. Holen, 335–355. Univ. of Kansas Publications in Anthropology 19.

———. 1992. Recognition and interpretation of Folsom technological variability on the Southern Plains. In *Ice Age hunters of the Rockies,* ed. D.J. Stanford and J.S. Day, 193–224. Niwot: Univ. Press of Colorado.

———. 1994. Paleoindian aggregations on the Great Plains. *Journal of Anthropological Archaeology* 13:341–370.

———. 1995. Dating Folsom occupations on the Southern Plains: The Lipscomb and Waugh sites. *Journal of Field Archaeology* 22:421–437.

———. 1996. Early hunter-gatherers of the Central Great Plains: Paleoindian and Mesoindian (Archaic) Cultures. In *Archaeology and paleoecology of the central Great Plains,* ed. J. Hofman, 41–100. Arkansas Archaeological Survey Research Series no. 48.

———. 1999a. Folsom fragments, site types, and assemblage formation. In *Folsom lithic technology: Explorations in structure and variation,* ed. D.S. Amick, 122–143. International Monographs in Prehistory, Archaeological Series 12. Ann Arbor, MI.

———. 1999b. Unbounded hunters: Folsom bison hunting on the Southern Plains, circa 10500 BP, the lithic evidence. In *Le bison: Gibier et moyen de subsistance de hommes du paleolithique aux paleoindiens des grandes plaines,* ed. J. Jaubert, J. Brugal, F. David, and J. Enloe, 383–415. Antibes, France: Editions APDCA.

Hofman, J.L., and R. Graham. 1998. The Paleoindian cultures of the Great Plains. In *Archaeology of the Great Plains,* ed. W.R. Wood, 87–139. Lawrence: Univ. Press of Kansas.

Hofman, J.L., and L. Todd. 1990. The Lipscomb bison quarry: 50 years of research. *Transactions of the 25th Regional Archaeological Symposium for Southeastern New Mexico and Western Texas,* 43–58.

———. 2001. Tyranny in the archaeological record of specialized hunters. In *People and wildlife in northern America: Essays in honor of R. Dale Guthrie,* ed. S.C. Gerlach and M.S. Murray, 200–215. BAR International Series 944.

Hofman, J.L., D. Amick, and R. Rose. 1990. Shifting sands: A Folsom-Midland assemblage from a campsite in western Texas. *Plains Anthropologist* 35:221–253.

Hofman, J.L. Todd, and M. Collins. 1991. Identification of central Texas Edwards chert at the Folsom and Lindenmeier sites. *Plains Anthropologist* 36:297–308.

Holliday, V.T. 1995. Stratigraphy and paleoenvironments of Late Quaternary valley fills on the Southern High Plains. *Geological Society of America Memoir* 186.

———. 1997. Paleoindian geoarchaeology of the southern High Plains. Austin: University of Texas Press.

———. 2000a. Folsom drought and episodic drying on the southern High Plains from 10,900–10,200 ^{14}C yr B.P. *Quaternary Research* 53, 1–12.

———. 2000b. The evolution of Paleoindian geochronology and typology on the Great Plains *Geoarchaeology* 15: 227–290.

———. 2001. Stratigraphy and geochronology of upper Quaternary eolian sand on the Southern High Plains of Texas and New Mexico, U.S.A. *Geological Society of America Bulletin* 113:88–108.

Holliday, V.T. and A. Anderson. 1993. "Paleoindian," "Clovis" and "Folsom": A brief etymology. *Current Research in the Pleistocene* 10:79–81.

Holliday, V.T. and C. Welty. 1981. Lithic tool resources of the eastern Llano Estacado. *Bulletin of the Texas Archeological Society* 52:201–214.

Holmes, W.H. 1890. A quarry workshop of the flaked stone implement makers in the District of Columbia. *American Anthropologist* 3:1–26.

———. 1892. Modern quarry refuse and the Palaeolithic theory. *Science* 20:295–297.

———. 1893a. Are there traces of man in the Trenton gravels? *Journal of Geology* 1:15–37.

———. 1893b. Traces of glacial man in Ohio. *Journal of Geology* 1:147–163.

———. 1893c. Vestiges of early man in Minnesota. *American Geologist* 11:219–224.

———. 1894. A natural history of flaked stone implements. In *Memoirs of the International Congress of Anthropology*, ed. C. Wake, 120–139. Chicago: Schulte Publishing.

———. 1897a. Primitive man in the Delaware Valley. *Science* 6:824–829.

———. 1897b. Stone implements of the Potomac-Chesapeake Tidewater province. *Bureau of American Ethnology Annual Report* 15:3–152

———. 1899. Preliminary revision of evidence relating to Auriferous gravel man in California. *American Anthropologist* 1:107–121, 614–645.

———. 1902. Fossil human remains found near Lansing, Kansas. *American Anthropologist* 4:743–752.

———. 1918. On the antiquity of man in America. *Science* 47:561–562

———. 1919. Handbook of aboriginal American antiquities. Part I. *Bureau of American Ethnology Bulletin* 60.

———. 1925. The antiquity phantom in American archaeology. *Science* 62:256–258.

Hooton, E.A. 1937. *Apes, men, and morons.* New York: G.P. Putnam's Sons.

Hoppe, K., P. Koch, R. Carlson, and S.D. Webb. 1999. Tracking mammoths and mastodons: Reconstruction of migratory behavior using strontium isotope ratios. *Geology* 27: 439–442.

Howard, E.B. 1935. Evidence of early man in North American. *The Museum Journal, Univ. of Pennsylvania Museum* 24(2–3).

———. 1936a. An outline of the problem of man's antiquity in North America. *American Anthropologist* 38:394–413.

———. 1936b. The association of a human culture with an extinct fauna in New Mexico. *American Naturalist* 70: 314–323

———. ed. 1936c. Early man in America with particular reference to the southwestern United States. *American Naturalist* 70:313–371.

———. 1938. Minutes of the International Symposium on Early Man held at the Academy of Natural Science of Philadelphia. *Proceedings of the Academy of Natural Science of Philadelphia* 89:439–447.

———. 1943. The Finley site. Discovery of Yuma points, in situ, near Eden, Wyoming. *American Antiquity* 8:224–234.

Hrdlička, A. 1902. The crania of Trenton, New Jersey, and their bearing upon the antiquity of man in that region. *Bulletin American Museum of Natural History* 16:23–62.

———. 1903. The Lansing skeleton. *American Anthropologist* 5:323–330.

———. 1907. Skeletal remains suggesting or attributed to early man in North America. *Bureau of American Ethnology Bulletin* 33.

———. 1912. Early man in South America. In collaboration with W. Holmes, B. Willis, F. Wright and C. Fenner. *Bureau of American Ethnology Bulletin* 52.

———. 1918. Recent discoveries attributed to early man in America. *Bureau of American Ethnology Bulletin* 66.

———. 1926. The race and antiquity of the American Indian. *Scientific American* 135:7–9

———. 1928. The origin and antiquity of man in America. *New York Academy of Medicine Bulletin* 4(7):802–816.

———. 1937. Early man in America: What have the bones to say? In *Early Man*, ed. G.G. MacCurdy, 93–104. Philadelphia: Lippincott.

———. 1942. The problem of man's antiquity in America. *Proceedings of the Eigth American Scientific Congress* II:53–55.

Huang, Y., F. Street-Perrott, S. Metcalfe, M. Brenner, M. Moreland, and K. Freeman. 2001. Climate change as the dominant control on glacial-interglacial variations in C_3 and C_4 plant abundance. *Science* 293:1647–1651.

Hubricht, L. 1983. Five new species of land snails from the southeastern United States with notes on other species. *Gastropodia* 2:13–19.

———. 1985. The distributions of the native land mollusks of the eastern United States. *Fieldiana Zoology* N.S. 24. Chicago: Field Museum of Natural History.

Huckell, L. 1998. Appendix D. In *Testing for additional Folsom-age components at the Folsom type site, 29CX1, New Mexico. Report on archaeological fieldwork conducted under State of New Mexico Permit no. AE–74*, by D.J. Meltzer. Report on file, Office of Cultural Affairs, State of New Mexico, Santa Fe.

Hudson, R.J., and S. Frank. 1987. Foraging ecology of bison in aspen boreal habitats. *Journal of Range Management* 40:71–75.

Hughen, K., J. Southon, S. Lehman, S., and J. Overpeck. 2000. Synchronous radiocarbon and climate shifts during the last deglaciation. *Science* 290:1951–1954.

Hull, D. 1988. *Science as a process: An evolutionary account of the social and conceptual development of science.* Chicago: Univ. of Chicago Press.

Humphrey, J.D., and C.R. Ferring. 1994. Evidence for Latest Pleistocene and Holocene climate change in North-Central Texas. *Quaternary Research* 41:200–213.

Hunt, C. 1967. *Physiography of the United States.* San Francisco: Freeman.

Hunt, A., S. Lucas, and B. Kues. 1987. Third-day road log, from Clayton to Des Moines, Capulin, Caoulin Mountain National Monument, Folsom, Raton and the Cretaceous/Tertiary boundary on Goat Hill. In *Northeastern New Mexico*, 41–54. New Mexico Geological Society Guidebook, 38th Field Conference,

Ingbar, E. 1992. The Hanson site and Folsom on the northwestern Plains. In *Ice Age hunters of the Rockies*, ed. D.J. Stanford and J.S. Day, 169–192. Niwot: Univ. Press of Colorado.

———. 1994. Lithic raw material selection and technological organization. In *The organization of North American prehistoric chipped stone tool technologies*, ed. P. Carr, ed. International Monographs in Prehistory, Archaeological Series 7. Ann Arbor, MI.

Ingbar, E., and J. Hofman. 1999. Folsom fluting fallacies. In *Folsom lithic technology: explorations in structure and variation*, ed. D.S. Amick, 98–110. International Monographs in Prehistory, Archaeological Series 12. Ann Arbor, MI.

Irving, W. 1985. Context and chronology of early man in the Americas. *Annual Review of Anthropology* 14:529–555.

Irving, W., A. Jopling, and B. Beebe. 1986. Indications of pre-Sangamon humans near Old Crow, Yukon, Canada. In *New evidence for the Pleistocene peopling of the Americas*, ed. A. Bryan, 49–63. Orono, ME: Center for the Study of Early Man.

Ivey, R. 1995. *Flowering plants of New Mexico*. 3rd ed. Albuquerque: Rio Rancho.

Jahren, A., L. C. Todd, and R. Amundson. 1998. Stable isotope dietary analysis of bison bone samples from the Hudson-Meng bonebed: Effects of paleotopography. *Journal of Archaeological Science* 25:465–475.

Jenness, D., ed. 1933. *The American Aborigines: Their origin and antiquity*, ed. D. Jenness, 9–45. Toronto: Univ. of Toronto Press.

Jodry, M. 1987. Stewart's Cattle Guard site: A Folsom site in southern Colorado. A report of the 1981 and 1983 field seasons. Unpublished master's thesis, Univ. of Texas, Austin.

———. 1999a. Folsom technological and socioeconomic strategies: views from Stewart's Cattle Guard and the Upper Rio Grande Basin, Colorado. Unpublished diss., American Univ.

———. 1999b. Paleoindian stage. In *Colorado prehistory: A context for the Rio Grande Basin*, ed. M. Martorano, T. Hoefer, M. Jodry, V. Spero, and M. Taylor, 45–114. Golden: Colorado Council of Professional Archaeologists.

Johnson, E., ed. 1987. *Lubbock Lake: Late Quaternary studies on the southern High Plains*. College Station: Texas A&M Press.

Jones, G. T., D. Grayson, and C. Beck. 1983. Artifact class richness and sample size in archaeological surface assemblages. In *Lulu Linear Punctated: essays in honor of George Irving Quimby*, ed. R. C. Dunnell and D. K. Grayson. *Anthropological Papers* 72:55–73. Museum of Anthropology, Univ. of Michigan.

Jones, G. T., C. Beck, E. Jones, and R. Hughes. 2003. Lithic source use and Paleoarchaic foraging territories in the Great Basin. *American Antiquity* 68:5–38.

Jones, J., D. Armstrong, and J. Choate. 1985. *Guide to mammals of the Plains states*. Lincoln: Univ. of Nebraska Press.

Jorgensen, D. G., J. Downey, A. Dutton, and R. Maclay. 1988. Region 16, Central Nonglaciated Plains. In *Hydrogeology*, ed. W. Back, J. S. Rosenhein, and P. R. Seaber. Vol. O-2 in *The geology of North America*, 141–156. Boulder, CO: Geological Society of America.

Judd, N. M. 1967. *The Bureau of American Ethnology: a partial history*. Norman: Univ. of Oklahoma Press.

Judge, W. J. 1970. Systems analysis and the Folsom-Midland question. *Southwestern Journal of Anthropology* 26:40–51.

———. 1973. *Paleoindian occupation of the central Rio Grande Valley in New Mexico*. Albuquerque:Univ. of New Mexico Press.

Karkanas, P., N. Kyparissi-Apostolika, O. Bar-Yosef, and S. Weiner. 1999. Mineral assemblages in Theopetra, Greece: A framework for understanding diagenesis in a prehistoric cave. *Journal of Archaeological Science* 26:1171–1180.

Karkanas, P., O. Bar–Yosef, P. Goldberg, and S. Weiner. 2000. Diagenesis in prehistoric caves: The use of minerals that form *in situ* to assess the completeness of the archaeological record. *Journal of Archaeological Science* 27:915–929.

Kay, M. 1996. Microwear analysis of some Clovis and experimental chipped stone tools. In *Stone tools: Theoretical insights into human prehistory*, ed. G. Odell, 315–344. New York: Plenum Press.

Keeley, L. 1982. Hafting and retooling: Effects on the archaeological record. *American Antiquity* 47:798–809.

Kelly, R. L. 1988. Three sides of a biface. *American Antiquity* 53: 717–734.

———. 1992. Mobility/sedentism: Concepts, archaeological measures, and effects. *Annual Review of Anthropology* 21: 43–66.

———. 1995. *The foraging spectrum*. Washington, DC: Smithsonian Institution Press.

———. 2003. Colonization of new land by hunter–gatherers: expectation and implications based on ethnographic data. In *Colonization of unfamiliar landscapes: The archaeology of adaptation*, ed. M. Rockman and J. Steele, 44–58. London: Routledge.

Kelly, R. L., and L. C. Todd. 1988. Coming into the country: Early Paleoindian hunting and mobility. *American Antiquity* 53:231–244.

Kennett, J. P., K. Elmstrom, and N. Penrose. 1985. The last deglaciation in Orca Basin, Gulf of Mexico: High-resolution planktonic foraminiferal changes. *Palaeogeography, Palaeoclimatology, Palaeoecology* 50:189–216.

Kidder, A. V. 1924. An introduction to the study of southwestern archaeology, with a preliminary account of the excavations at Pecos. *Papers of the Southwestern Expedition, Phillips Academy*, no. 1. New Haven, CT: Yale Univ. Press.

———. 1927. Early man in America. *The Masterkey* 1(5): 5–13.

———. 1936. Speculations on New World prehistory. In *Essays in anthropology*, ed. R. Lowie, 143–151. Berkeley: Univ. of California Press.

Kidder, A. V., and R. Terry. 1927. Section H. Anthropology. *Science* 65:111–112.

Kilmer, L. C. 1987. Water-bearing characteristics of geologic formations in northeastern New Mexico—southeastern Colorado. In *Northeastern New Mexico*, 275–279. New Mexico Geological Society Guidebook, 38th Field Conference.

Kindscher, K. 1987. *Edible wild plants of the prairie: An ethnobotanical guide*. Lawrence: Univ. Press of Kansas.

King, F. B., and R. W. Graham. 1981. Effects of ecological and paleoecological patterns on subsistence and paleoenvironmental reconstructions. *American Antiquity* 46:128–142.

Kitagawa, H., and J. van der Plicht. 1998. Atmospheric radiocarbon calibration to 45,000 yr B.P.: Late Glacial fluctuations and cosmogenic isotope production: *Science* 279: 1187–1190.

Knapp, A., J. Bair, J. Briggs, S. Collins, D. Harnett, L. Johnson, and E. Towne. 1999. The keystone role of Bison in North American tallgrass prairie. *BioScience* 49:39–50.

Knight, P. 1987. The vegetation of northeastern new Mexico. In *Northeastern New Mexico*, 2–6. New Mexico Geological Society Guidebook, 38th Field Conference.

Knox, J. C. 1983. Responses of river systems to Holocene climate. In *Late-Quaternary environments of the United States, Vol. 2, The Holocene*, ed. H. E. Wright, 26–41. Minneapolis: Univ. of Minnesota Press.

Koch, P. L., N. Diffenbaugh, and K. Hoppe. 2004. The effects of late Quaternary climate and pCO_2 change on C_4 plant abundance in the south-central United States. *Palaeogeography, Palaeoclimatology, Palaeoecology* 207:331–357.

Kraft, K. C. 1997. The distribution of Alibates silicified dolomite clasts along the Canadian River. *Current Research in the Pleistocene* 14:106–109.

Kreutzer, L. E. 1988. Megafaunal butchering at Lubbock Lake, Texas: A taphonomic reanalysis. *Quaternary Research* 30:221–231.

———. 1992. Bison and deer bone mineral densities: Comparisons and implications for the interpretation of archaeological faunas. *Journal of Archaeological Science* 19: 271–294.

———. 1996. Taphonomy of the Mill Iron site Bison bonebed. In *The Mill Iron Site*, ed. G. C. Frison, 101–143. Albuquerque: Univ. of New Mexico Press.

Krieger, A. 1947. Certain projectile points of the early American hunters. *Bulletin of the Texas Archeological and Paleontological Society* 18:7–27.

———. 1953. New World culture history: Anglo-America. In *Anthropology today: An encyclopedic inventory*, 238–264. Prepared under the chairmanship of A. L. Kroeber. Chicago: Univ. of Chicago Press.

———. 1964. Early man in the New World. In *Prehistoric man in the New World*, ed. J. Jennings and E. Norbeck, 23–81. UChicago: niv. of Chicago Press.

Kroeber, A. 1940. Conclusions: The present status of Americanistic problems. In *The Maya and their neighbors*, ed. C. Hay, S. Lothrop, R. Linton, H. Shapiro and G. Valliant, 460–476. New York: D. Appleton.

———. 1952. *The nature of culture*. Chicago: Univ. of Chicago Press.

Krueger, H., and C. Sullivan. 1984. Models for carbon isotope fractionation between diet and bone. In *Stable isotopes in nutrition*, ed. J. Turland and P. Johnson. *American Chemical Society Symposium Series* 258:205–220.

Kudo, A. M. 1976. A review of the volcanic history and stratigraphy of northeastern New Mexico. *New Mexico Geological Society Guidebook* 27:109–111.

Kues, B., and S. Lucas. 1987. Cretaceous stratigraphy and paleontology in the Dry Cimarron valley, New Mexico, Colorado, and Oklahoma. In *Northeastern New Mexico*, 167–198. New Mexico Geological Society Guidebook, 38th Field Conference.

Kutzbach, J. 1987. Model simulations of the climatic patterns during the deglaciation of North America. In *North America and adjacent oceans during the last deglaciation*, ed. W. F. Ruddiman and H. E. Wright, 425–446. Boulder, CO: Geological Society of America.

Kutzbach, J., and W. F. Ruddiman. 1993. Model description, external forcing, and surface boundary conditions. In *Global climates since the last glacial maximum*, ed. H. E. Wright, J. E. Kutzbach, T. Webb, W. F. Ruddiman, F. A. Street-Perrott, and P. J. Bartlein, 12–23. Minneapolis: Univ. of Minnesota Press.

Kutzbach, J., P. J. Guetter, P. J. Behling, and R. Selin. 1993. Simulated climatic changes: Results of the COHMAP climate-model experiments. In *Global climates since the last glacial maximum*, ed. H. E. Wright, J. E. Kutzbach, T. Webb, W. F. Ruddiman, F. A. Street-Perrott, and P. J. Bartlein, 24–93. Minneapolis: Univ. of Minnesota Press.

LaBelle, J. M. 2000. Review of *Bison Hunting at Cooper Site: Where Lightning Bolts Drew Thundering Herds*, by Leland Bement. *Geoarchaeology* 15(4):375–377.

LaBelle, J. 2002. Slim Arrow, the long-forgotten Yuma-type site in eastern Colorado. *Current Research in the Pleistocene* 19: 52–55.

LaBelle, J. M. 2005. Hunter-gatherer foraging variability during the Early Holocene of the Central Plains of North America. Unpublished Ph.D. diss., Southern Methodist Univ.

LaBelle, J., J. Seebach, and B. Andrews. 2003. Folsom site structure and settlement organization in the Great Plains and Rocky Mountains. Part A. Folsom and ethnographic site structure— 1:2000 scale. Part B. Folsom settlement at the foraging and regional scales. Part C. An atlas of Folsom site structure—1:2000 scale. Poster session presented at the 68th Annual Meeting of the Society for American Archaeology, Milwaukee, WI.

Lam, Y., X. Chen ,and O. Pearson. 1999. Intertaxonomic variability in patterns of bone density and the differential representation of bovid, cervid, and equid elements in the archaeological record. *American Antiquity* 64:343–362.

Lanner, R. 1981. *The piñon: A natural and cultural history*. Reno: Univ. of Nevada Press.

Largent, F., M. Waters, and D. Carlson. 1991. The spatiotemporal distribution and characteristics of Folsom projectile points in Texas. *PA* 36:323–341.

LaRocque, A. 1966. Pleistocene mollusca of Ohio. *Ohio Geological Survey Bulletin* 62(1):1–112.

Larson, F. 1940. The role of Bison in maintaining the short grass Plains. *Ecology* 21:113–121.

Larson, R., L. Todd, E. Kelly, and J. Welker. 2001. Carbon stable isotopic analysis of bison dentition. *Great Plains Research* 11: 25–64.

Lauenroth, W., I. Burke, and M. Gutman. 1999. The structure and function of ecosystems in the central North American grassland region. *Great Plains Research* 9:223–259.

Lavender, D. 1954. *Bent's Fort*. New York: Doubleday.

Laves, G. 1928. Reconnaissance at the Folsom site. Unpublished report on file, Anthropology Archives, Department of Anthropology, American Museum of Natural History, New York.

Leakey, L., R. Simpson, and T. Clements. 1968. Archaeological excavations in the Calico Mountains, California: Preliminary report. *Science* 160:1022–1023.

Lécolle, P. 1985. The oxygen isotope composition of landsnail shells as a climatic indicator: Applications to hydrogeology and paleoclimatology. *Chemical Geology* 58:157–181.

Lee, R. B. 1979. *The !Kung San. Men, women, and work in a foraging society*. Cambridge: Cambridge Univ. Press.

Leonard, A. B. 1950. A Yarmouthian molluscan fauna in the midcontinent region of the United States. *University of Kansas Paleontological Contributions, Mollusca* Article 3:1–48

———. 1959. *Handbook of gastropods in Kansas*. Miscellaneous Publication no. 20. Lawrence, KS: Museum of Natural History.

Leonard, R., and G. T. Jones, eds. 1989. *Quantifying diversity in archaeology*. Cambridge: Cambridge Univ. Press.

Lepper, B. 1983. Fluted point distributional patterns in the eastern United States: A contemporary phenomena. *Midcontinental Journal of Archaeology* 8:269–285.

LeTourneau, P. 1998a. The 'Folsom Problem.' In *Unit issues in archaeology: Measuring time, space, and material*, ed. A. Ramenofsky and A. Steffen, 52–73. Salt Lake City: Univ. of Utah Press.

———. 1998b. Folsom use of eastern New Mexico look-alike cherts on the Llano Estacado. *Current Research in the Pleistocene* 15:77–79.

———. 2000. Folsom toolstone procurement in the Southwest and Southern Plains. Unpublished Ph.D. diss., Univ. of New Mexico, Albuquerque.

Leuenberger M., U. Siegenthaler, and C. Langway. 1992. Carbon isotope composition of atmospheric CO_2 from an Antarctic ice core. *Nature* 357:488–490.

Lewis, H. C. 1881. The antiquity and origin of the Trenton gravel. In *Primitive industry*, ed. C. C. Abbott, 521–551. Salem, OR: George A. Bates.

Leyden, J., and G. Oetelaar. 2001. Carbon and nitrogen isotopes in archaeological bison remains as indicators of paleoenvironmental change in southern Alberta. *Great Plains Research* 11:3–23.

Little, C. 1947. The Folsom man in Colorado. *Rocky Mountain Empire Magazine*, February 2, 1947.

Lovvorn, M., G. Frison, and L. Tieszen. 2001. Paleoclimate and Amerindians: Evidence from stable isotopes and atmospheric circulation. *Proceedings of the National Academy of Sciences* 98: 2485–2490.

Lucas, S., A. Hunt, and S. Hayden. 1987. The Triassic system in the Dry Cimarron valley, New Mexico, Colorado, and Oklahoma. In *Northeastern New Mexico*, 97–117. New Mexico Geological Society Guidebook, 38th Field Conference.

Lyell, C. 1830–1833. *Principles of geology*. Vols. 1–3. London: John Murray.

Lyman, R. L. 1985. Bone frequencies: Differential transport, in situ destruction, and the MGUI. *Journal of Archaeological Science* 12:221–236.

———. 1994. *Vertebrate taphonomy*. Cambridge: Cambridge Univ. Press.

Lyman, R. L., M. J. O'Brien, and R. C. Dunnell. 1997. *The rise and fall of culture history*. New York: Plenum Press.

Lynch, T. 1991. Lack of evidence for glacial-age settlement of South America: Reply to Dillehay and Collins and to Gruhn and Bryan. *American Antiquity* 56:348–355.

MacArthur, R. 1972. *Geographical ecology: Patterns in the distribution of species*. New York: Harper and Row.

MacCurdy, G. G., ed. 1937. *Early man*. Philadelphia: Lippincott.

MacDonald, D. H. 1998. Subsistence, sex, and cultural transmission in Folsom culture. *Journal of Anthropological Archaeology* 17:217–239.

———. 1999. Modeling Folsom mobility, technological organization, and mating strategies in the northern Plains. *Plains Anthropologist* 44:141–161.

Madole, R. F. 1995. Spatial and temporal patterns of late Quaternary eolian deposition, eastern Colorado, U.S.A. *Quaternary Science Reviews* 14:155–177.

Maher, L. J., Jr. 1963. Pollen analysis of surface materials from the Southern San Juan Mountains, Colorado. *Geological Society of America Bulletin* 74:1485–1504.

Mallouf, R. 1989. Quarry hunting with Jack Hughes: Tecovas jasper in the south basin of the Canadian River, Oldham County, Texas. In *In the light of past experience: Papers in honor of Jack T. Hughes*, ed. B. C. Roper, 307–326. Clarendon, TX: Aquamarine.

Mandel, R. 1992. Soils and Holocene landscape evolution in central and southwestern Kansas: Implications for archaeological research. In *Soils in archaeology: Landscape evolution and human occupation*, ed. V. T. Holliday, 41–100. Washington, DC: Smithsonian Institution Press.

Mann, D. H. 2003. Fluvial stratigraphy and late Quaternary climates in the Dry Cimarron headwaters, northeastern New Mexico. In *Quest Archaeological Research Fund: Summary of field and laboratory research, 2002–2003*, D. J. Meltzer, ed. Dallas, TX: Report on file, Department of Anthropology, SMU.

———. 2004. Late Glacial and Holocene fluvial geomorphology of the Upper Dry Cimarron River, New Mexico. In *Quest Archaeological Research Fund: Summary of field and laboratory research, 2003–2004*, D. J. Meltzer. Dallas, TX: Report on file, Department of Anthropology, SMU.

Marcy, R. 1859. *The prairie traveler: A handbook for overland expedition*. New York: Harper and Brothers.

Marean, C., and C. Frey. 1997. Animal bones from caves to cities: Reverse utility curves as methodological artifacts. *American Antiquity* 62:698–711.

Marean, C., and S. Kim. 1998. Mousterian large mammal remains from Kobeh Cave. *Current Anthropology* 39:S79–113.

Marean, C., and L. Spencer. 1991. Impact of carnivore ravaging on zooarchaeological measures of element abundance. *American Antiquity* 56:645–658.

Marean, C., Y. Abe, P. Nilssen, and E. Stone. 2001. Estimating the minimum number of skeletal elements (MNE) in zooarchaeology: A review and a new image-analysis GIS approach. *American Antiquity* 66:333–348.

Margaritz, M., J. Heller, and M. Volokita. 1981. Land-air boundary environment as recorded by the $^{18}O/^{16}O$ and $^{13}C/^{12}C$ isotope ratios in the shells of land snails. *Earth and Planetary Science Letters* 52:101–106.

Markgraf, V. 1980. Pollen dispersal in a mountain area. *Grama* 19:127–146.

Markgraf, V., and L. Scott. 1981. Lower timberline in central Colorado during the past 15,000 yr. *Geology* 9:231–234.

Marshall, S., and G. Clarke. 1999. Modeling North American freshwater runoff through the last glacial cycle. *Quaternary Research* 52:300–315.

Martin, A. C., and W. D. Barkley. 1973. *Seed identification manual*. Berkeley: Univ. of California Press.

Mason, J. 2001. Transport direction of Peoria loess in Nebraska and implications for loess sources on the central Great Plains. *Quaternary Research* 56:79–86.

Mateer, N. 1987. The Dakota Group of northeastern New Mexico and southern Colorado. In *Northeastern New Mexico*, 223–236. New Mexico Geological Society Guidebook, 38th Field Conference.

Matheus, P. 1997. Paleoecology and ecomorphology of the giant short-faced bear in eastern Beringia. Unpublished Ph.D. diss., Univ. of Alaska, Fairbanks.

Matthew, W., and H. Cook. 1909. A Pliocene fauna from western Nebraska. *American Museum of Natural History* Bulletin 26: 361–414.

Mayer, J., T. Surovell, N. Waguespack, M. Kornfeld, R. Reider, and G. Frison. 2005. Paleoindian environmental change in Barger Gulch, Middle Park, Colorado. *Geoarchaeology*. Forthcoming.

Mayewski, P. A. L. Meeker, S. Whitlow, M. Twickler, M. Morrison, R. Alley, P. Bloomfield, and K. Taylor. 1993. The atmosphere during the Younger Dryas. *Science* 261:195–197.

Mayewski, P. A., L. Meeker, S. Whitlow, M. Twickler, M. Morrison, P. Bloomfield, G. Bond, R. Alley, A. Gow, P. Grootes, D. Meese, M. Ram, K. Taylor, and W. Wumkes. 1994. Changes in atmospheric circulation and ocean ice cover over the North Atlantic during the last 41,000 years. *Science* 263: 1747–1751.

McCabe, P. J. 1991. Geology of coal: Environments of deposition. In *Economic geology, U.S.*, ed. H. Gluskoter, D. Rice, and R. Taylor. Vol. P-2 in *The geology of North America*, 469–482. Boulder, CO: Geological Society of America.

McDonald, J. N. 1981. *North American bison: Their classification and evolution*. Berkeley: Univ. of California Press.

McElwain, J., F. Mayle, and D. Beerling. 2002. Stomatal evidence for a decline in atmospheric CO_2 concentration during the Younger Dryas stadial: A comparison with Antarctic ice core records. *Journal of Quaternary Science* 17:21–29.

McGee, WJ 1888. Paleolithic man in America: His antiquity and his environment. *Popular Science Monthly* 34:20–36.

———. 1893a. Man and the glacial period. *American Anthropologist* 6:85–95.

———. 1893b. Anthropology at the Madison meeting. *American Anthropologist* 6:435–448.

———. 1897. Anthropology at Detroit and Toronto. *American Anthropologist* 10:317–345.

McKinnon, N. A. 1986. Paleoenvironments and cultural dynamics at Head-Smashed-In Buffalo jump, Alberta: The carbon isotope record. Unpublished M.A. thesis, Univ. of Calgary, Alberta.

McNaghten, A. 1988. The flood as experienced by Allcutt McNaghten. In *Folsom 1888–1988. Then and now*, 32–33. Folsom, NM: Centennial Book Committee.

Meltzer, D. J. 1983. The antiquity of man and the development of American archaeology. *Advances in Archaeological Method and Theory* 6:1–51.

———. 1984. On stone procurement and settlement mobility in eastern fluted point groups. *North American Archaeologist* 6:1–24.

———. 1987. Paleoindian occupation of Texas: Results of the TAS survey. *Bulletin of Texas Archeological Society* 56:27–68.

———. 1989a. Why don't we know when the first people came to North America? *American Antiquity* 54(3): 471–490.

———. 1989b. Was stone exchanged among eastern North American Paleoindians? In *Eastern Paleo-indian lithic resource procurement and processing*, ed. C. Ellis and J. Lothrop, 11–39. Boulder, CO: Westview Press.

———. 1991a. Altithermal archaeology and paleoecology at Mustang Springs, on the Southern High Plains of Texas. *American Antiquity* 56:236–267.

———. 1991b. On "Paradigms" and "Paradigm bias" in controversies over human antiquity in America. In *The first Americans: Search and research*, ed. T. Dillehay and D. J. Meltzer, 13–49. Boa Raton, FL: CRC Press.

———. 1993. *Search for the first Americans*. Washington, DC: Smithsonian Books/Montreal: St. Remy's.

———. 1994. The discovery of deep time: A history of views on the peopling of the Americas. In *Method and theory for investigating the peopling of the Americas*, ed. R. Bonnichsen and D. G. Steele, 7–26. Corvallis, OR: Center for the Study of the First Americans.

———. 1999. Human responses to Middle Holocene (Altithermal) climates on the North American Great Plains. *Quaternary Research* 52:404–416.

———. 2000. Renewed investigations at the Folsom Paleoindian type site. *Antiquity* 74:35–36.

———. 2001. Late Pleistocene cultural and technological diversity of Beringia: A view from down under. *Arctic Anthropology* 38(2):206–213.

———. 2002. What do you do when no one's been there before? Thoughts on the exploration and colonization of new lands. In *The first Americans: The Pleistocene colonization of the New World*, ed. N. Jablonski. *Memoirs of the California Academy of Sciences* 27:25–56.

———. 2003. Lessons in landscape learning. In *Colonization of unfamiliar landscapes: The archaeology of adaptation*, ed. M. Rockman and J. Steele, 222–241. London: Routledge.

———. 2004. Modeling the initial colonization of the Americas: Issues of scale, demography, and landscape learning. In *The settlement of the American continents: A multidisciplinary approach to human biogeography*, ed. C. M. Barton, G. A. Clark, D. R. Yesner, and G. A. Pearson, 123–137. Tucson: Univ. of Arizona Press.

———. 2005. The seventy-year itch: Controversies over human antiquity and their resolution. *Journal of Anthropological Research* 61:433–468.

———. 2006. Archaeological research on the origins, antiquity, and adaptations of the first Americans, 1862–1997. *Handbook of North American Indians, Vol. 3, Environment, origins & population*, gen. ed. W. C. Sturtevant. Washington, DC: Government Printing Office. Forthcoming.

Meltzer, D. J., and M. Bever. 1995. Paleoindians of Texas: An update on the Texas Clovis fluted point survey. *Bulletin of Texas Archeological Society* 66:17–51.

Meltzer, D. J., and M. Collins. 1987. Prehistoric water wells on the southern High Plains: Clues to Altithermal climate. *Journal of Field Archaeology* 14(1):9–28.

Meltzer, D. J., D. K. Grayson, G. Ardila, A. Barker, D. Dincauze, C. V. Haynes, F. Mena, L. Núñez, and D. J. Stanford. 1997. On the Pleistocene antiquity of Monte Verde, southern Chile. *American Antiquity* 62:659–663.

Meltzer, D. J., D. Mann, and J. M. LaBelle. 2004. A *Bison antiquus* from Archuleta Creek, Folsom, New Mexico. *Current Research in the Pleistocene*. 21:107–109.

Meltzer, D. J., J. D. Seebach, and R. M. Byerly. 2006. The Hot Tubb Folsom-Midland site (41 CR 10), Texas. *Plains Anthropologist*. Forthcoming.

Meltzer, D. J., L. C. Todd, and V. T. Holliday. 2002. The Folsom (Paleoindian) type site: Past investigations, current studies. *American Antiquity* 67:5–36.

Menounos, B., and M. Reasoner. 1997. Evidence for cirque glaciation in the Colorado Front Range during the Younger Dryas Chronozone. *Quaternary Research* 48:38–47.

Merriam, J. C. 1914. Preliminary report on the discovery of human remains in an asphalt deposit at Rancho La Brea. *Science* 40:198–203.

———. 1924. Present status of investigations concerning antiquity of man in California. *Science* 60:1–2.

Merton, R. 1968. The Matthew effect in science: The reward and communication system in science. *Science* 199:55–63.

Meserve, F., and E. Barbour. 1932. Association of an arrow point with *Bison occidentalis* in Nebraska. *Nebraska State Museum Bulletin* 27:239–242.

Metcalf, A. L. 1984. Land Snails (Gastropoda: Pulmonata) from Cimarron County, Oklahoma. *Texas Journal of Science* 36(1):53–64.

———. 1997. Land snails of New Mexico from a historical zoogeographic point of view. In *land snails of New Mexico*, ed. A. Metcalf and R. A. Smartt, Bulletin 10, 71–108. Albuquerque: New Mexico Museum of Natural History and Science.

Metcalf, A. L., and R. A. Smartt. 1997. Land snails of New Mexico: A systematic review. In *Land Snails of New Mexico*, ed. A. Metcalf and R. A. Smartt, Bulletin 10, 1–69. Albuquerque: New Mexico Museum of Natural History and Science.

Metcalf, D., and K. Barlow. 1992. A model for exploring the optimal tradeoff between field processing and transport. *American Anthropologist* 94:340–356.

Meyer, H. W. 1992. Lapse rates and other variables applied to estimating paleoaltitudes from fossil floras. *Palaeogeography, Palaeoclimatology, Palaeoecology* 99, 71–99.

Miller, J.P. 1958. Problems of the Pleistocene in Cordilleran North America as related to reconstruction of environmental changes that affected Early Man. In *Climate and man in the Southwest*, ed. T.L. Smiley, 19–49. Contribution 6. Tucson: Univ. of Arizona Program in Geochronology.

Mole, S., A. Joern, M. O'Leary, and S. Madhavan. 1994. Spatial and temporal variation in carbon isotope discrimination in prairie graminoids. *Oecologia* 97:316–321.

Monnin, E., A. Indermuhle, A. Dallenbach, J. Flückiger, B. Stauffer, T. Stocker, D. Raynaud, and J.-M. Barnola. 2001. Atmospheric CO_2 concentrations over the last glacial termination. *Science* 291:112–114.

Moodie, D.W., and A.J. Ray. 1976. Buffalo migrations in the Canadian Plains. *Plains Anthropologist* 21:45–52.

Moorehead, W.K. 1893. The meeting of the American Association for the Advancement of Science. *Archaeologist* 1: 170–172.

Morgan, R.G. 1980. Bison movement patterns on the Canadian Plains: An ecological analysis. *Plains Anthropologist* 25:143–160.

Morlan, R. 1994. Oxbow bison procurement as seen from the Harder site, Saskatchewan. *Journal of Archaeological Science* 21: 757–777.

Morrow, T., and J. Morrow. 1999. On the fringe: Folsom points and performs in Iowa. In *Folsom lithic technology: Explorations in structure and variation*, ed. D.S. Amick, 65–81. International Monographs in Prehistory, Archaeological Series 12. Ann Arbor, MI.

Muehlberger, W.R. 1955. Relative age of Folsom man and Capulin Mountain eruption, Colfax and Union counties, New Mexico (abstract). *Geological Society of America Bulletin* 66:1600–1601.

Muehlberger, W., B. Baldwin, and R. Foster. 1961. *High Plains northeastern New Mexico*. Socorro: New Mexico State Bureau of Mines and Mineral Resources.

Muhs, D.R., T.W. Stafford, S.D. Cowherd, S.A. Mahan, R. Kihl, P.B. Maat, C.A. Bush, and J. Nehring. 1996. Origin of the late Quaternary dune fields of northeastern Colorado. *Geomorphology* 17:129–149.

Muhs, D.R., T.W. Stafford, J.B. Swinehart, S.D. Cowherd, S.A. Mahan, C.A. Bush, R.F. Madole, and P.B. Maat. 1997. Late Holocene eolian activity in the mineralogically mature Nebraska Sand Hills. *Quaternary Research* 48:162–176.

Murtaugh, W.J. 1976. *The National Register of Historic Places*. Washington, DC: U.S. Department of the Interior, National Park Service.

Musil, A.F. 1978. *Identification of crop and weed seeds*. Agricultural Handbook no. 219. Washington, DC: U.S. Department of Agriculture.

Nami, H. 1999. The Folsom biface reduction sequence: Evidence from the Lindenmeier collection. In *Folsom lithic technology: Explorations in structure and variation*, ed. D.S. Amick, 82–97. International Monographs in Prehistory, Archaeological Series 12. Ann Arbor, MI.

Nativ, R., and Riggio, R. 1990a. Meteorologic and isotopic characteristics of precipitation events with implications for ground-water recharge, Southern High Plains. In *Geological framework and regional hydrology: Upper Cenozoic Blackwater Draw and Ogallala formations, Great Plains*, ed. T. Gustavson, 152–179. Bureau of Economic Geology: Univ. of Texas,

———. 1990b. Precipitation in the Southern High Plains— Meteorologic and isotopic features. *Journal of Geophysical Research* 95:22,559–22,564.

Neck, R. 1986. Molluscan remains. In *Stratigraphy and paleoenvironments of Late Quaternary valley fills on the Southern High Plains*, ed. V.T. Holliday. Geological Society of America, Memoir 186.

Nelson, N.C. 1918. Review of "Additional Studies in the Pleistocene at Vero, Florida." *Science* 47:394–395.

———. 1928a. Pseudo-artifacts from the Pliocene of Nebraska. *Science* 67:316–317.

———. 1928b. Discussion of "The Origin and Antiquity of Man in America by A. Hrdlička." *New York Academy of Medicine Bulletin* 4(7):820–823.

———. 1933. The antiquity of man in America in the light of archaeology. In *The American Aborigines: Their origin and antiquity*, ed. D. Jenness, 87–130. Toronto: Univ. of Toronto Press.

Nelson, R. 1969. *Hunters of the northern ice*. Chicago: Univ. of Chicago Press.

Neuhauser, K., S. Lucas, J.S. de Albuquerque, R. Louden, S. Hayden, K. Kietzke, W. Oakes, and D. des Marais. 1987. Stromatolites of the Morrison formation (Upper Jurassic) Union County, New Mexico. In *Northeastern New Mexico*, 153–160. New Mexico Geological Society Guidebook, 38th Field Conference.

Nordt, L.C., T.W. Boutton, J.S. Jacob, and R.D. Mandel. 2002. C_4 plant productivity and climate–CO_2 variations in south-central Texas during the Late Quaternary. *Quaternary Research* 58:182–188.

Nowak, R. 1999. *Walker's Mammals of the world*. 6th ed. Baltimore: John Hopkins Univ. Press.

O'Connell, J.F. 1987. Alyawara site structure and its archaeological implications. *American Antiquity* 52:74–108.

———. 1995. Ethnoarchaeology needs a general theory of behavior. *Journal of Archaeological Research* 3:205–255.

O'Connell, J.F., K. Hawkes, and N.G. Blurton-Jones. 1988. Hadza hunting, butchering, and bone transport and their archaeological implications. *Journal of Anthropological Research* 44:113–161.

———. 1990. Reanalysis of large mammal body part transport among the Hadza. *Journal of Archaeological Science* 17: 301–316.

———. 1992. Patterns in the distribution, site structure and assemblage composition of Hadza kill-butchering sites. *Journal of Archaeological Science* 19:319–345.

Ode, D., L. Tieszen and J. Lerman. 1980. The seasonal contribution of C_3 and C_4 plant species to primary production in a mixed prairie. *Ecology* 61:1304–1311.

Odell, G. 1994. Prehistoric hafting and mobility in the North American midcontinent: Examples from Illinois. *Journal of Anthropological Archaeology* 13:51–73.

Odell, G., and F. Cowan. 1986. Experiments with spears and arrows on animal targets. *Journal of Field Archaeology* 13:195–212.

Oldroyd, D. 1990. *The Highlands controversy: Constructing geological knowledge through fieldwork in nineteenth-century Britain*. Chicago: Univ. of Chicago Press.

Opler, M. 1936. A summary of Jicarilla Apache culture. *American Anthropologist* 28:202–223.

Osborn, A. 1999. From global models to regional patterns: Possible detereminants of Folsom hunting weapon design, diversity and complexity. In *Folsom lithic technology: Explorations in structure and variation*, ed. D.S. Amick, 188–213. International Monographs in Prehistory, Archaeological Series 12. Ann Arbor, MI.

Osborn, H. F. 1922. *Hesperopithecus*, the first anthropoid primate found in America. *American Museum Novitates* no. 37.

Osterkamp, W., M. Fenton, T. Gustavson, R. Hadley, V. T. Holliday, R. Morrison, and T. J. Toy. 1987. Great Plains. In *Geomorphic systems of North America*, ed. W. Graf, 163–210. Centennial Special Vol. 2. Boulder, CO: Geological Society of North America.

Owen, T. 1951. Tom Owen describes men who first uncovered bones of Folsom bison. *Raton Daily Range,* 28 June.

———. 1988. George McJunkin. In *Folsom 1888–1988. Then and now*, 27. Folsom, NM: Centennial Book Committee.

Paruelo, J. M., and W. K. Lauenroth. 1996. Relative abundance of plant functional types in grasslands and shrublands of North America. *Ecological Applications* 6:1212–1224.

Peden, D. G. 1976. Botanical composition of bison diets on shortgrass Plains. *American Midland Naturalist* 96:225–229.

Peden, D. G., G. Van Dyne, R. Rice, and R. Hansen. 1974. The trophic ecology of *Bison bison* L. on shortgrass Plains. *Journal of Applied Ecology* 11:489–498.

Peteet, D. 1995. Global Younger Dryas? *Quaternary International* 28:93–104.

———. 2000. Sensitivity and rapidity of vegetational response to abrupt climate change. *Proceedings of the National Academy of Sciences* 97:1359–1361.

Phillips, R. J., D. Wiens, K. Detwiler, K. Larsen, R. Popelka, S. Robertson, and E. Wilson. 1997. Seismic and resistivity studies of the Folsom site. Dallas, TX: Reports on file, Department of Anthropology, Southern Methodist University.

Pierce, H. 1987. The gastropods, with notes on other invertebrates. In *Lubbock Lake: Late Quaternary studies on the southern High Plains*, ed. E. Johnson, 41–48. College Station: Texas A&M Press.

Pilsbry, H. A. 1939. Land Mollusca of North America (North of Mexico). *Academy of Natural Sciences of Philadelphia Monographs* no. 3, Vol. 1, Pt. 1.

———. 1948. Land Mollusca of North America (North of Mexico). *Academy of Natural Sciences of Philadelphia. Monographs* no. 3, Vol. 2, Pt. 1.

Plumb, G., and J. Dodd. 1993. Foraging ecology of bison and cattle on a mixed prairie: Implications for natural area management. *Ecological Applications* 3:631–643.

Polyak, V., J. Rasmussen, and Y. Asmerom. 2004. Prolonged wet period in the southwestern United States through the Younger Dryas. *Geology* 32:5–8.

Porter, S., K. Pierce, and T. Hamilton. 1983. Late Wisconsin mountain glaciation in the western United States. In *Late-Quaternary environments of the United States, Vol. 1, The Late Pleistocene, ed.* H. Wright and S. Porter, 71–111. Minneapolis: Univ. of Minnesota Press.

Preston, D. 1997. Fossils and the Folsom cowboy. *Natural History* 106:16–22.

Preston, F. 1948. The commonness and rarity of species. *Ecology* 29:254–283.

Prestwich, J. 1860. On the occurrence of flint implements, associated with the remains of extinct mammalia, in undisturbed beds of a late geological period. *Proceedings of the Royal Society of London* 10:50–59.

Putnam, F. W., compiler. 1888. Symposium on "Paleolithic Man in Eastern and Central North America." *Proceedings Boston Society of Natural History* 23:419–449.

———. 1889. Symposium on "Paleolithic Man in Eastern and Central North America" (Part III). *Proceedings Boston Society of Natural History* 24:141–165.

Rabbitt, M. 1980. *Minerals, lands, and geology for the common defence and general welfare, Vol. 2, 1879–1904.* Washington, DC: U.S. Government Printing Office.

Ramsey, C. B. 2003. OxCal radiocarbon calibration program, version 3.9. Available at: http://www.rlaha.ox.ac.uk/orau/oxcal.html

Raynaud, D., J.-M. Barnola, J. Chappellaz, T. Blunier, A. Indermuhle, and B. Stauffer. 2000. The ice record of greenhouse gases: A view in the context of future changes. *Quaternary Science Reviews* 19:9–17.

Reasoner, M., and M. Jodry. 2000. Rapid response of alpine timberline vegetation to the Younger Dryas climate oscillation in the Colorado Rocky Mountains, USA. *Geology* 28:51–54.

Reed, E. K. 1940. Archaeological site report on Folsom, New Mexico (Theme I: Early man). Unpublished manuscript prepared for the Historic Sites Survey, Region III. Santa Fe, NM: National Park Service.

Reher, C., and G. Frison. 1980. The Vore site, 48CK302, a stratified buffalo jump in the Wyoming Black Hills. *Plains Anthropologist Memoir* 16.

Renaud, E. B. 1931. Prehistoric flaked points from Colorado and neighboring districts. *Proceedings of the Colorado Museum Natural History* 10(2).

———. 1932. Yuma and Folsom artifacts. *Proceedings of the Colorado Museum Natural History* 11(2).

———. 1934. The first thousand Yuma-Folsom artifacts. Denver: Department of Anthropology, Univ. of Denver.

Reynolds, H. G., W. Clary, and P. Folliot. 1970. Gambel oak for southwestern wildlife. *Journal of Forestry* 68:545–547.

Reynolds, H. W., R. Hansen, and D. Peden. 1978. Diets of the Slave River lowland bison herd, Northwest Territories, Canada. *Journal of Wildlife Management* 42:581–590.

Richerson, P. J., R. Boyd, and R. Bettinger. 2001. Was agriculture impossible during the Pleistocene but mandatory during the Holocene? A climate change hypothesis. *American Antiquity* 66:387–411.

Richmond, G. 1965. Glaciation of the Rocky Mountains. In *The Quaternary of the United States,* ed. H. E. Wright and D. G. Frey, 217–230. Princeton, NJ: Princeton Univ. Press.

———. 1986. Stratigraphy and correlation of glacial deposits of the Rocky Mountains, the Colorado Plateau and the ranges of the Great Basin. *Quaternary Science Reviews* 5:99–127.

Roberts, F. H. H. 1934. Scientist describes true Folsom points. *Literary Digest* 118(4):18.

———. 1935. A Folsom Complex: Preliminary report on investigations at the Lindenmeier site in northern Colorado. *Smithsonian Miscellaneous Collections* 94(4):1–35.

———. 1936. Additional information on the Folsom Complex: Report on the second season's investigations at the Lindenmeier site in northern Colorado. *Smithsonian Miscellaneous Collections* 95(10):1–38.

———. 1937. The Folsom Problem in American archaeology. In *Early man*, ed. G. G. MacCurdy, 153–162. Philadelphia: Lippincott.

———. 1939. The Folsom Problem in American archaeology. *Smithsonian Institution Annual Report for 1938*, 531–546.

———. 1940. Developments in the problem of the North American Paleo-indian. *Smithsonian Miscellaneous Collections* 100:51–116.

———. 1951. Radiocarbon dates and early man. *American Antiquity Memoir* 7:20–22.

Rogers, A. 2000. On equifinality in faunal analysis. *American Antiquity* 65:709–723.

Rogers, R., and L. Martin. 1984. The 12 Mile Creek site: A reinvestigation. *American Antiquity* 49:757–764.

———. 1986. Replication and the history of Paleoindian studies. *Current Research in the Pleistocene* 3: 43–44.

———. 1987. The Folsom discovery and the concept of breakthrough sites in Paleoindian studies. *Current Research in the Pleistocene* 4:81–82.

Roe, F. G. 1951. *The North American buffalo: A critical study of the species in its wild state*. Toronto: Univ. of Toronto Press.

Romer, A. S. 1933. Pleistocene vertebrates and their bearing upon the problem of human antiquity in America. In *The American Aborigines: Their origin and antiquity*, ed. D. Jenness, 49–83. Toronto: Univ. of Toronto Press.

Roosa, W. 1977. Great Lakes Paleoindians: The Parkhill site, Ontario. *Annals of the New York Academy of Sciences* 288: 349–354.

Roosa, W., and D. B. Deller. 1982. The Parkhill complex and eastern Great Lakes Paleoindian. *Ontario Archaeology* 37:3–15.

Root, M. 2000. Intrasite comparisons. In *The archaeology of the Bobtail Wolf site: Folsom occupation of the Knife River Flint Quarry Area, North Dakota*, ed. M. Root, 347–362. Pullman: Washington State Univ. Press.

Root, M., J. William, and A. Emerson. 2000. Stone tools and flake debris. In *The archaeology of the Bobtail Wolf site: Folsom occupation of the Knife River Flint Quarry Area, North Dakota*, ed. M. Root, 223–308. Pullman: Washington State Univ. Press.

Rudwick, M. 1985. *The great Devonian controversy: The shaping of scientific knowledge among gentlemanly specialists*. Chicago: Univ. of Chicago Press.

Salisbury, R. 1892a. Review of *"Man and the Glacial Period."* *Chicago Tribune*, October 22.

———. 1892b. Wright's *"Man and the Glacial Period."* *The Nation* 55:496–497.

———. 1893a. Man and the glacial period. *American Geologist* 11:13–20.

———. 1893b. Distinct glacial epochs, and the criteria for their recognition. *Journal of Geology* 1:61–84.

Saucier, R. 1994. Evidence of Late Glacial runoff in the lower Mississippi Valley. *Quaternary Science Reviews* 13:973–981.

Sayre, W., and M. Ott. 1999. A geologic study of Capulin Volcano National Monument and surrounding areas. Final Report on Cooperative Agreement CA7029–2–0017. On file, Capulin Volcano National Monument, Capulin, NM.

Schiegl, S., P. Goldberg, O. Bar-Yosef, and S. Weiner. 1996. Ash deposits in Hayonim and Kebara caves, Israel: Macroscopic, microscopic and mineralogical observations, and their archaeological implications. *Journal of Archaeological Science* 23:763–781.

Schoeller, D. 1999. Isotope fractionation: Why aren't we what we eat? *Journal of Archaeological Science* 26:667–673.

Schopmeyer, C. S. 1974. *Seeds of woody plants in the Unites States*. Agricultural Handbook no. 450. Washington, DC: U.S. Department of Agriculture, Forest Service.

Schultz, C. B. 1932. Association of artifacts and extinct mammals in Nebraska. *Nebraska State Museum Bulletin* 33:271–282.

———. 1943. Some artifacts of early man in the Great Plains and adjacent areas. *American Antiquity* 8:242–249.

———. 1983. Early man and the Quaternary: Initial research in Nebraska. *Transactions of the Nebraska Academy of Sciences* 11:129–136.

Schumm, S., and R. Lichty. 1963. Channel widening and floodplain construction along Cimarron River in southwestern Kansas. *United States Geological Survey Professional Paper* 352–D.

Schwarcz, H., T. Dupras, and S. Fairgrieve. 1999. $\delta^{15}N$ enrichment in the Sahara: In search of a global relationship. *Journal of Archaeological Science* 26:629–636.

Schwartz, C. C., and J. Ellis. 1981. Feeding ecology and niche separation in some native ungulates on the shortgrass prairie. *Journal of Applied Ecology* 18:343–353.

Scott, G., and C. Pillmore. 1993. Geologic and structure-contour map of the Raton 30′ × 60′ quadrangle, Colfax and Union counties, New Mexico, and Las Animas County, Colorado. *United States Geological Survey, Miscellaneous Investigations Series, Map* I–2266.

Scott, L. J. 1972. Folsom Cave palynological analysis. Manuscript on file, Univ. of Colorado Museum, Boulder.

Secord, J. A. 1986. *Controversies in Victorian geology: The Cambrian-Silurian dispute*. Princeton, NJ: Princeton Univ. Press.

Sellards, E. H. 1952. *Early man in America*. Austin: Univ. of Texas Press.

Sellet, F. 1999. A dynamic view of Paleoindian assemblages at the hell gap site, Wyoming: Reconstructing lithic technological systems. Unpublished Ph.D. diss., Southern Methodist Univ.

———. 2001. A changing perspective on Paleoindian chronology and typology: A view from the Northern Plains. *Arctic Anthropology* 38:48–63.

———. 2004. Beyond the point: Projectile manufacture and behavioral inference. *Journal of Archaeological Science*. 31: 1553–1566.

Severinghaus, J., and E. Brook. 1999. Abrupt climate change at the end of the last glacial period inferred from trapped air in polar ice. *Science* 286:930–934.

Shaler, N. 1876. On the age of the Delaware gravel beds containing chipped pebbles. *Peabody Museum Annual Report* 10:44–47.

Shea, J. 1988. Spear points from the Middle Paleolithic of the Levant. *Journal of Field Archaeology* 15:441–450.

Shelford, V. 1963. *The ecology of North America*. Urbana: Univ. of Illinois Press.

Shetrone, H. 1936. The Folsom phenomena as seen from Ohio. *Ohio State Archaeological and Historical Quarterly* 45:240–256.

Shipman, P. 1981. *Life history of a fossil: An introduction to taphonomy and paleoecology*. Cambridge, MA: Harvard Univ. Press.

Shott, M. 1986. Settlement mobility and technological organization: An ethnographic examination. *Journal of Anthropological Research* 42:15–51.

———. 2002. Sample bias in the distribution and abundance of Midwestern fluted bifaces. *Midcontinental Journal of Archaeology* 27:89–123.

Shuman, B., T. Webb, P. Bartlein, and J. W. William. 2002. The anatomy of a climatic oscillation: Vegetation change in eastern North America during the Younger Dryas chronozone. *Quaternary Science Reviews* 21:1777–1791.

Silberbauer, G. 1981. *Hunter and habitat in the central Kalahari desert*. Cambridge: Cambridge Univ. Press.

Simmons, V. 2000. *The Ute Indians of Utah, Colorado, and New Mexico*. Boulder: Univ. Press of Colorado,

Simpson, R., L. Patterson, and C. Singer. 1986. Lithic technology of the Calico Mountains site, southern California. In *New evidence for the Pleistocene peopling of the Americas*, ed. A. Bryan, 89–105. Orono, ME: Center for the Study of Early Man.

Skinner, M., and O. Kaisen. 1947. The fossil *Bison* of Alaska and preliminary revision of the genus. *American Museum of Natural History Bulletin* 89:123–256.

Skinner, M., S. Skinner, and R. Gooris. 1977. Stratigraphy and biostratigraphy of late Cenozoic deposits in central Sioux County, western Nebraska. *American Museum of Natural History Bulletin* 158:271–367.

Smith, B. C. 1996. *Fundamentals of Fourier transform infrared spectroscopy*. Boca Raton, FL: CRC Press.

Smith, B., and S. Epstein. 1971. Two categories of $^{13}C/^{12}C$ ratios for higher plants. *Plant Physiology* 47:380–384.

Smith, E. A. 1991. *Inujjuamiut foraging strategies: Evolutionary ecology of an Arctic hunting economy*. New York: Aldine de Gruyter.

Sollberger, J. 1977. On fluting Folsom. *Bulletin of the Texas Archeological Society* 48:47–52.

———. 1985. A technique for Folsom fluting. *Lithic Technology* 14:41–50.

———. 1989. Comment on "A Stone Age Pressure Method of Folsom Fluting," by Eugene M. Gryba. *Plains Anthropologist* 34:63–64.

Sollberger, J., and L. Patterson. 1980. Attributes of experimental Folsom points and channel flakes. *Bulletin of the Texas Archeological Society* 51:289–299.

Spero, H., and D. Lea. 2002. The cause of carbon isotope minimum events on glacial terminations. *Science* 296:2438–2441.

Spero, H. J., and D. F. Williams. 1990. Evidence for seasonal low salinity surface waters in the Gulf of Mexico over the last 16,000 years. *Paleoceanography* 5:963–975.

Speth, J. 1983. *Bison kills and bone counts*. Chicago: Univ. of Chicago Press.

Speth, J., and K. Spielmann. 1983. Energy source, protein metabolism, and hunter-gatherer subsistence strategies. *Journal of Anthropological Archaeology* 2:1–31.

Spielmann, K. 1983. Late prehistoric exchange between the southwest and southern Plains. *Plains Anthropologist* 28:257–272.

Spier, L. 1928a. Concerning man's antiquity at Frederick, Oklahoma. *Science* 67:160–161.

———. 1928b. A note on reputed ancient artifacts from Frederick, Oklahoma. *Science* 68:184.

Sponheimer, M., and J. Lee–Thorp. 1999. Oxygen isotopes in enamel carbonate and their ecological significance. *Journal of Archaeological Science* 26:723–728.

Stafford, T. W. 1984. Quaternary stratigraphy, geochronology, and carbon isotope geology of alluvial deposits in the Texas Panhandle. Unpublished Ph.D. diss., Department of Geosciences, Univ. of Arizona, Tucson.

———. 1998. Radiocarbon chronostratigraphy. In *Wilson-Leonard. An 11,00-year archaeological record of hunter-gatherers in central Texas, Vol. IV, Archaeological features and technical analyses*, ed. M. B. Collins. Studies in Archaeology 31. Austin: Texas Archeological Research Laboratory, Univ. of Texas.

Stafford, T. W., K. Brendel, and R. C. Duhamel. 1988. Radiocarbon, ^{13}C and ^{15}N analysis of fossil bone: Removal of humates with XAD–2 resin. *Geochimica et Cosmochimica Acta* 52:2257–2267.

Stafford, T. W., P. E. Hare, L. Vurrie, A. J. T. Jull, and D. Donahue. 1991. Accelerator radiocarbon dating at the molecular level. *Journal of Archaeological Science* 18:35–72.

Stafford, T. W., A. J. T. Jull, K. Brendel, R. C. Duhamel, and D. Donahue. 1987. Study of bone radiocarbon dating accuracy at the University of Arizona NSF Accelerator Facility for Radioisotope Analysis. *Radiocarbon* 29:24–44.

Stanford, D. J. 1999. Paleoindian archaeology and Late Pleistocene environments in the Plains and Southwestern United States. In *Ice Age peoples of North America*, ed. R. Bonnichsen and K. Turnmire, 281–339. Center for the Study of the First Americans, Corvallis, Oregon.

Steen, C. 1955. *Prehistoric man in the Arkansas-White-Red River Basins*. Washington, DC: National Park Service.

Steuter, A., E. Steinauer, G. Hill, P. Bowers, and L. Tieszen. 1995. Distribution and diet of bison and pocket gophers in a sand-hill prairie. *Ecological Applications* 5:756–766.

Stevens, R. and R. E. Hedges. 2004. Carbon and nitrogen stable isotope analysis of northwest European horse bone and tooth collagen, 40,000 BP–present: Palaeoclimatic interpretations. *Quaternary Science Reviews* 23:977–991.

Steward, J., and F. Setzler. 1938. Function and configuration in archaeology. *American Antiquity* 4:4–10.

Stiner, M. C., S. L. Kuhn, T. A. Surovell, P. Goldberg, L. Meignen, S. Weiner, and O. Bar-Yosef. 2001. Bone preservation in Hayonim Cave (Israel): A macroscopic and mineralogical study. *Journal of Archaeological Science*. 28:643–659.

Stocking, G. 1968. *Race, culture, and evolution*. New York: Free Press.

———. 1987. *Victorian anthropology*. New York: Free Press.

Stormer, J. 1972. Mineralogy and petrology of the Raton-Clayton volcanic field, northeastern New Mexico. *Geological Society of America Bulletin* 83:3299–3322.

Stott, L. 2002. The influence of diet on the $\delta^{13}C$ of shell carbon in the pulmonate snail Helix aspersa. *Earth and Planetary Science Letters* 195:249–259.

Stowe, L., and J. Teeri. 1978. The geographic distribution of C_4 species of the Dicotyledonae in relation to climate. *American Naturalist* 112:609–623.

Stubbendieck, J., S. Hatch, and C. Butterfield. 1992. *North American range plants*. 4th ed. Lincoln: Univ. of Nebraska Press.

Stuiver, M. P., J. Reimer, and R. Reimer. 2000. CALIB radiocarbon calibration, version 4.4. Available at: http://radiocarbon.pa.qub.ac.uk/calib/

Taylor, D. W. 1960. Late Cenozoic molluscan faunas from the High Plains. *United States Geological Survey Professional Paper* 337.

———. 1965. The study of Pleistocene nonmarine mollusks in North America. In *The Quaternary of the United States*, ed. H. E. Wright and D. G. Frey, 597–611. Princeton, NJ: Princeton Univ. Press.

Taylor, D. W., and C. Hibbard. 1955. *A new Pleistocene Fauna from Harper County, Oklahoma*. Circular 37. Norman: Oklahoma Geological Survey.

Taylor, K., P. Mayewski, R. Alley, E. Brook, and G. Zielinski. 1997. The Holocene–Younger Dryas transition recorded at Summit, Greenland. *Science* 278:825–827.

Taylor, R. E. 1980. Radiocarbon dating of Pleistocene bone: Toward criteria for the selection of samples. *Radiocarbon* 22: 969–979.

Taylor, R. E., M. Stuiver, and P. J. Reimer. 1996. Development and extension of the radiocarbon time scale: Archaeological applications. *Quaternary Science Reviews* 15:655–668.

Teeri, J. A., and L. Stowe. 1976. Climatic patterns and the distribution of C$_4$ grasses in North America. *Oecologia* 23: 1–12.

Teller, J. T., D. W. Leverington, and J. D. Mann. 2002. Freshwater outbursts to the oceans from glacial Lake Agassiz and their role in climate change during the last deglaciation. *Quaternary Science Reviews* 21:879–887.

Telfer, E. S., and J. P. Kelsall. 1984. Adaptation of some large North American mammals for survival in snow. *Ecology* 65: 1828–1834.

Theler, J. L., D. G. Wyckoff, and B. J. Carter. 2004. The Southern Plains Gastropod Survey: The distribution of land snail populations in an American grassland environment. *American Malacological Bulletin* 18:1–20.

Thomas, C. 1898. *Introduction to the study of North American archaeology*. Cincinatti, OH: Robert Clarke.

Thompson, B. C., P. J. Crist, J. S. Prior-Magee, R. A. Deitner, C. L. Garber, and M. A. Hughes. 1996. *Gap analysis of biological diversity conservation in New Mexico using geographic information systems*, 1–94. Las Cruces: New Mexico Cooperative Fish and Wildlife Research Unit.

Thompson, F. 1967. Folsom find claimed by Canadian cowboy. *The Raton Daily Range*, 11 November, 1, 4.

Thompson, R. S., K. Anderson, and P. Bartlein. 1999a. Atlas of relations between climatic parameters and distributions of important trees and shrubs in North America—Introduction and conifers. *United States Geological Survey Professional Paper* 1650–A.

———. 1999b. Atlas of relations between climatic parameters and distributions of important trees and shrubs in North America—Hardwoods. *United States Geological Survey Professional Paper* 1650–B.

———. 2000. Atlas of relations between climatic parameters and distributions of important trees and shrubs in North America—Additional conifers, hardwoods, and monocots. *United States Geological Survey Professional Paper* 1650–C.

Thompson, R. S., C. Whitlock, P. J. Bartlein, S. P. Harrison, and W. G. Spaulding. 1993. Climatic changes in the western United States since 18,000 yr B.P. In *Global climates since the Last Glacial Maximum*, ed. H. E. Wright, J. E. Kutzbach, T. Webb, W. F. Ruddiman, F. A. Street-Perrott, and P. J. Bartlein, 468–513. Minneapolis: Univ. of Minnesota Press.

Tieszen, L. 1991. Natural variations in the carbon isotope values of plants: Implications for archaeology, ecology, and paleoecology. *Journal of Archaeological Science* 18:227–248.

———. 1994. Stable isotopes on the Plains: Vegetation analyses and diet determinations. In *Skeletal biology in the Great Plains*, ed. D. Owsley and R. Jantz, 261–282. Washington, DC: Smithsonian Institution Press.

Tieszen, L., K. Reinhard, and D. Forshoe. 1997a. Application of stable isotopes in analysis of dietary patterns. In Bioarchaeology of the north central United States, ed. D. Owsley and J. Rose. *Arkansas Archaeological Survey Research Series* 49:248–256.

———. 1997b. Appendix C. Stable isotopes in the central and northern Great Plains. In Bioarchaeology of the north central United States, ed. D. Owsley and J. Rose. *Arkansas Archaeological Survey Research Series* 49:329–336.

Tiller, V. E. 1983a. *The Jicarilla Apache tribe: A history, 1846–1970*. Lincoln: Univ. of Nebraska Press.

———. 1983b. Jicarilla Apache. In *Handbook of North American Indians, Vol. 10, Southwest*, ed. A. Ortiz, 440–461. Washington, DC: Smithsonian Institution Press.

Titmus, G., and J. Woods. 1991. A closer look at margin "grinding" on Folsom and Clovis points. *Journal of California and Great Basin Anthropology* 13:194–203.

Todd, L. C. 1987a. Taphonomy of the Horner II bone bed. In *The Horner site: The type site of the Cody cultural complex*, ed. G. C. Frison and L. C. Todd, 107–198. New York: Academic Press.

———. 1987b. Analysis of kill-butchery bonebeds and interpretation of Paleoindian hunting. In *The evolution of human hunting*, ed. M. Nitecki, 225–266. New York: Plenum Press.

———. 1991. Seasonality studies and Paleoindian subsistence strategies. In *Human predators & prey mortality*, ed. M. Stiner, 217–238. Boulder, CO: Westview Press.

———. 1993. Body parts, butchery, and post-depositional processes at Round Spring: Formational analysis of Body Size III (Brain) bovids. In *Hogan Pass: Final report on archaeological investigations along Forest Highway 10 (State Highway 72, Sevier County, Utah)*. Vol. III, ed. M. Metcalf et al., 341–386. Eagle, CO: Metcalf Archaeological Consultants.

Todd, L. C., and J. L. Hofman. 1991. Variation in Folsom Age Bison Assemblages: Implications for the Interpretation of Human Action. Paper presented at the 56th Annual Meeting of the Society for American Archaeology, New Orleans, LA.

Todd, L. C., and D. Rapson. 1999. Formational analysis of bison bonebeds and interpretation of Paleoindian subsistence. In *Le bison: Gibier et moyen de subsistance de hommes du paleolithique aux paleoindiens des grandes plaines*, ed. J. Jaubert, J. Brugal, F. David, and J. Enloe, 479–499. Antibes: Editions APDCA Antibes.

Todd, L. C., M. Hill, D. Rapson, and G. Frison. 1997. Cutmarks, impacts, and carnivores at the Casper site bison bonebed. In *Proceedings of the 1993 Bone Modification Conference, Hot Springs, South Dakota*, ed. L. A. Hannus, L. Rossum, and R. P. Winham, 136–157. Occasional Publication no. 1. Sioux Falls, SD: Archaeology Laboratory, Augustana College.

Todd, L. C., J. L. Hofman and C. B. Schultz. 1990. Seasonality of the Scottsbluff and Lipscomb bison bonebeds: Implications for modeling Paleoindian subsistence. *American Antiquity* 55: 813–827.

———. 1992. Faunal analysis and Paleoindian studies: A reexamination of the Lipscomb bison bonebed. *Plains Anthropologist* 37:137–165.

Todd, L. C., D. Rapson, and J. L. Hofman. 1996. Dentition studies of the Mill Iron and other early Paleoindian site Bison bonebed sites. In *The Mill Iron Site*, ed. G. C. Frison, 145–175. Albuquerque: Univ. of New Mexico Press.

Todd, L. C., R. Witter, and G. C. Frison. 1987. Excavation and documentation of the Princeton and Smithsonian Horner site assemblages. In *The Horner site: The type site of the Cody cultural complex*, G. C. Frison and L. C. Todd, 39–91. New York: Academic Press.

Tompa, A. 1976. Fossil eggs of the land snail genus *Vallonia* (Pulmonata: Valloniidae). *Nautilus* 90(1):5–7.

Trimble, D. 1990. *The geologic story of the Great Plains*. Medora, ND: Theodore Roosevelt Nature and History Association.

Tunnell, C. 1977. Fluted projectile point production as revealed by lithic specimens from the Adair-Steadman site in northwest Texas. *Museum Journal* 17:140–168.

Tunnell, C., and L. Johnson. 2000. Comparing dimensions for Folsom points and their by-products from Adair-Steadman and Lindenmeier sites and other localities. *Texas Historical Commission Archaeological Reports Series* no. 1.

Turgeon, D. D., J. F. Quinn, Jr., A. E. Bogan, E. V. Coan, F. G. Hockberg, W. G. Lyons, P. M. Mikkelsen, R. J. Neves, C. F. E. Roper, G. Rosenberg, B. Roth, A. Scheltema, F. G. Thompson, M. Vecchione, and J. D. Williams. 1998. *Common and scientific names of aquatic invertebrates from the United States and Canada: Mollusks.* 2nd ed. Special Publication 26. Bethesda, MD: American Fisheries Society.

Van Devender, T. R. 1990. Late Quaternary vegetation and climate of the Chihuahuan Desert, United States and Mexico. In *Packrat Middens*, J. L. Betancourt, T. R. Van Devender, and P. S. Martin, eds., 134–165. Tucson: Univ. of Arizona Press.

Van De Water, P., S. Leavitt, and J. Betancourt. 1994. Trends in stomatal density and $^{13}C/^{12}C$ ratios of *Pinus flexilis* needles during the last glacial-interglacial transition. *Science* 264:239–243.

Van Dyne, G. M., N. Brockington, Z. Szocs, J. Duek, and C. Ribic. 1980. Large herbivore system. In *Grasslands, systems analysis, and man*, ed. A. I. Breymeyer and G. M. Van Dyne, 269–537. Cambridge: Cambridge Univ. Press.

Van Klinken, G. 1999. Bone collagen quality indicators for paleodietary and radiocarbon measurements. *Journal of Archaeological Science* 26:687–695.

Van Vuren, D. 1984. Summer diets of bison and cattle in southern Utah. *Journal of Range Management* 37:260–261.

Verbicky-Todd, E. 1984. Communal buffalo hunting among the Plains Indians. *Archaeological Survey of Alberta, Occasional Paper* no. 24.

Vierling, L. A. 1998. Palynological evidence for late- and postglacial environmental change in central Colorado. *Quaternary Research* 49:222–232.

Vinton, M., D. Hartnett, E. Finck, and J. Briggs. 1993. Interactive effects of fire, Bison (*Bison bison*), grazing and plant community composition in tallgrass prairie. *American Midland Naturalist* 129:10–18.

Voorhies, M. 1969. Taphonomy population dynamics of an early Pliocene fauna, Knox County, Nebraska. *Univ. of Wyoming Contributions to Geology Special Paper* 1.

Walker, D. 1980. The Vore site local fauna. In The Vore site, 48CK302, a stratified buffalo jump in the Wyoming Black Hills, by C. Reher and G. Frison. 1980. *Plains Anthropologist Memoir* 16:154–169.

Wallace, A. R. 1887. The antiquity of man in North America. *Nineteenth Century* 22:667–679.

Waller, S., and J. K. Lewis. 1978. Occurrence of C_3 and C_4 photosynthetic pathways in North American grasses. *Journal of Range Management* 32:12–28.

Wandsnider, L. 1997. The roasted and the boiled: Food composition and heat treatment with special emphasis on pit-hearth cooking. *Journal of Anthropological Archaeology* 16.

Warnica, J. 1961. The Elida site, evidence of a Folsom occupation in Roosevelt County, eastern New Mexico. *Bulletin of Texas Archaeological Society* 30:209–215.

Weber, W. 1976. *Rocky Mountain Flora*. Boulder: Colorado Associated Univ. Press.

Weeks, J. B., and E. Gutentag. 1988. Region 17, High Plains. In *Hydrology*, ed. W. Back, J. S. Rosenshein, P. R. Seaber. Vol. O-2 in *The geology of North America*, 157–164. Boulder, CO: Geological Society of America.

Weiner, S., P. Goldberg, and O. Bar-Yosef. 1993. Bone preservation in Kebara Cave, Israel using on-site Fourier–transform infrared spectrometry. *Journal of Archaeological Science* 20: 613–627.

Weiner, S., S. Scheigl, P. Goldberg, and O. Bar-Yosef. 1995. Mineral sssemblages in Kebara and Hayonim caves, Israel: Excavation strategies, bone preservation, and wood ash remains. *Israel Journal of Chemistry* 35:143–154.

Weissner, P. 1984. Reconsidering the behavioral basis of style: a case study among the Kalahari San. *Journal of Anthropological Archaeology* 3:190–234.

Wells, P. V., and J. D. Stewart. 1987. Cordilleran-Boreal Taiga and fauna on the central Great Plains of North America, 14,000–18,000 Years Ago. *American Midland Naturalist* 118(1): 94–106.

Wendorf, F., and J. Hester. 1962. Early man's utilization of the Great Plains environment. *American Antiquity* 28: 159–171.

———. 1975. *Late Pleistocene environments of the Southern High Plains*. Publication 9. Dallas, TX: Fort Burgwin Research Center, Southern Methodist Univ.

Wendorf, F., and A. Krieger. 1959. New light on the Midland discovery. *American Antiquity* 25:66–78.

Wendorf, F., A. Krieger, C. Albritton and T. D. Stewart. 1955. *The Midland discovery*. Austin: Univ. of Texas Press.

Wheat, J. B. 1972. The Olsen-Chubbuck site: A Paleo–Indian bison kill. *Society for American Archaeology Memoir* 26.

———. 1979. The Jurgens site. *Plains Anthropologist, Memoir* 15.

Whitlock, C., and L. Grigg. 1999. Paleoecological evidence of Milankovitch climate variations in the western U.S. during the late Quaternary. In *Mechanisms of global climate change at millennial time scales*, ed. P. Clark, R. S. Webb, and L. Keigwin, 227–241. Washington, DC: American Geophysical Union.

Whittaker, J. C. 1994. *Flintknapping: Making and understanding stone tools*. Austin: Univ. of Texas Press.

Whittaker, J. C., and M. Stafford. 1999. Replicas, fakes, and art: The twentieth-century stone age and its effects on archaeology. *American Antiquity* 64:203–214.

Whittlesey, C. 1869. On the evidences of the antiquity of man in the United States. *Proceedings American Association for the Advancement of Science* 16:268–288.

Willey, G., and J. Sabloff. 1980. *A history of American archaeology*. 2nd ed. San Francisco: W. H. Freeman.

William, J. 2000. Folsom stone tools. In *The Big Black site (32DU955C): A Folsom workshop in the Knife River Flint Quarry Area*, ed. J. D. William, 169–231. Pullman: Washington State Univ. Press.

Williams, J., B. Shuman, and T. Webb. 2001. Dissimilarity analyses of late Quaternary vegetation of climate in eastern North America. *Ecology* 82:3346–3362.

Williston, S. 1902. An arrow-head found with the bones of Bison occidentalis, Lucas, in western Kansas. *American Geologist* 30:313–315.

Wilmsen, E. 1965. An outline of early man studies in the United States. *American Antiquity* 31:172–192.

Wilmsen, E. and F. H. H. Roberts. 1978. *Lindenmeier, 1934–1974. Concluding report on investigations.* Smithsonian Contributions to Anthropology, no. 24. Washington, DC: Smithsonian Institution.

Winchell, N. 1893a. Some recent criticisms. *American Geologist* 11:110–112.

———. 1893b. Professor Wright's book a service to science. *American Geologist* 11:194.

Winfrey, J. 1990. An event tree analysis of Folsom point failure. *Plains Anthropologist* 35:263–272.

Wissler, C. 1910. Material Culture of the Blackfoot Indians. American Museum of Natural History, *Anthropological Papers* 5(1).

———. 1928. Archaeological explorations at Folsom, N.M. Unpublished report on file, Anthropology Archives, Department of Anthropology, American Museum of Natural History, New York.

Woodhouse, C. A., and J. T. Overpeck. 1998. 2000 years of drought variability in the central United States. *Bulletin of the American Meteorological Society* 79:2693–2714.

Wormington, H. M. 1939. *Ancient man in North America.* 1st ed. Denver: Colorado Museum of Natural History.

———. 1944. *Ancient man in North America.* 2nd ed. Denver: Colorado Museum of Natural History.

———. 1948. A proposed revision of Yuma point terminology. *Proceedings of the Colorado Museum of Natural History* 18(2).

———. 1949. *Ancient man in North America.* 3rd ed. Denver: Denver Museum of Natural History.

———. 1957. *Ancient man in North America.* 4th ed. Denver: Denver Museum of Natural History.

Worster, D. 2001. *A river running west: The life of John Wesley Powell.* Oxford Univ. Press. Oxford.

Wright, G. F. 1881. On the age of the Trenton gravel. *Proceedings Boston Society of Natural History* 21:137–145.

———. 1883. Glacial phenomena in Ohio. *Science* 1:269–271.

———. 1888. The age of the Ohio gravel-beds. *Proceedings Boston Society of Natural History* 23:427–436.

———. 1889a. *The Ice Age in North America and its bearing upon the antiquity of man.* New York: D. Appleton.

———. 1889b. The age of the Philadelphia Red gravel. *Proceedings Boston Society of Natural History* 24:152–157.

———. 1890. Report. In *Discovery of a Paleolithic implement at New Comerstown, Ohio,* by W. C. Mills and G. F. Wright, 5–14. Western Reserve Historical Society, Tract 75.

———. 1892. *Man and the glacial period.* New York: D. Appleton.

Wyckoff, D. 1993. Gravel sources of knappable Alibates silicified dolomite. *Geoarchaeology* 8:35–58.

———. 1999. Southern Plains Folsom lithic technology: A view from the edge. In *Folsom lithic technology: Explorations in structure and variation,* ed. D. S. Amick, 39–64. International Monographs in Prehistory, Archaeological Series 12. Ann Arbor, MI.

Wyckoff, D., B. J. Carter, P. Flynn, L. D. Martin, B. A. Branson, and J. L. Theler. 1992. Interdisciplinary Studies of the Hajny Mammoth Site, Dewey County, Oklahoma. *Studies in Oklahoma's Past* 17. Norman: Oklahoma Archeological Survey.

Wyckoff, D., J. Theler, and B. Carter. 1997. Southern Great Plains gastropods: Modern occurrences, prehistoric implications. Final Report to the National Geographic Society, Grant 5477–95.

Youmans, W. J. 1893a. The insolence of office. *Popular Science Monthly* 42:841–842.

———. 1893b. The attack on Prof. Wright. *Popular Science Monthly* 43:412–413.

Yapp, C. 1979. Oxygen and carbon isotope measurements of land snail shell carbonate. *Geochimica et Cosmochimica Acta* 43:629–635.

Yu, Z. 2000. Ecosystem response to late Glacial and early Holocene climate oscillations in the Great Lakes region of North America. *Quaternary Science Reviews* 19:1723–1747.

Yu, Z., and H. E. Wright. 2001. Response of interior North America to abrupt climate oscillations in the North Atlantic region during the last deglaciation. *Earth Science Reviews* 52: 333–369.

Zar, J. 1999. *Biostatistical analysis.* 4th ed. Upper Saddle River, NJ: Prentice Hall.

Zedeno, N., and R. Stoffle. 2003. Tracking the role of pathways in the evolution of a human landscape. In *Colonization of unfamiliar landscapes: The archaeology of adaptation,* ed. M. Rockman and J. Steele, 59–80. London: Routledge.

INDEX